APPLIED PROBABILITY AND STOCHASTIC PROCESSES

SECOND EDITION

Applied Probability and Stochastic Processes

Second Edition

Frank Beichelt

University of the Witwatersrand

Johannesburg, South Africa

CRC Press
Taylor & Francis Group
Boca Raton London New York

CRC Press is an imprint of the
Taylor & Francis Group, an **informa** business

A CHAPMAN & HALL BOOK

CRC Press
Taylor & Francis Group
6000 Broken Sound Parkway NW, Suite 300
Boca Raton, FL 33487-2742

First issued in paperback 2020

© 2016 by Taylor & Francis Group, LLC
CRC Press is an imprint of Taylor & Francis Group, an Informa business

No claim to original U.S. Government works

ISBN-13: 978-1-4822-5764-9 (hbk)
ISBN-13: 978-0-367-65849-6 (pbk)

Visit the Taylor & Francis Web site at
http://www.taylorandfrancis.com

and the CRC Press Web site at
http://www.crcpress.com

CONTENTS

PREFACE

SYMBOLS AND ABBREVIATIONS

INTRODUCTION

PART I   PROBABILITY THEORY

1   RANDOM EVENTS AND THEIR PROBABILITIES

1.1	RANDOM EXPERIMENTS	7
1.2	RANDOM EVENTS	8
1.3	PROBABILITY	11
	1.3.1 Classical Definition of Probability	12
	1.3.2 Geometric Definition of Probability	15
	1.3.3 Axiomatic Definition of Probability	17
	1.3.4 Relative Frequency	20
1.4	CONDITIONAL PROBABILITY AND INDEPENDENCE OF RANDOM EVENTS	22
	1.4.1 Conditional Probability	22
	1.4.2 Total Probability Rule and Bayes' Theorem	25
	1.4.3 Independent Random Events	28
1.5	EXERCISES	32

2   ONE-DIMENSIONAL RANDOM VARIABLES

2.1	MOTIVATION AND TERMINOLOGY	39
2.2	DISCRETE RANDOM VARIABLES	43
	2.2.1 Probability Distributions and Distribution Parameters	43
	2.2.2 Important Discrete Probability Distributions	48
2.3	CONTINUOUS RANDOM VARIABLES	59
	2.3.1 Probability Distributions	59
	2.3.2 Distribution Parameters	63
	2.3.3 Important Continuous Probability Distributions	73
	2.3.4 Nonparametric Classes of Probability Distributions	86

2.4	MIXTURES OF RANDOM VARIABLES	92
2.5	GENERATING FUNCTIONS	95
	2.5.1 z-Transformation	96
	2.5.2 Laplace Transformation	99
2.6	EXERCISES	106

3 MULTIDIMENSIONAL RANDOM VARIABLES

3.1	TW0-DIMENSIONAL RANDOM VARIABLES	117
	3.1.1 Discrete Components	117
	3.1.2 Continuous Components	120
	3.1.2.1 Probability Distribution	120
	3.1.2.2 Conditional Probability Distribution	127
	3.1.2.3 Bivariate Normal Distribution	131
	3.1.2.4 Bivariate Exponential Distributions	132
	3.1.3 Linear Regression and Correlation Analysis	133
3.2	n-DIMENSIONAL RANDOM VARIABLES	144
3.3	EXERCISES	149

4 FUNCTIONS OF RANDOM VARIABLES

4.1	FUNCTIONS OF ONE RANDOM VARIABLE	155
	4.1.1 Probability Distribution	155
	4.1.2 Random Numbers	163
4.2	FUNCTIONS OF SEVERAL RANDOM VARIABLES	169
	4.2.1 Introduction	169
	4.2.2 Mean Value	170
	4.2.3 Product of Two Random Variables	172
	4.2.4 Ratio of Two Random Variables	173
	4.2.5 Maximum of Random Variables	175
	4.2.6 Minimum of Random Variables	177
4.3	SUMS OF RANDOM VARIABLES	179
	4.3.1 Sums of Discrete Random Variables	179
	4.3.2 Sums of Continuous Random Variables	181
	4.3.2.1 Sum of Two Random Variables	181
	4.3.2.2 Sum of Random Variables	186
	4.3.3 Sums of a Random Number of Random Variables	194
4.4	EXERCISES	196

5 INEQUALITIES AND LIMIT THEOREMS

5.1	INEQUALITIES	199
	5.1.1 Inequalities for Probabilities	199
	5.1.2 Inequalities for Moments	202
5.2	LIMIT THEOREMS	204
	5.2.1 Convergence Criteria for Sequences of Random Variables	204
	5.2.2 Laws of Large Numbers	206
	5.2.2.1 Weak Laws of Large Numbers	206
	5.2.2.2 Strong Laws of Large Numbers	207
	5.2.3 Central Limit Theorem	208
	5.2.4 Local Limit Theorems	214
5.3	EXERCISES	217

PART II STOCHASTIC PROCESSES

6 BASICS OF STOCHASTIC PROCESSES

6.1	MOTIVATION AND TERMINOLOGY	221
6.2	CHARACTERISTICS AND EXAMPLES	225
6.3	CLASSIFICATION OF STOCHASTIC PROCESSES	230
6.4	TIME SERIES IN DISCRETE TIME	237
	6.4.1 Introduction	237
	6.4.2 Smoothing of Time Series	239
	6.4.3 Trend Estimation	243
	6.4.4 Stationary Discrete-Time Stochastic Processes	246
6.5	EXERCISES	252

7 RANDOM POINT PROCESSES

7.1	BASIC CONCEPTS	255
7.2	POISSON PROCESSES	261
	7.2.1 Homogeneous Poisson Processes	261
	7.2.1.1 Definition and Properties	261
	7.2.1.2 Homogeneous Poisson Process and Uniform Distribution	267
	7.2.2 Nonhomogeneous Poisson Processes	274
	7.2.3 Mixed Poisson Processes	278
	7.2.4 Superposition and Thinning of Poisson Processes	284
	7.2.4.1 Superposition	284
	7.2.4.2 Thinning	285
	7.2.5 Compound Poisson Processes	287

	7.2.6	Applications to Maintenance	289
	7.2.7	Application to Risk Analysis	292
7.3		RENEWAL PROCESSES	299
	7.3.1	Definitions and Examples	299
	7.3.2	Renewal Function	302
		7.3.2.1 Renewal Equations	302
		7.3.2.2 Bounds on the Renewal Function	308
	7.3.3	Asymptotic Behavior	311
	7.3.4	Recurrence Times	315
	7.3.5	Stationary Renewal Processes	318
	7.3.6	Alternating Renewal Processes	319
	7.3.7	Compound Renewal Processes	324
		7.3.7.1 Definition and Properties	324
		7.3.7.2 First Passage Time	329
7.4		EXERCISES	332

8 DISCRETE-TIME MARKOV CHAINS

8.1		FOUNDATIONS AND EXAMPLES	339
8.2		CLASSIFICATION OF STATES	350
	8.2.1	Closed Sets of States	350
	8.2.2	Equivalence Classes	351
	8.2.3	Periodicity	353
	8.2.4	Recurrence and Transience	354
8.3		LIMIT THEOREMS AND STATIONARY DISTRIBUTION	360
8.4		BIRTH AND DEATH PROCESSES	364
	8.4.1	Introduction	364
	8.4.2	General Random Walk with Two Absorbing Barriers	365
	8.4.3	General Random Walk with One Absorbing Barrier	368
8.5		DISCRETE-TIME BRANCHING PROCESSES	370
	8.5.1	Introduction	370
	8.5.2	Generating Function and Distribution Parameters	371
	8.5.3	Probability of Extinction and Examples	373
8.6		EXERCISES	376

9 CONTINUOUS-TIME MARKOV CHAINS

9.1	BASIC CONCEPTS AND EXAMPLES	383
9.2	TRANSITION PROBABILITIES AND RATES	387
9.3	STATIONARY STATE PROBABILITIES	396

9.4	SOJOURN TIMES IN PROCESS STATES	399
9.5	CONSTRUCTION OF MARKOV SYSTEMS	401
9.6	BIRTH AND DEATH PROCESSES	405
	9.6.1 Birth Processes	405
	9.6.2 Death Processes	408
	9.6.3 Birth and Death Processes	410
	9.6.3.1 Time-Dependent State Probabilities	410
	9.6.3.2 Stationary State Probabilities	418
	9.6.3.3 Nonhomogeneous Birth and Death Processes	421
9.7	APPLICATIONS TO QUEUEING MODELS	425
	9.7.1 Basic Concepts	425
	9.7.2 Loss Systems	428
	9.7.2.1 $M/M/\infty$-System	428
	9.7.2.2 $M/M/s/0$-System	428
	9.7.2.3 Engset's Loss System	430
	9.7.3 Waiting Systems	431
	9.7.3.1 $M/M/s/\infty$-System	431
	9.7.3.2 $M/G/1/\infty$-System	434
	9.7.3.3 $G/M/1/\infty$-System	438
	9.7.4 Waiting-Loss-Systems	439
	9.7.4.1 $M/M/s/m$-System	439
	9.7.4.2 $M/M/s/\infty$-System with Impatient Customers	441
	9.7.5 Special Single-Server Queueing Systems	442
	9.7.5.1 System with Priorities	442
	9.7.5.2 $M/M/1/m$-System with Unreliable Server	445
	9.7.6 Networks of Queueing Systems	447
	9.7.6.1 Introduction	447
	9.7.6.2 Open Queueing Networks	447
	9.7.6.3 Closed Queueing Networks	454
9.8	SEMI-MARKOV CHAINS	458
9.9	EXERCISES	465

10 MARTINGALES

10.1	DISCRETE-TIME MARTINGALES	475
	10.1.1 Definition and Examples	475
	10.1.2 Doob-Type Martingales	479
	10.1.3 Martingale Stopping Theorem and Applications	486
10.2	CONTINUOUS-TIME MARTINGALES	489
10.3	EXERCISES	492

11   BROWNIAN MOTION

11.1   INTRODUCTION   495

11.2   PROPERTIES OF THE BROWNIAN MOTION   497

11.3   MULTIDIMENSIONAL AND CONDITIONAL DISTRIBUTIONS   501

11.4   FIRST PASSAGE TIMES   504

11.5   TRANSFORMATIONS OF THE BROWNIAN MOTION   508
    11.5.1 Identical Transformations   508
    11.5.2 Reflected Brownian Motion   509
    11.5.3 Geometric Brownian Motion   510
    11.5.4 Ornstein-Uhlenbeck Process   511
    11.5.5 Brownian Motion with Drift   512
        11.5.5.1 Definitions and First Passage Times   512
        11.5.5.2 Application to Option Pricing   516
        11.5.5.3 Application to Maintenance   522
    11.5.6 Integrated Brownian Motion   524

11.6   EXERCISES   526

12   SPECTRAL ANALYSIS OF STATIONARY PROCESSES

12.1   FOUNDATIONS   531

12.2   PROCESSES WITH DISCRETE SPECTRUM   533

12.3   PROCESSES WITH CONTINUOUS SPECTRUM   537
    12.3.1 Spectral Representation of the Covariance Function   537
    12.3.2 White Noise   544

12.4   EXERCISES   547

REFERENCES   549

INDEX   553

PREFACE TO THE SECOND EDITION

The book is a self-contained introduction into elementary probability theory and stochastic processes with special emphasis on their applications in science, engineering, finance, computer science and operations research. It provides theoretical foundations for modeling time-dependent random phenomena in these areas and illustrates their application through the analysis of numerous, practically relevant examples. As a non-measure theoretic text, the material is presented in a comprehensible, application-oriented way. Its study only assumes a mathematical maturity which students of applied sciences acquire during their undergraduate studies in mathematics. The study of stochastic processes and its fundament, probability theory, as of any other mathematically based science, requires less routine effort, but more creative work on one's own. Therefore, numerous exercises have been added to enable readers to assess to which extent they have grasped the subject. Solutions to many of the exercises can be downloaded from the website of the Publishers or the exercises are given together with their solutions. A complete solutions manual is available to instructors from the Publishers. To make the book attractive to theoretically interested readers as well, some important proofs and challenging examples and exercises have been included. 'Starred' exercises belong to this category. The chapters are organized in such a way that reading a chapter usually requires knowledge of some of the previous ones. The book has been developed in part as a course text for undergraduates and for self-study by non-statisticians. Some sections may also serve as a basis for preparing senior undergraduate courses.

The text is a thoroughly revised and supplemented version of the first edition so that it is to a large extent a new book: The part on probability theory has been completely rewritten and more than doubled. Several new sections have been included in the part about stochastic processes as well: Time series analysis, random walks, branching processes, and spectral analysis of stationary stochastic processes. Theoretically more challenging sections have been deleted and mainly replaced with a comprehensive numerical discussion of examples. All in all, the volume of the book has increased by about a third.

This book does not extensively deal with data analysis aspects in probability and stochastic processes. But sometimes connections between probabilistic concepts and the corresponding statistical approaches are established to facilitate the understanding. The author has no doubt the book will help students to pass their exams and practicians to apply stochastic modeling in their own fields of expertise.

The author is thankful for the constructive feedback from many readers of the first edition. Helpful comments to the second edition are very welcome as well and should be directed to: Frank.Beichelt@wits.ac.za.

Johannesburg, March 2016   *Frank Beichelt*

SYMBOLS AND ABBREVIATIONS

□ ■ ●	symbols after an example, a theorem, a definition
$f(t) \equiv c$	$f(t) = c$ <u>for all</u> t being element of the domain of definition of f
$f * g$	convolution of two functions f and g
$f^{*(n)}$	n th convolution power of f
$\hat{f}(s)$, $L\{f, s\}$	Laplace transform of a function f
$o(x)$	Landau order symbol
δ_{ij}	Kronecker symbol

Probability Theory

X, Y, Z	random variables
$E(X)$, $Var(X)$	mean (expected) value of X, variance of X
$f_X(x)$, $F_X(x)$	probability density function, (cumulative probability) distribution function of X
$F_Y(y|x)$, $f_Y(y|x)$	conditional distribution function, density of Y given $X = x$
X_t, $F_t(x)$	residual lifetime of a system of age t, distribution function of X_t
$E(Y|x)$	conditional mean value of Y given $X = x$
$\lambda(x)$, $\Lambda(x)$	failure rate, integrated failure rate (hazard function)
$N(\mu, \sigma^2)$	normally distributed random variable (normal distribution) with mean value μ and variance σ^2
$\varphi(x)$, $\Phi(x)$	probability density function, distribution function of a standard normal random variable $N(0, 1)$
$f_{\mathbf{X}}(x_1, x_2, \dots, x_n)$	joint probability density function of $\mathbf{X} = (X_1, X_2, \dots, X_n)$
$F_{\mathbf{X}}(x_1, x_2, \dots, x_n)$	joint distribution function of $\mathbf{X} = (X_1, X_2, \dots, X_n)$
$Cov(X, Y)$, $\rho(X, Y)$	covariance, correlation coefficient between X and Y
$M(z)$	z-transform (moment generating function) of a discrete random variable or of its probability distribution, respectively

Stochastic Processes

$\{X(t), t \in \mathbf{T}\}$, $\{X_t, t \in \mathbf{T}\}$	continuous-time, discrete-time stochastic process with parameter space $\mathbf{T}$
$\mathbf{Z}$	state space of a stochastic process
$f_t(x)$, $F_t(x)$	probability density, distribution function of $X(t)$
$f_{t_1, t_2, \dots, t_n}(x_1, x_2, \dots, x_n)$, $F_{t_1, t_2, \dots, t_n}(x_1, x_2, \dots, x_n)$	
	joint density, distribution function of $(X(t_1), X(t_2), \dots, X(t_n))$
$m(t)$	trend function of a stochastic process
$C(s, t)$	covariance function of a stochastic process
$C(\tau)$	covariance function of a stationary stochastic process

$C(t)$, $\{C(t),\ t \geq 0\}$	compound random variable, compound stochastic process
$\rho(s,t)$	correlation function of a stochastic process
$\{T_1, T_2, ...\}$	random point process
$\{Y_1, Y_2, ...\}$	sequence of interarrival times, renewal process
N	integer-valued random variable, discrete stopping time
$\{N(t),\ t \geq 0\}$	(random) counting process
$N(s,t)$	increment of a counting process in $(s,\ t]$
$H(t)$, $H_1(t)$	renewal function of an ordinary, delayed renewal processss
$A(t)$	forward recurrence time, point availability
$B(t)$	backward recurrence time
$R(t)$, $\{R(t),\ t \geq 0\}$	risk reserve, risk reserve process
A, $A(t)$	stationary (long-run) availability, point availability
p_{ij}, $p_{ij}^{(n)}$	one-step, n-step transition probabilities of a homogeneous, discrete-time Markov chain
$p_{ij}(t)$; q_{ij}, q_i	transition probabilities; conditional, unconditional transition rates of a homogeneous, continuous-time Markov chain
$\{\pi_i;\ i \in \mathbf{Z}\}$	stationary state distribution of a homogeneous Markov chain
π_0	extinction probability, vacant probability (sections 8.5, 9.7)
λ_j, μ_j	birth, death rates
λ, μ, ρ	arrival rate, service rate, traffic intensity λ/μ (in queueing models)
μ_i	mean sojourn time of a semi-Markov process in state i
μ	drift parameter of a Brownian motion process with drift
W	waiting time in a queueing system
L	lifetime, cycle length, queue length, continuous stopping time
$L(x)$	first-passage time with regard to level x
$L(a,b)$	first-passage time with regard to level $\min(a,b)$
$\{B(t),\ t \geq 0\}$	Brownian motion (process)
σ^2, σ	$\sigma^2 = Var(B(1))$ variance parameter, volatility
$\{S(t),\ t \geq 0\}$	seasonal component of a time series (section 6.4), standardized Brownian motion (chapter 11).
$\{\bar{B}(t),\ 0 \leq t \leq 1\}$	Brownian bridge
$\{D(t),\ t \geq 0\}$	Brownian motion with drift
$M(t)$	absolute maximum of the Brownian motion (with drift) in $[0, t]$
M	absolute maximum of the Brownian motion (with drift) in $[0, \infty)$
$\{U(t),\ t \geq 0\}$	Ornstein-Uhlenbeck process, integrated Brownian motion process
ω, w	circular frequency, bandwidth
$s(\omega)$, $S(\omega)$	spectral density, spectral function (chapter 12)

Introduction

Is the world a well-ordered entirety,
or a random mixture,
which nevertheless is called world-order?

Marc Aurel

Random influences or phenomena occur everywhere in nature and social life. Their consideration is an indispensable requirement for being successful in natural, econ-omical, social, and engineering sciences. Random influences partially or fully contri-bute to the variability of parameters like wind velocity, rainfall intensity, electromag-netic noise levels, fluctuations of share prices, failure time points of technical units, timely occurrences of births and deaths in biological populations, of earthquakes, or of arrivals of customers at service centers. Random influences induce *random events*. An event is called *random* if on given conditions it can occur or not. For instance, the events that during a thunderstorm a certain house will be struck by lightning, a child will reach adulthood, at least one shooting star appears in a specified time interval, a production process comes to a standstill for lack of material, a cancer patient survives chemotherapy by 5 years are random. Border cases of random events are the *deterministic events*, namely the *certain event* and the *impossible event*. On given conditions, a deterministic (impossible) event will always (never) occur. For instance, it is absolutely sure that lead, when heated to a temperature of over $327.5\,^0C$ will become liquid, but that lead during the heating process will turn to gold is an impossible event. Random is the shape, liquid lead assumes if poured on an even steel plate, and random is also the occurrence of events which are predicted from the form of these castings to the future. Even if the reader is not a lottery, card, or dice player, she/he will be confronted in her/his daily routine with random influences and must take into account their implications: When your old coffee machine fails after an unpredictable number of days, you go to the supermarket and pick a new one from the machines of your favorite brand. At home, when trying to make your first cup of coffee, you realize that you belong to the few unlucky ones who picked by chance a faulty machine. A car driver, when estimating the length of the trip to his destination, has to take into account that his vehicle may start only with delay, that a traffic jam could slow down the progress, and that scarce parking opportunities may cause further delay. Also, at the end of a year the overwhelming majority of the car drivers realize that having taken out a policy has only enriched the insurance compa-ny. Nevertheless, they will renew their policy because people tend to prefer moderate regular cost, even if they arise long-term, to the risk of larger unscheduled cost. Hence it is not surprising that insurance companies belonged to the first institutions that had a direct practical interest in making use of methods for the quantitative evaluation of random influences and gave in turn important impulses for the development of such methods. It is the probability theory, which provides the necessary mathematical tools for their work.

> *Probability theory deals with the investigation of regularities random events are subjected to.*

The existence of such *statistical* or *stochastic regularities* may come as a surprise to philosophically less educated readers, since at first glance it seems to be paradoxical to combine regularity and randomness. But even without philosophy and without probability theory, some simple regularities can already be illustrated at this stage:

1) When throwing a fair die once, then one of the integers from 1 to 6 will appear and no regularity can be observed. But if a die is thrown repeatedly, then the fraction of throws with outcome 1, say, will tend to 1/6, and with increasing number of throws this fraction will converge to the value 1/6. (A die is called *fair* if each integer has the same chance to appear.)

2) If a specific atom of a radioactive substance is observed, then the time from the beginning of its observation to its disintegration cannot be predicted with certainty, i.e., this time is random. On the other hand, one knows the *half-life period* of a radioactive substance, i.e., one can predict with absolute certainty after which time from say originally 10 gram (trillions of atoms) of the substance exactly 5 gram is left.

3) Random influences can also take effect by superimposing purely deterministic processes. A simple example is the measurement of a physical parameter, e.g., the temperature. There is nothing random about this parameter when it refers to a specific location at a specific time. However, when this parameter has to be measured with sufficiently high accuracy, then, even under always the same measurement conditions, different measurements will usually show different values. This is, e.g., due to the degree of inaccuracy, which is inherent to every measuring method, and to subjective moments. A statistical regularity in this situation is that with increasing number of measurements, which are carried out independently and are not biased by systematic errors, the arithmetic mean of these measurements converges towards the true temperature.

4) Consider the movement of a tiny particle in a container filled with a liquid. It moves along zig-zag paths in an apparently chaotic motion. This motion is generated by the huge number of impacts the particle is exposed to with surrounding molecules of the fluid. Under average conditions, there are about 10^{21} collisions per second between particle and molecules. Hence, a deterministic approach to modeling the motion of particles in a fluid is impossible. This movement has to be dealt with as a random phenomenon. But the pressure within the container generated by the vast number of impacts of fluid molecules with the sidewalls of the container is constant.

Examples 1 to 4 show the nature of a large class of statistical regularities:

> *The superposition of a large number of random influences leads under certain conditions to deterministic phenomena.*

Deterministic regularities (law of falling bodies, spreading of waves, Ohm's law, chemical reactions, theorem of Pythagoras) can be verified in a single experiment if the underlying assumptions are fulfilled. But, although statistical regularities can be proved in a mathematically exact way just as the theorem of Pythagoras or the rules of differentiation and integration of real functions, their experimental verification requires a huge number of repetitions of one and the same experiment. Even leading scientists spared no expense to do just this. The *Comte de Buffon* (1707 – 1788) and the mathematician *Karl Pearson* (1857 – 1936) had flipped a fair coin several thousand times and recorded how often 'head' had appeared. The following table shows their results (n number of total flippings, m number of outcome 'head'):

Scientist	n	m	m/n
Buffon	4040	2048	0.5080
Pearson	12000	6019	0.5016
Pearson	24000	12012	0.5005

Thus, the more frequently a coin is flipped, the more approaches the ratio m/n the value 1/2 (compare with example 1 above). In view of the large number of flippings, this principal observation is surely not a random result, but can be confirmed by all those readers who take pleasure in repeating these experiments. However, nowadays the experiment 'flipping a coin' many thousand times is done by a computer with a 'virtual coin' in a few seconds. The ratio m/n is called the *relative frequency* of the occurrence of the random event 'head appears.'

Already the expositions made so far may have convinced many readers that random phenomena are not figments of human imagination, but that their existence is objective reality. There have been attempts to deny the existence of random phenomena by arguing that if all factors and circumstances, which influence the occurrence of an event are known, then an absolutely sure prediction of its occurrence is possible. In other words, the protagonists of this thesis consider the creation of the concept of randomness only as a sign of 'human imperfection.' The young *Pierre Simeon Laplace* (1729 – 1827) believed that the world is down to the last detail governed by deterministic laws. Two of his famous statements concerning this are: 'The curve described by a simple molecule of air in any gas is regulated in a manner as certain as the planetary orbits. The only difference between them lies in our ignorance.' And: 'Give me all the necessary data, and I will tell you the exact position of a ball on a billiard table' (after having been pushed). However, this view has proved futile both from the philosophical and the practical point of view. Consider, for instance, a biologist who is interested in the movement of animals in the wilderness. How on earth is he supposed to be in a position to collect all that information, which would allow him to predict the movements of only one animal in a given time interval with absolute accuracy? Or imagine the amount of information you need and the corresponding software to determine the exact path of a particle, which travels in a fluid, when there are 10^{21} collisions with surrounding molecules per second. It is an unrealistic and impossible task to deal with problems like that in a deterministic way. The physicist *Marian von Smoluchowski* (1872 – 1917) wrote in a paper published in 1918 that 'all theories are inadequate, which consider randomness as an unknown partial cause of an event. The chance of the occurrence of an event can only depend on the conditions, which have influence on the event, but not on the degree of our knowledge.'

Already at a very early stage of dealing with random phenomena the need arose to quantify the *chance*, the *degree of certainty*, or the *likelihood* for the occurrence of random events. This had been done by defining the probability of random events and by developing methods for its calculation. For now the following explanation is given: The probability of a random event is a number between 0 and 1. The imposs- ible event has probability 0, and the certain event has probability 1. The probability of a random event is the closer to 1, the more frequently it occurs. Thus, if in a long series of experiments a random event *A* occurs more frequently than a random event *B*, then *A* has a larger probability than *B*. In this way, assigning probabilities to random events allows comparisons with regard to the frequency of their occurrence under identical conditions. There are other approaches to the definition of probabili- ty than the classical (frequency) approach, to which this explanation refers. For beginners the frequency approach is likely the most comprehensible one.

Gamblers, in particular dice gamblers, were likely the first people, who were in need of methods for comparing the chances of the occurrence of random events, i.e., the chances of winning or losing. Already in the medieval poem *De Vetula* of *Richard de Fournival* (ca 1200–1250) one can find a detailed discussion about the total number of possibilities to achieve a certain number, when throwing 3 dice. *Geronimo Cardano* (1501 – 1576) determined in his book *Liber de Ludo Aleae* the number of possibilities to achieve the total outomes 2, 3, ..,12, when two dice are thrown. For instance, there are two possibilities to achieve the outcome 3, namely (1,2) and (2,1), whereas 2 will be only then achieved, when (1,1) occurs. (The notation (i,j) means that one die shows an i and the other one a j.) *Galileo Galilei* (1564 – 1642) proved by analogous reasoning that, when throwing 3 dice, the probability to get the (total) outcome 10 is larger than the probability to get a 9. The gamblers knew this from their experience, and they had asked Galilei to find a mathematical proof. The *Chevalier de Méré* formulated three problems related to games of chance and asked the French mathematician *Blaise Pascal* (1623 – 1662) for solutions:

1) What is more likely, to obtain at least one 6 when throwing a die four times, or in a series of 24 throwings of two dice to obtain at least once the outcome (6,6)?

2) How many time does one have to throw two dice at least so that the probability to achieve the outcome (6,6) is larger than 1/2?

3) In a game of chance, two equivalent gamblers need each a certain number of points to become winners. How is the stake to fairly divide between the gamblers, when for some reason or other the game has to be prematurely broken off ? (This problem of the *fair division* had been already formulated before *de Méré*, e.g., in the *De Vetula*.)

Pascal sent these problems to *Pierre Fermat* (1601 – 1665) and both found their solutions, although by applying different methods. It is generally accepted that this work of *Pascal* and *Fermat* marked the beginning of the development of probability theory as a mathematical discipline. Their work has been continued by famous scientists as *Christian de Huygens* (1629 – 1695), *Jakob Bernoulli* (1654 – 1705), *Abraham de Moivre* (1667 – 1754), *Carl Friedrich Gauss* (1777 – 1855), and last but not least by *Simeon Denis de Poisson* (1781 – 1840). However, probability theory was out of its infancy only in the thirties of the twentieth century, when the Russian mathematician *Andrej Nikolajewič Kolmogorov* (1903 – 1987) found the solution of one of the famous Hilbert problems, namely to put probability theory as any other mathematical discipline on an axiomatic foundation.

Nowadays, probability theory together with its applications in science, medicine, engineering, economy et al. are integrated in the field of *stochastics*. The linguistic origin of this term can be found in the Greek word *stochastikon*. (Originally, this term denoted the ability of seers to be correct with their forecasts.) Apart from probability theory, *mathematical statistics* is the most important part of stochastics. A key subject of it is to infer by probabilistic methods from a sample taken from a set of interesting objects, called among else *sample space* or *universe,* to parameters or properties of the sample space (*inferential statistics*). Let us assume we have a lot of 10 000 electronic units. To obtain information on what percentage of these units is faulty, we take a sample of 100 units from this lot. In the sample, 4 units are faulty. Of course, this figure does not imply that there are exactly 400 faulty units in the lot. But inferential statistics will enable us to construct lower and upper bounds for the percentage of faulty units in the lot, which limit the 'true percentage' with a given high probability. Problems like this led to the development of an important part of mathematical statistics, the *statistical quality control*. Phenomena, which depend both on random and deterministic influences, gave rise to the theory of *stochastic processes*. For instance, meteorological parameters like temperature and air pressure are random, but obviously also depend on time and altitude. Fluctuations of share prices are governed by chance, but are also driven by periods of economic up and down turns. Electromagnetic noise caused by the sun is random, but also depends on the periodical variation of the intensity of sunspots.

Stochastic modeling in operations research comprises disciplines like queueing theory, reliability theory, inventory theory, and decision theory. All of them play an important role in applications, but also have given many impulses for the theoretical enhancement of the field of stochastics. *Queueing theory* provides the theoretical fundament for the quantitative evaluation and optimization of queueing systems, i.e., service systems like workshops, supermarkets, computer networks, filling stations, car parks, and junctions, but also military defense systems for 'serving' the enemy. *Inventory theory* helps with designing warehouses (storerooms) so that they can on the one hand meet the demand for goods with sufficiently high probability, and on the other hand keep the costs for storage as small as possible. The key problem with dimensioning queueing systems and storage capacities is that flows of customers, service times, demands, and delivery times of goods after ordering are subject to random influences. A main problem of *reliability theory* is the calculation of the reliability (survival probability, availability) of a system from the reliabilities of its subsystems or components. Another important subject of reliability theory is modelling the aging behavior of technical systems, which incidentally provides tools for the survival analysis of human beings and other living beings. Chess automats got their intelligence from the *game theory*, which arose from the abstraction of games of chance. But opponents within this theory can also be competing economic blocs or military enemies. Modern communication would be impossible without *information theory*. This theory provides the mathematical foundations for a reliable transmission of information although signals may be subject to noise at the transmitter, during transmission, and at the receiver. In order to verify stochastic regularities, nowadays no scientist needs to manually repeat thousands of experiments. Computers do this job much more efficiently. They are in a position to virtually replicate the operation of even highly complex systems, which are subjected to random influences, to any degree of accuracy. This process is called *(Monte Carlo) simulation*. More and very fruitful applications of stochastic (probabilistic) methods exist in fields like physics (kinetic gas theory, thermodynamics, quantum theory), astronomy (stellar statistics), biology (genetics, genomics, population dynamic), artificial intelligence (inference under undertainty), medicine, genomics, agronomy and forestry (design of experiments, yield prediction) as well as in economics (time series analysis) and social sciences. There is no doubt that probabilistic methods will open more and more possibilities for applications, which in turn will lead to a further enhancement of the field of stochastics.

More than 300 hundreds years ago, the famous Swiss mathematician *Jakob Bernoulli* proposed in his book *Ars Conjectandi* the recognition of *stochastics* as an independent new science, the subject of which he introduced as follows:

To conjecture about something is to measure its probability: The Art of conjecturing or the Stochastic Art is therefore defined as the art of measuring as exactly as possible the probability of things so that in our judgement and actions we always can choose or follow that which seems to be better, more satisfactory, safer and more considered.

In line with Bernoulli's proposal, an independent science of stochastics would have to be characterized by two features:
1) The subject of stochastics is uncertainty caused by randomness and/or ignorance.
2) Its methods, concepts, and language are based on mathematics.

But even now, in the twenty-first century, an independent science of stochastics is still far away from being officially established. There is, however, a powerful support for such a move by internationally leading academics; see *von Collani* (2003).

PART I

Probability Theory

There is no credibility in sciences in which no mathematical theory can be applied, and no credibility in fields which have no connections to mathematics.

Leonardo da Vinci

CHAPTER 1

Random Events and Their Probabilities

1.1 RANDOM EXPERIMENTS

If water is heated up to $100^0 C$ at an air pressure of $101\ 325\ Pa$, then it will inevitably start boiling. A motionless pendulum, when being pushed, will start swinging. If ferric sulfate is mixed with hydrochloric acid, then a chemical reaction starts, which releases hydrogen sulfide. These are examples for experiments with deterministic outcomes. Under specified conditions they yield an outcome, which had been known in advance.

Somewhat more complicated is the situation with *random experiments* or *experiments with random outcome.* They are characterized by two properties:

1. Repetitions of the experiment, even if carried out under identical conditions, generally have different outcomes.

2. The possible outcomes of the experiment are known.

Thus, the outcome of a random experiment cannot be predicted with certainty. This implies that the study of random experiments makes sense only if they can be repeated sufficiently frequently under identical conditions. Only in this case *stochastic* or *statistical regularities* can be found.

Let Ω be the set of possible outcomes of a random experiment. This set is called *sample space, space of elementary events,* or *universe.* Examples of random experiments and their respective sample spaces are:

1) Counting the number of traffic accidents a day in a specified area: $\Omega = \{0, 1, ...\}$.

2) Counting the number of cars in a parking area with maximally 200 parking bays at a fixed time point: $\Omega = \{0, 1, ..., 200\}$.

3) Counting the number of shooting stars during a fixed time interval: $\Omega = \{0, 1, ...\}$.

4) Recording the daily maximum wind velocity at a fixed location: $\Omega = [0, \infty)$.

5) Recording the lifetimes technical systems or organisms: $\Omega = [0, \infty)$.

6) Determining the number of faulty parts in a set of 1000: $\Omega = \{0, 1, ..., 1000\}$.

7) Recording the daily maximum fluctuation of a share price: $\Omega = [0, \infty)$.

8) The total profit sombody makes with her/his financial investments a year.
This 'profit' can be negative, i.e. any real number can be the outcome: $\Omega = (-\infty, +\infty)$.

9) Predicting the outcome of a wood reserve inventory in a forest stand: $\Omega = [0, \infty)$.

10) a) Number of eggs a sea turtle will bury at the beach: $\Omega = \{0, 1, ...\}$.

b) Will a baby turtle, hatched from such an egg, reach the water? $\Omega = \{0, 1\}$ with meaning 0: no, 1: yes.

As the examples show, in the context of a random experiment, the term 'experiment' has a more general meaning than in the customary sense.

A random experiment may also contain a deterministic component. For instance, the measurement of a physical quantity should ideally yield the exact (deterministic) parameter value. But in view of random measurement errors and other (subjective) influences, this ideal case does not materialize. Depending on the degree of accuracy required, different measurements, even if done under identical conditions, may yield different values of one and the same parameter (length, temperature, pressure, amperage,...).

1.2 RANDOM EVENTS

A possible outcome ω of a random experiment, i.e. any $\omega \in \Omega$, is called an *elementary event* or a *simple event.*

1) The sample space of the random experiment 'throwing two dice consists of 36 simple elements: $\Omega = \{(i,j), i,j = 1, 2, \cdots, 6\}$. The gambler wins if the sum $i+j$ is at least 10. Hence, the 'winning simple events' are $(5,5)$, $(5,6)$, $(6,5)$, and $(6,6)$.

2) In a delivery of 100 parts some may be defective. A subset (sample) of $n = 12$ parts is taken, and the number N of defective parts in the sample is counted. The elementary events are 0,1,...,12 (possible numbers of defective parts in the sample). The delivery is rejected if $N \geq 4$.

3) In training, a hunter shoots at a cardboard dummy. Given that he never fails the dummy, the latter is the sample space Ω, and any possible impact mark at the dummy is an elementary event. Crucial subsets to be hit are e.g. 'head' or 'heart.'

Already these three examples illustrate that often not single elementary events are interesting, but sets of elementary events. Hence it is not surprising that concepts and results from set theory play a key role in formally establishing probability theory. For this reason, next the reader will be reminded of some basic concepts of set theory.

Basic Concepts and Notation from Set Theory A set is given by its *elements.* We can consider the set of all real numbers, the set of all rational numbers, the set of all people attending a performance, the set of buffalos in a national park, and so on. A set is called *discrete* if it is a finite or a *countably infinite* set. By definition, a countably infinite set can be written as a sequence. In other words, its elements can be numbered. If a set is infinite, but not countably infinite, then it is called *nondenumerable.* Nondenumerable sets are for instance the whole real axis, the positive half-axis, a finite subinterval of the real axis, or a geometric object (area of a circle, target).

Let A and B be two sets. In what follows we assume that all sets $A, B, ...$ considered are subsets of a 'universal set' Ω. Hence, for any set A, $A \subseteq \Omega$.

A is called a *subset* of B if each element of A is also an element of B.

Symbol: $A \subseteq B$.

The *complement* of B with regard to A contains all those elements of B which are not element of A.

Symbol: $B \backslash A$

In particular, $\overline{A} = \Omega \backslash A$ contains all those elements which are not element of A.

The *intersection* of A and B contains all those elements which belong both to A and B.

Symbol: $A \cap B$

The *union* of A and B contains all those elements which belong to A or B (or to both).

Symbol: $A \cup B$

These relations between two sets are illustrated in Figure 1.1 (Venn diagram). The whole shaded area is $A \cup B$.

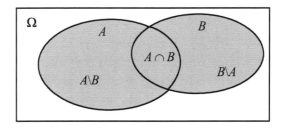

Figure 1.1 Venn diagram

For any sequence of sets $A_1, A_2, \cdots, A_n$, intersection and union are defined as

$$\bigcap_{i=1}^{n} A_i = A_1 \cap A_2 \cap \cdots \cap A_n, \quad \bigcup_{i=1}^{n} A_i = A_1 \cup A_2 \cup \cdots \cup A_n.$$

De Morgan Rules for 2 Sets

$$\overline{A \cup B} = \overline{A} \cap \overline{B}, \quad \overline{A \cap B} = \overline{A} \cup \overline{B}. \tag{1.1}$$

De Morgan Rules for n Sets

$$\overline{\bigcup_{i=1}^{n} A_i} = \bigcap_{i=1}^{n} \overline{A}_i, \quad \overline{\bigcap_{i=1}^{n} A_i} = \bigcup_{i=1}^{n} \overline{A}_i. \tag{1.2}$$

Random Events A *random event* (briefly: *event*) A is a subset of the set Ω of all possible outcomes of a random experiment, i.e. $A \subseteq \Omega$.

> *A random event A is said to have* occurred *as a result of a random experiment if the observed outcome ω of this experiment is an element of A: $\omega \in A$.*

The empty set $\varnothing$ is the *impossible event* since, for not containing any elementary event, it can never occur. Likewise, Ω is the *certain event,* since it comprises all possible outcomes of the random experiment. Thus, there is nothing random about the events $\varnothing$ and Ω. They are actually *deterministic events.* Even before having completed a random experiment, we are absolutely sure that Ω will occur and $\varnothing$ will not.

Let A and B be two events. Then the set-theoretic operations introduced above can be interpreted in terms of the occurrence of random events as follows:

$A \cap B$ is the event that both A and B occur,

$A \cup B$ is the event that A or B (or both) occur,

If $A \subseteq B$ (A is a subset of B), then the occurrence of A implies the occurrence of B.

$A \backslash B$ is the set of all those elementary events which are elements of A, but not of B. Thus, $A \backslash B$ is the event that A occurs, but not B. Note that (see Figure 1.1)

$$A \backslash B = A \backslash (A \cap B). \tag{1.3}$$

The event $\overline{A} = \Omega \backslash A$ is called the *complement of A*. It consists of all those elementary events, which are not in A.

Two events A and B are called *disjoint* or *(mutually) exclusive* if their joint occurrence is impossible, i.e. if $A \cap B = \varnothing$. In this case the occurrence of A implies that B cannot occur and vice versa. In particular, A and $\overline{A}$ are disjoint for any event $A \subseteq \Omega$.

Short Terminology

$A \cap B$	A and B
$A \cup B$	A or B
$A \subseteq B$	A implies B, B follows from A
$A \backslash B$	A but not B
$\overline{A}$	A not

Example 1.1 Let us consider the random experiment 'throwing a die' with sample space $\Omega = \{1, 2, \cdots, 6\}$ and the random events $A = \{2, 3\}$ and $B = \{3, 4, 6\}$. Then, $A \cap B = \{3\}$ and $A \cup B = \{2, 3, 4, 6\}$. Thus, if a 3 had been thrown, then both the events A and B have occurred. Hence, A and B are not disjoint. Moreover, $A \backslash B = \{2\}$, $B \backslash A = \{4, 6\}$, and $\overline{A} = \{1, 4, 5, 6\}$. □

Example 1.2 Two dice D_1 and D_2 are thrown. The sample space is

$$\Omega = \{(i_1, i_2), i_1, i_2 = 1, 2, \cdots, 6\}.$$

Thus, an elementary event ω consists of two integers indicating the results i_1 and i_2 of D_1 and D_2, respectively. Let $A = \{i_1 + i_2 \leq 3\}$ and $B = \{i_1/i_2 = 2\}$. Then,

$$A = \{(1, 1), (1, 2), (2, 1)\}, \quad B = \{(2, 1), (4, 2), (6, 3)\}.$$

Hence,

$$A \cap B = \{(2, 1)\}, \quad A \cup B = \{(1, 1), (1, 2), (2, 1), (4, 2), (6, 3)\}$$

and

$$A \backslash B = \{(1, 1), (1, 2)\}. \qquad □$$

Example 1.3 A company is provided with power by three generators G_1, G_2, and G_3. The company has sufficient power to maintain its production if only two out of the three generators are operating. Let A_i be the event that generator G_i, $i = 1, 2, 3$, is operating, and B be the event that at least two generators are operating. Then,

$$B = A_1 A_2 A_3 \cup A_1 A_2 \overline{A_3} \cup A_1 \overline{A_2} A_3 \cup \overline{A_1} A_2 A_3. \qquad □$$

1.3 PROBABILITY

The aim of this section consists in constructing rules for determining the probabilities of random events. Such a rule is principally given by a function P on the set E of all random events A: $P = P(A)$, $A \in E$.

Note that in this context A is an element of the set E so that the notation $A \subseteq E$ would not be correct. Moreover, not all subsets of Ω need to be random events, i.e., the set E need not necessarily be the set of all possible subsets of Ω.

The function P assigns to every event A a number $P(A)$, which is its probability. Of course, the construction of such a function cannot be done arbitrarily. It has to be done in such a way that some obvious properties are fulfilled. For instance, if A implies the occurrence of the event B, i.e. $A \subseteq B$, the B occurs more frequently than A so that the relation $P(A) \leq P(B)$ should be valid. If in addition the function P has properties $P(\emptyset) = 0$ and $P(\Omega) = 1$, then the probabilities of random events yield indeed the desired information about their degree of uncertainty: The closer $P(A)$ is to 0, the more unlikely is the occurrence of A, and the closer $P(A)$ is to 1, the more likely becomes the occurrence of A.

To formalize this intuitive approach, let for now $P = P(A)$ be a function on E with properties

I) $P(\emptyset) = 0, \quad P(\Omega) = 1,$ II) If $A \subseteq B$, then $P(A) \leq P(B)$.

As a corollary from these two properties we get the following property of P:

III) For any event $A, \ 0 \leq P(A) \leq 1.$

1.3.1 Classical Definition of Probability

The classical concept of probability is based on the following two assumptions:

1) The space Ω of the elementary events is finite.

2) As a result of the underlying random experiment, each elementary event has the same probability to occur.

A random experiment with properties 1 and 2 is called a *Laplace random experiment*.

Let n be the total number of elementary events (i.e. the cardinality of Ω). Then any random event $A \subseteq \Omega$ consisting of m elementary events has probability

$$P(A) = m/n. \tag{1.4}$$

Let $\Omega = \{a_1, a_2, \cdots, a_n\}$. Then every elementary event has probability

$$P(a_i) = 1/n, \quad i = 1, 2, ..., n.$$

Obviously, this definition of probability satisfies the properties I, II, and III listed above. The integer m is said to be the number of *favorable cases* (for the occurrence of A), and n is the number of *possible cases*.

The classical definition of probability arose in the Middle Ages to be able to determine the chances to win in various games of chance. Then formula (1.4) is applicable given that the players are honest and do not use marked cards or manipulated dice. For instance, what is the probability of the event A that throwing a die yields an even number? In this case, $A = \{2, 4, 6\}$ so that $m = 3$ and $P(A) = 3/6 = 0.5$.

Example 1.4 When throwing 3 dice, what is more likely, to achieve the total sum 9 (event A_9) or the total sum 10 (event A_{10})? The corresponding sample space is

$$\Omega = \{(i, j, k), \ 1 \leq i, j, k \leq 6\} \text{ with } n = 6^3 = 216$$

possible outcomes. The integers 9 and 10 can be represented a as sum of 3 positive integers in the following ways:

$$9 = 3 + 3 + 3 = 4 + 3 + 2 = 4 + 4 + 1 = 5 + 2 + 2 = 5 + 3 + 1 = 6 + 2 + 1,$$
$$10 = 4 + 3 + 3 = 4 + 4 + 2 = 5 + 3 + 2 = 5 + 4 + 1 = 6 + 2 + 2 = 6 + 3 + 1.$$

The sum 3+3+3 corresponds to the event $A_{333} =$ 'every die shows a 3' $= \{(3, 3, 3)\}$. The sum 4+3+2 corresponds to the event A_{432} that one die shows a 4, another die a 3, and the remaining one a 2:

$$A_{432} = \{(2,3,4),(2,4,3),(3,2,4),(3,4,2),(4,2,3),(4,3,2)\}.$$

Analogously,

$$A_{441} = \{(1,4,4),(4,1,4),(4,4,1)\}, \quad A_{522} = \{(2,2,5),(2,5,2),(5,5,2),$$

$$A_{531} = \{(1,3,5),(1,5,3),(3,1,5),(3,5,1),(5,1,3),(5,3,1)\},$$

$$A_{621} = \{(1,2,6),(1,6,2),(2,1,6),(2,6,1),(6,1,2),(6,2,1)\}.$$

Corresponding to the given sum representations for 9 and 10, the numbers of favorable elementary events belonging to the events A_9 and A_{10}, respectively, are

$$m_A = 1+6+3+3+6+6 = 25, \quad m_B = 2+3+6+6+3+6 = 27.$$

Hence, the desired probabilities are:

$$P(A_9) = 25/216 = 0.116, \quad P(A_{10}) = 27/216 = 0.125.$$

The dice gamblers of the Middle Ages could not mathematically prove this result, but from their experience they knew that $P(A_9) < P(A_{10})$.   □

Example 1.5 d dice are thrown at the same time.

What is the smallest number $d = d^*$ with property that the probability of the event $A = $ 'no die shows a 6' does not exceed 0.1?

The problem makes sense, since with increasing d the probability $P(A)$ tends to 0, and if $d = 1$, then $P(A) = 5/6$. For $d \geq 1$, the corresponding space of elementary events Ω has $n = 6^d$ elements, namely the vectors $(i_1, i_2, \cdots, i_d)$, where the i_k are integers between 1 and 6. Amongst the 6^d elementary events those are favorable for the occurrence of A, where the i_k only assume integers between 1 and 5. Hence, for the occurrence of A exactly 5^d elementary events are favorable:

$$P(A) = 5^d/6^d.$$

The inequality $5^d/6^d \leq 0.1$ is equivalent to

$$d\,(\ln 5/6) \leq \ln(0.1) \quad \text{or} \quad d\,(-0.1823) \leq -2.3026 \quad \text{or} \quad d \geq \frac{2.3026}{0.1823} = 12.63.$$

Hence, $d^* = 13$.   □

Binomial Coefficient and Faculty For solving the next examples, we need a result from elementary combinatorics: The number of possibilities to select subsets of k different elements from a set of n different elements, $1 \leq k \leq n$, is given by the *binomial coefficient* $\binom{n}{k}$, which is defined as

$$\binom{n}{k} = \frac{n\,(n-1)\cdots(n-k+1)}{k!}, \quad 1 \leq k \leq n, \tag{1.5}$$

where $k!$ is the *faculty* of k: $k! = 1 \cdot 2 \cdots k$. By agreement

$$\binom{n}{0} = 1 \quad \text{and} \quad 0! = 1.$$

The faculty of a positive integer has its own significance in combinatorics:

| *There are k! different possibilities to order a set of k different objects.*

Example 1.6 An optimist buys one ticket in a '6 out of 49' lottery and hopes for hitting the jackpot. What are his chances? There are

$$\binom{49}{6} = \frac{49 \cdot 48 \cdot 47 \cdot 46 \cdot 45 \cdot 44}{6!} = 13\,983\,816$$

different possibilities to select 6 numbers out of 49. Thus, one has to fill in almost 14 million tickets to make absolutely sure that the winning one is amongst them. It is $m = 1$ and $n = 13\,983\,816$. Hence, the probability p_6 of having picked the six 'correct' numbers is

$$p_6 = \frac{1}{13\,983\,816} = 0.0000000715.$$ □

The classical definition of probability satisfies the properties $P(\emptyset) = 0$ and $P(\Omega) = 1$, since the impossible event $\emptyset$ does not contain any elementary events ($m = 0$) and the certain event Ω comprises all elementary events ($m = n$).

Now, let A and B be two events containing m_A and m_B elementary events, respectively. If $A \subseteq B$, then $m_A \le m_B$ so that $P(A) \le P(B)$. If the events A and B are disjoint, then they have no elementary events in common so that $A \cup B$ contains $m_A + m_B$ elementary events. Hence

$$P(A \cup B) = \frac{m_A + m_B}{n} = \frac{m_A}{n} + \frac{m_B}{n} = P(A) + P(B)$$

or $$P(A \cup B) = P(A) + P(B) \quad \text{if} \quad A \cap B = \emptyset.$$ (1.6)

More generally, if $A_1, A_2, \cdots, A_r$ are pairwise disjoint events, then

$$P(A_1 \cup A_2 \cup \cdots \cup A_r) = P(A_1) + P(A_2) + \cdots + P(A_r), \quad A_i \cap A_k = \emptyset, \ i \ne k. \quad (1.7)$$

Example 1.7 When participating in the lottery '6 out of 49' with one ticket, what is the probability of the event A to have at least 4 correct numbers?

Let A_i be the event of having got i numbers correct. Then,

$$A = A_4 \cup A_5 \cup A_6.$$

A_4, A_5, and A_6 are pairwise disjoint events. (It is impossible that there are on one and the same ticket, say, exactly 4 and exactly 5 correct numbers.) Hence,

$$P(A) = P(A_4) + P(A_5) + P(A_6).$$

There are $\binom{6}{4} = 15$ possibilities to choose 4 numbers from the 6 'correct' ones. To each of these 15 choices there are

$$\binom{49 - 6}{6 - 4} = \binom{43}{2} = 903$$

possibilities to pick 2 numbers from the 43 'wrong' numbers. Therefore, favorable for the occurrence of A_4 are $m_4 = 15 \cdot 903 = 13\,545$ elementary events. Hence,

$$p_4 = P(A_4) = 13\,545/13\,983\,616 = 0.0009686336.$$

Analogously,

$$p_5 = P(A_5) = \frac{\binom{6}{5}\binom{49-6}{6-5}}{\binom{49}{6}} = \frac{6 \cdot 43}{\binom{49}{6}} = 0.0000184499.$$

Together with the result of example 1.6, $P(A) = p_4 + p_5 + p_6 = 0.0009871552$, i.e., almost 10 000 tickets have to be bought to achieve the desired result.   □

1.3.2 Geometric Definition of Probability

The geometric definition of probability is subject to random experiments, in which every outcome has the same chance to occur (as with Laplace experiments), but the sample space Ω is a bounded subset of the one, two or three dimensional Euklidian space (real line, plain, space). Hence, in each case Ω is a nondenumerable set. In most applications, Ω is a finite interval, a rectangular, a circle, a cube or a sphere.

Let $A \subseteq \Omega$ be a random event. Then we denote by $\mu(A)$ the *measure* of A. For instance, if Ω is a finite interval, then $\mu(\Omega)$ is the length of this interval. If A is the union of disjoint subintervals of Ω, then $\mu(A)$ is the total length of these subintervals. (We do not consider subsets like the set of all irrational numbers in a finite interval.) If Ω is a rectangular and A is a circle embedded in this rectangular, then $\mu(A)$ is the area of this circle and so on. If μ is defined in this way, then

$$A \subseteq B \subseteq \Omega \text{ implies } \mu(A) \leq \mu(B) \leq \mu(\Omega).$$

Under the assumptions stated, a probability is assigned to every event $A \subseteq \Omega$ by

$$P(A) = \frac{\mu(A)}{\mu(\Omega)}. \tag{1.8}$$

For disjoint events A and B, $\mu(A \cup B) = \mu(A) + \mu(B)$ so that formulas (1.6) and (1.7) are true again. Analogously to the classical probability, $\mu(A)$ can be interpreted as the measure of all elementary events, which are favorable to the occurrence of A. With the given interpretation of the measure $\mu(\cdot)$, every elementary event, i.e. every point in Ω, has measure and probability 0 (different to the Laplace random experiment). (A point, whether at a line, in a plane or space has always extension 0 in all directions.) But the assumption "every elementary event has the same chance to occur" is not equivalent to the fact that every elementary event has probability 0. Rather, this assumption has to be understood in the following sense:

| *All those random events, which have the same measure, have the same probability.*

Thus, never mind where the events (subsets of Ω) with the same measure are located in Ω and however small their measure is, the outcome of the random experiment will be in any of these events with the same probability, i.e., no area in Ω is preferred with regard to the occurrence of elementary events.

Example 1.8 For the sake of a tensile test, a wire is clamped at its ends so that the free wire has a length of 400 *cm*. The wire is supposed to be homogeneous with regard to its physical parameters. Under these assumptions, the probability p that the wire will tear up between 0 and 40 *cm* or 360 and 400 *cm* is

$$p = \frac{40 + 40}{400} = 0.2.$$

Repeated tensile tests will confirm or reject the assumption that the wire is indeed homogeneous.   □

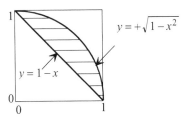

Figure 1.2 Illustration to example 1.9

Example 1.9 Two numbers x and y are randomly picked from the interval $[0, 1]$. What is the probability that x and y satisfy both the conditions

$$x + y \geq 1 \quad \text{and} \quad x^2 + y^2 \leq 1?$$

Note: In this context, 'randomly' means that every number between 0 and 1 has the same chance of being picked.

In this case the sample space is the unit square $\Omega = [0 \leq x \leq 1, 0 \leq y \leq 1]$, since an equivalent formulation of the problem is to pick at random a point out of the unit square, which is favorable for the occurrence of the event

$$A = \{(x,y); x + y \geq 1, x^2 + y^2 \leq 1\}.$$

Figure 1.2 shows the area (hatched) given by A, whereas the 'possible area' Ω is left white, but also includes the hatched area. Since $\mu(\Omega) = 1$ and $\mu(A) = \pi/4 - 0.5$ (area of a quarter of a circle with radius 1 minus the area of the half of a unit square),

$$P(A) = \mu(A) \approx 0.2854.$$ □

Example 1.10 (***Buffon's needle problem***) At an even surface, parallel straight lines are drawn at a distance of a *cm*. At this surface a needle of length L is thrown, $L < a$. What is the probability of the event A that the needle and a parallel intersect?

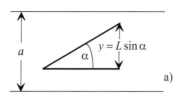
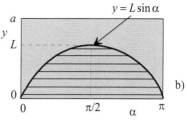

Figure 1.3 Illustration to example 1.10

The position of the needle at the surface is fully determined by its distance of its 'lower' endpoint to the 'upper' parallel and by its angle of inclination α to the parallels (Figure 1.3a), since a shift of the needle parallel to the lines obviously has no influence on the desired probability. Thus, the sample space is given by the rectangle

$$\Omega = \{(y, \alpha),\, 0 \le y \le a,\, 0 \le \alpha \le \pi\}$$

with area $\mu(\Omega) = a\pi$ (Figure 1.3b). Hence, Buffon's needle problem formally consists in randomly picking elementary events given by (y, α) from the rectangle Ω. Since the needle and the upper parallel intersect if and only if $y < L \sin \alpha$, the favorable area for the occurrence of A is given by the hatched part in Figure 1.3b. The area of this part is

$$\mu(A) = \int_0^\pi L \sin \alpha \, d\alpha = L\,[-\cos \alpha]_0^\pi = L[1+1] = 2\,L.$$

Hence, the desired probability is $P(A) = 2\,L/a\pi$.   □

1.3.3 Axiomatic Definition of Probability

The classical and the geometric concepts of probability are only applicable to very restricted classes of random experiments. But these concepts have illustrated which general properties a universally applicable probability definition should have:

Definition 1.1 A function $P = P(A)$ on the set of all random events E with $\varnothing \in E$ and $\Omega \in E$ is called *probability* if it has the following properties:

I) $P(\Omega) = 1.$

II) For any $A \in E,\ 0 \le P(A) \le 1.$

III) For any sequence of disjoint events $A_1, A_2, \ldots$, i.e., $A_i \cap A_j = \varnothing$ for $i \neq j$,

$$P\left(\bigcup_{i=1}^{\infty} A_i\right) = \sum_{i=1}^{\infty} P(A_i). \qquad (1.9)$$

●

Property III makes sense only if with $A_i \in E$ the union $\bigcup_{i=1}^{\infty} A_i$ is also an element of E. Hence we assume that the set of all random events E is a σ-algebra:

Definition 1.2 Any set of random events E is called a $\sigma-algebra$ if

1) $\Omega \in E$.

2) If $A \in E$, then $\overline{A} \in E$. In particular, $\overline{\Omega} = \varnothing \in E$.

3) For any sequence $A_1, A_2, \ldots$ with $A_i \in E$, the union $\bigcup_{i=1}^{\infty} A_i$ is also a random event, i.e.,

$$\bigcup_{i=1}^{\infty} A_i \in E.$$

$[\Omega, E]$ is called a *measurable space*, and $[\Omega, E, P]$ is called a *probability space*.   ●

Note: In case of a finite or a countably infinite set Ω, the set E is usually the *power set* of Ω, i.e. the set of all subsets of Ω. A power set is, of course, always a σ–algebra. In this book, taking into account its applied orientation, specifying explicitly the underlying σ– algebra is usually not necessary. $[\Omega, E]$ is called a *measurable space,* since to any random event $A \in E$ a *measure*, namely its probability, can be assigned. In view of the de Morgan rules (1.1): If A and B are elements of E, then $A \cap B$ as well.

Given that E is a σ–algebra, properties I–III of definition 1.1 imply all the properties of the probability functions, which we found useful in sections 1.3.1 and 1.3.2:

a) Let $A_i = \varnothing$ for $i = n + 1,\ n + 2, \cdots$. Then, from III),

$$P(\textstyle\bigcup_{i=1}^{n} A_i) = \sum_{i=1}^{n} P(A_i), \quad A_i \cap A_j = \varnothing,\ i \ne j,\ i,j = 1, 2, \cdots, n. \tag{1.10}$$

In particular, letting $n = 2$ and $A = A_1,\ B = A_2,$ this formula implies

$$P(A \cup B) = P(A) + P(B) \ \text{if} \ A \cap B = \varnothing. \tag{1.11}$$

With $B = \overline{A},$ taking into account $\Omega = A \cup \overline{A}$ and $P(\Omega) = 1$, formula (1.11) yields

$$P(A) + P(\overline{A}) = 1 \ \text{or} \ P(\overline{A}) = 1 - P(A). \tag{1.12}$$

Applying (1.12) with $A = \Omega$ yields $P(\Omega) + P(\varnothing) = 1$, so that

$$P(\varnothing) = 0, \quad P(\Omega) = 1. \tag{1.13}$$

Note that $P(\Omega) = 1$ is part of definition 1.1.

b) If A and B are two events with $A \subseteq B,$ then B can be represented as $B = A \cup (B \backslash A).$ Since A and $B \backslash A$ are disjoint, by (1.11), $P(B) = P(A) + P(B \backslash A)$ or, equivalently,

$$P(B \backslash A) = P(B) - P(A) \ \text{if} \ A \subseteq B. \tag{1.14}$$

Therefore,
$$P(A) \le P(B) \ \text{if} \ A \subseteq B. \tag{1.15}$$

c) For any events A and B, the event $A \cup B$ can be represented as follows (Figure 1.1)

$$A \cup B = \{A \backslash A \cap B)\} \cup \{B \backslash (A \cap B)\} \cup (A \cap B).$$

In this representation, the three events combined by '$\cup$' are disjoint. Hence, by (1.10) with $n = 3$:

$$PA \cup B) = P(\{A \backslash A \cap B)\}) + P(\{B \backslash (A \cap B)\}) + P(A \cap B).$$

On the other hand, since $(A \cap B) \subseteq A$ and $(A \cap B) \subseteq B$, from (1.14),

$$P(A \cup B) = P(A) + P(B) - P(A \cap B). \tag{1.16}$$

Given any 3 events A, B, and C, the probability of the event $A \cup B \cup C$ can be determined by replacing in (1.16) A with $A \cup B$ and B with C. This yields

$$P(A \cup B \cup C) = P(A) + P(B) + P(C) - P(A \cap B) - P(A \cap C) - P(B \cap C)$$
$$+ P(A \cap B \cap C) \tag{1.17}$$

d) For any n events $A_1, A_2, ..., A_n$ one obtains by repeated application of (1.16) (more exactly, by induction) the *Inclusion-Exclusion Formula* or the *Formula of Poincaré* for the probability of the event $A_1 \cup A_2 \cup \cdots \cup A_n$:

$$P(A_1 \cup A_2 \cup \cdots \cup A_n) = \sum_{k=1}^{n} (-1)^{k+1} R_k \tag{1.18}$$

with
$$R_k = \sum_{(i_1 < i_2 < \cdots < i_k)}^{n} P(A_{i_1} \cap A_{i_2} \cap \cdots \cap A_{i_k}),$$

where the summation runs over all k-dimensional vectors $(i_1, i_2, ..., i_k)$ out of the set $\{1, 2, ..., n\}$ with $1 \leq i_1 < i_2 < \cdots < i_k \leq n$ and $k = 1, 2, ..., n$. The sum representing R_k has exactly $\binom{n}{k}$ terms, so that the total number of terms in (1.18) is

$$\sum_{k=1}^{n} \binom{n}{k} = 2^n - 1.$$

For instance, if $n = 3$, then the R_k in (1.18) are

$$R_1 = P(A_1) + P(A_2) + P(A_3),$$
$$R_2 = P(A_1 \cap A_2) + P(A_1 \cap A_3) + P(A_2 \cap A_3),$$
$$R_3 = P(A_1 \cap A_2 \cap A_3).$$

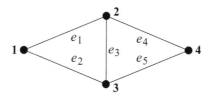

Figure 1.4 Computer network with 4 computers

Example 1.11 Figure 1.4 shows a simple local computer network. Computers are located at nodes 1, 2, 3, and 4. The transmission of data between the computers is possible via cables $e_1, e_2, \cdots, e_5$, which link the four computers. Cable e_i is available, i.e. in a position to transfer information, with probability p_i and unavailable (e.g. under maintenance, waiting for maintenance, waiting for replacement for having been stolen) with probability $q_i = 1 - p_i$, $i = 1, 2, ..., 5$.

What is the probability of the event A that the computer at node 1 can transfer data to the computer at node 4 via one or more paths (chains) of available edges which connect node 1 to node 4? There are four potential candidates for such paths:

$$w_1 = \{e_1, e_4\}, \quad w_2 = \{e_2, e_5\}, \quad w_3 = \{e_1, e_3, e_5\}, \quad w_4 = \{e_2, e_3, e_4\}.$$

Let A_i be the event that all edges in path w_i are available, $i = 1, 2, 3, 4$. Then event A occurs when at least one of these four events occurs. Hence, A can be represented as

$$A = A_1 \cup A_2 \cup A_3 \cup A_4.$$

The A_i are not disjoint. Hence we apply the inclusion-exclusion formula (1.11) for representing A:

$$P(A) = P(A_1 \cup A_2 \cup A_3 \cup A_4) = R_1 - R_2 + R_3 - R_4$$

with

$$R_1 = P(A_1) + P(A_2) + P(A_3) + P(A_4),$$
$$R_2 = P(A_1 \cap A_2) + P(A_1 \cap A_3) + P(A_1 \cap A_4) + P(A_2 \cap A_3) + P(A_2 \cap A_4)$$
$$+ P(A_2 \cap A_4) + P(A_3 \cap A_4),$$
$$R_3 = P(A_1 \cap A_2 \cap A_3) + P(A_1 \cap A_2 \cap A_4) + P(A_1 \cap A_3 \cap A_4) + P(A_2 \cap A_3 \cap A_4),$$
$$R_4 = P(A_1 \cap A_2 \cap A_3 \cap A_4).$$

The event $A_1 \cap A_2$ means that both the cables in A_1 and in A_2 are operating. Thus, to the event $A_1 \cap A_2$ there belongs the set of cables $\{e_1, e_2, e_4, e_5\}$. Hence, the notation $P(A_1 \cap A_2) = p_{1245}$ will be used. To the event $A_1 \cap A_2 \cap A_3$ there belongs the set of cables $\{e_1, e_2, e_3, e_4, e_5\}$: $P(A_1 \cap A_2 \cap A_3) = p_{12345}$. If this way of notation is applied to all other probabilities occurring in the R_i, then

$$R_1 = p_{14} + p_{25} + p_{135} + p_{234},$$

$$R_2 = p_{1245} + p_{1345} + p_{1234} + p_{1235} + p_{2345} + p_{12345},$$

$$R_3 = p_{12345} + p_{12345} + p_{12345} + p_{12345}, \quad R_4 = p_{12345}.$$

The desired probability is

$$P(A) = p_{14} + p_{25} + p_{135} + p_{234} - p_{1245} - p_{1345} - p_{1234} - p_{1235} - p_{2345} + 3p_{12345}.$$

In section 1.4.2, an additional assumption on the operation modus of the cables will be imposed which enables the calculation of $P(A)$ only on the basis of the p_i.   □

1.3.4 Relative Frequency

The probabilities of random events are usually unknown. However, they can be estimated by their relative frequencies. If in a series of n repetitions of one and the same random experiment the event A has been observed exactly $m = m(A)$ times, then the *relative frequency* of A is given by

$$\hat{p}_n(A) = \frac{m(A)}{n}. \tag{1.19}$$

Generally, the relative frequency of A tends to $P(A)$ as n increases:

$$\lim_{n \to \infty} \hat{p}_n(A) = P(A). \tag{1.20}$$

Thus, the probability of A can be estimated with any required level of accuracy from its relative frequency by sufficiently frequently repeating the random experiment (for the theoretical background see section 5.2.2). Empirical verifications of the limit relation (1.20) were aleady given in the introduction by the coin experiments of *Buffon* and *Pearson*. Without the validity of (1.20) the gamblers in the Middle Ages would not have been in a position to empirically verify that, when throwing three dice, the chance to obtain sum 9 is lower than the chance to obtain sum 10 (example 1.4).

It is interesting that the relationship (1.20) in connection with Buffon's needle problem (example 1.10) allows to estimate the number π with any desired degree of accuracy. To do this, in the formula $P(A) = 2L/\pi a$ the probability $P(A)$ is replaced with the relative frequency $\hat{p}_n(A)$ for the occurrence of A in a series of n needle throwings. This gives for π the estimate

$$\hat{\pi}_n = \frac{2L}{a\hat{p}_n(A)}.$$

Lazzarini (1901) threw the needle $n = 3408$ times and got for π the estimate

$$\hat{\pi}_{3408} = 3.141529,$$

i.e., the first six figures are the exact ones. The approximate calculation of π was one of the first examples how to solve deterministic problems by probabilistic methods. Nowadays, nobody needs to throw a needle manually several tousand times. Computers 'simulate' random experiments of this simple structure many thousand times in a twinkling of an eye.

The reader may object that the approximate calculation of probabilities of all random events by their relative frequency is practically not possible, in particular, if the sample space is not finite. However, depending on the respective random experiment, the probabilities of all its elementary events are frequently given by a unifying mathematical pattern (model). For instance, the probability that the random number of traffic accidents occurring in a specific area during a year is equal to k can frequently be determined by the formula

$$p_k = \frac{\lambda^k}{k!} e^{-\lambda}; \quad k = 0, 1, \dots,$$

where λ is the average number of traffic accidents which occur a year in that area. Hence, for determining all infinitely many probabilities $p_0, p_1, \dots$, only the parameter λ has to be estimated. This is done by counting the number x_i of traffic accidents occurring in year i over a period of n years and determining the arithmetic mean

$$\hat{\lambda} = \frac{1}{n} \sum_{i=1}^{n} x_i.$$

Defining and discussing mathematical models for the calculation of the probabilities of random events is the subject of chapter 2.

1.4 CONDITIONAL PROBABILITY AND INDEPENDENCE
OF RANDOM EVENTS

1.4.1 Conditional Probability

Two random events A and B can depend on each other in the following sense: The occurrence of B will change the probability of the occurrence of A and vice versa. Hence, the additional piece of information 'B has occurred' should be used in order to predict the probability of the occurrence of A more precisely. If one has to determine the probability that a device does not fail during its guarantee period (event A), then this probability may depend on the manufacturer of the device (event B) if there are several of them who produce the same type. The probability of having a sunny day on 21 August (event A) will increase if there is a sunny day on 20 August (event B) in view of the inertia of weather patterns. The probability of attracting a certain disease (event A) will usually be larger than average if there was/is a family member, who had suffered from this disease (event B). If A is the random event to spot a leopard in a certain area of a National Park during a safari, then the probability of A increases if it is known that there are baboons in this area (event B).

Let us now consider some numerical examples to illustrate how to define the probability of the occurrence of an event A given that another event B has occurred.

Example 1.12 A gambler throws the dice 1 and 2 simultaneously. What is the probability that die 1 shows a 6 (event A) on condition that both dice showed an even number (event B). This probability will be denoted as $P(A|B)$. The sample space is

$$\Omega = \{(i,j); \ i,j = 1, 2, ..., 6\}.$$

In terms of the elementary events (i,j), the events A and B are given by

$$A = \{(6,1), (6,2), (6,3), (6,4), (6,5), (6,6)\},$$
$$B = \{(2,2), (2,4), (2,6), (4,2), (4,4), (4,6), (6,2), (6,4), (6,6)\}.$$

Hence,

$$P(A) = 6/36 \ \text{ and } \ P(B) = 9/36.$$

On condition 'B has occurred' the sample space Ω reduces to the 9 elementary events given by B. From these 9, only the 3 elementary events in the conjunction

$$A \cap B = \{(6,2), (6,4), (6,6)\}$$

are favorable for the occurrence of A: Therefore,

$$P(A|B) = 3/9.$$

The following representation shows the general structure of $P(A|B)$:

$$P(A|B) = 1/3 = \frac{3/36}{9/36} = \frac{P(A \cap B)}{P(B)}. \qquad \square$$

Example 1.13 In a bowl there are two white and two red marbles. The numbers 1 and 2 are assigned to the white marbles and the numbers 3 and 4 are assigned to the red marbles. Two marbles are one after the other randomly picked from the bowl. Find the probability of the event A that one of the drawn marbles is white and the other red given the event B that the first drawn marble is white.

The sample space consists of $4 \cdot 3 = 12$ elementary events:

$$\Omega = \{(i,j);\ i \neq j,\ i,j = 1,2,3,4\}.$$

The events A and B are given by the following sets of elementary events:

$$A = \{(1,3),(1,4),\ (2,3),\ (2,4),\ (3,1),\ (3,2),\ (4,1),\ (4,2)\},$$
$$B = \{(1,2),(1,3),\ (1,4),\ (2,1),\ (2,3),\ (2,4)\}.$$

Hence,

$$P(A) = 8/12 = 2/3 \quad \text{and} \quad P(B) = 6/12 = 1/2.$$

Since it is known that event B has happened, the space of possible elementary events is given by B. Hence, the elementary events which are favorable for the occurrence of event A are given by the conjunction

$$A \cap B = \{(1,3),\ (1,4),\ (2,3),\ (2,4)\}.$$

This yields

$$P(A|B) = \frac{4}{6} = \frac{2}{3} = \frac{4/12}{6/12} = \frac{P(A \cap B)}{P(B)}.$$

For the sake of arriving at the general structure of $P(A|B)$, solution of the problem had been unnecessarily complicated. The problem is namely quickly solved as follows: If the first drawn marble is white (event B), then there are one white and two red marbles left in the bowl. Event A occurs if one of the red marbles will be drawn, i.e., $P(A|B) = 2/3$. □

Example 1.14 The lifetimes of $n = 1000$ electronic units had been tested. 205 units failed in the interval $[0,\ 500\,h)$, 180 units failed in the interval $[500,\ 600\,h)$, and the remaining 615 units failed after $600\,h$. Let A be the event that a unit fails in the interval $[500,\ 600\,h)$, and B be the event that a unit fails after a lifetime of at least $500\,h$. By formula (1.19) with $n = 1000$, the relative frequencies for the occurrence of events A and B are

$$\hat{p}_n(A) = \frac{m(A)}{n} = \frac{180}{1000}, \quad \hat{p}_n(B) = \frac{m(B)}{n} = \frac{1000 - 205}{1000} = 0.795.$$

What is the relative frequency $\hat{p}_n(A|B)$ of the event A on condition that event B has occurred?

Under this condition, only the 795 units, which have survived the first $500\,h$, need to be taken into account. From these 795 units, 180 fail in $[500,\ 600\,h)$. Therefore,

$$\hat{p}_n(A|B) = \frac{180}{795} = 0.2264.$$

Since $A \subseteq B$, i.e. the occurrence of A implies the occurrence of B, event A satisfies $A = A \cap B$. Hence, the 'conditional relative frequency' $\hat{p}_n(A|B)$ can be written as

$$\hat{p}_n(A|B) = \frac{m(A \cap B)}{m(B)} = \frac{\frac{m(A \cap B)}{n}}{\frac{m(B)}{n}}. \qquad (1.21)$$

By (1.20), the relative frequencies $\frac{m(A \cap B)}{n}$ and $\frac{m(B)}{n}$ tend to $P(A \cap B)$ and $P(B)$ as $n \to \infty$, respectively. Thus, the conditional probability of A given B has again the structure we know from the previous examples:

$$\lim_{n \to \infty} \hat{p}_n(A|B) = P(A|B) = \frac{P(A \cap B)}{P(B)}. \qquad \square$$

Now it is no longer surprising that the probability of 'A on condition B' or, equivalently, the probability of 'A given B' is defined as follows.

Definition 1.3 Let A and B be two events with $P(B) > 0$. Then the *probability* of A on *condition B* is given by

$$P(A|B) = \frac{P(A \cap B)}{P(B)}. \qquad (1.22)$$

$\bullet$

Note: $P(A|B)$ is also denoted as the *probability of A given B, the conditional probability of A on condition B, or the conditional probability of A given B*. Of course, in (1.22) the roles of A and B can be changed.

If A and B are arbitrary random events, formula (1.22) implies a *product formula* for the probability $P(A \cap B)$ of the joint occurrence of arbitrary events A and B:

$$P(A \cap B) = P(A|B) P(B) \quad \text{or} \quad P(A \cap B) = P(B|A) P(A). \qquad (1.23)$$

Example 1.15 In a bowl there are three white and two red marbles. Two marbles are randomly taken out one after the other. What is the probability that both of these marbles are red?

Let be A and B be the events that the first and the second, respectively, of the chosen marbles are red. Hence, the probability $P(A \cap B)$ has to be determined. The probability of A is equal to $P(A) = 2/5$. On condition A, there are 3 white and 1 red marble in the bowl. Hence, $P(B|A) = 1/4$ so that

$$P(A \cap B) = P(B|A) P(A) = \frac{1}{4} \cdot \frac{2}{5} = 0.1. \qquad \square$$

Example 1.16 In a study, data from a sample of 12 000 persons had been collected. 4800 persons in this sample were obese and 3600 suffered from diabetes 2. From the diabetes sufferers, 2700 were obese. A person is randomly selected from the sample of 12 000 persons. It happens to be Max. Let A be the event that Max is obese, and B be the event that Max has diabetes 2. Then

$$P(A) = 0.4, \ P(B) = 0.3, \ \text{and} \ P(A|B) = 2700/3600 = 0.75.$$

Hence, the probability that Max is both obese and a diabetes 2 sufferer is, by (1.22),

$$P(A \cap B) = P(A|B)P(B) = 0.75 \cdot 0.3 = 0.225.$$

2) To see whether being obese increases the probability of contracting diabetes 2, the probability $P(B|A)$ has to be determined: From the right equation of (1.23),

$$P(A \cap B) = 0.225 = P(B|A)P(A) = P(B|A) \cdot 0.4.$$

Hence, $P(B|A) = 0.5625$. Thus, based on this study, being obese increases the probability of contracting diabetes 2.   □

1.4.2 Total Probability Rule and Bayes' Theorem

Frequently several mutually exclusive conditions have influence on the occurrence of a random event A. The whole of these conditions are known, but it is not known, which of these conditions is taking effect. However, the probabilities are known which of these conditions affects the occurrence of A at the time point of interest. Under these assumptions, a formula for the occurrence of A will be derived. But next the procedure is illustrated by an example.

Example 1.17 A machine is subject to two stress levels 1 (event B_1) and 2 (event B_2) with respective probabilities 0.8 and 0.2. Stress levels can be determined e.g. by different production conditions as speed, pressu,re or humidity. It is supposed that the stress level does not change during a fixed working period (hour, day). Given stress level 1 or 2, the machine will fail during a working period with probability 0.3 or 0.6, respectively. Hence,

$$P(A|B_1) = 0.3, \ P(A|B_2) = 0.6.$$

Since the events B_1 and B_2 are disjoint (mutually exclusive) and $\Omega = B_1 \cup B_2$ is the certain event, A can be represented as

$$A = A \cap \Omega = A \cap (B_1 \cup B_2) = (A \cap B_1) \cup (A \cap B_2).$$

The events $A \cap B_1$ and $A \cap B_2$ are disjoint so that by formula (1.11)

$$P(A) = P(A \cap B_1) + P(A \cap B_2).$$

By applying (1.23) to each of the two terms on the right-hand side of this formula,

$$P(A) = P(A|B_1)P(B_1) + P(A|B_2)P(B_2)$$

$$= 0.3 \cdot 0.8 + 0.6 \cdot 0.2 = 0.36.$$

Thus, without information on the respective stress level, the failure probability of the machine in the working period is 0.36.   □

Now the principle, illustrated by this example, is formulated more generally:

Definition 1.4 The set of random events $\{B_1, B_2, ..., B_n, n \leq \infty\}$ is an *exhaustive set of random events* for Ω if

$$\Omega = \bigcup_{i=1}^{n} B_i,$$

and it is a mutually disjoint set of events if

$$B_i \cap B_j = \varnothing, \quad i \neq j, \quad i, j = 1, 2, ..., n.$$

A mutually disjoint and exhaustive (for Ω) set of events is called a *partition* of Ω. ●

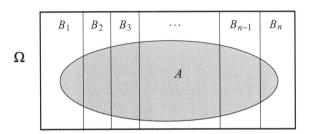

Figure 1.5 Partition of a sample space

Let $\{B_1, B_2, ..., B_n\}$ be an exhaustive and mutually disjoint set of events with property $P(B_i) > 0$ for all $i = 1, 2, ..., n$, and let A be an event with $P(A) > 0$. Then A can be represented as follows (see Figure 1.5):

$$P(A) = \bigcup_{i=1}^{n} (A \cap B_i).$$

Since the B_i are disjoint, the conjunctions $A \cap B_i$ are disjoint as well. Formula (1.10) is applicable and yields $P(A) = \sum_{i=1}^{n} P(A \cap B_i)$. Now formula (1.23) applied to all n probabilities $P(A \cap B_i)$ yields

$$P(A) = \sum_{i=1}^{n} P(A|B_i) P(B_i). \tag{1.24}$$

This result is called the *Formula of total probability* or the *Total probability rule.*

Moreover, formulas (1.22) and (1.23) yield

$$P(B_i|A) = \frac{P(B_i \cap A)}{P(A)} = \frac{P(A \cap B_i)}{P(A)} = \frac{P(A|B_i) P(B_i)}{P(A)}.$$

If $P(A)$ is replaced with its representation (1.24), then

$$P(B_i|A) = \frac{P(A|B_i) P(B_i)}{\sum_{i=1}^{n} P(A|B_i) P(B_i)}, \quad i = 1, 2, ..., n. \tag{1.25}$$

Formula (1.25) is called *Bayes' theorem* or *Formula of Bayes.* For obvious reasons, the probabilities $P(B_i)$ are called *a priori probabilities* and the conditional probabilities $P(B_i|A)$ *a posteriori probabilities.*

Example 1.18 The manufacturers M_1, M_2, and M_3 delivered to a supermarket a total of 1000 fluorescent tubes of the same type with shares 200, 300, and 500, respectively. In these shares, there are in this order 12, 9, and 5 defective tubes.

1) What is the probability that a randomly chosen tube is not defective?

2) What is the probability that a defective tube had been produced by M_i, $i = 1, 2, 3$?

Let events A and B_i be introduced as follows:

A = 'A tube, randomly chosen from the whole delivery, is not defective.'

B_i = 'A tube, randomly chosen from the whole delivery, is from M_i, $i = 1, 2, 3$.'

According to the figures given:

$$P(B_1) = 0.2, \quad P(B_2) = 0.3, \quad P(B_3) = 0.5,$$

$$P(A|B_1) = 12/200 = 0.06, \quad P(A|B_2) = 9/300 = 0.03, \quad P(A|B_3) = 5/500 = 0.01.$$

$\{B_1, B_2, B_3\}$ is a set of exhaustive and mutually disjoint events, since there are no other manufacturers delivering tubes of this brand to that supermarket and no two manufacturers can have produced one and the same tube.

1) Formula (1.23) yields

$$P(A) = 0.06 \cdot 0.2 + 0.03 \cdot 0.3 + 0.01 \cdot 0.5 = 0.026.$$

2) Bayes' theorem (1.25) gives the desired probabilities:

$$P(B_1|A) = \frac{P(A|B_1)P(B_1)}{P(A)} = \frac{0.06 \cdot 0.2}{0.026} = 0.4615,$$

$$P(B_2|A) = \frac{P(A|B_2)P(B_2)}{P(A)} = \frac{0.03 \cdot 0.3}{0.026} = 0.3462,$$

$$P(B_3|A) = \frac{P(A|B_3)P(B_3)}{P(A)} = \frac{0.01 \cdot 0.5}{0.026} = 0.1923.$$

Thus, despite having by far the largest proportion of tubes in the delivery, the high quality of tubes from manufacturer M_3 guarantees that a defective tube is most likely not produced by this manufacturer. □

Example 1.19 1% of the population in a country are HIV-positive. A test procedure for diagnosing whether a person is HIV-positive indicates with probability 0.98 that the person is HIV-positive if indeed he/she is HIV-positive, and with probability 0.96 that this person is not HIV-positve if he/she is not HIV-positive.

1) What is the probability that a test person is HIV-positive if the test indicates that?

To solve the problem, random events A and B are introduced:

A = 'The test indicates that a person is HIV-positive.'

B = 'A test person is HIV-positive.'

Then, from the figures given,

$$P(B) = 0.01, \quad P(\overline{B}) = 0.99$$

$$P(A|B) = 0.98, \quad P(\overline{A}|B) = 0.02, \quad P(\overline{A}|\overline{B}) = 0.96, \quad P(A|\overline{B}) = 0.04.$$

Since $\{B, \overline{B}\}$ is an exhaustive and disjoint set of events, the total probability rule (1.23) is applicable to determining $P(A)$:

$$P(A) = P(A|B) P(B) + P(A|\overline{B}) P(\overline{B}) = 0.98 \cdot 0.01 + 0.04 \cdot 0.99 = 0.0494.$$

Bayes' theorem (1.24) yields the desired probability $P(B|A)$:

$$P(B|A) = \frac{P(A|B) P(B)}{P(A)} = \frac{0.98 \cdot 0.01}{0.0494} = 0.1984.$$

Although the initial parameters of the test look acceptable, this result is quite unsatisfactory: In view of $P(\overline{B}|A) = 0.8016$, about 80% HIV-negative test persons will be shocked to learn that the test procedure indicates they are HIV-positive. In such a situation the test has to be repeated several times. The reason for this unsatisfactory numerical result is that only a small percentage of the population is HIV-positive.

2) The probability that a person is HIV-negative if the test procedure indicates this is

$$P(\overline{B}|\overline{A}) = \frac{P(\overline{A}|\overline{B}) P(\overline{B})}{P(\overline{A})} = \frac{0.96 \cdot 0.99}{1 - 0.0494} = 0.99979.$$

This result is, of course, an excellent feature of the test.   □

1.4.3 Independent Random Events

If a die is thrown twice, then the result of the first throw does not influence the result of the second throw and vice versa. If you have not won in the weekly lottery during the past 20 years, then this bad luck will not increase or decrease your chance to win in the lottery the following week. An aircraft crash over the Pacific for technical reasons has no connection to the crash of an aircraft over the Atlantic for technical reasons the same day. Thus, there are random events which do not at all influence each other. Events like that are called independent (of each other). Of course, for a quantitative probabilistic analysis a more accurate definition is required.

If the occurrence of a random event B has no influence on the occurrence of a random event A, then the probability of the occurrence of A will not be changed by the additional information that B has occurred, i.e.

$$P(A) = P(A|B) = \frac{P(A \cap B)}{P(B)}. \tag{1.26}$$

This motivates the definition of independent random events:

Definition 1.5: Two random events A and B are called *independent* if

$$P(A \cap B) = P(A) P(B). \tag{1.27}$$

●

This is the *product formula* for independent events A and B. Obviously, (1.27) is also valid for $P(B) = 0$ and/or $P(A) = 0$. Hence, defining independence of two random events by (1.27) is preferred to defining independence by formula (1.26).

If A and B are independent random events, then the pairs A and $\overline{B}$, $\overline{A}$ and B, as well as $\overline{A}$ and $\overline{B}$ are independent, too. That means relation (1.27) implies, e.g.,

$$P(A \cap \overline{B}) = P(A) P(\overline{B}).$$

This can be proved as follows:

$$P(A \cap \overline{B}) = P(A \cap (\Omega \backslash B)) = P((A \cap \Omega) \backslash (A \cap B)) = P(A \backslash (A \cap B))$$
$$= P(A) - P(A \cap B) = P(A) - P(A)P(B)$$
$$= P(A)[1 - P(B)] = P(A) P(\overline{B}).$$

The generalization of the independence property to more than two random events is not obvious. The pairwise independence between $n \geq 2$ events is defined as follows: The events $A_1, A_2, ..., A_n$ are called *pairwise independent* if for each pair (A_i, A_j)

$$P(A_i \cap A_j) = P(A_i) P(A_j), \quad i \neq j, \ i,j = 1, 2, ..., n.$$

A more general definition of the independence of n events is the following one:

Definition 1.6 The random events $A_1, A_2, ..., A_n$ are called *completely independent* or simply *independent* if for all $k = 2, 3, ..., n$,

$$P(A_{i_1} \cap A_{i_2} \cap \cdots \cap A_{i_k}) = P(A_{i_1}) P(A_{i_2}) \cdots P(A_{i_k}) \tag{1.28}$$

for any subset $\{A_{i_1}, A_{i_2}, ..., A_{i_k}\}$ of $\{A_1, A_2, ..., A_n\}$ with $1 \leq i_1 < i_2 < \cdots < i_k \leq n$. ●

Thus, to verify the complete independence of n random events, one has to check

$$\sum_{k=2}^{n} \binom{n}{k} = 2^n - n - 1$$

conditions. Luckily, in most applications it is sufficient to verify the case $k = n$:

$$P(A_1 \cap A_2 \cap \cdots \cap A_n) = P(A_1) P(A_2) \cdots P(A_n). \tag{1.29}$$

The complete independence is a stronger property than the pairwise independence. For this reason it is interesting to consider an example, in which the $A_1, A_2, ..., A_n$ are pairwise independent, but not complete independent.

Example 1.20 The dice D_1 and D_2 are thrown. The corresponding sample space consists of 36 elementary events: $\Omega = \{(i,j); \ i,j = 1, 2, ..., 6\}$. Let

$A_1 = 'D_1$ shows a $1' = \{(1,1), (1,2), (1,3), (1,4), (1,5), (1,6)\}$,

$A_2 = 'D_2$ shows a $1' = \{(1,1), (2,1), (3,1), (4,1), (5,1), (6,1)\}$,

$A_3 = $ 'both D_1 and D_2 show the same number' $= '\{(i,i), i = 1, 2, ..., 6)\}.'$

Since the A_i each contain 6 elementary events,

$$P(A_1) = P(A_2) = P(A_3) = 1/6.$$

The A_i have only one elementary event in common, namely $(1, 1)$. Hence,

$$P(A_1 \cap A_2) = P(A_1 \cap A_3) = P(A_2 \cap A_3) = \frac{1}{6} \cdot \frac{1}{6} = \frac{1}{36}.$$

Therefore, the A_i are pairwise independent. However, there is

$$A_1 \cap A_2 \cap A_3 = \{(1, 1)\}.$$

Hence,

$$P(A_1 \cap A_2 \cap A_3) = \frac{1}{36} \neq P(A_1)P(A_2)P(A_3) = \frac{1}{6} \cdot \frac{1}{6} \cdot \frac{1}{6} = \frac{1}{216}. \qquad \square$$

Example 1.21 (*Chevalier de Méré*) What is more likely: 1) to get at least one 6, when throwing four dice simultaneously (event A), or 2) to get the outcome (6,6) at least once, when throwing two dice 24 times simultaneously (event B)?

The complementary events to A and B are:

$\overline{A}$ = 'none of the dice shows a 6, when four dice are thrown simultaneously,'

$\overline{B}$ = 'the outcome (6,6) does not occur, when two dice are thrown 24 times.'

1) Both the four results obtained by throwing four or two dice and the results by repeatedly throwing two dice are independent of each other. Hence, since the probability to get no 6, when throwing one die, is 5/6, formula (1.29) with $n = 4$ yields

$$P(\overline{A}) = (5/6)^4.$$

The probability, not to get the result (6,6) when throwing two dice, is 35/36. Hence, formula (1.29) yields with $n = 24$ the probability

$$P(\overline{B}) = (35/36)^{24}.$$

Thus, the desired probabilities are

$$P(A) = 1 - (5/6)^4 \approx 0.518, \quad P(B) = 1 - (35/36)^{24} \approx 0.491. \qquad \square$$

Example 1.22 In a set of traffic lights, the color 'red' (as well as green and yellow) is indicated by two bulbs which operate independently of each other. Color 'red' is clearly visible if at least one bulb is operating.

What is the probability that in the time interval [0, 200 *hours*] color 'red' is visible if it is known that a bulb survives this interval with probability 0.95 ?

To answer this question, let

A = 'bulb 1 does not fail in [0, 200],' B = 'bulb 2 does not fail in [0, 200].'

The event of interest is

$$C = A \cup B = \text{'red light is clearly visible in } [0, 200].\text{'}$$

By formula (1.16),

$$P(C) = P(A \cup B) = P(A) + P(B) - P(A \cap B).$$

Since A and B are independent,

$$P(C) = P(A) + P(B) - P(A)P(B) = 0.95 + 0.95 - (0.95)^2.$$

Thus, the desired probability is

$$P(C) = 0.9975.$$

Another possibility of solving this problem is to apply the *Rules of de Morgan* (1.1):

$$P(\overline{C}) = P(\overline{A \cup B}) = P(\overline{A} \cap \overline{B}) = P(\overline{A})P(\overline{B}) = (1 - 0.95)(1 - 0.95)$$

$$= 0.0025$$

so that $P(C) = 1 - P(\overline{C}) = 0.9975.$ □

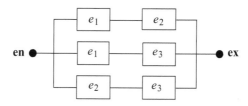

Figure 1.6 Diagram of a '2 out of 3-system'

Example 1.23 ('2 out of 3 system') A system S consists of 3 independently operating subsystems S_1, S_2, and S_3. The system operates if and only if at least 2 of its subsystems operate. Figure 1.6 illustrates the situation: S operates if there is at least one path with two operating subsystems (symbolized by rectangles) from the entrance node **en** to the exit node **ex**. As an application may serve the following one: The pressure in a high-pressure tank is indicated by 3 gauges. If at least 2 gauges show the same pressure, then this value can be accepted as the true one. (But for safety reasons the failed gauge has to be replaced immediately.)

At a given time point t_0, subsystem S_i is operating with probability p_i, $i = 1, 2, 3$. What is the probability p_S that the system S is operating at time point t_0?

Let A_S be the event that S is working at time point t_0, and A_i be the event that S_i is operating at time point t_0. Then,

$$A_S = (A_1 \cap A_2) \cup (A_1 \cap A_3) \cup (A_2 \cap A_3).$$

With $A = A_1 \cap A_2$, $B = A_1 \cap A_3$, and $C = A_2 \cap A_3$, formula (1.17) can be directly applied and yields the following representation of A_S:

$$P(A_S) = P(A_1 \cap A_2) + P(A_1 \cap A_3) + (A_2 \cap A_3) - 2P(A_1 \cap A_2 \cap A_3).$$

In view of the independence of the A_1, A_2, and A_3, this probability can be written as

$$P(A_S) = P(A_1)P(A_2) + P(A_1)P(A_3) + P(A_2)P(A_3) - 2P(A_1)P(A_2)P(A_3).$$

or

$$P(A_S) = p_1p_2 + p_1p_3 + p_2p_3 - 2p_1p_2p_3.$$

In particular, if $p = p_i$, $i = 1, 2, 3$, then

$$P(A_S) = (3 - 2p)p^2. \qquad \square$$

Disjoint and *independent* random events are causally not connected. Nevertheless, sometimes there is confusion about their meaning and use. This may be due to the formal analogy between their properties:

If the random events $A_1, A_2, ..., A_n$ are disjoint, then, by formula (1.10),

$$P(A_1 \cup A_2 \cup \cdots \cup A_n) = P(A_1) + P(A_2) + \cdots + P(A_n).$$

If the random events $A_1, A_2, ..., A_n$ are independent, then, by formula (1.29),

$$P(A_1 \cap A_2 \cap \cdots \cap A_n) = P(A_1) \cdot P(A_2) \cdots P(A_n).$$

1.5 EXERCISES

Sections 1.1–1.3

1.1) A random experiment consists of simultaneously flipping three coins.

(1) What is the corresponding sample space?

(2) Give the following events in terms of elementary events:
A = 'head appears at least two times,' B = 'head appears not more than once,' and C = 'no head appears.'

(3) Characterize verbally the complementary events of A, B, and C.

1.2) A random experiment consists of flipping a die to the first appearance of a '6'. What is the corresponding sample space?

1.3) Castings are produced weighing either 1, 5, 10, or 20 kg. Let A, B, and C be the events that a casting weighs 1 or 5kg, exactly 10kg, and at least 10kg, respectively. Characterize verbally the events $A \cap B$, $A \cup B$, $A \cap \overline{C}$, and $(\overline{A} \cup \overline{B}) \cap C$.

1.4) Three randomly chosen persons are to be tested for the presence of gene g. Three random events are introduced:

A = 'none of them has gene g,'
B = 'at least one of them has gene g,'
C = 'not more than one of them has gene g'.

Determine the corresponding sample space Ω and characterize the events $A \cap B$, $B \cup \overline{C}$, and $\overline{B \cap C}$ by elementary events.

1.5) Under which conditions are the following relations between events A and B true:
(1) $A \cap B = \Omega$, (2) $A \cup B = \Omega$, (3) $A \cup B = A \cap B$?

1.6) Visualize by a Venn diagram whether the following relations between random events A, B, and C are true:
(1) $A \cap (B \cup C) = (A \cap B) \cup (A \cap C)$,
(2) $(A \cap B) \cup (A \cap \overline{B}) = A$,
(3) $A \cup B = B \cup (A \cap \overline{B})$.

1.7) (1) Verify by a Venn diagram that for three random events A, B, and C the following relation is true: $(A \backslash B) \cap C = (A \cap C) \backslash (B \cap C)$.
(2) Is the relation $(A \cap B) \backslash C = (A \backslash C) \cap (B \backslash C)$ true as well?

1.8) The random events A and B belong to a σ–algebra E.
What other events, generated by A and B, must belong to E (see definition 1.2)?

1.9) Two dice D_1 and D_2 are simultaneously thrown. The respective outcomes of D_1 and D_2 are ω_1 and ω_2. Thus, the sample space is $\Omega = \{(\omega_1, \omega_2); \omega_1, \omega_2 = 1, 2, ..., 6\}$.

Let the events A, B, and C be defined as follows:

A = 'The outcome of D_1 is even and the outcome of D_2 is odd',

B = "The outcomes of D_1 and D_2 are both even".

What is the smallest σ–algebra E generated by A and B ('smallest' with regard to the number of elements in E)?

1.10) Let A and B be two disjoint random events, $A \subset \Omega$, $B \subset \Omega$.
Check whether the set of events $\{A, B, A \cap \overline{B}, \text{ and } \overline{A} \cap B\}$ is (1) an exhaustive and (2) a disjoint set of events (Venn diagram).

1.11) A coin is flipped 5 times in a row. What is the probability of the event A that 'head' appears at least 3 times one after the other?

1.12) A die is thrown. Let $A = \{1, 2, 3\}$ and $B = \{3, 4, 6\}$ be two random events.
Determine the probabilities $P(A \cup B)$, $P(A \cap B)$, and $P(B \backslash A)$.

1.13) A die is thrown 3 times. Determine the probability of the event A that the resulting sequence of three integers is strictly increasing.

1.14) Two dice are thrown simultaneously. Let (ω_1, ω_2) be an outcome of this random experiment, $A =$ '$\omega_1 + \omega_2 \leq 10$' and $B =$ '$\omega_1 \cdot \omega_2 \geq 19$.'
Determine the probability $P(A \cap B)$.

1.15) What is the probability p_3 to get 3 numbers right with 1 ticket in the '6 out of 49' number lottery?

1.16) A sample of 300 students showed the following results with regard to physical fitness and body weight:

weight [kg]

		$60 <$	$[60\text{-}80]$	$80 >$
fitness	good	48	64	11
	satisfactory	22	42	29
	bad	19	17	48

One student is randomly chosen. It happens to be *Paul.*

(1) What is the probability that the fitness of Paul is satisfactory?

(2) What is the probability that the weight of Paul is greater than 80 *kg?*

(3) What is the probability that the fitness of Paul is bad and that his weight is less than 60 *kg?*

1.17) Paul writes four letters and addresses the four accompanying envelopes. After having had a bottle of whisky, he puts the letters randomly into the envelopes. Determine the probabilities p_k that k letters are in the 'correct' envelopes, $k = 0, 1, 2, 3$.

1.18) A straight stick is broken at two randomly chosen positions. What is the probability that the resulting three parts of the stick allow the construction of a triangle?

1.19) Two hikers climb to the top of a mountain from different directions. Their arrival time points are between 9:00 and 10:00 a.m., and they stay on the top for 10 and 20 minutes, respectively. For each hiker, every time point between 9 and 10:00 has the same chance to be the arrival time. What is the probability that the hikers meet on the top?

1.20) A fence consists of horizontal and vertical wooden rods with a distance of 10 *cm* between them (measured from the center of the rods). The rods have a circular sectional view with a diameter of 2*cm*. Thus, the arising squares have an edge length of 8*cm*. Children throw balls with a diameter of 5*cm* horizontally at the fence. What is the probability that a ball passes the fence without touching the rods?

1.21) Determine the probability that the quadratic equation
$$x^2 + 2\sqrt{a}\,x = b - 1$$
does not have a real solution if the pair (a,b) is randomly chosen from the quarter circle $\{(a, b); a, b > 0, a^2 + b^2 < 1\}$.

1.22) Let A and B be disjoint events with $P(A) = 0.3$ and $P(B) = 0.45$. Determine the probabilities $P(A \cup B)$, $P(\overline{A \cup B})$, $P(\overline{A} \cup \overline{B})$, and $P(\overline{A} \cap B)$.

1.23) Let $P(A \cap \overline{B}) = 0.3$ and $P(\overline{B}) = 0.6$. Determine $P(A \cup B)$.

1.24) Is it possible that for two events A and B with $P(A) = 0.4$ and $P(B) = 0.2$ the relation $P(A \cap B) = 0.3$ is true?

1.25) Check whether for 3 arbitrary random events A, B, and C the following constellations of probabilities can be true:

(1) $P(A) = 0.6$, $P(A \cap B) = 0.2$, and $P(A \cap \overline{B}) = 0.5$,

(2) $P(A) = 0.6$, $P(B) = 0.4$, $P(A \cap B) = 0$, and $P(A \cap B \cap C) = 0.1$,

(3) $P(A \cup B \cup C) = 0.68$ and $P(A \cap B) = P(A \cap C) = 1$.

1.26) Show that for two arbitrary random events A and B the following inequalities are true: $P(A \cap B) \le P(A) \le P(A \cup B) \le P(A) + P(B)$.

1.27) Let A, B, and C be 3 arbitrary random events.

(1) Express the event 'A occurs, but B and C do not occur' in terms of suitable relations between these events and their complements.

(2) Prove: the probability of the event 'exactly one of the events A, B, or C occurs' is

$$P(A) + P(B) + P(C) - 2P(A \cap B) - 2P(A \cap C) - 2P(B \cap C) + 3P(A \cap B \cap C).$$

Section 1.4

1.28) Two dice are simultaneously thrown. The result is (ω_1, ω_2). What is the probability p of the event '$\omega_2 = 6$' on condition that '$\omega_1 + \omega_2 = 8$?'

1.29) Two dice are simultaneously thrown. By means of formula (1.24) determine the probability p that the dice show the same number.

1.30) A publishing house offers a new book as standard or luxury edition and with or without a CD. The publisher analyzes the first 1000 orders:

		luxury edition	
		yes	no
with CD	yes	324	82
	no	48	546

Let A (B) the random event that a book, randomly choosen from these 1000, is a luxury one (comes with a CD). (1) Determine the probabilities

$$P(A), \; P(B), \; P(A \cup B), \; P(A \cap B), \; P(A|B), \; P(B|A), \; P(A \cup B|\overline{B}), \; \text{and} \; P(\overline{A} \,|\, \overline{B}).$$

(2) Are the events A and B independent?

1.31) A manufacturer equips its newly developed car of type *Treekill* optionally with or without a tracking device and with or without speed limitation technology and analyzes the first 1200 orders:

		speed limitation	
		yes	no
tracking device	yes	74	642
	no	48	436

Let A (B) the random event that a car, randomly chosen from these 1200, has speed limitation (comes with a tracking device).

(1) Calculate the probabilities $P(A)$, $P(B)$, and $P(A \cap B)$ from the figures in the table.

(2) Based on the probabilities determined under a), only by using the rules developed in section 1.3.3, determine the probabilities

$$P(A \cup B), \ P(A|B), \ P(B|A), \ P(A \cup B|\overline{B}), \ \text{and} \ P(\overline{A}|\overline{B}).$$

1.32) A bowl contains m white marbles and n red marbles. A marble is taken randomly from the bowl and returned to the bowl together with r marbles of the same color. This procedure continues to infinity.

(1) What is the probability that the second marble taken is red?

(2) What is the probability that the first marble taken is red on condition that the second marble taken is red? (This is a variant of *Pólya's urn problem.*)

1.33) A test procedure for diagnosing faults in circuits indicates no fault with probability 0.99 if the circuit is faultless. It indicates a fault with probability 0.90 if the circuit is faulty. Let the probability of a circuit to be faulty be 0.02.

(1) What is the probability that a circuit is faulty if the test procedure indicates a fault?

(2) What is the probability that a circuit is faultless if the test procedure indicates that it is faultless?

1.34) Suppose 2% of cotton fabric rolls and 3% of nylon fabric rolls contain flaws. Of the rolls used by a manufacturer, 70% are cotton and 30% are nylon.

a) What is the probability that a randomly selected roll used by the manufacturer contains flaws?

b) Given that a randomly selected roll used by the manufacturer does not contain flaws, what is the probability that it is a nylon fabric roll?

1.35) A group of 8 students arrives at an examination. Of these students 1 is very well prepared, 2 are well prepared, 3 are satisfactorily prepared, and 2 are insufficiently prepared. There is a total of 16 questions. A very well prepared student can answer all of them, a well prepared 12, a satisfactorily prepared 8, and an insuffi-

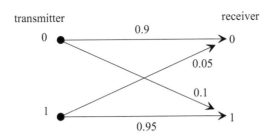

ciently prepared 4. Each student has to draw randomly 4 questions. Student *Frank* could answer all the 4 questions. What is the probability that Frank

(1) was very well prepared,
(2) was satisfactorily prepared,
(3) was insufficiently prepared?

1.36) Symbols 0 and 1 are transmitted independently from each other in proportion $1:4$. Random noise may cause transmission failures: If a 0 was sent, then a 1 will arrive at the sink with probability 0.1. If a 1 was sent, then a 0 will arrive at the sink with probability 0.05 (figure).

(1) What is the probability that a received symbol is '1'?

(2) '1' has been received. What is the probability that '1' had been sent?

(3) '0' has been received. What is the probability that '1' had been sent?

1.37) The companies 1, 2, and 3 have 60, 80, and 100 employees with 45, 40, and 25 women, respectively. In every company, employees have the same chance to be retrenched. It is known that a woman had been retrenched (event B).

What is the probability that she had worked in company 1, 2, and 3, respectively?

1.38) John needs to take an examination, which is organized as follows: To each question 5 answers are given. But John knows the correct answer only with probability 0.6. Thus, with probability 0.4 he has to guess the right answer. In this case, John guesses the correct answer with probability 1/5 (that means, he chooses an answer by chance). What is the probability that John knew the answer to a question given that he did answer the question correctly?

1.39) A delivery of 25 parts is subject to a quality control according to the following scheme: A sample of size 5 is drawn (without replacement of drawn parts). If at least one part is faulty, then the delivery is rejected. If all 5 parts are o.k., then they are returned to the lot, and a sample of size 10 is randomly taken from the original 25 parts. The delivery is rejected if at least 1 part out of the 10 is faulty.

Determine the probabilities that a delivery is accepted on condition that

(1) the delivery contains 2 defective parts,
(2) the delivery contains 4 defective parts.

1.40) The random events $A_1, A_2, ..., A_n$ are assumed to be independent. Show that
$$P(A_1 \cup A_2 \cup \cdots \cup A_n) = 1 - (1 - P(A_1))(1 - P(A_2)) \cdots (1 - P(A_n)).$$

1.41) n hunters shoot at a target independently of each other, and each of them hits it with probability 0.8. Determine the smallest n with property that the target is hit with probability 0.99 by at least one hunter.

1.42) Starting a car of type *Treekill* is successful with probability 0.6. What is the probability that the driver needs no more than 4 start trials to be able to leave?

1.43) Let A and B be two subintervals of $[0, 1]$. A point x is randomly chosen from $[0, 1]$. Now A and B can be interpreted as random events, which occur if $x \in A$ or $x \in B$, respectively. Under which condition are A and B independent?

1.44) A tank is shot at by 3 independently acting anti-tank helicopters with one anti-tank missile each. Each missile hits the tank with probability 0.6. If the tank is hit by 1 missile, it is put out of action with probability 0.8. If the tank is hit by at least 2 missiles, it is put out of action with probability 1.

What is the probability that the tank is put out of action by this attack?

1.45) An aircraft is targeted by two independently acting ground-to-air missiles. Each missile hits the aircraft with probability 0.6 if these missiles are not being destroyed before. The aircraft will crash with probability 1 if being hit by at least one missile. On the other hand, the aircraft defends itself by firing one air-to-air missile each at the approaching ground-to-air missiles. The air-to-air missiles destroy their respective targets with probablity 0.5.

(1) What is the probability that p the aircraft will crash as a result of this attack?

(2) What is the probability that the aircraft will crash if two independent air-to-air missiles are fired at each of the approaching ground-to-air-missiles?

1.46) The liquid flow in a pipe can be interrupted by two independent valves V_1 and V_2, which are connected in series (figure). For interrupting the liquid flow it is sufficient if one valve closes properly. The probability that an interruption is achieved when necessary is 0.98 for both valves. On the other hand, liquid flow is only possible if both valves are open. Switching from 'closed' to 'open' is successful with probability 0.99 for each of the valves.

(1) Determine the probability to be able to interrupt the liquid flow if necessary.

(2) What is the probability to be able to resume liquid flow if both valves are closed?

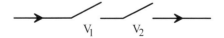

CHAPTER 2

One-Dimensional Random Variables

2.1 MOTIVATION AND TERMINOLOGY

Starting point of chapter 1 is a random experiment with sample space Ω, which is the set of all possible outcomes of the random experiment under consideration, and the set (σ–algebra) E of all random events, where a random event $A \in E$ is a subset of the sample space: $A \subseteq \Omega$. In this way, together with a probability function P defined on E, the probability space $[\Omega, E, P]$ is given. In many cases, the outcomes (elementary events) of random experiments are real numbers (throwing a die, counting the number of customers arriving per unit time at a service station, counting of wildlife in a specific area, total number of goals in a soccer match, or measurement of lifetimes of organisms and technical products). In these cases, the outcomes of a series of identical random experiments allow an immediate quantitative analysis. However, when the outcomes are not real numbers, i.e. Ω is not a subset of the real axis or the whole real axis, then such an immediate numerical analysis is not possible. To overcome this problem, a real number z is assigned to the outcome ω by a given real-valued function g defined on Ω: $z = g(\omega)$, $\omega \in \Omega$.

Examples for situations like that are:

1) When flipping a coin, the two possible outcomes are ω_1 = '*head*' and ω_2 = '*tail*'. A '1' is assigned to *head* and a '0' to *tail* (for instance).

2) An examination has the outcomes ω_1 = '*with distinction*', ω_2 = '*very good*', ω_3 = '*good*', ω_4 = 'satisfactory', and ω_5 = 'not passed'. The figures '5', '4', $\cdots$, '1' (for instance) are assigned to these verbal evaluations.

3) Even if the outcomes are real numbers, you may be more interested in figures derived from these numbers. For instance, the outcome is the number n of items you have produced during a workday. For first item you get a financial reward of $10, for the second of $11, for the third $12, and so on. Then you are first of all interested in your total income per working day.

4) If the outcomes of random experiments are vectors of real numbers, it may be opportune to assign a real number to these vectors. For instance, if you throw four dice simultaneously, you get a vector with four components. If you win, when the total sum exceeds a certain amount, then you are not in the first place interested in the four individual results, but in their sum. In this way, you reduce the complexity of the ran- dom experiment.

5) The random experiment consists in testing the quality of 100 spare parts taken randomly from a delivery. A '1' is assigned to a spare part which meets the requirements,

and a '0' otherwise. The outcome of this experiment is a vector $\vec{\omega} = (\omega_1, \omega_2, \cdots, \omega_{100})$, the components ω_i of which are 0 or 1. Such a vector is not tractable, so you want to assign a summarizing quality parameter to it to get a random experiment, which has a one-dimensional result. This can be, e.g., the relative frequency of those items in the sample, which meet the requirements:

$$z = g(\vec{\omega}) = \frac{1}{100} \sum_{k=1}^{100} \omega_k. \tag{2.1}$$

Basically, application of a real function to the outcomes of a random experiment does not change the 'nature' of the random experiment, but simply replaces the 'old' sample space with a 'new' one, which is more suitable for the solution of directly interesting numerical problems. In the cases 1 and 3 – 5 listed above:

1) The sample space {*tail, head*} is replaced with {0, 1}.

3) The sample space {0, 1, 2, 3, 4, ...} is replaced with {0, 10, 21, 33, 46,...}.

4) The sample space {$(\omega_1, \omega_2, \omega_3, \omega_4)$; $\omega_i = 1, 2, ..., 6$}, which consists of $6^4 = 1296$ elementary events of the structure $\omega = (\omega_1, \omega_2, \omega_3, \omega_4)$, is replaced with the sample space {6, 7, ..., 24}.

5) The sample space consisting of the 2^{100} elementary events $\omega = (\omega_1, \omega_2, ..., \omega_{100})$ with ω_k is 0 or 1 is reduced by the relative frequency function g given by (2.1) to a sample space with 101 elementary events:

$$\{0, \frac{1}{100}, \frac{2}{100}, \cdots, \frac{99}{100}, 1\}.$$

Since the outcome ω of a random experiment is not predictable, it is also random which value the function $g(\omega)$ will assume after the random experiment. Hence, functions on the sample space of a random experiment are called *random variables*. In the end, the concept of a random variable is only a somewhat more abstract formulation of the concept of a random experiment. But the terminology has changed: One says on the one hand that as a result of a random experiment *an elementary event has occurred*, and on the other hand, *a random variable has assumed a value*. In this book (apart from Chapter 12) only real-valued random variables are considered. As it is common in literature, random variables will be denoted by capital Latin letters, e.g. X, Y, Z or by Greek letters as ζ, ξ, η.

Let X be a random variable: $X = X(\omega)$, $\omega \in \Omega$. The *range* R_X of X is the set of all possible values X can assume. Symbolically: $R_X = X(\Omega)$. The elements of R_X are called the *realizations* of X or their *values*. If there is no doubt about the underlying random variable, the range is simply denoted as R.

> A random variable X is a real function on the sample space Ω of a random experiment. This function generates a new random experiment, whose sample space is given by the range R_X of X. The probabilistic structure of the new random experiment is determined by the probabilistic structure of the original one.

When discussing random variables, the original, application-oriented random experiment will play no explicit role anymore. Thus, a random variable can be considered to be an abstract formulation of a random experiment. With this in mind, the probability that X assumes a value out of a set A, $A \subseteq R$, is an equivalent formulation for the probability that the random event A occurs, i.e.

$$P(A) = P(X \in A) = P(\omega, X(\omega) \in A).$$

For one-dimensional random variables X, it is sufficient to know the *interval probabilities* $P(I) = P(X \in I)$ for all intervals: $I = [a, b)$, $a < b$, i.e.

$$P(X \in I) = P(a < X \le b) = P(\omega, a < X(\omega) \le b). \qquad (2.2)$$

If R is a finite or countably infinite set, then $I = [a, b)$ is simply the set of all those realizations of X, which belong to I.

Definition 2.1 The *probability distribution* or simply *distribution* of a one-dimensional random variable X is given by a rule $\boldsymbol{P}$, which assigns to every interval of the real axis $I = [a, b]$, $a < X \le b$, the probabilities (2.2). $\qquad\qquad\bullet$

Remark In view of definition 1.2, the probability distribution of any random variable X should provide probabilities $P(X \in A)$ for any random event A from the sigma algebra $\boldsymbol{E}$ of the underlying measurable space $[\Omega, \boldsymbol{E}]$, i.e. not only for intervals. This is indeed the case, since from measure theory it is known that a probability function, defined on all intervals, also provides probabilities for all those events, which can be generated by finite or countably infinite unions and conjunctions of intervals. For this reason, a random variable is called a *measurable function* with regard to $[\Omega, \boldsymbol{E}]$. This application-oriented text does not explicitly refer to this measure-theoretic background and is presented without measure-theoretic terminology.

> *A random variable X is fully characterized by its range R_X and by its probability distribution. If a random variable is multidimensional, i.e. its values are n-dimensional vectors, then the definition of its probability distribution is done by assigning probabilities to rectangles for $n = 2$ and to rectangular parallelepipeds for $n = 3$ and so on.*

In chapter 2, only one-dimensional random variables will be considered, i.e., their values are scalars.

The set of all possible values R_X, which a random variable X can assume, only plays a minor role compared to its probability distribution. In most cases, this set is determined by the respective applications; in other cases there prevails a certain arbitrariness. For instance, the faces of a die can be numbered from 7 to 12; a 3 (2) can be assigned to an operating (nonoperating) system instead of 1 or 0. Thus, the most important thing is to find the probability distribution of a random variable.

Fortunately, the probability distribution of a random variable X is fully characterized by one function, called its *(cumulative) distribution function* or its *probability distribution function*:

Definition 2.2 The *probability distribution function* (*cumulative distribution function or simply distribution function*) $F(x)$ of a random variable X is defined as

$$F(x) = P(X \le x), \quad -\infty \le x \le +\infty. \qquad \bullet$$

Any distribution function $F(x)$ has the following obvious properties:

1) $F(-\infty) = 0, \; F(+\infty) = 1,$ \hfill (2.3)

2) $F(x_1) \le F(x_2)$ if $x_1 \le x_2$. \hfill (2.4)

On the other hand, every function $F(x)$ satisfying the conditions (2.3) and (2.4) and being continuous on the left can be considered the distribution function of a random variable.

Given the distribution function of X, it must be possible to determine the interval probabilities (2.2). This can be done as follows:

For $a < b$, the event "$X \le b$" is given by the union of two disjoint events:

$$"X \le b" = "X \le a" \cup "a < X \le b".$$

Hence, by formula (1.11), $P(X \le b) = P(X \le a) + P(a < X \le b)$, or, equivalently,

$$P(a < X \le b) = F(b) - F(a). \qquad (2.5)$$

Thus, the cumulative distribution function contains all the information, specified in definition 2.1, about the probability distribution of a random variable. Note that definition 2.2 refers both to discrete and continuous random variables:

> *A random variable X is called* discrete *if it can assume only finite or countably infinite many values, i.e., its range R is a finite or a countably infinite set. A random variable X is called* continuous *if it can assume all values from the whole real axis, a real half-axis, or at least from a finite interval of the real axis or unions of finite intervals.*

Examples for discrete random variables are:

Number of flipping a coin to the first appearance of 'head', number of customers arriving at a service station per hour, number of served customers at service station per hour, number of traffic accidents in a specified area per day, number of staff being on sick leave a day, number of rhinos poached in the Krüger National park a year, number of exam questions correctly answered by a student, number of sperling errors in this chapter.

Examples for continuous random variables are:

Length of a chess match, service time of a customer at a service station, lifetimes of biological and technical systems, repair time of a failed machine, amount of rainfall per day at a measurement point, measurement errors, sulfur dioxide content of the air (with regard to time and location), daily stock market fluctuations.

2.2 DISCRETE RANDOM VARIABLES

2.2.1 Probability Distribution and Distribution Parameters

Let X be a discrete random variable with range $R = \{x_0, x_1, \cdots\}$. The probability distribution of X is given by a *probability mass function* $f(x)$. This function assigns to each realization of X its probability $p_i = f(x_i)$; $i = 0, 1, \ldots$. Without loss of generality it can be assumed that each p_i is positive. Otherwise, an x_i with $f(x_i) = 0$ could be deleted from R. Let $A_i = "X = x_i"$ be the random event that X assumes value x_i. The A_i are mutually disjoint events, since X cannot assume two different realizations at the same time. The union of all A_i,

$$\bigcup_{i=0}^{\infty} A_i,$$

is the certain event Ω, since X must take on any of its realizations. (A random experiment must have an outcome.) Taking into account (1.9), a probability mass function $f(x)$ has two characteristic properties:

$$1)\ f(x_i) > 0, \quad 2)\ \textstyle\sum_{i=0}^{\infty} f(x_i) = 1. \tag{2.6}$$

Every function $f(x)$ having these two properties can be considered to be the probability mass function of a discrete random variable. By means of $f(x)$, the probability distribution function of X, defined by (2.3), can be written as follows:

$$F(x) = \begin{cases} 0 & \text{if } x < x_0, \\ \sum\limits_{\{x_i, x_i \le x\}} f(x_i) & \text{if } x_0 \le x. \end{cases}$$

With $p_i = f(x_i)$, an equivalent representation of $F(x)$ is

$$F(x) = P(X \le x) = \begin{cases} 0 & \text{for } x < x_0, \\ \sum_{i=0}^{k} p_i & \text{for } x_k \le x < x_{k+1}, \quad k = 0, 1, 2, \cdots. \end{cases}$$

Figure 2.1 shows the typical graph of the distribution function of a discrete random variable X in terms of the *cumulative probabilities* s_i:

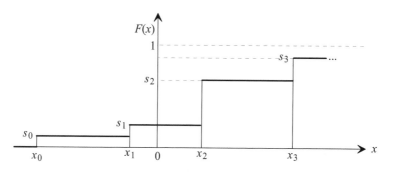

Figure 2.1 Graph of the distribution function of an arbitrary discrete random variable

$$s_k = p_0 + p_1 + \cdots + p_k; \quad k = 0, 1, \ldots,$$

or
$$s_k = F(x_k) = f(x_0) + f(x_1) + \cdots + f(x_k).$$

Thus, the distribution function of a discrete random variable is a piecewise constant function (*step function*) with jumps of sizes

$$p_i = P(X = x_i) = F(x_i) - F(x_i - 0), \quad i = 0, 1, \ldots.$$

The probability mass function of a random variable X as well as its distribution function can be identified with the probability distribution P_X of X.

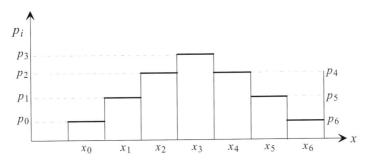

Figure 2.2 Probability histogram of a symmetric discrete distribution

Figure 2.2 shows the *probability histogram* of a discrete random variable. It graphically illustrates the *frequency distribution* of the occurrence of the values x_i of X. In this special case, the distribution is symmetric around x_3, i.e.

$$p_0 = p_6, \quad p_1 = p_5, \text{ and } p_2 = p_4.$$

Hint For technical reasons it is frequently practical to renumber the x_i and p_i and start with x_1 (p_1) instead of x_0 (p_0). In what follows, no further reference will be made regarding this. Moreover, the notation p_i will be preferred to $f(x_i)$.

Example 2.1 (*uniform distribution*) A random variable X is *uniformly distributed* over its range $R = \{1, 2, \ldots, m\}$ if it has the probability distribution

$$p_i = P(X = x_i) = \frac{1}{m}; \quad i = 1, 2, \ldots, m; \quad m < \infty.$$

The conditions (2.6) are fulfilled. Thus, X is the outcome of a Laplace random experiment (section 1.3), since every value of X has the same chance to occur. The cumulative probabilities are $s_i = i/m, \ i \le m$. The corresponding distribution function is

$$F(x) = P(X \le x) = \begin{cases} 0 & \text{for } x < 1, \\ i/m & \text{for } i \le x < i+1, \quad i = 1, 2, \cdots, m-1, \\ 1 & \text{for } m \le x. \end{cases} \qquad \square$$

Example 2.2 The leaves of *Fraxinus excelsior* (an ash tree) have an odd number of leaflets. This number varies from 3 to 11. A sample of $n = 300$ leaves had been taken from a tree. Let X be the number of leaflets of a randomly picked leaf from this sample. Then X is a random variable with range $R = \{3, 5, 7, 9, 11\}$.

Table 2.1 shows the probability distribution of X: The first column contains the possible number of leaflets i, the second column the number n_i of leaves with i leaflets, the third one the probability that a randomly choosen leaf from the sample has i leaflets: $p_i = n_i/n$. (In terms of mathematical statistics, p_i is the *relative frequency* of the occurrence of leaves with i leaflets in the sample.) The fourth column contains the cumulative probabilities s_k (cumulative frequencies).

i	n_i	p_i	s_i
3	8	0.0267	0.0267
5	36	0.1200	0.1467
7	108	0.3600	0.5067
9	118	0.3933	0.9000
11	30	0.1000	1

Table 2.1 Distribution of leaflets at leaves of *Fraxinus excelsior*

Figure 2.3 shows the distribution function and the probability histogram of X. For instance, $s_7 = 0.5607$ is the probability that a randomly selected leaf has at most 7 leaflets, and a randomly drawn leaf from the sample has most likely 9 leaflets.   ☐

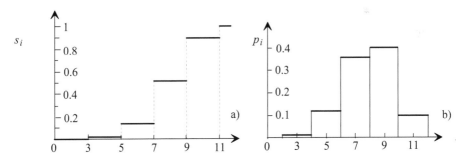

Figure 2.3 Distribution function a) and histogram b) for example 2.2

As pointed out before, the probability distribution and the range R contain all the information on X. However, to get quick information on essential features of a random variable, it is desirable to condense as much as possible of this information to some numerical parameters. In what follows, let the range of X be $R = \{x_0, x_1, \cdots\}$. If the range is finite, i.e., $R = \{x_0, x_1, \cdots, x_m; \ m < \infty\}$, the formulas to be given stay valid if letting $x_{m+1} = x_{m+2} = \cdots = 0$.

Mean Value If a random variable X has the finite range $R = \{x_0, x_1, ..., x_m\}$, then at first glance the average result of a random experiment with outcome X is

$$\bar{x} = \frac{1}{m+1} \sum_{i=0}^{m} x_i,$$

the *arithmetic mean* of all possible values of X. But this is only true if every value of X has the same chance to occur as this is the case with a uniformlydistributed random variable. Otherwise, those realizations of X contribute most to the average result (relatively to their absolute value), which occur more frequently than other realizations. To illustrate this, let us assume that in a series of n random experiments n_0 times x_0, n_1 times $x_1, \cdots$, and n_m times x_m occurred. Then there is $n = n_0 + n_1 + \cdots + n_m$, and the arithmetic mean of all observations is

$$\bar{x} = \frac{1}{n}(n_0 x_0 + n_1 x_1 + \cdots + n_m x_m) = \frac{n_0}{n} x_0 + \frac{n_1}{n} x_1 + \cdots + \frac{n_m}{n} x_m.$$

The ratio n_i/n is the relative frequency for the occurrence of x_i out of the total of n observations, which, as will be shown in section 5.2.2, tends for all $i = 0, 1, ..., m$ to the probability $p_i = P(X = x_i)$ as $n \to \infty$. Thus, the following definition is well motivated:

The *mean value*, or *expected value*, or simply the *mean* of a random variable X is

$$E(X) = \sum_{i=0}^{\infty} x_i p_i \quad \text{given that} \quad \sum_{i=0}^{\infty} |x_i| p_i < \infty. \tag{2.7}$$

Thus, the mean value of a discrete random variable X is the *weighted sum* of all its possible values, where the weights of the x_i are their respective probabilities. The convergence condition in (2.7) makes sure that $E(X)$ exists (i.e., is finite). Note that

$$E(|X|) = \sum_{i=0}^{\infty} |x_i| p_i. \tag{2.8}$$

A statistical motivation to the mean value of a random variable is the following one: If one and the same random experiment with outcome X is repeated n times and the results are $x_{i_1}, x_{i_2}, ..., x_{i_n}$, the arithmetic mean $\frac{1}{n} \sum_{k=1}^{n} x_{i_k}$ tends to $E(X)$ as $n \to \infty$.

If X is nonnegative with range $R = \{0, 1, 2, ...\}$, then its mean value can be written in the following way:

$$E(X) = \sum_{i=0}^{\infty} i p_i = \sum_{i=1}^{\infty} P(X \geq i) = \sum_{i=1}^{\infty} \sum_{k=i}^{\infty} p_k. \tag{2.9}$$

If $h(x)$ is a real function, then the mean value of the random variable $Y = h(X)$ is

$$E(Y) = \sum_{i=0}^{\infty} h(x_i) p_i. \tag{2.10}$$

In this formula, $y_i = h(x_i)$, $i = 0, 1, ...$ are the possible values which the random variable Y can take on. Since the y_i occur with the same probabilities as the x_i, namely p_i, (2.10) gives indeed the mean value of Y. As a special case, let $Y = X^n$. Then the mean value of X^n is given by (2.10) with $h(x_i) = x_i^n$:

$$E(X^n) = \sum_{i=0}^{\infty} x_i^n p_i, \quad n = 0, 1, ... $$

$E(X^n)$ is called the nth (*ordinary*) *moment* of X. Therefore, the mean value $E(X)$ is the first (ordiary) moment of X.

Variability In addition to its mean value $E(X)$, one is interested in the *variability* (*scatter, fluctuations*) of the outomes of a random experiment (given by X) in series of repetitions of this experiment. These fluctuations are measured by the absolute distances of the values x_i from $E(X)$: $|x_i - E(X)|$. This leads to the *mean absolute linear deviation* of a random variable X from its mean value:

$$E(|X - E(X)|) = \sum_{i=0}^{\infty} |x_i - E(X)| p_i. \tag{2.11}$$

The mean absolute linear deviation of X is a special case of the nth *absolute central (ordinary) moment* of X:

$$M_n = E(|X - E(X)|^n) = \sum_{i=0}^{\infty} |x_i - E(X)|^n p_i, \quad n = 0, 1, ... $$

For pactical and theoretical reasons, one usually prefers to work with the squared deviation of the x_i from $E(X)$: $(x_i - E(X))^2$. The mean value of the squared deviation of a random variable X from its mean value $E(X)$ is called *variance* of X and denoted as $Var(X)$:

$$Var(X) = E(X - E(X))^2 = \sum_{i=0}^{\infty} (x_i - E(X))^2 p_i. \tag{2.12}$$

The variance is obviously equal to the second absolute central moment of X. The square root of the variance $\sqrt{Var(X)}$ is called the *standard deviation* of X. For any random variable X, the following notation is common:

$$\sigma^2 = Var(X), \quad \sigma = \sqrt{Var(X)}.$$

Note, in determining $Var(X)$, formula (2.10) has been used with $h(x_i) = (x_i - E(X))^2$. From (2.12), for any constant h,

$$Var(hX) = h^2 Var(X).$$

There is a useful relationship between the variance and the second moment of X:

$$Var(X) = E(X - E(X))^2 = E(X^2) - 2 E[X E(X)] + E[(E(X)]^2$$

so that

$$Var(X) = E(X^2) - (E(X))^2. \tag{2.13}$$

The *coefficient of variation* of X is

$$V(X) = \sigma/|E(X)|.$$

Variance, standard deviation, and coefficient of variation are all measures for the variability of X. The coefficient of variation is most informative in this regard for taking into account not only the deviation of X from its mean value, but also relates this deviation to the mean value of X. For instance, if the variabilities of two random variables X and Y with equal variances $Var(X) = Var(Y) = 5$, but with different mean values $E(X) = 10$ and $E(Y) = 100$, have to be compared, then it is already intuitively obvious that the scatter behavior of X is more distinct than that of Y:

$$V(X) = 0.5, \quad V(Y) = 0.05.$$

Continuation of Example 2.2 The mean number of leaflets is

$$E(X) = 3 \cdot 0.0267 + 5 \cdot 0.1200 + 7 \cdot 0.3600 + 9 \cdot 0.3933 + 11 \cdot 0.1000 = 7.8398.$$

The variance of the number of leaflets is

$$Var(X) = (3 - 7.8398)^2 \cdot 0.0267 + (5 - 7.8398)^2 \cdot 0.12 + (7 - 7.8398)^2 \cdot 0.36$$
$$+ (9 - 7.8398)^2 \cdot 0.3933 + (11 - 7.8398)^2 \cdot 0.1 = 3.3751.$$

Altogether,

$$E(X) = 7.8398, \ \ Var(X) = 3.3751, \ \ \sqrt{Var(X)} = 1.8371, \ \ V(X) = 0.2343.$$

It is interesting to compare the standard deviation to the mean absolute linear deviation, since one expects that $E(|X - E(X)|)$ is somewhere in the order of the standard deviation: From (2.14),

$$E(|X - E(X)|) = |3 - 7.8398| \cdot 0.0267 + |5 - 7.8398| \cdot 0.12 + |7 - 7.8398| \cdot 0.36$$
$$+ |9 - 7.8398| \cdot 0.3933 + |11 - 7.8398| \cdot 0.1 = 1.5447.$$

Thus, $$E(|X - E(X)|) = 1.5447 < \sqrt{Var(X)} = 1.8371. \qquad \square$$

2.2.2 Important Discrete Probability Distributions

In this section, the following finite and infinite series are needed:

$$\sum_{i=0}^{n} i = \frac{n(n+1)}{2} \tag{2.14}$$

$$\sum_{i=0}^{n} i^2 = \frac{n(n+1)(2n+1)}{6} \tag{2.15}$$

$$\sum_{i=0}^{\infty} x^i = \frac{1}{1-x}, \qquad 0 \le x < 1 \qquad (geometric\ series) \tag{2.16}$$

$$\sum_{i=0}^{\infty} i x^i = \frac{x}{(1-x)^2}, \qquad 0 \le x < 1 \tag{2.17}$$

$$\sum_{i=0}^{n} x^i = \frac{1-x^{n+1}}{1-x}, \qquad x \ne 1 \tag{2.18}$$

$$\sum_{i=0}^{\infty} \frac{x^i}{i!} = e^x, \qquad |x| < +\infty \qquad (exponential\ series) \tag{2.19}$$

$$\sum_{i=0}^{n} \binom{n}{i} x^i y^{n-i} = (x+y)^n \qquad (binomial\ series) \tag{2.20}$$

Note that in view of (2.6) every probability distribution $\{p_0, p_1, ...\}$ of a discrete random variable must fulfill the *normalizing condition*

$$\sum_{i=0}^{\infty} p_i = 1. \tag{2.21}$$

Uniform Distribution A random variable X with range $R = \{x_1, x_2, ..., x_n\}$ has a *discrete uniform distribution* if

$$p_i = P(X = x_i) = \frac{1}{n}; \quad i = 1, 2, ..., n.$$

Thus, each possible value has the same probability. The normalizing condition (2.21) is obviously fulfilled. Mean value and variance are

$$E(X) = \bar{x} = \frac{1}{n} \sum_{i=1}^{n} x_i, \quad Var(X) = \frac{1}{n} \sum_{i=1}^{n} (x_i - \bar{x})^2.$$

Thus, $E(X)$ is the arithmetic mean of all values which X can assume.

Particularly, if $x_i = i$ for $i = 1, 2, ..., n$, then the formulas (2.14) and (2.15) yield

$$E(X) = \frac{n+1}{2}, \quad Var(X) = \frac{(n-1)(n+1)}{12}. \tag{2.22}$$

For instance, if X is the outcome of 'rolling a die', then $\mathbf{R} = \{1, 2, ..., 6\}$ and $p_i = 1/6$ so that

$$E(X) = 3.5, \text{ and } Var(X) \approx 2.92, \sqrt{Var(X)} \approx 1.71, \ V(X) = 0.59 \stackrel{\wedge}{=} 59\%,$$

and $E(|X - E(X)|) = \frac{1}{6} |1 - 3.5| + |2 - 3.5| + \cdots + |6 - 3.5|) = 1.5$ so that

$$E(|X - E(X)|) = 1.5 < \sqrt{Var(X)} \approx 1.71.$$

Bernoulli Distribution A random variable X with range $\mathbf{R} = \{0, 1\}$ has a *Bernoulli distribution* with parameter p, $0 < p < 1$, if

$$p_0 = P(X = 0) = 1 - p, \quad p_1 = P(X = 1) = p. \tag{2.23}$$

Mean value and variance are

$$E(X) = p, \quad Var(X) = p(1 - p). \tag{2.24}$$

This is easily verified:

$$E(X) = 0 \cdot (1 - p) + 1 \cdot p = p$$

$$Var(X) = (0 - p)^2 (1 - p) + (1 - p)^2 p = p(1 - p).$$

The random experiment, which leads to the Bernoulli distribution, is called *Bernoulli trial.* It has two outcomes: event A and its complementary event $\bar{A}$. Event A occurs with probability p, and event $\bar{A}$ occurs with probability $1 - p$. The random variable X defined by (2.23) assigns a "1" to event A and a "0" to event $\bar{A}$:

$$X = \begin{cases} 1 & \text{if } A \text{ has occurred,} \\ 0 & \text{if } \bar{A} \text{ has occurred.} \end{cases} \tag{2.25}$$

The occurrence of A is frequently referred to as <u>*success*</u>. With this terminology, X is the *indicator variable* for the occurrence of a success or a failure, respectively. Generally, since X can only assume two values, it is called a *(random) binary variable.* Specifically, since the two possible values of X are 0 and 1, it is a $(0, 1)$-*variable.*

Geometric Distribution A random variable X with range $R = \{1, 2, ...\}$ has a *geometric distribution* with parameter p, $0 < p < 1$, if

$$p_i = P(X = i) = p(1 - p)^{i-1}, \quad i = 1, 2, ... \qquad (2.26)$$

In view of the geometrical series (2.21), the normalizing condition (2.26) is fulfilled. Mean value and variance are

$$E(X) = 1/p, \quad Var(X) = (1 - p)/p^2.$$

To verify these formulas, use the series (2.16) and (2.17) as well as formula (2.13). A more elegant derivation is given in section 2.5.1.

For instance, if X is the random integer indicating how frequently one has to toss a die to get for the first time a '6' (= success), then X has a geometric distribution with

$$p = 1/6, \quad E(X) = 6, \quad Var(X) = 30, \quad \text{and} \quad \sqrt{Var(X)} \approx 5.4772.$$

Generally, a geometrically distributed random variable X can be interpreted as the number of independent Bernoulli trials one has to carry out to have for the first time a 'success'.

The geometric distribution is also defined with range $R = \{0, 1, ...\}$ and

$$p_i = P(X = i) = p(1 - p)^i, \quad i = 0, 1, 2, ... \qquad (2.27)$$

In this case, mean value and variance are

$$E(X) = (1 - p)/p, \quad Var(X) = (1 - p)/p^2.$$

Example 2.3 (*'nonaging property' of the geometric distribution*) Let X be a geometrically with parameter p distributed random variable. For any integers $m \geq 0$ and $n \geq 1$ determine the conditional probability $P(X = m + n | X > m)$.

In view of the geometrical series (2.16) with $x = 1 - p$,

$$P(X > m) = \Sigma_{i=m+1}^{\infty} p(1 - p)^{i-1} = p(1 - p)^m \Sigma_{i=0}^{\infty}(1 - p)^i = (1 - p)^m.$$

By the formula of conditional probability (1.22) and since the event "$X = m + n$" implies the event "$X > m$",

$$P(X = m + n | X > m) = \frac{P((X = m + n) \cap (X > m))}{P(X > m)} = \frac{P(X = m + n)}{P(X > m)}$$

$$= \frac{p(1 - p)^{m+n-1}}{(1 - p)^m} = p(1 - p)^{n-1}.$$

Hence,

$$P(X = m + n | X > m) = P(X = n), \quad m, n = 1, 2, ... \qquad (2.28)$$

This result has an interesting interpretation: If X is the lifetime of a technical unit, which can only fail at time points $n = 1, 2...$, and which has survived m time units, then the *residual lifetime* of the unit has the same lifetime distribution as the unit at the start of its operation, i.e. as a new unit. Such a unit is called *nonaging*. $\square$

Binomial Distribution A random variable X with range $R = \{0, 1, ..., n\}$ has a *binomial distribution with parameters p and n* if

$$p_i = P(X = i) = \binom{n}{i} p^i (1 - p)^{n-i}, \quad i = 0, 1, ..., n. \tag{2.29}$$

Frequently the notation $p_i = b(i, n, p)$ is used.

In view of the binomial series (2.20) with $x = p$ and $y = 1 - p$, the normalizing condition (2.21) is fulfilled. Mean value and variance are

$$E(X) = np, \quad Var(X) = np(1 - p). \tag{2.30}$$

The proofs will be given in section 2.5.1.

The binomial distribution occurs in the following situation: A Bernoulli trial, whose outcome is the $(0,1)$-indicator variable for the occurrences of events A and $\overline{A}$ as given by (2.25), is independently repeated n times. (Independence in the sense of definition 1.5: The respective outcomes of the n Bernoulli are independent random events.) Let the outcome of the ith trial be X_i:

$$X_i = \begin{cases} 1 & \text{if } A \text{ has occurred,} \\ 0 & \text{if } \overline{A} \text{ has occurred,} \end{cases} \quad i = 1, 2, ..., n.$$

The outcome of a series of n Bernoulli trials is a random vector $\vec{X} = (X_1, X_2, ..., X_n)$, whose components X_i can take on values 0 or 1. The sum

$$X = \sum_{i=1}^{n} X_i$$

is equal to the random number of successes in a series of n Bernoulli trials. X has a binomial distribution with parameters n and p: In view of the product formula for independent events (1.29), the probability that in $\vec{X}$ a '1' occurs i times and a '0' occurs $(n - i)$ times in a specific order, is

$$p^i (1 - p)^{n-i}.$$

There are $\binom{n}{i}$ different possibilities to order the i '1's and $(n - i)$ '0's.

For instance, let $n = 3$. Then the probability that vector $(0, 1, 1)$ (first Bernoulli trial is a failure, the second and third trial are successes) occurs is $(1 - p)p^2$. But there are $\binom{3}{2} = 3$ vectors with 1 failure and 2 successes having probability $(1 - p)p^2$:

$$(1, 1, 0), (1, 0, 1), (0, 1, 1).$$

Hence, the probability that a series of three Bernoulli trials yields one failure and two successes is $3p^2(1 - p)$.

Example 2.4 A power station supplies power to 10 bulk consumers. They use power independently of each other and in random time intervals, which, for each customer, accumulate to 20% of the calendar time. What is the probability of the random event B that at a randomly chosen time point at least seven customers use power?

The problem leads to a Bernoulli trial, where the 'success event' A for every customer is 'using power'. By assumption, $p = P(A) = 0.2$. Let B_i be the event that exactly i customers simultaneously use power. Then the event of interest is

$$B = B_7 \cup B_8 \cup B_9 \cup B_{10}.$$

The B_i are disjoint so that

$$P(B) = \textstyle\sum_{i=7}^{10} P(B_i) = \sum_{i=7}^{10} \binom{10}{i}(0.2)^i (0.8)^{10-i}$$

$$= 7.864 \cdot 10^{-4} + 7.373 \cdot 10^{-5} + 4.096 \cdot 10^{-6} + 1.024 \cdot 10^{-7}$$

$$= 0.000864. \qquad\qquad \square$$

Example 2.5 From a large delivery of calculators a sample of size $n = 100$ is taken. The delivery will be accepted if there are at most 4 defective calculators in the sample. The average rate of defective calculators from the producer is known to be 2%.

1) What is the probability P_{risk} that the delivery will be rejected (producer's risk)?
2) What is the probability C_{risk} to accept the delivery although it contains 7% defective calculators (consumer's risk)?

1) Picking a defective calculator is declared a "success" (event A). The probability of this event is $P(A) = 0.02$. Thus, the underlying Bernoulli trial has parameters $p = 0.02$ and $n = 100$. The probabilities p_i that i from 100 calculators are defective are:

$$p_i = \binom{100}{i}(0.02)^i (0.98)^{100-i}, \quad i = 0, 1, ..., 100.$$

In particular,

$$p_0 = 0.1326, \; p_1 = 0.2706, \; p_2 = 0.2734, \; p_3 = 0.1823, \; p_4 = 0.0902$$

so that the producer's risk is

$$P_{risk} = 1 - p_0 - p_1 - p_2 - p_3 - p_4 = 0.0509.$$

2) Now a "success" (picking a defective calculator) has probability $p = P(A) = 0.07$ so that the probabilities p_i to have i defective calculators in a sample of 100 are

$$p_i = \binom{100}{i}(0.07)^i (0.93)^{100-i}, \quad i = 0, 1, ..., 100.$$

In particular,

$$p_0 = 0.0007, \; p_1 = 0.0053, \; p_2 = 0.0198, \; p_3 = 0.0486, \; p_4 = 0.0888.$$

Thus, the consumer's risk is $C_{risk} = p_0 + p_1 + p_2 + p_3 + p_4 = 0.1632$. Thus, the proposed acceptance/rejection plan favors the producer. $\qquad \square$

In examples like the previous one the successive calculation of the probabilities p_i can be efficiently done by using the following recursion formula:

$$p_{i+1} = \frac{n-i}{i+1} \cdot \frac{p}{1-p} \cdot p_i; \quad i = 0, 1, ..., n-1.$$

Negative Binomial Distribution A random variable X with range $R = \{0, 1, ...\}$ has a *negative binomial distribution* with parameters p and r, $0 < p < 1$, $r > 0$, if

$$p_i = P(X = i) = \binom{i-1+r}{i} p^i (1-p)^r; \quad i = 0, 1, ... \quad (2.31)$$

Equivalently,

$$p_i = P(X = i) = \binom{-r}{i} (-p)^i (1-p)^r; \quad i = 0, 1, ... $$

Mean value and variance are

$$E(X) = \frac{r}{p}, \quad Var(X) = \frac{(1-p)r}{p^2}. \quad (2.32)$$

If r is a positive integer, then X can be interpreted as the total number of trials in a series of independent Bernoulli trials till the rth success occurs. The geometric distribution is a special case of the negative binomial distribution if $r = 1$.

The negative binomial distribution is also called *Pascal distribution.*

Hypergeometric Distribution A random variable X with range

$$R = \{0, 1, ..., \min(n, M)\}$$

has a *hypergeometric distribution* with parameters M, N, and n, $M \le N$, $n \le N$, if

$$p_m = P(X = m) = \frac{\binom{M}{m} \binom{N-M}{n-m}}{\binom{N}{n}}; \quad m = 0, 1, ..., \min(n, M). \quad (2.33)$$

Mean value and variance are:

$$E(X) = n\frac{M}{N}, \quad Var(X) = n\frac{M}{N}\left(1 - \frac{M}{N}\right)\left(1 - \frac{n-1}{N-1}\right). \quad (2.34)$$

As an application, consider the lottery '6 out of 49'. In this case, $M = n = 6$, $N = 49$, and p_m is the probability that a gambler hits exactly m winning numbers with one coupon (see example 1.7). More generally, hypergeometrically distributed random variables occur in the following situations: In a set of N elements belong M elements to type 1 and $N - M$ elements to type 2. A sample of n elements is randomly taken from this set. What is the probability that there are m elements of type 1 (and, hence, $n - m$ elements of type 2) in this sample?

If X is the random number of type 1 elements in the sample, then X has the distribution (2.33): There are $\binom{M}{m}$ possibilities to select from M type 1-elements exactly m, and to each of these possibilities there are $\binom{N-M}{n-m}$ possibilities to select from $N - M$ type 2-elements exactly $n - m$. The product of both numbers is the number of favorable cases for the occurrence of the event '$X = m$'. Finally, there are $\binom{N}{n}$ possibilities to select n elements from a total of N elements. Problems of this kind are typical ones in statistical quality control.

Example 2.6 A customer knows that on average 4% of parts delivered by a manufacturer are defective and has accepted this percentage. To check whether the manufacturer exceeds this limit, the customer takes from each batch of 800 parts randomly a sample of size 80 and accepts the delivery if there are at most 3 defective parts in a batch. What is the probability that the customer accepts a batch, which contains 50 defective parts? In this case,

$$N = 800, \ M = 50, \ \text{and} \ n = 80.$$

Let X be the random number of defective parts in the sample. Then the probabilities $p_i = P(X = i)$ are

$$p_i = \frac{\binom{50}{i}\binom{800-50}{80-i}}{\binom{800}{80}}; \quad i = 0, 1, 2, 3.$$

The exact values are

$$p_0 = 0.00431, \ p_1 = 0.02572, \ p_2 = 0.07406, \ p_3 = 0.13734.$$

Thus, the acceptance probability C_{risk} of the delivery (consumer's risk) is

$$C_{risk} = p_0 + p_1 + p_2 + p_3 = 0.24143.$$

Note that according to agreement the average number of faulty parts in a batch is supposed to be 32. □

Remark When comparing examples 2.5 and 2.6, the reader will notice that despite the same type of problems, for their solution first the binomial disribution and then the hypergeometric distribution had been used. This is because in example 2.5 the size of the delivery, from which a sample was taken, had been assumed to be large compared to the sample size, whereas in example 2.6 the size of the set of parts, namely 50, is fairly small compared to the sample of size 5 taken from this lot. If a sample of moderate size is taken from a sufficiently large set of parts, then this will not significantly change the ratio between defective and nondefective parts in the set, and one can assume the probability p of picking a defective part stays approximately the same. In this case the binomial distribution will yield acceptable approximate values. But if you want to apply the binomial distribution to small lots of parts, then, after every test of a part, you have to return it to the lot. In this case the ratio between defective and nondefective parts in the lot will not change either. The policy 'with replacement' is not always applicable, since during a check a part is frequently 'tested to death'. Generally, when applying the binomial distribution (hypergeometric distribution) in quality control, then "sampling with replacement" ("sampling without replacement") refers.

Example 2.7 Let N be the unknown number of adult zebras in a large National Park. A number of $M = 100$ randomly selected zebras from the total population of this park had been marked. A year later, a second sample from the whole adult zebra population of this park was taken, this time of size $n = 50$. Amongst these there were $m = 7$ zebras marked a year ago. Construct an estimation $\hat{N}$ for N with property that for $N = \hat{N}$ the probability of the observed event '$X = 7$' is maximal.

This way of estimating $\hat{N}$ makes sense, since one does not assume to have observed by chance an unlikely event instead of a very likely one. In this case, the hypergeometrically distributed random variable X is the number of marked zebras in the second sample of size $n = 50$. Let $p_7(N) = P(X = 7|N)$ be the probability that there are 7 marked zebras in the second sample given that the whole zebra population is of size N. Then, by definition of $\hat{N}$, the following two inequalities must be true:

$$p_8(\hat{N}+1) = \frac{\binom{100}{7}\binom{\hat{N}+1-100}{50-7}}{\binom{\hat{N}+1}{50}} \le \frac{\binom{100}{7}\binom{\hat{N}-100}{50-7}}{\binom{\hat{N}}{50}} = p_7(\hat{N}), \qquad (2.35)$$

$$p_7(\hat{N}-1) = \frac{\binom{100}{7}\binom{\hat{N}-1-100}{50-7}}{\binom{\hat{N}-1}{50}} \le \frac{\binom{100}{7}\binom{\hat{N}-100}{50-7}}{\binom{\hat{N}}{50}} = p_7(\hat{N}). \qquad (2.36)$$

Inequality (2.35) is equivalent to

$$\binom{\hat{N}-99}{43}\binom{\hat{N}}{50} \le \binom{\hat{N}-100}{43}\binom{\hat{N}+1}{50}.$$

By making use of the representation (1.5) of the binomial coefficient (cancelling the factors which are equal at both sides), this inequality reduces to

$$(\hat{N}-99)(\hat{N}-49) \le (\hat{N}-142)(\hat{N}+1) \quad \text{or} \quad 4993 \le 7\hat{N} \quad \text{or} \quad 713.3 \le \hat{N}.$$

Inequality (2.36) is equivalent to

$$\binom{\hat{N}-101}{43}\binom{\hat{N}}{50} \le \binom{\hat{N}-100}{43}\binom{\hat{N}-1}{50}.$$

Again by using (1.5), this inequality simplifies to

$$(\hat{N}-143)\hat{N} \le (\hat{N}-100)(\hat{N}-50) \quad \text{or} \quad 7\hat{N} \le 5000 \quad \text{or} \quad \hat{N} \le 714.3.$$

Hence, $713.3 \le \hat{N} \le 714.3$, so that $\hat{N} = 714$. $\qquad\qquad\qquad\square$

If the probabilities p_m of the hypergeometric distribution have to be successively calculated, then the following recursion formula is useful:

$$p_{m+1} = \frac{(n-m)(M-m)}{(m+1)(N-M-n+m+1)}p_m; \quad m = 0, 1, ..., \min(n, M).$$

Poisson Distribution A random variable X with range $R = \{0, 1, ...\}$ has a *Poisson distribution* with parameter λ if

$$p_i = P(X = i) = \frac{\lambda^i}{i!} e^{-\lambda}, \quad \lambda > 0, \quad i = 0, 1, ... \tag{2.37}$$

In view of the exponential series (2.19), the normalizing condition (2.21) is fulfilled. Again by making use of (2.19),

$$E(X) = \sum_{i=0}^{\infty} i p_i = \sum_{i=1}^{\infty} i \frac{\lambda^i}{i!} e^{-\lambda} = \sum_{i=1}^{\infty} \frac{\lambda^i}{(i-1)!} e^{-\lambda}$$

$$= \lambda e^{-\lambda} \sum_{i=0}^{\infty} \frac{\lambda^i}{i!} = \lambda e^{-\lambda} e^{+\lambda} = \lambda .$$

In section 2.2.3 it will be proved that the variance of X is equal to λ as well. Thus,

$$E(X) = \lambda, \quad Var(X) = \lambda . \tag{2.38}$$

In the context of the Poisson distribution, X is frequently said to be the number of *Poisson events* (occurring in time or in a spacial area).

Example 2.8 Let X be the random number of staff of a company being on sick leave a day. Long-term observations have shown that X has a Poisson distribution with parameter $\lambda = E(X) = 10$.

What is the probability that the number of staff being on sick leave a day is 9, 10, or 11?

$$p_9 = \frac{10^9}{9!} e^{-10} = 0.1251,$$

$$p_{10} = \frac{10^{10}}{10!} e^{-10} = 0.1251,$$

$$p_{11} = \frac{10^{11}}{11!} e^{-10} = 0.1137.$$

Hence, the desired probability is

$$P(9 \leq X \leq 11) = p_9 + p_{10} + p_{11} = 0.3639. \qquad \square$$

With regard to applications, it is frequently more adequate to write the Poisson probabilities (2.37) in the following form:

$$p_i = P(X = i) = \frac{(\lambda t)^i}{i!} e^{-\lambda t}, \quad \lambda > 0, \quad t > 0; \quad i = 0, 1, ... \tag{2.39}$$

In this form, the Poisson distribution depends on the two parameters λ and t. The parameter t refers to the time span or to the size of a spacial area (1-, 2-, or 3-dimensional), and λ refers to the mean number of Poisson events occurring per unit time, per length unit, etc. Thus, t is a *scale parameter*.

Example 2.9 The number of trees per unit of area in a virgin forest stand with a stem diameter of at least $50\,cm$ (measured at a height of $1.3\,m$) follows a Poisson distribution with parameter $\lambda = 0.004\,[m^2]^{-1}$.

What are the probabilities that in any subarea of $1000\,m^2$ in this stand there are (1) none of such trees, and (2) exactly four of such trees?

Formula (2.39) is applied with $\lambda = 0.004\,[m^2]^{-1}$ and $t = 1000\,m^2$. The results are

$$p_0 = e^{-0.004 \cdot 1000} = e^{-4} \approx 0.0183,$$

$$p_4 = \frac{[(0.004) \cdot 1000]^4}{4!} e^{-0.004 \cdot 1000}$$

$$= \frac{4^4}{4!} e^{-4} \approx 0.1954. \qquad \square$$

If the 'Poisson probabilities' p_i have to be manually calculated, then the following recursion formula is useful:

$$p_{i+1} = \frac{\lambda}{i+1} p_i; \quad i = 0, 1, \ldots$$

Approximations In view of binomial coefficients involved in the definition of the binomial and particularly in the hypergeometric distribution, the following approximations are useful for numerical analysis with a calculator:

Poisson Approximation to the Binomial Distribution If n is sufficiently large and p is sufficiently small, then

$$\binom{n}{i} p^i (1-p)^{n-i} \approx \frac{\lambda^i}{i!} e^{-\lambda}; \quad \lambda = np, \quad n = 0, 1, \ldots. \qquad (2.40)$$

As a rule of thumb, the Poisson approximation is applicable if

$$np < 10 \text{ and } n > 1500p.$$

Binomial Approximation to the Hypergeometric Distribution

$$\frac{\binom{M}{m}\binom{N-M}{n-m}}{\binom{N}{n}} \approx \binom{n}{m} p^m (1-p)^{n-m} \text{ with } p = M/N; \quad m = 0, 1, \ldots, n. \qquad (2.41)$$

As a rule of thumb, the binomial approximation to the hypergeometric distribution is applicable if

$$0.1 < M/N < 0.9, \quad n > 10, \text{ and } n/N < 0.05.$$

This approximation is heuristically motivated by the remark after example 2.6.

Poisson Approximation to the Hypergeometric Distribution If n is sufficiently large and $p = M/N$ is sufficiently small, then

$$\frac{\binom{M}{m}\binom{N-M}{n-m}}{\binom{N}{n}} \approx \frac{\lambda^m}{m!} e^{-\lambda} \quad \text{with } \lambda = n \cdot \frac{M}{N}. \tag{2.42}$$

This relation combines the approximations (2.40) and (2.41). As a rule of thumb, the Poisson approximation is applicable if

$$M/N \le 0.1, \quad n > 30, \quad n/N < 0.05.$$

Example 2.10 On average, only 0.01% of trout eggs will develop into adult fish. What is the probability $p_{\ge 3}$ that at least three adult fish arise from 40 000 eggs?

Let X be the random number of eggs out of 40 000 which develop into adult fish. It is assumed that the eggs develop independently of each other. Then X has a binomial distribution with parameters $n = 40\,000$ and $p = 0.0001$. Thus,

$$p_i = P(X = i) = \binom{40\,000}{i} (0.0001)^i (0.9999)^{40\,000-i},$$

where $i = 0, 1, ..., 40\,000$. Since n is large and p is small, the Poisson distribution with parameter $\lambda = np = 4$ can be used to approximately calculate the p_i:

$$p_i = \frac{4^i}{i!} e^{-4}; \quad i = 0, 1, ...$$

The desired probability is

$$p_{\ge 3} = 1 - p_0 - p_1 - p_2 = 1 - 0.0183 - 0.0733 - 0.1465$$

$$= 0.7619. \qquad \square$$

Continuation of Example 2.6 The binomial and the Poisson approximations to the hypergeometric distribution are applied with

$$N = 800, \quad M = 50, \quad \text{and } n = 80.$$

Table 2.2 compares the exact values to the ones obtained from approximations. The third condition in the corresponding 'rule of thumbs', namely $n/N < 0.05$, is not fulfilled. $\qquad \square$

	p_0	p_1	p_2	p_3	C_{risk}
Exact	0.00431	0.02572	0.07406	0.13734	0.24143
Binomial	0.00572	0.03053	0.08039	0.13934	0.25598
Poisson	0.00673	0.03369	0.08422	0.14037	0.26501

Table 2.2 Comparison of exact probabilities to its approximative values (example 2.6)

2.3 CONTINUOUS RANDOM VARIABLES

2.3.1 Probability Distribution

The probability distribution of a discrete random variable Y is given by assigning to each possible value of Y its probability according to the probability mass function of Y. This approach is no longer feasible for random variables, which can assume non-countably many values. To illustrate the situation, let us recall the geometric distribution over the interval $[0, T]$ (page 15). This distribution defines the probability distribution of a random variable X with noncountable, but finite, range $R = [0, T]$ in the following way: The probability that X takes on a value out of an interval $[a, b]$ with $0 \leq a < b \leq T < \infty$ is

$$P(a \leq X < b) = (b - a)/T.$$

If $b \to a$, then length of this 'interval probability' tends to 0: $P(X = a) = 0$. However, to assign to each value of X the probability 0 cannot be the way to define the probability distribution of a random variable with noncountably many values. Moreover, a noncountable range R does not exclude the possibility that there exists a finite or countably infinite set of values of X which actually have positive probabilities. Hence, the probability distribution of X will be defined via the distribution function of X (definition 2.2) as suggested in section 2.1:

$$F(x) = P(X \leq x), \quad x \in R. \tag{2.43}$$

As shown there (formula 2.5), the interval probabilities for any interval $I = [a, b]$ with $a < b$ and $a, b \in R$ are given in terms of the distribution function by

$$P(a < X \leq b) = F(b) - F(a). \tag{2.44}$$

To exclude the case that $F(x)$ has jumps for some $x \in R$ (i.e. $F(x)$ has points of discontinuity), a *continuous random variable* is defined as follows:

> *A random variable is called* continuous *if its distribution function $F(x)$ has a first derivative $f(x) = F'(x)$.*

Equivalently, a random variable is called *continuous* if there is a function $f(x)$ so that

$$F(x) = \int_{-\infty}^{x} f(u)\, du. \tag{2.45}$$

The function

$$f(x) = F'(x) = dF(x)/dx, \quad x \in \mathbf{R}, \tag{2.46}$$

is called *probability density function, probability density,* or briefly *density* of X. Sometimes the term *probability mass function* is used. A density has properties

$$f(x) \geq 0, \quad \int_{-\infty}^{+\infty} f(x)\, dx = 1. \tag{2.47}$$

Conversely, every function $f(x)$ with properties (2.47) can be interpreted as the density of a continuous random variable (Figure 2.4).

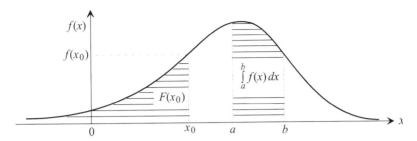

Figure 2.4 Relationship between distribution function and density

Note If a random variable X has a density $f(x)$, then its distribution function need not exist in an explicit form. This is the case if $f(x)$ is not integrable. Then, if no tables are available, the values of $F(x)$ have to be calculated by numerical integration of (2.45).

The range of X coincides with the set of all those x for which its density is positive: $R = \{x, f(x) > 0\}$ (Figure 2.4). In terms of the density, the interval probability (2.44) has the form

$$P(a < X \le b) = \int_a^b f(x)\, dx. \tag{2.48}$$

Thus, the probability that X assumes a value between a and b is equal to the area below $f(x)$ and above the x-axis between a and b (Figure 2.4). This implies the larger $f(x)$ is in an environment of x, the larger is the probability that X assumes a value out of this environment.

Example 2.11 A popular example for a continuous probability distribution is the *exponential distribution* with parameter λ: It has distribution function and density (see Figure 2.5 a) and b))

$$F(x) = \begin{cases} 1 - e^{-\lambda x}, & x > 0, \\ 0, & x \le 0, \end{cases} \qquad f(x) = \begin{cases} \lambda e^{-\lambda x}, & x > 0, \\ 0, & x \le 0. \end{cases} \tag{2.49}$$

A random variable with this distribution cannot take on negative values since

$$F(0) = P(X \le 0) = 0.$$

By (2.44), if $\lambda = 1$, $a = 1$, and $b = 2$, the probability that X takes on a value between 1 and 2 is $P(1 < X \le 2) = F(2) - F(1) = (1 - e^{-2}) - (1 - e^{-1}) = 0.2325$. □

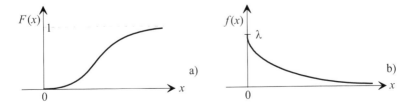

Figure 2.5 Distribution function a) and density b) of the exponential distribution

A motivation of the term 'probability density' follows from the definition of $f(x)$ as

$$f(x) = \lim_{\Delta x \to 0} \frac{F(x + \Delta x) - F(x)}{\Delta x}$$

so that, for small Δx,

$$f(x) \approx \frac{F(x + \Delta x) - F(x)}{\Delta x} \quad \text{or} \quad f(x)\Delta x \approx F(x + \Delta x) - F(x). \tag{2.50}$$

Hence, $f(x)$ is indeed a probability per unit of x, and $f(x)\Delta x$ is approximately the probability that X takes on a value in the interval $[x, x + \Delta x]$. This is the reason why for some heuristic derivations it is useful to interpret $f(x)\,dx$ as the probability that X takes on value x. Of course, for continuous random variables this probability is 0:

$$P(X = x) = \lim_{\Delta x \to 0} [F(x + \Delta x) - F(x)] = F(x) - F(x) = 0.$$

Example 2.12 The weights of 60 balls for ball bearings of the same type have been measured. Normally, one would expect that all balls have the same weight as prescribed by the standard for this type of ball bearings. In view of unavoidable technological fluctuations and measurement errors, this is not a realistic expectation. Table 2.3 shows the results of the measurements [in g]:

5.77	5.82	5.70	5.78	5.70	5.62	5.66	5.66	5.64	5.76
5.73	5.80	5.76	5.76	5.68	5.66	5.62	5.72	5.70	5.78
5.76	5.67	5.70	5.72	5.81	5.79	5.78	5.66	5.76	5.72
5.70	5.78	5.76	5.70	5.76	5.76	5.62	5.68	5.74	5.74
5.81	5.66	5.72	5.74	5.64	5.79	5.72	5.82	5.74	5.73
5.81	5.77	5.60	5.72	5.78	5.76	5.74	5.70	5.64	5.78

Table 2.3 Sample of 60 weight measurements of balls for ball bearings of the same type

The data fluctuate between 5.60 and 5.82. This interval is called the *range* of the sample. Of course, the weights of the balls can principally assume any value within the range, but the accuracy of the measurement method applied is restricted to two decimals after the point. To get an idea on the frequency distribution of the data, they are partitioned into *class intervals* (or *cells*). In Table 2.4, the integer n_i denotes the number of measurements which belong to class i, and $p_i = n_i/n$ with $n = 60$ is the relative frequency of the random event $A_i =$ 'a measurement is in class interval i'. A ball is randomly selected from the data set. Let X be the number of the class which the weight of this ball belongs to. Then X is a discrete random variable with range $\{1, 2, \ldots, 6\}$ and probability distribution

$$p_i = P(X = i) = n_i/n, \quad i = 1, 2, \ldots, 6.$$

The corresponding *cumulative probabilities* s_i are

$$s_i = p_1 + p_2 + \cdots + p_i, \quad i = 1, 2, \ldots, 6, \quad s_6 = 1.$$

X	class	n_i	p_i	s_i
1	[5.59-5.63)	4	0.0667	0.0667
2	[5.63-5.67)	8	0.1333	0.2000
3	[5.67-5.71)	10	0.1667	0.3667
4	[5.71-5.75)	13	0.2167	0.5834
5	[5.75-5.79)	17	0.2833	0.8667
6	[5.79-5.83)	8	0.1333	1

Table 2.4 Probability distribution of X for example 2.12

Now we essentially are in the same situation as in example 2.3. In Table 2.4 the notation $[a_i, a_{i+1})$ means that the left end point a_i belongs to the class interval, but the
right end point a_{i+1} does not.

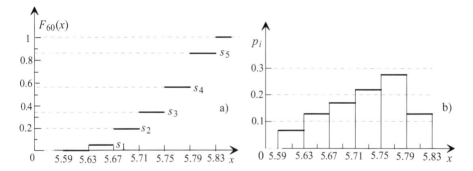

Figure 2.6 Distribution function a) and probability histogram b) of X (example 2.12)

The jump size of the distribution function between the i th and the $(i+1)$ th class is
determined by the data belonging to the i th class, i.e., by the probabilities $p_i = n_i/n$:

$$F_{60}(x) = P(X \le x) = \begin{cases} 0 & \text{for} & x < 5.63 \\ 0.0667 & \text{for } 5.63 \le x < 5.67 \\ 0.2000 & \text{for } 5.67 \le x < 5.71 \\ 0.3667 & \text{for } 5.71 \le x < 5.75 \\ 0.5834 & \text{for } 5.75 \le x < 5.79 \\ 0.8667 & \text{for } 5.79 \le x < 5.83 \\ 1 & \text{for} & 5.83 \le x \end{cases} .$$

The histogram is an approximation to the probability density of the random weight Y
of the balls, which actually is a continuous random variable, for the following reason:

If the length of the class intervals is scheduled to be one, what can always be done by scaling the x-axis accordingly (see Figure 2.10, page 70), then the area of the column over this interval is the probability $p_i = n_i/n$ that Y takes on a value from this interval. This corresponds to the interval probabilities (2.48) given by a density. By comparing the probability histogram with the theoretical densities proposed in section 2.3.4, one gets a first hint at the type of the probability distribution of Y. For instance, when comparing the histogram (Figure 2.6 b) with the density of the exponential distribution (Figure 2.5 b), this distribution can immediately be excluded as a suitable model.   ☐

By partitioning in the previous example the 60 ball weights in classes, information about the probability distribution of the ball weights was lost. No information is lost when defining an *empirical distribution function* $F_n(x)$ of Y based on a sample of size n (i.e., the results of n repetitions of a random experiment with outcome Y have been registered) as follows:

$$F_n(x) = \frac{n(x)}{n},$$

where $n(x)$ is the number of values in the sample, which are equal or smaller than x.

Theorem of I. V. Glivenko: $F_n(x)$ tends to $F(x) = P(Y \le x)$ as $n \to \infty$ in the following sense: If $G_n = \sup_{x \in \mathbf{R}} |F_n(x) - F(x)|$, where $\mathbf{R}$ is the range of Y, then

$$P(\lim_{n \to \infty} G_n = 0) = 1.$$

Note that $F_n(x)$ has jumps of size $1/n$ at each sample value.

2.3.2 Distribution Parameters

The probability distribution function and/or the density of a continuous random variable X contain all the information on X. But, as with discrete random variables, to get fast information on essential features of a random variable or its probability distribution, it is desirable to condense as much as possible of this information into some numerical parameters. Their interpretation is the same as for discrete random variables. Remember that a random variable X can be interpreted as the outcome of a random experiment. The *mean value* gives information on the average outcome of the random experiment in a long series of repetitions. The characteristic feature of the *median* is that, in a long series of repetitions of the random experiment, on average 50% of its outcomes are to the left of the median and 50% to the right. Hence, mean value and median characterize the *central tendency* of X.

Mean Value The *mean value* (*mean, expected value*) of X is defined as

$$E(X) = \int_{-\infty}^{+\infty} x f(x) \, dx \tag{2.51}$$

on condition that $\int_{-\infty}^{+\infty} |x| f(x) \, dx < \infty$.

The condition makes sure that the integral (2.51) exists. Note that

$$E(|X|) = \int_{-\infty}^{+\infty} |x| f(x) dx.$$

Formula (2.51) can be derived from the definition of the mean value of a discrete random variable (2.7): For simplicity of notation, let the range of X be $\mathbf{R} = [0, \infty)$. $\mathbf{R}$ is partitioned in intervals I_k of length Δx as follows:

$$I_k = (k\Delta, (k+1)\Delta], \; k = 0, 1, \dots.$$

Let $\tilde{X}$ be a discrete random variable, which takes on a value x_k from each I_k with probability $p_k = F((k+1)\Delta) - F(k\Delta); \; k = 0, 1, \dots$ Then, by (2.7) and (2.50), as $\Delta \to 0$,

$$E(\tilde{X}) = \sum_{k=0}^{\infty} x_k p_k = \sum_{k=0}^{\infty} \int_{k\Delta}^{(k+1)\Delta} x_k f(x) dx$$

$$\to \int_0^{\infty} x f(x) dx = E(X).$$

For nonnegative continuous random variables, the analogue to formula (2.10) is

$$E(X) = \int_0^{\infty} [1 - F(x)] dx. \tag{2.52}$$

This formula is verified by partial integration as follows:

$$E(X) = \int_0^{+\infty} x f(x) dx = \lim_{t \to \infty} \int_0^t x f(x) dx$$

$$= \lim_{t \to \infty} \left[t F(t) - \int_0^t F(x) dx \right] = \lim_{t \to \infty} \int_0^t [F(t) - F(x)] dx$$

$$= \int_0^{\infty} [1 - F(x)] dx.$$

From (2.51) one gets analogously by partial integration the mean value of a random variable X with range $\mathbf{R} = (-\infty, +\infty)$ as

$$E(X) = \int_0^{\infty} [1 - F(x)] dx - \int_{-\infty}^0 F(x) dx.$$

If $h(x)$ is a real function and X any continuous random variable with density $f(x)$, then the mean value of the random variable $Y = h(X)$ can directly be obtained from the density of X:

$$E(h(X)) = \int_{-\infty}^{+\infty} h(x) f(x) dx. \tag{2.53}$$

If $h(x) = ax + b$ with constants a and b, then $Y = aX + b$ and

$$E(aX + b) = a E(X) + b. \tag{2.54}$$

If both X and $h(x)$ are nonnegative, one obtains by partial integration of (2.53) a formula for $E(h(X))$, which generalizes formula (2.52):

$$E(h(X)) = \int_0^{\infty} [1 - F(x)] dh(x) = \int_0^{\infty} [1 - F(x)] h'(x) dx, \tag{2.55}$$

where $h'(x)$ denotes the first derivative of $h(x)$ (assuming its existence).

Moments By specifying $h(x)$, formula (2.53) yields the *moments of X*:

The (ordinary) nth *moment of X* is the mean value of X^n:

$$\mu_n = E(X^n) = \int_{-\infty}^{+\infty} x^n f(x)\,dx; \ n = 0, 1, \dots. \tag{2.56}$$

In particular, $\mu_0 = 1$ and $\mu_1 = E(X)$.

The nth (*ordinary*) *central moment of X* is

$$m_n = E((X - E(X))^n) = \int_{-\infty}^{+\infty} (x - E(X))^n f(x)\,dx, \ n = 0, 1, \dots, \tag{2.57}$$

and the nth *absolute central moment of X* is

$$M_n = E(|X - E(X)|^n) = \int_{-\infty}^{+\infty} |x - E(X)|^n f(x)\,dx, \ n = 0, 1, \dots. \tag{2.58}$$

Median The *median* of a continuous random variable X with distribution function $F(x)$ is defined as that value $x_{0.5}$ of X which satisfies $F(x_{0.5}) = 0.5$.

Hence, in a long series of experiments with outcome X about 50% of the results will be to the left of $x_{0.5}$ and 50% to the right of $x_{0.5}$ (Figure 2.7). One may expect that $x_{0.5} = E(X)$. But this is not generally true as the following example shows.

Example 2.13 Let X have an exponential distribution with parameter λ (see example 2.11), i.e., $F(x) = 1 - e^{-\lambda x}$, $x \geq 0$. Then, by formula (2.52),

$$E(X) = \int_0^\infty e^{-\lambda x}\,dx = 1/\lambda.$$

Now, let $h(x) = x^2$. Then, by (2.55), the second moment of X becomes

$$E(X^2) = \int_0^\infty e^{-\lambda x} 2x\,dx = 2\int_0^\infty x e^{-\lambda x}\,dx = -\frac{2}{\lambda^2}\left[e^{-\lambda x}(\lambda x + 1)\right]_0^\infty$$

$$= -\frac{2}{\lambda^2}[0 - 1] = 2/\lambda^2.$$

The median $x_{0.5}$ is solution of the equation $1 - e^{-\lambda x_{0.5}} = 0.5$ so that

$$x_{0.5} = 0.6931/\lambda.$$

Thus, for the exponential distribution, $x_{0.5} < E(X)$ and $E(X^2) > [E(X)]^2$. ☐

Percentile The α-*percentile* (also denoted as α-*quantile*) of a continuous random variable X is defined as that value x_α of X which satisfies

$$F(x_\alpha) = \alpha, \ 0 < \alpha < 1. \tag{2.59}$$

Hence, in a long series of experiments with outcome X, about α% of the results will be to the left of x_α and $(1 - \alpha)$% to the right of x_α (Figure 2.7). Thus, the median is the 0.5-percentile of X or of its probability distribution, respectively.

Percentiles are important criteria in quality control. For instance, for an exponentially distributed lifetime, what should the mean life of an electronic part be so that 95% of these parts operate at least 5 years without failure? The mean life is $\mu = 1/\lambda$ so that μ must satisfy $P(X > 5) = e^{-5/\mu} \geq 0.95$. Therefore, $\mu \geq 97.5\,years$.

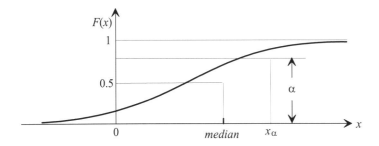

Figure 2.7 Illustration of the percentiles

Mode A *mode* x_m of a continuous random variable X with density $f(x)$ is a value at which $f(x)$ assumes a relative maximum. $f(x)$ is *unimodal* if it has exactly one mode. Otherwise it is called *multimodal*.

A density may have an uncountably infinite set of modes. This happens when the density takes on a (relative) maximum over a whole interval. For a unimodal density (in this case $f(x)$ assumes its absolute maximum at x_m), the most outcomes during a long series of experiments will be in an environment of x_m.

A function $f(x)$ is said to be *symmetric* with *symmetry center* x_s if for all x

$$f(x_s - x) = f(x_s + x).$$

It is quite obvious that for a random variable X with a unimodal and symmetric probability density $f(x)$, median, mode and symmetry center coincide. If, in addition, the mean value of X is finite, then

$$E(X) = x_{0.5} = x_m = x_s.$$

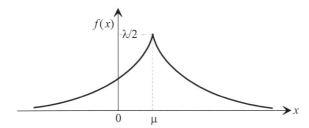

Figure 2.8 Density of the Laplace distribution

Example 2.14 The *Laplace distribution*, also called *doubly exponential distribution*, has a symmetric density with symmetry center at $x_s = \mu$ (Figure 2.8):

$$f(x) = \tfrac{1}{2}\lambda e^{-\lambda |x-\mu|}, \quad -\infty < x < \infty.$$

This density assumes its maximum at $x_m = \mu$, namely $f(\mu) = \lambda/2$. Thus,

$$E(X) = x_{0.5} = x_m = \mu.$$

□

In what follows, formulas for the *measures of variability*, introduced in section 2.2 for discrete random variables, are given for continuous random variables. Their interpretation does not change.

Variance The *variance* of X is the mean value of the squared deviation of X from its mean value $E(X)$, i.e. the mean value of the random variable $Y = (X - E(X))^2$:

$$Var(X) = E(X - E(X))^2.$$

The calculation of this mean value does not require knowledge of the density of Y, but can be done by (2.53) with $h(x) = (x - E(X))^2$:

$$Var(X) = \int_{-\infty}^{+\infty} (x - E(X))^2 f(x)\,dx. \tag{2.60}$$

Thus, the variance of X is its 2 nd central moment (equation 2.57).

If with constants a and b the random variable $aX + b$ is of interest, then $h(x)$ becomes

$$h(x) = (ax + b - aE(X) - b)^2 = a^2(x - E(X))^2$$

so that

$$Var(aX + b) = a^2 Var(X). \tag{2.61}$$

There is an important relationship between the variance and the second moment of X:

$$Var(X) = E(X^2) - [E(X)]^2. \tag{2.62}$$

The proof is identical to the one for the corresponding relationship for discrete random variables (see formula 2.17).

Standard Deviation The *standard deviation* of X is the square root of $Var(X)$. It is frequently denoted as σ:

$$\sigma = \sqrt{Var(X)}.$$

Coefficient of Variation The *coefficient of variation* of X is defined as the ratio

$$V(X) = \sigma / E(X).$$

It follows from formulas (2.54) and (2.61) that X and aX have the same coefficient of variation. More generally, since the coefficient of variation considers the values of X in relation to their average size, this coefficient allows to compare the variability of different random variables.

An important measure of the variability is also the *mean absolute linear deviation* of X from its mean value:

$$E(|x - E(X)|) = \int_{-\infty}^{+\infty} |x - E(X)| f(x)\,dx. \tag{2.63}$$

This is the 1st absolute central moment of X as defined by (2.58):

$$M_1 = E(|x - E(X)|).$$

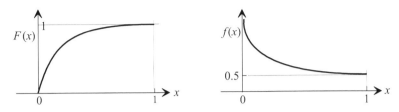

Figure 2.9 Distribution function and density for example 2.15

Example 2.15 Let X be the random emission of SO_2 [in $100\,kg/h$] of a chemical factory. Its distribution function $F(x)$ (density $f(x)$) over one day, starting at midnight, has been found to be (Figure 2.9)

$$F(x) = \begin{cases} 0 & \text{for} & x < 0, \\ \sqrt{x} & \text{for} & 0 \le x \le 1, \\ 1 & \text{for} & 1 < x. \end{cases} \qquad f(x) = \begin{cases} 0 & \text{for} & x < 0, \\ 0.5\,x^{-0.5} & \text{for} & 0 \le x \le 1, \\ 0 & \text{for} & 1 < x. \end{cases}$$

The graph of the density shows that the bulk of the (illegal) emissions occurs immediately after midnight. Later the emissions tend to the accepted values.

By (2.52), the mean value of X is

$$E(X) = \int_0^1 (1 - \sqrt{x})\,dx = [x - \tfrac{2}{3} \cdot x^{3/2}]_0^1 = 1/3 \ [100\,kg/h].$$

This result and formulas (2.56) and (2.62) yield the second moment and the variance:

$$E(X^2) = \int_0^1 x^2\, 0.5\,x^{-1/2}\,dx = 0.5 \int_0^1 x^{1.5}\,dx = 0.2,$$

$$\sigma^2 = Var(X) = 0.2 - (1/3)^2 \approx 0.0889.$$

Standard deviation and coefficient of variation are

$$\sigma = \sqrt{Var(X)} \approx 0.2981, \quad V(X) = \sigma/E(X) \approx 0.8943 \,\hat{=}\, 89,43\%.$$

The 1st absolute central moment of X is

$$M_1 = E(|X - 1/3|) = \int_0^1 |x - 1/3|\, 0.5\,x^{-0.5}\,dx$$

$$= \int_0^{1/3} (1/3 - x)\, 0.5\,x^{-0.5}\,dx + \int_{1/3}^1 (x - 1/3)\, 0.5\,x^{-0.5}\,dx = 0.1283 + 0.1283$$

so that $E(|X - 1/3|) = 0.2566 \ [100\,kg/h]$. $\qquad\qquad\qquad\qquad\qquad\qquad\qquad$ $\square$

Continuation of Example 2.12 a) The probabilities p_i in example 2.12 are actually assigned to the class numbers 1, 2, ...,6. To be able to get quantitative information on the ball weights, now the p_i are assigned to the middle points of the class intervals. That means the original range of X, namely $\{1, 2, \dots, 6\}$, is replaced with the range $\{5.605, 5.645, 5.68.5, 5.725, 5.765, 5.805\}$. The choice of the middle points takes into account that the classes do not contain their upper limit.

In this way, a discrete random variable has been generated, which approximates the original continuous one, the weight of the balls. Mean value and variance of X are

$$E(X) = 5.605 \cdot 0.0667 + 5.645 \cdot 0.1333 + 5.685 \cdot 0.1667 + 5.725 \cdot 0.2167,$$

and

$$Var(X) = (5.605 - 5.722)^2 \cdot 0.0667 + (5.645 - 5.722)^2 \cdot 0.1333$$
$$+ (5.685 - 5.722)^2 \cdot 0.1667 + (5.725 - 5.722)^2 \cdot 0.2167$$
$$+ (5.765 - 5.722)^2 \cdot 0.2833 + (5.805 - 5.722)^2 \cdot 0.1333$$

so that

$$E(X) = 5.722, \quad Var(X) = 0.00343, \quad \sqrt{Var(X)} = 0.05857.$$

For the sake of comparison, the first absolute central moment is calculated:

$$E(|X - E(X)|) = |5.605 - 5.722| \cdot 0.0667 + |5.645 - 5.722| \cdot 0.1333$$
$$+ |5.685 - 5.722| \cdot 0.1667 + |5.725 - 5.722| \cdot 0.2167$$
$$+ |5.765 - 5.722| \cdot 0.2833 + |5.805 - 5.722| \cdot 0.1333$$
$$= 0.0481.$$

By representing several values of the original data set by their average value, the numerical effort is reduced, but some of the information contained in the data set is lost. Based on the data set given, maximal information on the mean value and on the variance of X give the *arithmetic mean* $\bar{x}$ and the *empirical variance* s^2, respectively, which are calculated from the individual $n = 60$ values provided by Table 2.2:

$$\bar{x} = \frac{1}{n} \sum_{i=1}^{n} x_i \quad \text{and} \quad s^2 = \frac{1}{n-1} \sum_{i=1}^{n} (x_i - \bar{x})^2 = \frac{1}{n-1} \sum_{i=1}^{n} x_i^2 - \frac{n}{n-1} \bar{x}^2 . \quad (2.64)$$

The numerical results are, including the *empirical standard deviation* $s = \sqrt{s^2}$:

$$\bar{x} = 5.727, \quad s^2 = 0.0032, \quad \text{and} \quad s = 0.0566.$$

Directly from the data set, the *empirical mean absolute deviation* is given by

$$\frac{1}{n} \sum_{i=1}^{n} |x_i - \bar{x}| = \frac{1}{60} \sum_{i=1}^{60} |x_i - 5.727| = 0.0475.$$

b) The frequency histogram of Figure 2.5 suggests a suitable empirical density $f_{60}(y)$ with respect to class intervals of length 1:

$$f_{60}(y) = \begin{cases} 0 & \text{if } y < 2/3, \\ \frac{3}{145}(3y - 2) & \text{if } 2/3 \leq y < 5.5, \\ \frac{3}{55}(-3y + 22) & \text{if } 5.5 \leq y \leq 22/3, \\ 0 & \text{if } 22/3 < y. \end{cases}$$

Having assigned length 1 to all class intervals formally means that the variables x and y in Figure 2.10 are related by the linear transformation $y = 25x - 138.75$, or, in terms of the corresponding variables Y and X:

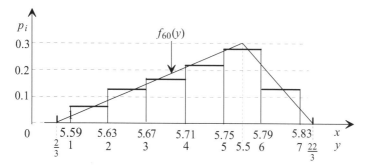

Figure 2.10 Probability histogram and empirical density for example 2.12

$$Y = 25X - 138.75 \quad \text{or} \quad X = 0.04Y + 5.55. \tag{2.65}$$

First of all, it has to be shown that $f_{60}(y)$ is indeed a probability density, i.e,. it has to be shown that the area $\mathbf{A}$ of the triangle is equal to 1: Since it is composed of 2 rectangular triangles, there is no need for integration:

$$\mathbf{A} = \tfrac{1}{2}0.3 \cdot (5.5 - 2/3) + \tfrac{1}{2}0.3 \cdot (22/3 - 5.5) = 1.$$

This empirical density allows the calculation of estimates for the distribution parameters by the formulas given in this section.

The mean value of Y is

$$E(Y) = \int_{2/3}^{22/3} y f_{60}(y)dy = \frac{3}{145}\int_{2/3}^{5.5} y(3y-2)dy + \frac{3}{55}\int_{5.5}^{22/3} y(-3y+22)dy$$

$$= \frac{3}{145}\left[y^3 - y^2\right]_{2/3}^{5.5} + \frac{3}{55}\left[-y^3 + 11y^2\right]_{5.5}^{22/3}$$

$$= 4.4965.$$

By formulas (2.54) and (2.65), $E(X) = 0.04 E(Y) + 5.55$ so that

$$E(X) = 0.04 \cdot 4.4965 + 5.55 = 5.729.$$

By formula (2.60), an estimate of the variance of Y is

$$Var(Y) = \int_{2/3}^{22/3} y^2 f_{60}(y)dy - [E(Y)]^2$$

$$= \frac{3}{145}\int_{2/3}^{5.5} y^2(3y-2)dy + \frac{3}{55}\int_{5.5}^{22/3} y^2(-3y+22)dy - [4.4965]^2$$

$$= \frac{3}{145}\left[y^3\left(\tfrac{3}{4}y - \tfrac{2}{3}\right)\right]_{2/3}^{5.5} + \frac{3}{55}\left[y^3\left(-\tfrac{3}{4}y + \tfrac{22}{3}\right)\right]_{5.5}^{22/3} - [4.4965]^2$$

$$= 2.0083.$$

Hence, by formulas (2.59) and (2.60),

$$Var(X) = 0.04^2 Var(Y) = 0.003213.$$

By (2.63), the mean absolute linear deviation of Y from $E(Y)$ is

$$E(|Y - E(Y)|) = \int_{2/3}^{22/3} |y - 4.4965| f_{60}(y)\, dy$$

$$= \frac{3}{145} \int_{2/3}^{4.4965} (4.4965 - y)(3y - 2)\, dy + \frac{3}{145} \int_{4.4965}^{5.5} (y - 4.4965)(3y - 2)\, dy$$

$$+ \frac{3}{55} \int_{5.5}^{22/3} (y - 4.4965)(-3y + 22)\, dy$$

$$= 0.58111 + 0.14060 + 0.44402 = 1.16573.$$

Hence, $E(|X - E(X)|) = 0.04 E(|Y - E(Y)|) = 0.04663.$ □

Truncation Most of the probability distributions for random variables have ranges $[0, \infty)$ or $(-\infty, +\infty)$, respectively. If, however, in view of whatever reasons a random variable, which is supposed to have distribution function $F(x)$, can only take on values from an interval $[c, d]$, then a *truncation* of the range of X or its distribution, respectively, makes sense. This is being done by replacing $F(x) = P(X \le x)$ with the conditional distribution function $F_{[c,d]}(x) = F(X \le x | c \le X \le d)$. By formula (1.22),

$$F_{[c,d]}(x) = \begin{cases} 0 & \text{if } x < c, \\ \frac{F(x) - F(c)}{F(d) - F(c)} & \text{if } c \le x \le d, \\ 1 & \text{if } d < x. \end{cases} \tag{2.66}$$

For instance, when the exponential distribution (example 2.10) is truncated with regard to the interval $[c, d]$, then

$$F_{[c,d]}(x) = \begin{cases} 0 & \text{if } x < c, \\ \frac{e^{-\lambda c} - e^{-\lambda x}}{e^{-\lambda c} - e^{-\lambda d}} & \text{if } c \le x \le d, \\ 1 & \text{if } d < x. \end{cases} \tag{2.67}$$

Most important is the special case $c = 0$. Then,

$$F_{[0,d]}(x) = \begin{cases} 0 & \text{if } x < 0, \\ \frac{1 - e^{-\lambda x}}{1 - e^{-\lambda d}} & \text{if } 0 \le x \le d, \\ 1 & \text{if } d < x. \end{cases} \tag{2.68}$$

Truncation is actually a very adequate tool to tailor probability distributions to the respective application. Although, as mentioned above, most of the common probability distributions have unbounded ranges (at least to the right), unbounded random variables are unrealistic (even impossible) outcomes of random experiments like determining life-, repair-, and service times or measurement errors.

Standardization A random variable S (discrete or continuous) with

$$E(S) = 0 \text{ and } Var(S) = 1$$

is called a *standard random variable*.

In view of formulas (2.54) and (2.59), for any random variable X with finite mean value $\mu = E(X)$ and variance $\sigma^2 = Var(X)$, the linear transformation of X given by

$$S = \frac{X - \mu}{\sigma} \qquad (2.69)$$

or, equivalently, by

$$S = \frac{1}{\sigma} X - \frac{\mu}{\sigma}$$

is a standard random variable. S is called the *standardization* or *normalization* of X.

Skewness In case of a continuous random variable, its distribution is symmetric if and only if its density is a symmetric function. The *skewness* of a distribution measures the degree of asymmetry of arbitrary probability distributions, including discrete ones. (Remember the skewness of a discrete probability distribution is visualized by its histogram.) The two most popular skewness criterions are *Charlier's skewness* γ_C and *Pearson's skewness* γ_P:

$$\gamma_C = \frac{m_3}{\sigma^3}, \quad \gamma_P = \frac{\mu - x_m}{\sigma},$$

where μ, m_3, x_m, and σ are in this order mean value, third central moment (see formula 2.57), mode, and standard deviation of X. For symmetric distributions both criteria are equal to 0. They are negative if the density is skewed to the right ('long tail' of the density to the right (Figure 2.11)) and positive if the density is skewed to the left ('long tail' of the density to the left).

Charlier's skewness is invariant to the linear transformation (2.69), i,.e., invariant to standardization. That means, X and its standardization $(X - E(X))/\sigma$ have the same skewness if measured by γ_C.

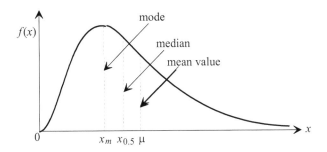

Figure 2.11 Asymmetric density skewed to the right

2.3.3 Important Continuous Probability Distributions

In this section some important probability distributions of continuous random variables X will be listed. If the distribution function is not explicitly given, it can only be
represented as integral over the density.

Note: In what follows, the areas where the distribution function is 0 or 1 or, equivalently, the
density is 0, are no longer explicitly taken into account when specifying the domains of definition of these functions.

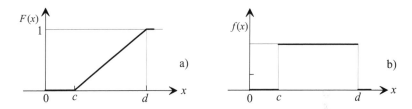

Figure 2.12 Distribution function a) and density b) for the uniform distribution

Uniform Distribution A random variable X has a *uniform distribution* over the finite
interval (range) $[c,d]$ with $c < d$ if it has distribution function and density

$$F(x) = \frac{x-c}{d-c}, \quad c \le x \le d, \quad f(x) = \frac{1}{d-c}, \quad c \le x \le d.$$

Thus, for any subinterval $[a,b]$ of $[c,d]$, the corresponding interval probability is

$$P(a < X \le b) = \frac{b-a}{d-c}.$$

This probability depends only on the length of the interval $[a,b]$, but not on its position within the interval $[c,d]$, i.e., all subintervals of $[c,d]$ of the same length have
the same chance that X takes on a value out of it.

Mean value and variance of X are

$$E(X) = \frac{c+d}{2}, \quad Var(X) = \frac{1}{12}(d-c)^2.$$

Power Distribution A random variable X has a *power distribution* with finite range
$[0,\tau]$ if it has distribution function and density (Figure 2.13)

$$F(x) = \left(\frac{x}{\tau}\right)^\alpha, \quad f(x) = \frac{\alpha}{\tau}\left(\frac{x}{\tau}\right)^{\alpha-1}, \quad \alpha > 0, \ \tau > 0, \ 0 \le x \le \tau.$$

Mean value and variance are

$$E(X) = \frac{\alpha\tau}{\alpha+1}, \quad Var(X) = \frac{\alpha\tau^2}{(\alpha+1)^2(\alpha+2)}, \quad \alpha > 0, \ \tau > 0.$$

The uniform distribution with range $[0,\tau]$ is seen to be a special case if $\alpha = 1$.

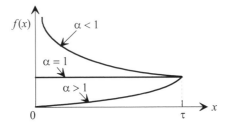

Figure 2.13 Density of the power distribution

Note τ is a *scale parameter*, i.e., without loss of generality $\tau = 1$ can be chosen as measurement unit. α is the *shape* or *form parameter* of this distribution, since α determines the shape of the graph of the density.

Pareto Distribution A random variable X has a *Pareto distribution* with range $[\tau, \infty)$ if it has distribution function and density

$$F(x) = 1 - \left(\frac{\tau}{x}\right)^\alpha, \quad f(x) = \frac{\alpha}{\tau}\left(\frac{\tau}{x}\right)^{\alpha+1}, \quad x \ge \tau > 0, \ \alpha > 0.$$

Mean value and variance are

$$E(X) = \frac{\alpha\tau}{\alpha-1}, \quad \alpha > 1, \ Var(X) = \frac{\alpha\tau^2}{(\alpha-1)^2(\alpha-2)}, \quad \alpha > 2.$$

For $\alpha < 1$ and $\alpha < 2$ mean value and variance, respectively, do not exist, i.e., they are not finite.

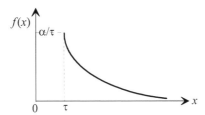

Figure 2.14 Density of the Pareto distribution

Cauchy Distribution A random variable X has a *Cauchy distribution* with parameters λ and μ if it has density

$$f(x) = \frac{\lambda}{\pi[\lambda^2 + (x-\mu)^2]}, \quad -\infty < x < \infty, \ \lambda > 0, \ -\infty < \mu < \infty.$$

This distribution is symmetric with symmetry center μ. Mean value and variance are infinite.

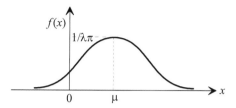

Figure 2.15 Density of the Cauchy distribution

Exponential Distribution A random variable X has an *exponential distribution* with (scale) parameter λ if it has distribution function and density (Figure 2.5, page 60)

$$F(x) = 1 - e^{-\lambda x}, \quad f(x) = \lambda e^{-\lambda x}, \quad \lambda > 0, \quad x \geq 0. \tag{2.70}$$

Mean value and variance are

$$E(X) = 1/\lambda, \quad Var(X) = 1/\lambda^2. \tag{2.71}$$

Erlang Distribution A random variable X has an *Erlang distribution* with parameters λ and n if it has distribution function and density

$$F(x) = 1 - e^{-\lambda x} \sum_{i=0}^{n-1} \frac{(\lambda x)^i}{i!} = e^{-\lambda x} \sum_{i=n}^{\infty} \frac{(\lambda x)^i}{i!}, \tag{2.72}$$

$$f(x) = \lambda \frac{(\lambda x)^{n-1}}{(n-1)!} e^{-\lambda x}; \quad x \geq 0, \; \lambda > 0, \; n = 1, 2, ... \tag{2.73}$$

Mean value and variance are

$$E(X) = n/\lambda, \quad Var(X) = n/\lambda^2.$$

The exponential distribution is a special case of the Erlang distribution for $n = 1$. The relationship between the Erlang distribution and the Poisson distribution with parameter λ is obvious, since the right-hand side of (2.72) is the probability that at least n Poisson events occur in the interval $[0, x]$ (formula (2.39), page 56).

Gamma Distribution A random variable X has a *gamma distribution* with parameters α and β if it has density (Figure 2.16)

$$f(x) = \frac{\beta^\alpha}{\Gamma(\alpha)} x^{\alpha-1} e^{-\beta x}, \quad x > 0, \; \alpha > 0, \; \beta > 0, \tag{2.74}$$

where the *gammafunction* $\Gamma(y)$ is defined by

$$\Gamma(y) = \int_0^\infty t^{y-1} e^{-t} dt, \quad y > 0. \tag{2.75}$$

Mean value, variance, mode and Charlier's skewness are

$$E(X) = \alpha/\beta, \quad Var(X) = \alpha/\beta^2, \quad x_m = (\alpha - 1)/\beta, \quad \gamma_C = 2/\sqrt{\alpha}. \tag{2.76}$$

Special cases: Exponential distribution for $\alpha = 1$ and $\beta = \lambda$, Erlang distribution for $\alpha = n$ and $\beta = \lambda$.

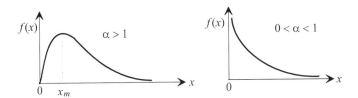

Figure 2.16 Densities of the gamma distribution

Beta Distribution A random variable X has a *beta distribution* with range (c,d) and parameters α and β if it has density

$$f(x) = \frac{(d-c)^{1-\alpha-\beta}}{B(\alpha,\beta)}(x-c)^{\alpha-1}(d-x)^{\beta-1}, \quad c < x < d, \; \alpha > 0, \; \beta > 0,$$

where the *beta function* $B(x,y)$ is defined as

$$B(\alpha,\beta) = \int_0^1 x^{\alpha-1}(1-x)^{\beta-1}dx.$$

An equivalent representation of the beta function is

$$B(x,y) = \frac{\Gamma(x)\Gamma(y)}{\Gamma(x+y)}; \quad x > 0, \; y > 0.$$

Mean value and variance are

$$E(X) = c + (d-c)\frac{\alpha}{\alpha+\beta}, \quad Var(X) = \frac{(d-c)^2\alpha\beta}{(\alpha+\beta)^2(\alpha+\beta+1)}.$$

The mode of this distribution is

$$x_m = c + (d-c)\frac{\alpha-1}{\alpha+\beta-2} \quad \text{for } \alpha \geq 1, \; \beta \geq 1, \text{ and } \alpha+\beta > 2.$$

A special case is the uniform distribution in $[c,d]$ if $\alpha = \beta = 1$.

If X has a beta distribution on the interval (c,d), then $Y = (X-c)/(d-c)$ has a beta distribution on the interval $(0,1)$. Hence, it is sufficient to consider the beta distribution with range $(0,1)$. The corresponding density is

$$f(x) = \frac{1}{B(\alpha,\beta)}x^{\alpha-1}(1-x)^{\beta-1}, \quad 0 < x < 1, \; \alpha > 0, \; \beta > 0.$$

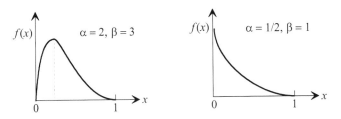

Figure 2.17 Densities of the beta-distribution over $(0, 1)$

Weibull Distribution A random variable X has a *Weibull distribution* with scale parameter θ and shape parameter β (*2-parameter Weibull distribution*) if it has distribution function and density (Figure 2.18)

$$F(x) = 1 - e^{(x/\theta)^\beta}, \quad f(x) = \frac{\beta}{\theta}\left(\frac{x}{\theta}\right)^{\beta-1} e^{(x/\theta)^\beta}; \quad x > 0, \ \beta > 0, \ \theta > 0. \qquad (2.77)$$

Mean value and variance are

$$E(X) = \theta \Gamma\left(\frac{1}{\beta} + 1\right), \quad Var(X) = \theta^2\left[\Gamma\left(\frac{2}{\beta} + 1\right) - \left(\Gamma\left(\frac{1}{\beta} + 1\right)\right)^2\right]. \qquad (2.78)$$

Special cases: Exponential distribution if $\theta = 1/\lambda$ and $\beta = 1$. Rayleigh distribution if $\beta = 2$. Distribution function, density, and parameters of the Rayleigh distribution are

$$F(x) = 1 - e^{(x/\theta)^2}, \quad f(x) = \frac{2x}{\theta^2} e^{(x/\theta)^2}; \quad x > 0, \ \theta > 0. \qquad (2.79)$$

$$E(X) = \theta\sqrt{\pi/4}, \quad Var(X) = \theta^2(1 - \pi/4). \qquad (2.80)$$

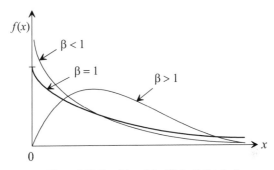

Figure 2.18 Densities of the Weibull distribution

3-parameter Weibull distribution A random variable X has a *3-parameter Weibull distribution* with parameters $\alpha, \beta,$ and θ if it has distribution function and density

$$F(x) = \begin{cases} 0 & \text{for } x < \alpha, \\ 1 - e^{-\left(\frac{x-\alpha}{\theta}\right)^\beta} & \text{for } \alpha \leq x, \end{cases}$$

$$f(x) = \begin{cases} 0 & \text{for } x < \alpha, \\ \frac{\beta}{\alpha}\left(\frac{x-\alpha}{\theta}\right)^{\beta-1} e^{-\left(\frac{x-\alpha}{\theta}\right)^\beta} & \text{for } \alpha \leq x. \end{cases}$$

α is a *parameter of location*, since X cannot assume values smaller than α.

Remark The Weibull distribution was found by the German mining engineers *E. Rosin* and *E. Rammler* in the late twenties of the past century when investigating the distribution of the size of stone, coal, and other particles after a grinding process (see, for example, *Rosin, Rammler* (1931)). Hence, in the mining engineering literature, the Weibull distribution is called *Rosin-Rammler distribution*. The Swedish engineer *W. Weibull* came across this distribution type when investigating mechanical wear in the early thirties of the past century.

Example 2.16 By a valid standard, the useful life X of front tires of a certain type of trucks comes to an end if their tread depth has decreased to $5\,mm$. From a large sample of $n = 120$ useful lifes of front tires, taken under average usage conditions, the mean useful life had been determined to be $2\,years$. The histogram of the same sample also justifies to assume that X has a Rayleigh distribution.

a) What is the probability of the random event A that the useful life of a tire exceeds 2.4 years?

By (2.77), the unknown parameter θ of the Rayleigh distribution can be obtained from the equation $E(X) = 2 = \theta \sqrt{\pi/4}$. It follows $\theta = 2.25676$. Hence,

$$P(A) = P(X > 2.4) = e^{-(2.4/\theta^2)} = 0.34526.$$

b) What is the probability of A on condition that a tire has not yet reached the end of its useful life after 2 years of usage? From the formula of the conditional probability (1.22), the desired probability is

$$P(A|X > 2) = P(X > 2.4|X > 2)$$

$$= \frac{1 - F(2.4)}{1 - F(2)} = \frac{e^{-(2.4/\theta^2)}}{e^{-(2/\theta^2)}}$$

$$= e^{-0.4/2.25676^2} = 0.83757. \qquad \square$$

Normal Distribution A random variable X has a *normal* (or *Gaussian*) *distribution* with parameters μ and σ^2 if it has density (Figure 2.19)

$$f(x) = \frac{1}{\sqrt{2\pi}\,\sigma} e^{-\frac{1}{2}\left(\frac{x-\mu}{\sigma}\right)^2}, \quad -\infty < x < +\infty, \; -\infty < \mu < +\infty, \; \sigma > 0. \quad (2.81)$$

The corresponding distribution function can only be given as an integral, since there exists no function the first derivative of which is $f(x)$:

$$F(x) = \frac{1}{\sqrt{2\pi}\,\sigma} \int_{-\infty}^{x} e^{-\frac{(y-\mu)^2}{2\sigma^2}} \, dy, \quad -\infty < x < +\infty. \quad (2.82)$$

As the notation of the parameters indicates, mean value and variance are

$$E(X) = \mu, \quad Var(X) = \sigma^2. \quad (2.83)$$

The mean absolute deviation of X from $E(X)$ is

$$E(|X - E(X)|) = \sqrt{2/\pi}\,\sigma \approx 0.798\sigma. \quad (2.84)$$

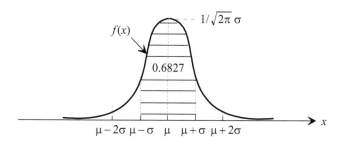

Figure 2.19 Density of the normal distribution (Gaussian bell curve)

This can be seen as follows: The substitution $y = (x - \mu)/\sigma$ in

$$E(|X - E(X)|) = \int_{-\infty}^{+\infty} |x - \mu| \frac{1}{\sqrt{2\pi}\,\sigma} e^{-(x-\mu)^2/2\sigma^2} dx$$

yields

$$E(|X - E(X)|) = \int_{-\infty}^{+\infty} |y| \frac{1}{\sqrt{2\pi}} e^{-y^2/2} \sigma\, dy$$

$$= \frac{\sigma}{\sqrt{2\pi}} \left[\int_{-\infty}^{0} (-y) e^{-y^2/2}\, dy + \int_{0}^{\infty} y e^{-y^2/2}\, dy \right]$$

$$= \frac{2\sigma}{\sqrt{2\pi}} \int_{0}^{\infty} y e^{-y^2/2}\, dy = \frac{2\sigma}{\sqrt{2\pi}} \left[-e^{-y^2/2} \right]_{0}^{\infty} = \frac{2\sigma}{\sqrt{2\pi}}.$$

The density $f(x)$ is positive at the whole real axis. It is symmetric with symmetry center $x_s = \mu$ and has points of inflection at $x_1 = \mu - \sigma$ and $x_2 = \mu + \sigma$.
In the intervals $[\mu - k\sigma, \mu + k\sigma]$, $k = 1, 2, 3$, X assumes values with probabilities:

$$P(\mu - \sigma \leq X \leq \mu + \sigma) \quad = 0.6827,$$
$$P(\mu - 2\sigma \leq X \leq \mu + 2\sigma) = 0.9545,$$
$$P(\mu - 3\sigma \leq X \leq \mu + 3\sigma) = 0.9973.$$

In particular, if a random experiment with outcome X is repeated many times, then 99.73% of the values of X will be in the '3σ-interval' $[\mu - 3\sigma, \mu + 3\sigma]$. Therefore, only 0.27% of all outcomes will be outside the 3σ-interval. In view of the symmetry of $f(x)$, this implies that for $\mu \geq 3\sigma$ negative values of X occur only with probability

$$\tfrac{1}{2}(1 - 0.9973) = 0.000135 \hat{=} 0.0135\%.$$

Thus, in case $\mu \geq 3\sigma$ the normal distribution can approximately serve as probability distribution for a nonnegative random variable. If $\mu < 3\sigma$, then a truncation with regard to $x = 0$ is recommended according to formula (2.68) with $c = 0$ and $d = \infty$. This makes sure that negative values cannot occur. The truncated normal distribution is a favorite model for lifetimes of systems subject to wear out. Generally, for reasons to be substantiated later (section 5.2.3, page 208), the normal distribution is a suitable probability distribution of random variables, which are generated by additive superposition of numerous effects.

A normally distributed random variable X with parameters μ and σ^2 is denoted as

$$X = N(\mu, \sigma^2).$$

Generally, the standardization S of a random variable X as given by (2.70) does not have the same distribution type as X. But the standardization

$$S = \frac{X - \mu}{\sigma}$$

of a normally distributed random variable $X = N(\mu, \sigma)$ is again normally distributed. This can be seen as follows:

$$F_S(x) = P(S \le x) = P\left(\frac{X-\mu}{\sigma} \le x\right) = P(X \le \sigma x + \mu).$$

From (2.82), substituting there $u = \frac{y-\mu}{\sigma}$,

$$F_S(x) = \frac{1}{\sqrt{2\pi}\,\sigma} \int_{-\infty}^{\sigma x + \mu} e^{-\frac{(y-\mu)^2}{2\sigma^2}} \, dy = \frac{1}{\sqrt{2\pi}} \int_{-\infty}^{x} e^{-u^2/2} du.$$

By comparison with (2.82), the right integral in this line is seen to be the distribution function of a normally distributed random variable with mean value 0 and variance 1. This implies the desired result, namely $S = N(0, 1)$. S is said to be *standard normal*. Its distribution function is denoted as $\Phi(x)$:

$$\Phi(x) = P(N(0, 1) \le x) = \frac{1}{\sqrt{2\pi}} \int_{-\infty}^{x} e^{-u^2/2} du, \quad -\infty < x < \infty. \tag{2.85}$$

The corresponding density $\varphi(x) = \Phi'(x)$ is

$$\varphi(x) = \frac{1}{\sqrt{2\pi}} e^{-x^2/2}, \quad -\infty < x < \infty. \tag{2.86}$$

$\Phi(x)$ or $\varphi(x)$, respectively, determins the *standard normal distribution*.

$\Phi(x)$ is closely related to the *Gaussian error integral* Erf(x), which led *C. F. Gauss* to the normal distribution:

$$\text{Erf}(x) = \int_0^x e^{-u^2/2} du.$$

Simple transformations, taking into account $\Phi(0) = 1/2$, yield

$$\Phi(x) = \frac{1}{2} + \frac{1}{\sqrt{\pi}} \text{Erf}\left(\frac{x}{\sqrt{2}}\right) \quad \text{and} \quad \text{Erf}(x) = \sqrt{\pi}\left(\Phi(\sqrt{2}\,x) - \frac{1}{2}\right).$$

Since $\varphi(x)$ is symmetric with symmetry center $x_s = 0$ (Figure 2.20),

$$\Phi(x) = 1 - \Phi(-x).$$

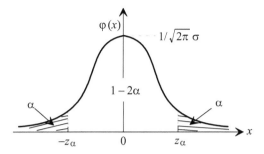

Figure 2.20 Density and percentiles of the standardized normal distribution

From this another useful formula results:

$$P(-x \le N(0, 1) \le +x) = \Phi(x) - \Phi(-x) = 2\Phi(x) - 1. \qquad (2.87)$$

Hence, there is the following relationship between the α- and the $(1-\alpha)$-percentiles of the standardized normal distribution:

$$-x_\alpha = x_{1-\alpha}, \quad 0 < \alpha < 1/2.$$

This is the reason for introducing the following notation (Figure 2.20):

$$z_\alpha = x_{1-\alpha}, \quad 0 < \alpha < 1/2.$$

Hence, with α replaced by $\alpha/2$,

$$P(-z_{\alpha/2} \le N(0, 1) \le z_{\alpha/2}) = \Phi(z_{\alpha/2}) - \Phi(-z_{\alpha/2}) = 1 - \alpha.$$

The distribution function $F(x)$ of $X = N(\mu, \sigma^2)$ can be expressed in terms of $\Phi(x)$ as follows:

$$F(x) = P(X \le x) = P\left(\frac{X-\mu}{\sigma} \le \frac{x-\mu}{\sigma}\right) = P\left(N(0, 1) \le \frac{x-\mu}{\sigma}\right) = \Phi\left(\frac{x-\mu}{\sigma}\right).$$

Corollaries 1) The interval probabilities (2.5) are given for any normally distributed random variable $X = N(\mu, \sigma^2)$ by

$$P(a \le X \le b) = \Phi\left(\frac{b-\mu}{\sigma}\right) - \Phi\left(\frac{a-\mu}{\sigma}\right). \qquad (2.88)$$

2) If x_α denotes the α-percentile of $X = N(\mu, \sigma^2)$, then

$$\alpha = F(x_\alpha) = \Phi\left(\frac{x_\alpha - \mu}{\sigma}\right)$$

so that, for any $\alpha < 1/2$,

$$\frac{x_\alpha - \mu}{\sigma} = z_\alpha \text{ or } x_\alpha = \sigma z_\alpha + \mu.$$

Therefore, determining the percentiles of any normally distributed random variable can be done by a table of the percentiles of the standardized normal distribution.

Example 2.17 A company needs cylinders with a diameter of 20 mm. It accepts deviations of ± 0.5 mm. The manufacturer produces these cylinders with a random diameter X, which has a $N(20, \sigma^2)$-distribution.

a) What percentage of cylinders is accepted by the company if $\sigma^2 = 0.04$ mm?

Since the condition $\mu \geq 3\sigma$ is fulfilled ($\mu \geq 100\sigma$), X can be considered a positive random variable. By (2.89) and (2.88), the probability to accept a cylinder is

$$P(|X - 20| \leq 0.5) = P(19.5 \leq X \leq 20.5) = P\left(\frac{19.5 - 20}{0.2} \leq N(0, 1) \leq \frac{20.5 - 20}{0.2}\right)$$

$$= P(-2.5 \leq N(0, 1) \leq +2.5) = 2\,\Phi(2.5) - 1$$

$$= 2 \cdot 0.9938 - 1 = 0.9876.$$

Thus, 98.76% of the produced cylinders are accepted.

b) What is the value of σ^2 if the company would reject 4% of the cylinders?

$$P(|X - 20| > 0.5) = 1 - P(19.5 \leq X \leq 20.5)$$

$$= 1 - P\left(\frac{19.5 - 20}{\sigma} \leq N(0, 1) \leq \frac{20.5 - 20}{\sigma}\right)$$

$$= 1 - P\left(-\frac{0.5}{\sigma} \leq N(0, 1) \leq \frac{0.5}{\sigma}\right) = 1 - [2\,\Phi(0.5/\sigma) - 1]$$

$$= 2\,[1 - \Phi(0.5/\sigma)].$$

The term $2\,[1 - \Phi(0.5/\sigma)]$ is required to be equal to 0.04. This leads to the equation

$$\Phi(0.5/\sigma) = 0.98.$$

Now one takes from the table that value $x_{0.98}$ for which $\Phi(x_{0.98}) = 0.98$. In other words, one determines the 0.98-percentile of the standardized normal distribution. This percentile is seen to be $x_{0.98} = 2.06$. Hence, the desired σ must satisfy

$$0.5/\sigma = 2.06.$$

It follows $\sigma = 0.2427$.   □

Example 2.18 By a data set collected over 32 years, the monthly rainfall from November to February in an area has been found to be normally distributed with mean value 92 mm and variance 784 mm. (Again, the condition $\mu \geq 3\sigma$ is fulfilled.)

What are the probabilities of the 'extreme cases' that (1) the monthly rainfall during the given time period is between 0 and 30 mm, and (2) exceeds 150 mm?

(1) $\quad P(0 \leq X \leq 30) = P\left(\frac{0 - 92}{28} \leq N(0, 1) \leq \frac{30 - 92}{28}\right) = \Phi(-2.214) - \Phi(-3.286)$

$$\approx \Phi(-2.214) \approx 0.0135.$$

(2) $\quad P(X > 150) = P\left(N(0, 1) > \frac{150 - 92}{28}\right) = 1 - \Phi(2.071) \approx 1 - 0.981$

$$= 0.019.$$ □

The first four moments (2.56) of the normal distribution $N(\mu, \sigma^2)$ are

$$\mu_1 = \mu = E(X),$$
$$\mu_2 = \sigma^2 + \mu^2,$$
$$\mu_3 = 3\mu\sigma^2 + \mu^3,$$
$$\mu_4 = \mu^4 + 6\mu^2\sigma^2 + 3\sigma^4,$$

and its first four central moments (2.57) are

$$m_1 = 0, \quad m_2 = \sigma^2, \quad m_3 = 0, \quad m_4 = \mu^4 + 6\mu^2\sigma^2 + 3\sigma^4.$$

In view of the key role the normal distribution plays in probability theory, it is useful, particularly for applications, to know how well any other probability distribution can be approximated by the normal distribution. Information about this gives the *excess* γ_E defined for any probability distribution with second central moment m_2 and fourth central moment m_4:

$$\gamma_E = \frac{m_4}{(m_2)^2} - 3.$$

Since γ_E is 0 for $N(\mu, \sigma^2)$, the excess can serve as a measure for the deviation of the distribution of any random variable with mean μ and variance σ^2 from $N(\mu, \sigma^2)$ in an environment of μ.

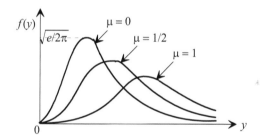

Figure 2.21 Densities of the logarithmic normal distribution

Logarithmic Normal Distribution A random variable Y has a *logarithmic normal distribution* or *log-normal distribution* with parameters μ and σ^2 if it has distribution function and density (Figure 2.21)

$$F(y) = \Phi\left(\frac{\ln y - \mu}{\sigma}\right); \qquad y > 0, \ \sigma > 0, \ -\infty < \mu < \infty,$$

$$f(y) = \frac{1}{\sqrt{2\pi}\ \sigma y}\ e^{-\frac{(\ln y - \mu)^2}{2\sigma^2}} \ ; \quad y > 0, \ \sigma > 0, \ -\infty < \mu < \infty.$$

Thus, Y has a log-normal distribution with parameters μ and σ^2 if it has structure $Y = e^X$ with $X = N(\mu, \sigma^2)$. Hence, if y_α is the α-percentile of the log-normal distribution and x_α the α-percentile of the $N(\mu, \sigma^2)$, then $y_\alpha = e^{x_\alpha}$, or, in terms of the α-percentile u_α of the standard normal distribution, $y_\alpha = e^{\sigma u_\alpha + \mu}$. Since $u_{0.5} = 0$, the median is $y_{0.5} = e^\mu$. The distribution is unimodal with mode $y_m = e^{\mu - \sigma^2}$.

Mean value and variance of X are

$$E(X) = e^{\mu + \sigma^2/2}, \quad Var(X) = [E(X)]^2 \left(e^{\sigma^2} - 1 \right).$$

The Charlier skewness and the excess are

$$\gamma_C = \left(\sqrt{e^{\sigma^2} - 1} \right) \left(e^{\sigma^2} + 2 \right), \quad \gamma_E = e^{4\sigma^2} + 2e^{3\sigma^2} + 3e^{2\sigma^2} - 6.$$

Example 2.19 As the Rosin-Rammler distribution, the logarithmic normal distribution is a favorite model for the particle size of stone and other materials after a grinding process. Statistical analysis has shown that the diameter of lava rock particles after a grinding process in a special mill has a logarithmic normal distribution with mean value $E(X) = 1.3002 \, mm$ and variance $Var(X) = 0.0778$.

What percentage of particles have their diameter in $I = [1.1, \, 1.5 \, mm]$?

Solving the system of equations $E(X) = 1.3002$, $Var(X) = 0.0778$ for μ and σ^2 gives $\mu = 0.24 \, mm$ and $\sigma^2 = 0.045$. Therefore,

$$P(1.1 \le X \le 1.5) = \Phi\left(\frac{\ln 1.5 - 0.24}{0.212} \right) - \Phi\left(\frac{\ln 1.1 - 0.24}{0.212} \right)$$

$$= \Phi(0.781) - \Phi(-0.683) = 0.783 - 0.246 = 0.537.$$

Thus, the corresponding percentage of particles is 53.7%.   □

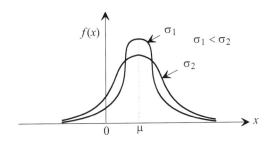

Figure 2.22 Density of the logistic distribution

Logistic Distribution A random variable X has a *logistic distribution* with parameters μ and σ if it has distribution function

$$F(x) = \frac{1}{1 + e^{-\frac{\pi}{\sqrt{3}\,\sigma}(x - \mu)}}, \quad -\infty < x < +\infty, \; \sigma > 0,$$

and density (Figure 2.22)

$$f(x) = \frac{\dfrac{\pi}{\sqrt{3}\,\sigma}\,e^{-\dfrac{\pi}{\sqrt{3}\,\sigma}(x-\mu)}}{\left(1 + e^{-\dfrac{\pi}{\sqrt{3}\,\sigma}(x-\mu)}\right)^2}, \quad -\infty < x < +\infty, \ \sigma > 0.$$

This distribution is symmetric with regard to μ. Mean value, variance, and excess are

$$E(X) = \mu, \quad Var(X) = \sigma^2, \quad \gamma_E = 1.2.$$

The denominator of $F(x)$ has the functional structure of a well-known growth curve originally proposed by *Verhulst* (1845). Generally, the logistic distribution proved to be a suitable probabilistic model for growth processes with saturation (i.e., not exceeding a given upper bound) of plants, in particular trees.

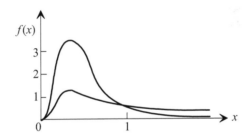

Figure 2.23 Densities of the inverse Gaussian distribution

Inverse Gaussian Distribution A random variable X has an *inverse Gaussian distribution* or a *Wald distribution* with positive parameters α and β if it has the density (Figure 2.23)

$$f(x) = \sqrt{\frac{\alpha}{2\pi x^3}}\, \exp\left(-\frac{\alpha(x-\beta)^2}{2\beta^2 x}\right), \quad x > 0. \tag{2.89}$$

Integration gives the corresponding distribution function

$$F(x) = \Phi\left(\frac{x-\beta}{\beta\sqrt{\alpha x}}\right) + e^{-2\alpha/\beta}\,\Phi\left(-\frac{x+\beta}{\beta\sqrt{\alpha x}}\right), \quad x > 0.$$

Mean value, variance, and mode are

$$E(X) = \beta, \quad Var(X) = \beta^3/\alpha, \quad x_m = \beta\left(\sqrt{1 + (3\beta/2\alpha)^2} - 3\beta/2\alpha\right). \tag{2.90}$$

Charlier's skewness and excess are

$$\gamma_C = \sqrt[3]{\beta/\alpha}, \gamma_E = 15\beta/\alpha.$$

The practical significance of the inverse Gaussian distribution is mainly due to the fact that it is the first passage time distribution of the Brownian motion process and some of its derivatives (pages 504, 513). This has made the inverse Gaussian distribution a favorite model for predicting time to failures of systems, which are subject to wearout.

2.3.4 Nonparametric Classes of Probability Distributions

This section is restricted to the class of nonnegative random variables. Lifetimes of technical systems and organisms are likely to be the most prominent members of this class. Hence, the terminology is tailored to this application. The lifetime of a system is the time from its starting up time point (birth) to its failure (death), where 'failure' is assumed to be an instantaneous event. In the engineering context, a failure of a system needs not be equivalent to the end of its useful life. If X is a lifetime of a system with distribution function $F(x) = P(X \leq x)$, then $F(x)$ is called its *failure probability* and $\overline{F}(x) = 1 - F(x)$ is its *survival probability*. $F(x)$ and $\overline{F}(x)$ are the respective probabilities that the system does or does not fail in the interval $[0, x]$.

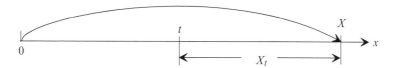

Figure 2.24 Illustration of the residual lifetime

Residual Lifetime Let $F_t(x)$ be the distribution function of the *residual lifetime* X_t of a system, which has already worked for t time units without failing (Figure 2.24):

$$F_t(x) = P(X_t \leq x) = P(X - t \leq x | X > t).$$

By the formula of the conditional probability (1.22)

$$F_t(x) = \frac{P(X - t \leq x \cap X > t)}{P(X > t)} = \frac{P(t < X \leq t + x)}{P(X > t)}$$

so that, by (2.44), page 59,

$$F_t(x) = \frac{F(t + x) - F(t)}{1 - F(t)}, \quad t > 0, \ x \geq 0. \tag{2.91}$$

The corresponding conditional survival probability $\overline{F}_t(x) = 1 - F_t(x)$ is

$$\overline{F}_t(x) = \frac{\overline{F}(t + x)}{\overline{F}(t)}, \quad t > 0, \ x \geq 0. \tag{2.92}$$

Hence, by using formula (2.52), the *mean residual lifetime* $\mu(t) = E(X_t)$ is seen to be

$$\mu(t) = \int_0^\infty \overline{F}_t(x) \, dx = \frac{1}{\overline{F}(t)} \int_t^\infty \overline{F}_t(x) \, dx. \tag{2.93}$$

Example 2.20 Let the lifetime X have a uniform distribution over $[0, T]$:

$$F(x) = x/T, \ 0 \le x \le T.$$

Then,

$$F_t(x) = \frac{x}{T-t}, \ 0 \le t < T, \ 0 \le x \le T - t.$$

Thus, X_t is uniformly distributed over the interval $[0, T-t]$, and for fixed x, the conditional failure probability is increasing with increasing age t of the system, $t < T$. $\square$

Example 2.21 Let X have an exponential distribution with parameter λ:

$$F(x) = 1 - e^{-\lambda x}, \ x \ge 0.$$

Then, for given $t > 0$ the conditional failure probability of the system in $[t, t+x]$ is

$$F_t(x) = \frac{(1 - e^{-\lambda(t+x)}) - (1 - e^{-\lambda t})}{e^{-\lambda t}} = 1 - e^{-\lambda x} = F(x), \ x \ge 0.$$

That means, if a system with exponentially distributed lifetime is known to have survived the interval $[0, t]$, then it is at time point t 'as good as new' from the point of view of its future failure behavior, since its residual lifetime X_t has the same failure probability as the system had at time point $t = 0$, when it started operating. In other words, systems with property

$$F_t(x) = F(x) \quad \text{for all } t \ge 0. \tag{2.94}$$

'do not age'. Thus, the exponential distribution is the continuous analogue to the geometric distribution (example 2.3). Its is, moreover, the only continuous distribution which has this so-called *memoryless property* or *lack of memory property*. Usually, systems (technical or biological ones) have this *nonaging property* only in certain finite subintervals of their useful life. These intervals start after the early failures have tapered off and last till wearout processes start. In the nonaging period failures or deaths are caused by purely random influences as natural catastrophes or accidents. In real life there is always some overlap of the early failure, nonaging, and wear out periods. $\square$

The fundamental relationship (2.94) is equivalent to the functional equation

$$\overline{F}(t+x) = \overline{F}(x) \cdot \overline{F}(t).$$

Only functions of type e^{ax} are solutions of this equation, where a is a constant.

The engineering and biological background of the conditional failure probability motivates the following definition:

Definition 2.3 A system is *aging (rejuvenating)* in the interval $[t_1, t_2]$, $t_1 < t_2$, if for an arbitrary but fixed x, $x > 0$, the conditional failure probability $F_t(x)$ is increasing (decreasing) with increasing t, $t_1 \le t \le t_2$. $\bullet$

Remark Here and in what follows the terms 'increasing' and 'decreasing' have the meaning of 'nondecreasing' and 'nonincreasing', respectively.

For technical systems periods of rejuvenation may be due to maintenance actions, and for human beings due to successful medical treatment or adopting a healthier lifestyle.

Provided the existence of the density $f(x) = F'(x)$, another approach to modeling the aging behavior of a system is based on the concept of its failure rate. To derive this rate, the conditional failure probability $F_t(\Delta t)$ of a system in the interval $[t, t + \Delta t]$ is considered relative to the length Δt of this interval. This gives a conditional failure probability per unit time, i.e. a 'failure probability rate':

$$\frac{1}{\Delta t} F_t(\Delta t) = \frac{1}{\overline{F}(t)} \cdot \frac{F(t + \Delta t) - F(t)}{\Delta t}.$$

If $\Delta t \to 0$, the second ratio on the right-hand side tends to $f(t)$. Hence,

$$\lim_{\Delta t \to 0} \frac{1}{\Delta t} F_t(\Delta t) = f(t)/\overline{F}(t). \tag{2.95}$$

This limit is called *failure rate* or *hazard function,* and it is denoted as $\lambda(t)$:

$$\lambda(t) = f(t)/\overline{F}(t). \tag{2.96}$$

In demography and actuarial science, $\lambda(t)$ is called *force of mortality.* Integration on both sides of (2.96) yields

$$F(x) = 1 - e^{-\int_0^x \lambda(t)dt} \quad \text{or} \quad \overline{F}(x) = e^{-\int_0^x \lambda(t)dt}, \quad x \ge 0. \tag{2.97}$$

By introducing the *integrated failure rate*

$$\Lambda(x) = \int_0^x \lambda(t)dt,$$

$F(x)$, $F_t(x)$ and the corresponding survival probabilities can be written as follows:

$$F(x) = 1 - e^{-\Lambda(x)}, \quad \overline{F}(x) = e^{-\Lambda(x)},$$

$$F_t(x) = 1 - e^{-[\Lambda(t+x) - \Lambda(t)]}, \quad \overline{F}_t(x) = e^{-[\Lambda(t+x) - \Lambda(t)]}, \quad x \ge 0, \ t \ge 0. \tag{2.98}$$

This representation of $F_t(x)$ implies an important property of the failure rate:

> A system ages in $[t_1, t_2]$, $t_1 < t_2$, if and only if its failure rate is increasing in this interval.

Formula (2.95) can be interpreted in the following way: For small Δt,

$$F_t(\Delta t) \approx \lambda(t)\,\Delta t. \tag{2.99}$$

Thus, for Δt sufficiently small, $\lambda(t)\,\Delta t$ is approximately the probability that the system fails 'shortly' after time point t if it has survived the interval $[0, t]$. Hence, the failure rate gives information on both the instantaneous tendency of a system to fail and its 'state of wear' at any age t.

The relationship (2.99) can be written more exactly in the form

$$F_t(\Delta t) = \lambda(t)\,\Delta t + o(\Delta t),$$

where $o(x)$ is the *Landau order symbol* with respect to $x \to 0$, i.e. $o(x)$ is any function of x with property

$$\lim_{x \to 0} \frac{o(x)}{x} = 0. \tag{2.100}$$

In the ratio of (2.100), both functions $y_1(x) = o(x)$ and the function $y_2(x) = x$ tend to 0 if $x \to 0$, but $y_1(x) = o(x)$ must approach 0 'much faster' than $y_2(x) = x$ if $x \to 0$. Otherwise (2.100) could not be true.

The relationship (2.99) can be used for the statistical estimation of $\lambda(t)$: At time $t = 0$, n identical systems start operating. Let $n(t)$ be the number of those systems, which have failed in the interval $[0, t]$. Then the number of systems which have survived $[0, t]$ is $n - n(t)$, and the number of systems which have failed in the interval $(t, t + \Delta t]$ is $n(t + \Delta t) - n(t)$. Then an estimate for the system failure rate in $(t, t + \Delta t]$ is

$$\hat{\lambda}(x) = \frac{1}{\Delta t} \frac{n(t + \Delta t) - n(t)}{n - n(t)}, \quad t < x \le t + \Delta t.$$

Based on the behavior of the conditional failure probability of systems, numerous nonparametric classes of probability distributions have been proposed and investigated during the past 60 years. Originally, they aimed at applications in reliability engineering, but nowadays these classes also play an important role in fields like demography, actuarial science, and risk analysis.

Definition 2.4 $F(x)$ is an *IFR (increasing failure rate)* or a *DFR (decreasing failure rate)* distribution if $F_t(x)$ is increasing or decreasing in t for fixed but arbitrary x, respectively. Briefly: $F(x)$ is *IFR (DFR)*.   ●

If the density $f(x) = F'(x)$ exists, then (2.98) implies the following corollary:

Corollary $F(x)$ is *IFR (DFR)* in the interval $[x_1, x_2]$, $x_1 < x_2$, if and only if the correponding failure rate $\lambda(x)$ is increasing (decreasing) in $[x_1, x_2]$.

The Weibull distribution shows that, within one and the same parametric class of probability distributions, a distribution may belong to different nonparametric probability distributions: From (2.77) and (2.97),

$$\Lambda(x) = (x/\theta)^\beta$$

so that

$$\lambda(x) = \frac{\beta}{\theta} \cdot \left(\frac{x}{\theta}\right)^{\beta-1}, \quad x \ge 0.$$

Hence, the Weibull distribution is *IFR* for $\beta > 1$ and *DFR* for $\beta < 1$. For $\beta = 1$ the failure rate is constant: $\lambda = \beta/\theta$ (exponential distribution). The exponential distribution is both *IFR* and *DFR*. This versatility of the Weibull distribution is one reason for being a favorite model in applications.

The failure rate (force of mortality) of human beings and other organisms is usually not strictly increasing. In short time periods, for instance, after having overcome a serious disease or another life-threatening situation, the failure rate will decrease, although the average failure rate will definitely increase. Actually, in view of the finite lifetimes of organism, their failure rates $\lambda(x)$ will tend to infinity as $x \to \infty$.

Analogously, technical systems, which operate under different, time-dependent stress levels (temperature, pressure, humidity, speed), will not have a strictly increasing failure rates, although in the long-run, their average failure rates are increasing. This motivates the following definition:

Definition 2.5 $F(x)$ is an *IFRA* (*increasing failure rate average*) distribution or a *DFRA* (*decreasing failure rate average*) distribution if

$$-\frac{1}{x} \ln \overline{F}(x)$$

is an increasing or a decreasing function in x, respectively.   ●

To justify the terminology, assuming the density $f(x) = F'(x)$ exists and taking the natural logarithm on both sides of the right equation in (2.97) yields

$$\ln \overline{F}(x) = -\int_0^x \lambda(t)\, dt.$$

Therefore,

$$\overline{\lambda}(x) = -\frac{1}{x} \ln \overline{F}(x) = \frac{1}{x} \int_0^x \lambda(t)\, dt$$

so that $-(1/x)\ln \overline{F}(x)$ turns out to be the average failure rate in $[0, x]$. An advantage of definitions 2.3 to 2.5 is that they do not require the existence of the density. But the existence of the density and, hence, the existence of the failure rate, motivates the terminology. Other intuitive proposals for nonparametric classes are based on the 'new better than used' concept or on the behavior of the mean residual lifetime $\mu(t)$; see *Lai, Xie* (2006) for a comprehensive survey.

Obviously, $F(x)$ being *IFR* (*DFR*) implies $F(x)$ being *IFRA* (*DFRA*):

$$IFR \Rightarrow IFRA, \quad DFR \Rightarrow DFRA.$$

Knowing the type of the nonparametric class $F(x)$ belongs to allows the construction of upper or lower bounds on $F(x)$ or $\overline{F}(x)$. For instance, if $\mu_n = E(X^n)$ is the nth moment of X and $F(x) = P(X \le x)$ is *IFR*, then

$$\overline{F}(x) \ge \begin{cases} \exp\{-x\,(n!/\mu_n)^{1/n}\} & \text{for } x \le \mu_n^{1/n}, \\ 0 & \text{otherwise.} \end{cases}$$

In particular, for $n = 1$ with $\mu = \mu_1 = E(X)$,

$$\overline{F}(x) \ge \begin{cases} \exp\{-x/\mu\} & \text{for } x \le \mu, \\ 0 & \text{otherwise.} \end{cases} \tag{2.101}$$

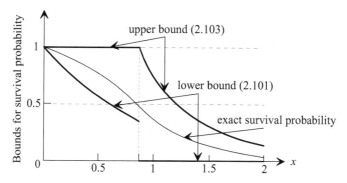

Figure 2.25 Upper and lower bounds for example 2.22

If $F(x)$ is *IFR*, then

$$\sup_x \left| \overline{F}(x) - e^{-x/\mu} \right| \leq 1 - \sqrt{2\gamma + 1} \qquad (2.102)$$

with

$$\gamma = \frac{\mu_2}{2\mu^2} - 1.$$

It can be shown that $\gamma \leq 0$ ($\gamma \geq 0$) if $F(x)$ is *IFR* (DFR).

If $F(x)$ is *IFRA*, then

$$\overline{F}(x) \leq \begin{cases} 1 & \text{for } x < \mu, \\ e^{-rx} & \text{for } x \geq \mu, \end{cases} \qquad (2.103)$$

where $r = r(x, \mu)$ is the unique solution of

$$1 - r\mu = e^{-r\mu x}.$$

Example 2.22 Let X have distribution function

$$F(x) = P(X \leq x) = 1 - e^{-x^2} , \ x \geq 0.$$

This is a Rayleigh distribution (page 77) so that $F(x)$ is *IFR* and X has mean value

$$\mu = E(X) = \sqrt{\pi/4} \ \text{ and second moment } \mu_2 = Var(X) + \mu^2 = 1$$

(see formulas (2.80)). Figure 2.25 compares the exact graph of the corresponding survival probability $\overline{F}(x)$ with the lower bound (2.101) and the upper bound (2.103). By (2.102), an upper bound for the maximum deviation of the exact graph of $\overline{F}(x)$ from the exponential survival probability with the same mean $\mu = \sqrt{\pi/4}$ as X is, since $\gamma = 2/\pi - 1 \approx -0.3634$,

$$\sup_x \left| \overline{F}(x) - e^{-x/\sqrt{\pi/4}} \right| = \sup_x \left| e^{-x^2} - e^{-x/\sqrt{\pi/4}} \right| \leq 0.4773. \qquad \square$$

2.4 MIXTURES OF RANDOM VARIABLES

The probability distribution P_X (definition 2.1) of any random variable X depends on one or more numerical parameters. To emphasize the dependency on a special parameter λ, in this section the notation $P_X(\lambda)$ instead of P_X is used. Equivalently, in terms of the distribution function and density of X,

$$F_X(x) = F_X(x, \lambda), \quad f_X(x) = f_X(x, \lambda).$$

Mixtures of random variables or, equivalently, their probability distributions arise from the assumption that the parameter λ is a realization (value) of a random variable L, and all the probability distributions belonging to the set $\{P_X(\lambda), \lambda \in R_L\}$, where R_L is the range of L, are mixed in a way to be explained as follows:

1. Discrete random parameter L Let L have range $R_L = \{\lambda_0, \lambda_1, ...\}$ and probability distribution

$$P_L = \{\pi_0, \pi_1, ...\} \text{ with } \pi_n = P(L = \lambda_n), \ n = 0, 1, ..., \ \textstyle\sum_{n=0}^{\infty} \pi_n = 1.$$

Then the mixture of the probability distributions of type $P_X(\lambda)$ in terms of the mixture of the corresponding probability distribution functions of type $F_X(x, \lambda)$, $\lambda \in R_L$, is defined as

$$G(x) = \textstyle\sum_{n=0}^{\infty} F_X(x, \lambda_n) \pi_n.$$

2. Continuous random parameter L Let L have range R_L with $R_L \subseteq (-\infty, +\infty)$ and probability density

$$f_L(\lambda), \ \lambda \in R_L.$$

Then the mixture of the probability distributions of type $P_X(\lambda)$ in terms of the distribution functions of type $F_X(x, \lambda)$ is defined as

$$G(x) = \int_{R_L} F_X(x, \lambda) f_L(\lambda) \, d\lambda.$$

Thus, if L is a discrete random variable, then $G(x)$ is the weighted sum of the distribution functions $F_X(x, \lambda_n)$ with weights π_n given by the probability mass function of L. If L is continuous, then $G(x)$ is the weighted integral of $F_X(x, \lambda)$ with *weight function* $f_L(x, \lambda)$. In either case, $G(x)$ has properties (2.3) and (2.4) so that it is the distribution function of a random variable Y, called a *mixed random variable*, and the probability distribution of Y is the *mixture of probability distributions* of type $P_X(\lambda)$.

If X is continuous and L discrete, then the density of Y is

$$g(x) = \textstyle\sum_{n=0}^{\infty} f_X(x, \lambda_n) \pi_n.$$

If X and L are continuous, then the density of Y is

$$g(x) = \int_{R_L} f_X(x, \lambda) f_L(\lambda) \, d\lambda.$$

Formally, $G(x)$ and $g(x)$ are the mean values of $F_X(x, L)$ and $f_X(x, L)$, respectively:

$$G(x) = E(F_X(x, L)), \quad g(x) = E(f_X(x, L)).$$

If L is discrete and X is discrete with probability distribution

$$\boldsymbol{P}_X(\lambda) = \{p_i(\lambda) = P(X = x, \lambda); \; i = 0, 1, ...\},$$

then the probability distribution of Y, given so far by its distribution function $G(x)$, can also be characterized by its individual probabilities:

$$P(Y = x_i) = \sum_{n=0}^{\infty} p_i(\lambda_n)\pi_n = E(p_i(L)); \quad i = 0, 1, ... \tag{2.104}$$

If L is continuous and X is discrete, then

$$P(Y = x_i) = \int_{R_L} p_i(\lambda) f_L(\lambda) d\lambda = E(p_i(L)). \tag{2.105}$$

The probability distribution of L is sometimes called *structure* or *mixing distribution*. Hence, the probability distribution $\boldsymbol{P}_Y$ of the '*mixed random variable*' Y is a *mixture of probability distributions of type* $\boldsymbol{P}_{X,\lambda}$ *with regard to a structure distribution* $\boldsymbol{P}_L$.

The mixture of probability distributions provides a method for producing types of probability distributions, which are specifically tailored to serve the needs of special applications.

Example 2.23 (*mixture of exponential distributions*) Assume X has an exponential distribution with parameter λ:

$$F_X(x, \lambda) = P(X \le x) = 1 - e^{-\lambda x}, \; x \ge 0.$$

This distribution is to be mixed with regard to a structure distribution $\boldsymbol{P}_L$, where L is exponentially distributed with density

$$f_L(\lambda) = \mu e^{-\mu\lambda}, \; \mu > 0.$$

Mixing yields the distribution function

$$G(x) = \int_0^\infty F_X(x, \lambda) f_L(\lambda) d\lambda = \int_0^\infty (1 - e^{-\lambda x}) \mu e^{-\mu\lambda} d\lambda = 1 - \frac{\mu}{x + \mu}.$$

Hence, mixing exponential distributions with regard to an exponential structure distribution gives the *Lomax distribution* with distribution function and density

$$G(x) = \frac{x}{x + \mu}, \quad g(x) = \frac{\mu}{(x + \mu)^2}, \quad x \ge 0, \; \mu > 0. \tag{2.106}$$

The Lomax distribution is also known as *Pareto distribution of the second kind*. □

Example 2.24 (*mixture of binomial distributions*) Let X have a binomial distribution with parameters n and p:

$$P(X = i) = \binom{n}{i} p^i (1 - p)^{n-i}, \quad i = 0, 1, 2, ..., n.$$

The parameter n is considered to be a value of a Poisson with parameter λ distributed random variable N:

$$P(N = n) = \frac{\lambda^n}{n!} e^{-\lambda}; \quad n = 0, 1, \dots \quad (\lambda \text{ fixed}).$$

Then, from (2.104), using

$$\binom{n}{i} = 0 \text{ for } n < i,$$

the mixture of binomial distributions $P_X(n)$, $n = 0, 1, \dots$, with regard to the Poisson structure distribution P_N is obtained as follows:

$$P(Y = i) = \sum_{n=0}^{\infty} \binom{n}{i} p^i (1-p)^{n-i} \frac{\lambda^n}{n!} e^{-\lambda}$$

$$= \sum_{n=i}^{\infty} \binom{n}{i} p^i (1-p)^{n-i} \frac{\lambda^n}{n!} e^{-\lambda}$$

$$= \frac{(\lambda p)^i}{i!} e^{-\lambda} \sum_{k=0}^{\infty} \frac{[\lambda(1-p)]^k}{k!} = \frac{(\lambda p)^i}{i!} e^{-\lambda} e^{\lambda(1-p)}.$$

Thus,

$$P(Y = i) = \frac{(\lambda p)^i}{i!} e^{-\lambda p}; \quad i = 0, 1, \dots.$$

This is a Poisson distribution with parameter λp. □

Mixed Poisson Distribution Let X have a Poisson distribution with parameter λ:

$$P_X(\lambda) = \{P(X = i) = \frac{\lambda^i}{i!} e^{-\lambda}; \ i = 0, 1, \dots; \lambda > 0\}.$$

A random variable Y with range $\{0, 1, \dots\}$ is said to have a *mixed Poisson distribution* if its probability distribution is a mixture of the Poisson distributions $P_X(\lambda)$ with regard to any structure distribution. For instance, if the structure distribution is given by the density $f_L(\lambda)$ of a positive random variable L (i.e., the parameter λ of the Poisson distribution is a realization of L), the distribution of Y is

$$P(Y = i) = \int_0^\infty \frac{\lambda^i}{i!} e^{-\lambda} f_L(\lambda) \, d\lambda, \quad i = 0, 1, \dots. \tag{2.107}$$

A mixed Poisson distributed random variable Y with any structure parameter L has the following properties

$$\left.\begin{array}{c} E(Y) = E(L) \\ Var(Y) = E(L) + Var(L) \\ P(Y > n) = \int_0^\infty \frac{\lambda^n}{n!} e^{-\lambda} \overline{F}_L(\lambda)) \, d\lambda \end{array}\right\} \tag{2.108}$$

where $F_L(\lambda) = P(L \le \lambda)$ is the distribution function of L and $\overline{F}_L(\lambda) = 1 - F_L(\lambda)$.

Example 2.25 (*mixed Poisson distribution, gamma structure distribution*) Let the random structure variable L have a gamma distribution with density

$$f_L(\lambda) = \frac{\beta^\alpha}{\Gamma(\alpha)} \lambda^{\alpha-1} e^{-\beta\lambda}, \quad \lambda > 0, \ \alpha > 0, \ \beta > 0.$$

The corresponding mixed Poisson distribution is obtained as follows:

$$P(Y = i) = \int_0^\infty \frac{\lambda^i}{i!} e^{-\lambda} \frac{\beta^\alpha}{\Gamma(\alpha)} \lambda^{\alpha-1} e^{-\beta\lambda} \, d\lambda$$

$$= \frac{1}{i!} \frac{\beta^\alpha}{\Gamma(\alpha)} \int_0^\infty \lambda^{i+\alpha-1} e^{-\lambda(\beta+1)} \, d\lambda$$

$$= \frac{1}{i!} \frac{\beta^\alpha}{\Gamma(\alpha)} \frac{1}{(\beta+1)^{i+\alpha}} \int_0^\infty x^{i+\alpha-1} e^{-x} \, dx$$

$$= \frac{1}{i!} \frac{\Gamma(i+\alpha)}{\Gamma(\alpha)} \frac{\beta^\alpha}{(\beta+1)^{i+\alpha}}.$$

Thus,

$$P(Y = i) = \binom{i-1+\alpha}{i} \left(\frac{1}{\beta+1}\right)^i \left(\frac{\beta}{\beta+1}\right)^\alpha; \quad \alpha > 0, \ \beta > 0, \quad i = 0, 1, \dots. \quad (2.109)$$

This is a negative binomial distribution with parameters $r = \alpha$ and $p = 1/(\beta+1)$ (see formula (2.31), page 53). In deriving this result, the following property of the gamma function with $x = i + \alpha$, $i = 1, 2, \dots$, had been used

$$\Gamma(x) = (x-1)\Gamma(x-1); \ x > 0. \qquad \square$$

2.5 GENERATING FUNCTIONS

Probability distributions or at least moments of random variables can frequently be obtained from special functions, called (*probability* or *moment*) *generating functions* of random variables or, equivalently, of their probability distributions. This is of importance, since it is in many applications of stochastic methods easier to determine the generating function of a random variable instead of directly its probability distribution. This will be in particular demonstrated in Part II of this book in numerous applications. The method of determining the probability distribution of a random variable from its generating function is mathematically justified, since to every probability distribution belongs exactly one generating function and vice versa.

Formally, going over from a probability distribution to its generating function is a *transformation* of this distribution. In this section, transformations are separately considered for discrete random variables (z-transformation) and for continuous random variables (Laplace transformation).

2.5.1 z-Transformation

The discrete random variable X has range $R = \{0, 1, ...\}$ and probability distribution

$$\{p_i = P(X = i); \ i = 0, 1, ...\}.$$

The *z-transform* $M(z)$ of X or, equivalently, of its probability distribution is for any real number z with $|z| \leq 1$ defined as the power series

$$M(z) = \sum_{i=0}^{\infty} p_i z^i.$$

Thus, the probability distribution of X has been transformed into a power series. In this book, the extension of this series to complex numbers z is not necessary.

To avoid misunderstandings, sometimes the notation $M_X(z)$ is used instead of $M(z)$.

From (2.10) with $h(z_i) = z^i$, $M(z)$ is seen to be the mean value of $Y = z^X$:

$$M(z) = E(z^X). \tag{2.110}$$

$M(z)$ converges absolutely for $|z| \leq 1$:

$$|M(z)| \leq \sum_{i=0}^{\infty} p_i |z^i| \leq \sum_{i=0}^{\infty} p_i = 1.$$

Therefore, $M(z)$ can be differentiated (as well as integrated) term by term:

$$M'(z) = \sum_{i=0}^{\infty} i p_i z^{i-1}.$$

Letting $z = 1$ yields

$$M'(1) = \sum_{i=0}^{\infty} i p_i = E(X).$$

Taking the second derivative of $M(z)$ gives

$$M''(z) = \sum_{i=0}^{\infty} (i-1) i p_i z^{i-2}.$$

Letting $z = 1$ yields

$$M''(1) = \sum_{i=0}^{\infty} (i-1) i p_i = \sum_{i=0}^{\infty} i^2 p_i - \sum_{i=0}^{\infty} i p_i.$$

Therefore, $M''(1) = E(X^2) - E(X)$. Thus, the first two moments of X are

$$E(X) = M'(1), \quad E(X^2) = M''(1) + M'(1). \tag{2.111}$$

Continuing in this way, all moments of X can be generated by derivatives of $M(z)$. Hence, the power series $M(z)$ is indeed a *moment generating function*. By (2.13),

$$E(X) = M'(1), \quad Var(X) = M''(1) + M'(1) - [M'(1)]^2. \tag{2.112}$$

$M(z)$ is also a *probability generating function,* since

$$p_0 = M(0), \ p_1 = M'(0), \ p_2 = \frac{1}{2!} M''(0), \ p_3 = \frac{1}{3!} M'''(0), \dots.$$

Generally,

$$p_n = \frac{1}{n!} \frac{d^n M(z)}{dz} \bigg|_{z=0}; \ n = 0, 1, \dots. \tag{2.113}$$

Otherwise, according to the definition of $M(z)$, developing a given z-transform with unknown underlying probability disribution into a power series yields the probabilities p_i simply as the coefficients of z^i.

Geometric Distribution Let X have a geometric distribution with parameter p (page 50):

$$p_i = P(X = i) = p(1-p)^i; \quad i = 0, 1, \dots .$$

Then,

$$M(z) = \sum_{i=0}^{\infty} p(1-p)^i z^i$$

$$= p \sum_{i=0}^{\infty} [(1-p)z]^i .$$

By the geometrical series (2.16) with $x = (1-p)z$,

$$M(z) = \frac{p}{1-(1-p)z} .$$

The first two derivatives are

$$M'(z) = \frac{p(1-p)}{[1-(1-p)z]^2}, \quad M''(z) = \frac{2p(1-p)^2}{[1-(1-p)z]^3} .$$

Hence, by (2.111) and (2.112),

$$E(X) = \frac{1-p}{p}, \quad E(X^2) = \frac{(1-p)(2-p)}{p^2}, \quad Var(X) = \frac{1-p}{p^2} .$$

Poisson Distribution Let X have a Poisson distribution with parameter λ (page 56):

$$p_i = P(X = i) = \frac{\lambda^i}{i!} e^{-\lambda}; \quad i = 0, 1, \dots .$$

Then, in view of the exponential series (2.19),

$$M(z) = \sum_{i=0}^{\infty} \frac{\lambda^i}{i!} e^{-\lambda} z^i = e^{-\lambda} \sum_{i=0}^{\infty} \frac{(\lambda z)^i}{i!} = e^{-\lambda} e^{+\lambda z} .$$

Hence,

$$M(z) = e^{\lambda(z-1)} .$$

The first two derivatives are

$$M'(z) = \lambda e^{\lambda(z-1)}, \quad M''(z) = \lambda^2 e^{\lambda(z-1)} .$$

Letting $z = 1$ yields

$$M'(1) = \lambda, \quad M''(1) = \lambda^2 .$$

Thus, mean value, second moment, and variance of X are

$$E(X) = \lambda, \quad E(X^2) = \lambda(\lambda+1), \quad Var(X) = \lambda .$$

Mixed Poisson Distribution The mixed Poisson distribution with density $f_L(\lambda)$ of its structure parameter L has the individual probabilities (formula (2.107))

$$P(Y = i) = \int_0^\infty \frac{\lambda^i}{i!} e^{-\lambda} f_L(\lambda) d\lambda, \quad i = 0, 1, \dots.$$

Hence, its z-transform is

$$M_Y(z) = \sum_{i=0}^\infty \left(\int_0^\infty \frac{\lambda^i}{i!} e^{-\lambda} f_L(\lambda) d\lambda \right) z^i = \int_0^\infty \sum_{i=0}^\infty \frac{(\lambda z)^i}{i!} e^{-\lambda} f_L(\lambda) d\lambda$$

so that

$$M_Y(z) = \int_0^\infty e^{\lambda(z-1)} f_L(\lambda) d\lambda.$$

This result can be interpreted as 'mixture of z-transforms of Poisson distributions'.

Binomial Distribution Let X have a binomial distribution with parameters n and p (page 51):

$$p_i = P(X = i) = \binom{n}{i} p^i (1-p)^{n-i}; \quad i = 0, 1, \dots, n.$$

Then,

$$M(z) = \sum_{i=0}^n p_i z^i = \sum_{i=0}^n \binom{n}{i} p^i (1-p)^{n-i} z^i$$

$$= \sum_{i=0}^n \binom{n}{i} (pz)^i (1-p)^{n-i}.$$

This is the binomial series (2.20) with $x = pz$ and $y = 1 - p$ so that

$$M(z) = [pz + 1 - p)]^n.$$

By differentiation,

$$M'(z) = np[pz + 1 - p)]^{n-1},$$

$$M''(z) = (n-1)np^2 [pz + 1 - p)]^{n-2}.$$

Hence,

$$M'(1) = np \quad \text{and} \quad M''(1) = (n-1)np^2$$

so that mean value, second moment, and variance of X are

$$E(X) = np, \quad E(X^2) = (n-1)np^2 + np, \quad Var(X) = np(1-p).$$

Convolution Let $\{p_0, p_1, \dots\}$ and $\{q_0, q_1, \dots\}$ be the respective probability distribution of the discrete random variables X and Y, and let a sequence $\{r_0, r_1, \dots\}$ be defined as follows

$$r_n = \sum_{i=0}^n p_i q_{n-i} = p_0 q_n + p_1 q_{n-1} + \dots + p_n q_0, \quad n = 0, 1, \dots. \tag{2.114}$$

The sequence $\{r_0, r_1, ...\}$ is called the *convolution* of the probability distributions of X and Y. The convolution is the probability distribution of a certain random variable Z since $\{r_0, r_1, ...\}$ fulfills the conditions of a discrete probability distribution (2.6):

$$\sum_{n=0}^{\infty} r_n = 1, \quad r_n \geq 0.$$

For deriving the z-transform of Z, *Dirichlet's formula* on how to change the order of summation in finite or infinite double sums is needed:

$$\sum_{n=0}^{\infty} \sum_{i=0}^{n} a_{in} = \sum_{i=0}^{\infty} \sum_{n=i}^{\infty} a_{in} . \tag{2.115}$$

Now,

$$M_Z(z) = \sum_{n=0}^{\infty} r_n z^n = \sum_{n=0}^{\infty} \sum_{i=0}^{n} p_i q_{n-i} z^n$$

$$= \sum_{i=0}^{\infty} p_i z^i \left(\sum_{n=i}^{\infty} q_{n-i} z^{n-i} \right)$$

$$= \left(\sum_{i=0}^{\infty} p_i z^i \right) \left(\sum_{k=0}^{\infty} q_k z^k \right).$$

Thus, the z-transform of the convolution of the probability distributions of two random variables X and Y is equal to the product of the z-transforms of the probability distributions of X and Y:

$$M_Z(z) = M_X(z) \cdot M_Y(z). \tag{2.116}$$

2.5.2 Laplace Transformation

The *Laplace transform* $\hat{f}(s)$ of a real function $f(x)$, $x \in (-\infty, +\infty)$, is defined as

$$\hat{f}(s) = \int_{-\infty}^{+\infty} e^{-sx} f(x)\, dx, \tag{2.117}$$

where the parameter s is a complex number.

The Laplace transform of a function need not exist. The following assumptions 1 and 2 make sure that this function exists for all s with $Re(s) > b$:

1) $f(x)$ is piecewise continuous.

2) There exist finite real constants a and b so that $f(x) \leq ae^{bx}$ for all $x > 0$.

Notation If $c = x + iy$ is any complex number (i.e., $i = \sqrt{-1}$ and x, y are real numbers), then $Re(c)$ denotes the *real part* of c: $Re(c) = x$. For the applications dealt with in this book, the parameter s can be assumed to be a real number.

If $f(t)$ is the density of a random variable X, then $\hat{f}(s)$ has a simple meaning:

$$\hat{f}(s) = E(e^{-sX}). \tag{2.118}$$

This formula is identical to (2.110) if there z is written in the form $z = e^{-s}$.

The n-fold derivative of $\hat{f}(s)$ with respect to s is

$$\frac{d^n \hat{f}(s)}{ds^n} = (-1)^n \int_{-\infty}^{+\infty} x^n e^{sx} f(x)\, dx.$$

Hence, the moments of all orders of X can be obtained from $E(X^0) = E(1) = 1$ and

$$E(X^n) = (-1)^n \left. \frac{d^n \hat{f}(s)}{ds^n} \right|_{s=0}, \quad n = 1, 2, \dots . \tag{2.119}$$

Sometimes it is more convenient to use the notation

$$\hat{f}(s) = L(f, s).$$

Partial integration in $\hat{f}(s)$ yields

$$L\left(\int_{-\infty}^x f(u)\, du,\ s \right) = \frac{1}{s} \hat{f}(s) \tag{2.120}$$

and, if $f(x) > 0$ for all $x \in (-\infty, +\infty)$ and $f^{(n)}(x)$ denotes the nth derivative of $f(x)$ with regard to x, then

$$\hat{f}^{(n)}(s) = s^n \hat{f}(s); \quad n = 1, 2, \dots . \tag{2.121}$$

Note This equation has to be modified for all $n = 1, 2, \dots$ if $f(x) = 0$ for $x < 0$:

$$\hat{f}^{(n)}(s) = s^n \hat{f}(s) - s^{n-1} f(0) - s^{n-2} f'(0) - \cdots - s^1 f^{(n-2)}(0) - f^{(n-1)}(0). \tag{2.122}$$

In particular, for $n = 1$,

$$L\left(\frac{df(x)}{dx},\ s \right) = s \hat{f}(s) - f(0). \tag{2.123}$$

Let $f_1, f_2, \dots, f_n$ be any n functions for which the corresponding Laplace transforms exist and $f = f_1 + f_2 + \cdots + f_n$. Then,

$$\hat{f}(s) = \hat{f}_1(s) + \hat{f}_2(s) + \cdots + \hat{f}(s). \tag{2.124}$$

Convolution The *convolution* $f_1 * f_2$ of two continuous functions f_1 and f_2, which are defined on $(-\infty, +\infty)$, is given by

$$(f_1 * f_2)(x) = \int_{-\infty}^{+\infty} f_1(x - u) f_2(u)\, du. \tag{2.125}$$

The convolution is a commutative operation, i.e.,

$$(f_1 * f_2)(x) = (f_2 * f_1)(x) = \int_{-\infty}^{+\infty} f_2(x - u) f_1(u)\, du.$$

If $f_1(x) = f_2(x) = 0$ for all $x < 0$, then

$$(f_1 * f_2)(x) = \int_0^x f_2(x - u) f_1(u)\, du = \int_0^x f_1(x - u) f_2(u)\, du. \tag{2.126}$$

The following formula is the 'continuous' analogue to (2.116):

$$L(f_1 * f_2, s) = L(f_1, s) \cdot L(f_2, s) = \hat{f}_1(s) \cdot \hat{f}_2(s). \tag{2.127}$$

A proof of this relationship is easily established:

$$L(f_1 * f_2, s) = \int_{-\infty}^{+\infty} e^{-sx} \int_{-\infty}^{+\infty} f_2(x-u) f_1(u) \, du \, dx$$

$$= \int_{-\infty}^{+\infty} e^{-su} f_1(u) \int_{-\infty}^{+\infty} e^{-s(x-u)} f_2(x-u) \, dx \, du$$

$$= \int_{-\infty}^{+\infty} e^{-su} f_1(u) \int_{-\infty}^{+\infty} e^{-sy} f_2(y) \, dy \, du$$

$$= L(f_1, s) \cdot L(f_2, s) = \hat{f}_1(s) \cdot \hat{f}_2(s).$$

In proving this relationship, the 'continuous version' of *Dirichlet's formula* (2.115) had been applied:

$$\int_{-\infty}^{z} \int_{-\infty}^{y} f(x,y) \, dx \, dy = \int_{-\infty}^{z} \int_{-\infty}^{z} f(x,y) \, dy \, dx.$$

Verbally, equation (2.126) means that the Laplace transform of the convolution of two functions is equal to the product of the Laplace transforms of these functions.

Retransformation The Laplace transform $\hat{f}(s)$ is called the *image* of $f(x)$, and $f(x)$ is the *preimage* of $\hat{f}(s)$. Finding the preimage of a given Laplace transform (*retransformation*) can be a difficult task. Properties (2.124) and (2.127) of the Laplace transformation suggest that Laplace transforms should be decomposed as far as possible into terms and factors (for instance, decomposing a fraction into partial fractions), because the retransformation of the arising less complex terms is usually easier than the retransformation of the original image.

Retransformation is facilitated by *contingency tables*. These tables contain important functions (preimages) and their Laplace transforms. Table 2.5 presents a selection of Laplace transforms, which are given by rational functions in s, and their preimages. There exists, moreover, an explicit formula for the preimages of Laplace transforms. Its application requires knowledge of complex calculus.

Example 2.26 Let X have an exponential distribution with parameter λ:

$$f(x) = \lambda e^{-\lambda x}, \quad x \geq 0.$$

The Laplace transform of $f(x)$ is

$$\hat{f}(s) = \int_0^{\infty} e^{-sx} \lambda e^{-\lambda x} \, dx = \lambda \int_0^{\infty} e^{-(s+\lambda)x} \, dx$$

so that

$$\hat{f}(s) = \frac{\lambda}{s+\lambda}.$$

The nth derivative of $\hat{f}(s)$ is

$$\frac{d^n \hat{f}(s)}{ds^n} = (-1)^n \frac{\lambda n!}{(s+\lambda)^{n+1}}.$$

Thus, the n th moment of X is

$$E(X^n) = \frac{n!}{\lambda^n}; \quad n = 0, 1, \dots.$$ □

Example 2.27 Let X have a normal distribution with density

$$f(x) = \frac{1}{\sqrt{2\pi}\,\sigma}\, e^{-\frac{(x-\mu)^2}{2\sigma^2}}; \quad x \in (-\infty, +\infty).$$

The Laplace transform of $f(x)$ is

$$\hat{f}(s) = \frac{1}{\sqrt{2\pi}\,\sigma} \int_{-\infty}^{+\infty} e^{-sx} e^{-\frac{(x-\mu)^2}{2\sigma^2}}\, dx.$$

This improper parameter integral exists for all s. Substituting $u = (x-\mu)/\sigma$ yields

$$\hat{f}(s) = \frac{1}{\sqrt{2\pi}}\, e^{-\mu s} \int_{-\infty}^{+\infty} e^{-\sigma s u} e^{-u^2/2}\, du = \frac{1}{\sqrt{2\pi}}\, e^{-\mu s + \frac{1}{2}\sigma^2 s^2} \int_{-\infty}^{+\infty} e^{-\frac{1}{2}(u+\sigma s)^2}\, du.$$

By substituting $y = u + \sigma s$, the second integral is seen to be $\sqrt{2\pi}$. Hence,

$$\hat{f}(s) = e^{-\mu s + \frac{1}{2}\sigma^2 s^2}.$$ (2.128)

□

Two important special cases of the Laplace transform are the *characteristic function* and the *moment generating function*.

Characteristic Function The *characteristic function*

$$\psi(y) = \int_{-\infty}^{+\infty} e^{iyx} f(x)\, dx$$

of a random variable with density $f(x)$ is a special case of its Laplace transform, namely if the parameter s is purely imaginary number, i.e. $s = iy$. Thus, the characteristic function is nothing else but the *Fourier transform* of $f(x)$. The advantage of the characteristic function to the Laplace transform is that it always exists:

$$|\psi(y)| = \left| \int_{-\infty}^{+\infty} e^{iyx} f(x)\, dx \right|$$

$$\leq \int_{-\infty}^{+\infty} |e^{iyx}| f(x)\, dx$$

$$= \int_{-\infty}^{+\infty} f(x)\, dx = 1.$$

As the z-transform and the Laplace transform, the characteristic function is moment and probability generating. Characteristic functions belong to the most important tools for solving theoretical and practical problems in probability theory.

Moment Generating Function Formally, the moment generating function $M(\cdot)$ is exactly defined as the Laplace transform $\hat{f}(s)$, namely by formula (2.117). The difference is that in case of the moment generating function the parameter s is always real and usually denoted as '$-t$' so that

$$M(t) = \int_{-\infty}^{+\infty} e^{tx} f(x)\, dx.$$

The key properties derived for Laplace transforms are of course also valid for the moment generating function. In particular, if $f(x)$ is a probability density, then

$$M(t) = E(e^{tX}).$$

The terminology is a bit confusing, since, as mentioned before, z-transform, the Laplace transform, and the characteristic function of a random variable are all moment- as well as probability generating.

Example 2.28 Let an image function be given by

$$\hat{f}(s) = \frac{s}{(s^2 - 1)^2}.$$

$\hat{f}(s)$ can be written as

$$\hat{f}(s) = \frac{s}{s^2 - 1} \cdot \frac{1}{s^2 - 1} = \hat{f}_1(s) \cdot \hat{f}_2(s).$$

The preimages of the factors can be found by means of Table 2.5:

$$f_1(x) = \cosh x = \tfrac{1}{2}(e^x + e^{-x})$$

and

$$f_2(x) = \sinh x = \tfrac{1}{2}(e^x - e^{-x}).$$

Let $f_1(x)$ and $f_2(x)$ be 0 for all $x < 0$. Then preimage $f(x)$ of $\hat{f}(s)$ is given by the convolution (2.126) of $f_1(x)$ and $f_2(x)$:

$$(f_1 * f_2)(x) = \tfrac{1}{4}\int_0^x (e^{(x-u)} + e^{-(x-u)})(e^u - e^{-u})\, du$$

$$= \tfrac{1}{4}\left[\int_0^x e^x(1 - e^{-2u})\, du + \int_0^x e^{-x}(e^{2u} - 1)\, du\right]$$

$$= \tfrac{1}{4}\left\{e^x\left[u + \tfrac{1}{2}e^{-2u}\right]_0^x + e^{-x}\left[\tfrac{1}{2}e^{2u} - u\right]_0^x\right\}$$

$$= \tfrac{1}{4}\left\{xe^x + \tfrac{1}{2}e^{-x} - \tfrac{1}{2}e^x + \tfrac{1}{2}e^x - xe^{-x} - \tfrac{1}{2}e^{-x}\right\}.$$

Thus,

$$f(x) = \frac{1}{2}x \sinh x.$$

This verifies the preimage given in Table 2.5 with $a = 1$. $\qquad\square$

Example 2.29 Let an image function be given by

$$\hat{f}(s) = \frac{s}{(s^2 - 1)(s+2)^2}.$$

The preimage cannot be taken from Table 2.5. But as in the previous example, it can be determined by factorization. But now the method of decomposition of $\hat{f}(s)$ into par- tial fractions is used: The denominator has the simple zeros $s = 1$, $s = -1$ and the doubly zero $s = -2$. Hence, $\hat{f}(s)$ can be written in the form

$$\hat{f}(s) = \frac{s}{(s^2 - 1)(s+2)^2} = \frac{A_1}{s-1} + \frac{A_2}{s+1} + \frac{B_1}{s+2} + \frac{B_2}{(s+2)^2}.$$

The coefficients A_1, A_2, B_1, and B_2 are determined by multiplying the equation by $(s^2 - 1)(s+2)^2$ and subsequent comparison of the coefficients of s^n; $n = 0, 1, 2, 3$; on both sides. This gives the equations

s^0 : $4A_1 - 4A_2 - 2B_1 - B_2 = 0$

s^1 : $8A_1 - B_1 = 1$

s^2 : $5A_1 + 3A_2 + 2B_1 + B_2 = 0$

s^3 : $A_1 + A_2 + B_1 = 0$

The solution is

$$A_1 = 1/18, \quad A_2 = 1/2, \quad B_1 = -5/9, \quad B_2 = -2/3.$$

Therefore,

$$\hat{f}(s) = \frac{1}{18} \cdot \frac{1}{s-1} + \frac{1}{2} \cdot \frac{1}{s+1} - \frac{5}{9} \cdot \frac{1}{s+2} + \frac{2}{3} \cdot \frac{1}{(s+2)^2}.$$

The preimage of the last term can be found in Table 2.5. If no table is available, then this term is represented as

$$\frac{1}{(s+2)^2} = \frac{1}{s+2} \cdot \frac{1}{s+2}.$$

The preimage of each factor is e^{-2x} so that the preimage of $1/(s+2)^2$ is equal to the convolution of e^{-2x} with itself:

$$e^{-2x} * e^{-2x} = \int_0^x e^{-2(x-y)} \cdot e^{-2y} dy$$

$$= \int_0^x e^{-2x} dy$$

$$= x e^{-2x}.$$

Now, by (2.124), retransformation of the image $\hat{f}(s)$ can be done term by term:

$$f(x) = \frac{1}{18} e^x + \frac{1}{2} e^{-x} - \frac{5}{9} e^{-2x} + \frac{2}{3} x e^{-2x}. \qquad \square$$

$\hat{f}(s)$	preimage	$\hat{f}(s)$	preimage
$\dfrac{1}{s}$	1	$\dfrac{1}{(s^2-a^2)^2}$	$\dfrac{1}{2a^2}(x\cosh ax - \dfrac{1}{a}\sinh ax)$
$\dfrac{1}{s^n},\ n\geq 1$	$\dfrac{1}{(n-1)!}x^{n-1}$	$\dfrac{s}{(s^2-a^2)^2}$	$\dfrac{1}{2a}x\sinh ax$
$\dfrac{1}{s+a}$	e^{-ax}	$\dfrac{s^2}{(s^2-a^2)^2}$	$\dfrac{1}{2a}(\sinh ax + ax\cosh ax)$
$\dfrac{1}{(s+a)^n}$	$\dfrac{1}{(n-1)!}x^{n-1}e^{-ax}$	$\dfrac{1}{(s+a)(s+b)}$	$\dfrac{1}{b-a}(e^{-ax}-e^{-bx})$
$\dfrac{s}{(s+a)^2}$	$(1-ax)e^{-ax}$	$\dfrac{s}{(s+a)(s+b)}$	$\dfrac{1}{b-a}(be^{-bx}-ae^{-ax})$
$\dfrac{s}{(s+a)^3}$	$(1-\dfrac{a}{2})xe^{-ax}$	$\dfrac{1}{(s+a)(s+b)^2}$	$\dfrac{1}{(b-a)^2}(e^{-ax}-e^{-bx}-(b-a)xe^{-bx})$
$\dfrac{s}{(s+a)^4}$	$\dfrac{1}{2}x^2 e^{-ax} - \dfrac{a}{6}x^3 e^{-ax}$	$\dfrac{s}{(s+a)(s+b)^2}$	$\dfrac{1}{(b-a)^2}\{-ae^{-ax}+[a+b(b-a)x]e^{-bx}\}$
$\dfrac{1}{s^2-a^2}$	$\dfrac{1}{a}\sinh(ax)$	$\dfrac{s^2}{(s+a)(s+b)^2}$	$\dfrac{1}{(b-a)^2}[(a^2e^{-ax}+b(b-2a-b^2x+abx)]e^{-bx}$
$\dfrac{1}{s^2+a^2}$	$\dfrac{1}{a}\sin(ax)$	$\dfrac{1}{s(s+a)^2}$	$\dfrac{1}{a^2}(1-e^{-ax}-axe^{-ax})$
$\dfrac{s}{s^2-a^2}$	$\cosh(ax)$	$\dfrac{1}{s(s+a)(s+b)}$	$\dfrac{1}{ab(a-b)}[a(1-e^{-bx})-b(1-e^{-ax})]$
$\dfrac{s}{s^2+a^2}$	$\cos ax$	$\dfrac{1}{(s+a)(s+b)(s+c)}$	$\dfrac{1}{(a-b)(b-c)(c-a)}[(c-b)e^{-ax}+$ $+(a-c)e^{-bx}+(b-a)e^{-cx}]$
$\dfrac{1}{s(s+a)}$	$\dfrac{1}{a}(1-e^{-ax})$	$\dfrac{s}{(s+a)(s+b)(s+c)}$	$\dfrac{1}{(a-b)(b-c)(c-a)}[a(b-c)e^{-ax}+$ $+b(c-a)e^{-bx}+c(a-b)e^{-cx}]$
$\dfrac{1}{s^2(s+a)}$	$\dfrac{1}{a^2}(e^{-ax}+ax-1)$	$\dfrac{s^2}{(s+a)(s+b)(s+c)}$	$\dfrac{1}{(a-b)(b-c)(c-a)}[-a^2(b-c)e^{-ax}$ $-b^2(c-a)e^{-bx}-c^2(a-b)e^{-cx}]$
$\dfrac{1}{(s^2+a^2)^2}$	$\dfrac{1}{2a^2}(\dfrac{1}{a}\sin ax - x\cos ax)$	$\dfrac{1}{(s+a)(s^2+b^2)}$	$\dfrac{1}{a^2+b^2}[e^{-ax}+\dfrac{a}{b}\sin bx - \cos bx]$
$\dfrac{s}{(s^2+a^2)^2}$	$\dfrac{1}{2a}x\sin ax$	$\dfrac{s}{(s+a)(s^2+b^2)}$	$\dfrac{1}{a^2+b^2}[-ae^{-ax}+a\cos bx+b\sin bx]$
$\dfrac{s^2}{(s^2+a^2)^2}$	$\dfrac{1}{2a}(\sin ax + ax\cos ax)$	$\dfrac{s^2}{(s+a)(s^2+b^2)}$	$\dfrac{1}{a^2+b^2}[a^2e^{-ax}-ab\sin bx+b^2\cos bx]$

Table 2.5 Images and the corresponding preimages of the Laplace transformation

2.6 EXERCISES

Sections 2.1 and 2.2

2.1) An ornithologist measured the weight of 132 eggs of helmeted guinea fowls [*gram*]:

number i	1	2	3	4	5	6	7	8	9	10
weight x_i	38	41	42	43	44	45	46	47	48	50
number of eggs n_i	4	6	7	10	13	26	33	16	10	7

There are no eggs weighing less than 38 and more than 50. Let X be the weight of a randomly picked egg from this sample.

(1) Determine the probability distribution of X.

(2) Draw the distribution function of X.

(3) Determine the probabilities $P(43 \le X \le 48)$ and $P(X > 45)$.

(4) Determine $E(X)$, $\sqrt{Var(X)}$, and $E(|X - E(X)|)$.

2.2) 114 nails are classified by length:

number i	1	2	3	4	5	6	7	
length (in mm) x_i	< 15.0	15.0	15.1	15.2	15.3	15.4	15.5	15.6	> 15.6
number of nails n_i	0	3	10	25	40	18	16	2	0

Let X denote the length of a nail selected randomly from this population.

(1) Determine the probability distribution of X.

(2) Determine the probabilities $P(X \le 15.1)$, and $P(15.0 < X \le 15.5)$.

(3) Determine $E(X)$, $m_3 = E(X - E(X))^3$, $\sigma = \sqrt{Var(X)}$, γ_C, and γ_P.

Interpret the skewness measures.

2.3) A set of 100 coins from an ongoing production process had been sampled and their diameters measured. The measurement procedure allows for a degree of accuracy of $\pm 0.04\, mm$. The table shows the measured values x_i and their numbers:

i	1	2	3	4	5	6	7
x_i	24.88	24.92	24.96	25.00	25.04	25.08	25.12
n_i	2	6	20	40	22	8	2

Let X be the diameter of a randomly from this set picked coin.

(1) Draw the distribution function of X.

(2) Determine $E(X)$, $E(|X - E(X)|)$, $Var(X)$, and $V(X)$.

2.4) 84 specimen copies of soft coal, sampled from the ongoing production in a colliery over a period of 7 days, had been analyzed with regard to ash and water content, respectively [in %]. Both ash and water content have been partitioned into 6 classes. The table shows the results:

water

		[16, 17)	[17, 18)	[18, 19)	[19, 20)	[20, 21)	[21, 22]
	[23, 24)	0	0	1	1	2	4
	[24, 25)	0	1	3	4	3	3
ash	[25, 26)	0	2	8	7	2	1
	[26, 27)	1	4	10	8	1	0
	[27, 28)	0	5	4	4	0	0
	[28, 29)	2	0	1	0	1	0

Let X be the water content and Y be the ash content of a randomly chosen specimen copy out of the 84 ones. Since the originally measured values are not given, it is assumed that the values, which X and Y can take on, are the centers of the given classes, i.e., 16.5, 17.5, ···, 21.5.

(1) Draw the distribution functions of X and Y.

(2) Determine $E(X)$, $Var(X)$, $E(Y)$, and $Var(Y)$.

2.5) It costs $ 50 to find out whether a spare part required for repairing a failed device is faulty or not. Installing a faulty spare part causes damage of $1000.

Is it on average more profitable to use a spare part without checking if

(1) 1% of all spare parts of that type,

(2) 3% of all spare parts of that type, and

(3) 10 % of all spare parts of that type are faulty?

2.6) Market analysts predict that a newly developed product in design 1 will bring in a profit of $ 500 000, whereas in design 2 it will bring in a profit of $ 200 000 with probability 0.4, and a profit of $ 800 000 with probability 0.6.

What design should the producer prefer?

2.7) Let X be the random number one has to throw a die, till for the first time a 6 occurs. Determine $E(X)$ and $Var(X)$.

2.8) 2% of the citizens of a country are HIV-positive. Test persons are selected at random from the population and checked for their HIV-status.

What is the mean number of persons which have to be checked till for the first time an HIV-positive person is found?

2.9) Let X be the difference between the number of *head* and the number of *tail* if a coin is flipped 10 times.

(1) What is the range of X?

(2) Determine the probability distribution of X.

2.10) A locksmith stands in front of a locked door. He has 9 keys and knows that only one of them fits, but he has otherwise no a priori knowledge. He tries the keys one after the other.

What is the mean number of trials till the door opens?

2.11) A submarine attacks a warship with 8 torpedoes. The torpedoes hit the warship independently of each other with probability 0.8. Any successful torpedo hits one of the 8 submerged chambers of the ship independently of other successful ones with probability 1/8. The chambers are isolated from each other. In case of one or more hits, a chamber fills up with water. The ship will sink if at least 3 chambers are hit by one or more torpedos. What is the probability that the attack sinks the warship?

2.12) Three hunters shoot at 3 partridges. Every hunter, independently of the others, takes aim at a randomly selected partridge and hits his/her target with probability 1. Thus, a partridge may be hit by several pellets, whereas lucky ones escape a hit.

Determine the mean $E(X)$ of the random number X of hit partridges.

2.13) A lecturer, for having otherwise no merits, claims to be equipped with extra-sensory powers. His students have some doubt about it and ask him to predict the outcomes of ten flippings of a fair coin. The lecturer is five times successful. Do you believe that, based on this test, the claim of the lecturer is justified?

2.14) Let X have a binomial distribution with parameters $n = 5$ and $p = 0.4$.

(1) Draw the distribution function of X.

(2) Determine the probabilities

$$P(X > 6), \ P(X < 2), \ P(3 \leq X < 7), \ P(X > 3 | X \leq 2), \ \text{and} \ P(X \leq 3 | X \geq 4).$$

2.15) Let X have a binomial distribution with parameters $n = 10$ and p.

Determine an interval $\mathbf{I}$ so that $P(X = 2) < P(X = 3)$ for all $p \in \mathbf{I}$.

2.16) The stop sign at an intersection is on average ignored by 4% of all cars. A car, which ignores the stop sign, causes an accident with probability 0.01. Assuming independent behavior of the car drivers:

(1) What is the probability that from 100 cars at least 3 ignore the stop sign?

(2) What is the probability that at least one of the 100 cars causes an accident due to ignoring the stop sign?

2.17) Tessa bought a dozen claimed to be fresh-laid farm eggs in a supermarket. There are 2 rotten eggs amongst them. For breakfast she boils 2 eggs.

What is the probability that her breakfast is spoilt if already one bad egg will have this effect?

2.18) A smart baker mixes 20 stale breads from the previous days with 100 freshly baked ones and offers this mixture for sale. Tessa randomly chooses 3 breads from the 120, i.e., she does not feel and smell them. What is the probability that she has bought at least one stale bread?

2.19) Some of the 270 spruces of a small forest stand are infested with rot (a fungus affecting first the core of the stems). Samples are taken from the stems of 30 random-ly selected trees.

(1) If 24 trees from the 270 are infested, what is the probability that there are less than 4 infested trees in the sample?

Determine this probability both by the binomial approximation and by the Poisson approximation to the hypergeometric distribution.

(2) If the sample contains six infested trees, what is the most likely number of infest-ed trees in the forest stand (see example 2.7)?

2.20) Because it happens that one or more airline passengers do not show up for their reserved seats, an airline would sell 602 tickets for a flight that holds only 600 pas-sengers. The probability that, for some reason or other, a passenger does not show up is 0.008.

What is the probability that every passenger who shows up will have a seat?

2.21) Flaws are randomly located along the length of a thin copper wire. The number of flaws follows a Poisson distribution with a mean of 0.15 flaws per *cm*. What is the probability $p_{\geq 2}$ of at least 2 flaws in a section of length $10\,cm$?

2.22) The random number of crackle sounds produced per hour by an old radio has a Poisson distribution with parameter $\lambda = 12$.

What is the probability that there is no crackle sound during the 4 minutes transmis-sion of a listener's favorite hit?

2.23) The random number of tickets car driver Odundo receives has a Poisson distri-bution with parameter $\lambda = 2$ a year. In the current year, Odundo had received his first ticket on the 31st of March.

What is the probability that he will receive another ticket in that year?

2.24) Let X have a Poisson distribution with parameter λ.

For which nonnegative integer n is the probability $p_n = P(X = n)$ maximal?

2.25) In $100kg$ of a low-grade molten steel tapping there are on average 120 impuri-ties. Castings weighing $1kg$ are manufactured from this raw material.

What is the probability that there are at least 2 impurities in a casting if the spacial distribution of the impurities in the raw material is assumed to be Poisson?

2.26) In a piece of fabric of length $100\,m$ there are on average 10 flaws. These flaws are assumed to be Poisson distributed over the length. The $100m$ of fabric are cut in pieces of length $4\,m$.

What percentage of the $4\,m$ cuts can be expected to be without flaws?

2.27) X have a binomial distribution with parameters n and p. Compare the following exact probabilities with the corresponding Poisson approximations and give reasons for possible larger deviations:

(1) $P(X=2)$ for $n=20, p=0.1$,

(2) $P(X=2)$ for $n=20, p=0.9$,

(3) $P(X=0)$ for $n=10, p=0.1$,

(4) $P(X=3)$ for $n=20, p=0.4$.

2.28) A random variable X has range $R=\{x_1,x_2,\cdots,x_m\}$ and probability distribution

$$\{p_k=P(X=x_k);\ k=1,2,...,m\},\quad \Sigma_{k=1}^{m}p_k=1.$$

A random experiment with outcome X is repeated n times. The outcome of the kth repetition has no influence on the outcome of the $(k+1)th$ one, $k=1,2,...,m-1$. Show that the probability of the event

$$\{x_1 \text{ occurs } n_1 \text{ times, } x_2 \text{ occurs } n_2 \text{ times, } \cdots, x_m \text{ occurs } n_m \text{ times}\}$$

is given by

$$\frac{n!}{n_1!n_2!\cdots n_m!}\, p_1^{n_1}p_2^{n_2}\cdots p_m^{n_m}\quad \text{with}\ \Sigma_{k=1}^{m}n_k=1.$$

This probability distribution is called the *multinomial distribution*. It contains as a special case the binomial distribution $(n=2)$.

2.29) A branch of the PROFIT-Bank has found that on average 68% of its customers visit the branch for routine money matters (type 1-visitors), 14% are there for invest-ment matters (type 2-visitors), 9% need a credit (type 3-visitors), 8% need foreign exchange service (type 4-visitors), and 1% only make a suspicious impression or even carry out a robbery (type 5-visitors).

(1) What is the probability that amongst 10 randomly chosen visitors 5, 3, 1, 1, and 0 are of type 1, 2, 3, 4, or 5, respectively?
(2) What is the probability that amongst 12 randomly chosen visitors 4, 3, 3, 1, and 1 are of type 1, 2, 3, 4, or 5, respectively?

Section 2.3

2.30) Let $F(x)$ and $f(x)$ be the respective distribution function and the probability density of a random variable X. Answer with *yes* or *no* the following questions:

(1) $F(x)$ and $f(x)$ can be arbitrary real functions.

(2) $f(x)$ is a nondecreasing function.

(3) $f(x)$ cannot have jumps.

(4) $f(x)$ cannot be negative.

(5) $F(x)$ is always a continuous function.

(6) $F(x)$ can assume values between -1 and $+1$.

(7) The area between the abscissa and the graph of $F(x)$ is always equal to 1.

(8) $f(x)$ must always be smaller than 1.

(9) The area between the abscissa and the graph of $f(x)$ is always equal to 1.

(10) The properties of $F(x)$ and $f(x)$ are all the same to me.

2.31) Check whether by suitable choice of the parameter a the following functions are densities of random variables. If the answer is *yes*, determine the respective distribution functions, mean values, variances, medians, and modes.

(1) $f(x) = a|x|, \ -3 \le x \le +3,$

(2) $f(x) = axe^{-x^2}, \ x \ge 0,$

(3) $f(x) = a \sin x, \ 0 \le x \le \pi,$

(4) $f(x) = a \cos x, \ 0 \le x \le \pi.$

2.32) (1) Show that $f(x) = \dfrac{1}{2\sqrt{x}}, \ 0 < x \le 1,$ is a probability density.

(2) Draw the graph of the corresponding distribution function and determine the corresponding 0.1, 0.5, and the 0.9-percentiles. Check whether the mean value exists.

2.33) Let X be a continuous random variable. Confirm or deny the following statements:

(1) The probability $P(X = E(X))$ is always positive.

(2) There is always $Var(X) \le 1$.

(3) $Var(X)$ can be negative if X can assume negative values.

(4) $E(X)$ is never negative.

2.34) The current which flows through a thin copper wire is uniformly distributed in the interval $[0, 10]$ (in mA). For safety reasons, the current should not fall below the crucial level of $4\,mA$.

What is the probability that at any randomly chosen time point the current is below $4\ mA$?

2.35) According to the timetable, a lecture begins at 8:15 a.m. The arrival time of Professor *Wisdom* in the venue is uniformly distributed between 8:13 and 8:20, whereas the arrival time of student *Sluggish* is uniformly distributed over the time interval from 8:05 to 8:30.

What is the probability that *Sluggish* arrives after *Wisdom* in the venue?

2.36) A road traffic light is switched on every day at 5:00 a.m. It always begins with *red* and holds this colour for two minutes. Then it changes to *yellow* and holds this colour for 30 seconds before it switches to *green* to hold this colour for 2.5 minutes. This cycle continues till midnight.

(1) A car driver arrives at this traffic light at a time point which is uniformly distributed between 9:00 and 9:10 a.m. What is the probability that the driver catches the green light period?

(2) Determine the same probability on condition that the driver's arrival time point has a uniform distribution over the interval [8:58, 9:08].

2.37) A continuous random variable X has the probability density

$$f(x) = \begin{cases} 1/4 \text{ for } 0 \le x \le 2, \\ 1/2 \text{ for } 2 < x \le 3. \end{cases}$$

Determine $\sqrt{Var(X)}$ and $E(|X-E(X)|)$.

2.38) A continuous random variable X has the probability density

$$f(x) = 2x, \ 0 \le x \le 1.$$

(1) Draw the corresponding distribution function.

(2) Determine and compare the measures of variability

$$E(|X-E(X)|) \text{ and } \sqrt{Var(X)} .$$

2.39) The lifetime X of a bulb has an exponential distribution with a mean value of $E(X) = 8000 \ hours$. Calculate the probabilities

$$P(X \le 4000), \ P(X > 12000), \ P(7000 \le X < 9000), \text{ and } P(X < 4000)$$

(time limits in *hours*).

2.40) The lifetimes of 5 identical bulbs are exponentially distributed with parameter $\lambda = 1.25 \cdot 10^{-4} [h^{-1}]$.

All of them are switched on at time $t = 0$ and will fail independently of each other.

(1) What is the probability that at time $t = 8000 \ hours$ a) all 5 bulbs and b) at least 3 bulbs have failed?

(2) What is the probability that at least one bulb survives 12 000 hours?

2.41) The period of employment of staff in a certain company has an exponential distribution with property that 92% of staff leave the company after only 16 months. What is the mean time an employee is with this company and the corresponding standard deviation?

2.42) The times between the arrivals of taxis at a rank are independent and have an exponential distribution with parameter $\lambda = 4 \; [h^{-1}]$. An arriving customer does not find an available taxi and the previous one left 3 minutes earlier. No other customers are waiting. What is the probability that the customer has to wait at least 5 minutes for the next free taxi?

2.43) A small branch of the *Profit Bank* has the two tellers 1 and 2. The service times at these tellers are independent and exponentially distributed with parameter $\lambda = 0.4 \; [\text{min}^{-1}]$. When Pumeza arrives, the tellers are occupied by a customer each. So she has to wait. Teller 1 is the first to become free, and the service of Pumeza starts immediately. What is the probability that the service of Pumeza is finished sooner than the service of the customer at teller 2?

2.44) Four weeks later Pumeza visits the same branch as in exercise 2.43. Now the service times at tellers 1 and 2 are again independent, but exponentially distributed with respective parameters $\lambda_1 = 0.4 \, [\text{min}^{-1}]$ and $\lambda_2 = 0.2 \, [\text{min}^{-1}]$.

(1) When Pumeza enters the branch, both tellers are occupied and no customer is waiting. What is the mean time Pumeza spends in the branch till the end of her service?

(2) When Pumeza enters the branch, both tellers are occupied, and another customer is waiting for service. What is the mean time Pumeza spends in the branch till the end of her service? (Pumeza does not get preferential service.)

2.45) An insurance company offers policies for fire insurance. Achmed holds a policy according to which he gets full refund for that part of the claim which exceeds $3000. He gets nothing for a claim size less than or equal to $ 3000. The company knows that the average claim size is $5642.

(1) What is the mean refund Achmed gets from the company for a claim if the claim size is exponentially distributed?

(2) What is the mean refund Achmed gets from the company for a claim if the claim size is Rayleigh-distributed?

2.46) Pedro runs a fruit shop. Mondays he opens his shop with a fresh supply of strawberries of s pounds, which is supposed to satisfy the demand for three days. He knows that for this time span the demand X is exponentially distributed with a mean value of 200 pounds. Pedro pays $ 2 for a pound and sells it for $ 4. So he will lose $ 2 for each pound he cannot sell, and he will make a profit of $ 2 out of each pound he sells. What amount $s = s^*$ of strawberries Pedro should stock for a period of three days to maximize his mean profit?

2.47) The probability density function of the random annual energy consumption X of an enterprise [in $10^8 kwh$] is

$$f(x) = 30(x-2)^2[1-2(x-2)+(x-2)^2], \quad 2 \le x \le 3.$$

(1) Determine the distribution function of X. What is the probability that the annual energy consumption exceeds 2.8?

(2) What is the mean annual energy consumption?

2.48) The random variable X is normally distributed with mean $\mu = 5$ and standard deviation $\sigma = 4$.

Determine the respective values of x which satisfy

$$P(X \le x) = 0.5, \quad P(X > x) = 0.95, P(x \le X < 9) = 0.2, \quad P(3 < X \le x) = 0.95,$$

$$P(-x \le X \le +x) = 0.99.$$

2.49) The response time of an average male car driver is normally distributed with mean value 0.5 and standard deviation 0.06 (in seconds).

(1) What is the probability that his response time is greater than 0.6 seconds?

(2) What is the probability that his response time is between 0.50 and 0.55 seconds?

2.50) The tensile strength of a certain brand of paper is modeled by a normal distribution with mean $24 psi$ and variance $9 [psi]^2$.

What is the probability that the tensile strength of a sample does not fall below the critical level of $20 psi$?

2.51) The total monthly sick leave time of employees of a small company has a normal distribution with mean $100 hours$ and standard deviation $20 hours$.

(1) What is the probability that the total monthly sick leave time will be between 50 and $80 hours$?

(2) How much time has to be budgeted for sick leave to make sure that the budgeted total amount for sick leave is only exceeded with a probability of less than 0.1?

2.52) The random variable X has a Weibull distribution with mean value 12 and variance 9.

(1) Calculate the parameters β and θ of this distribution.

(2) Determine the conditional probabilities $P(X > 10|X > 8)$ and $P(X \le 6|X > 8)$.

2.53) The random measurement error X of a meter has a normal distribution with mean 0 and variance σ^2, i.e., $X = N(0, \sigma^2)$. It is known that the percentage of measurements, which deviate from the 'true' value by more than $|0.4|$, is 80%. Use this piece of information to determine σ.

2.54) If sand from gravel pit 1 is used, then molten glass for producing armored glass has a random impurity content X which is $N(60, 16)$-distributed. But if sand from gravel pit 2 is used, then this content is $N(62, 9)$-distributed (μ and σ in 0.01%). The admissable degree of impurity should not exceed 0.64%.

Sand from which gravel pit should be used?

2.55) Let X have a geometric distribution with

$$P(X = i) = (1 - p)p^i; \quad i = 0, 1, ...; \quad 0 < p < 1.$$

By mixing these geometric distributions with regard to a suitable structure distribution density $f(p)$ show that

$$\sum_{i=0}^{\infty} \frac{1}{(i+1)(i+2)} = 1.$$

2.56) A random variable X has distribution function

$$F_\alpha(x) = e^{-\alpha/x}; \quad \alpha > 0, \ x > 0$$

(*Frechét distribution*).

What distribution type arises when mixing this distribution with regard to the exponential structure distribution density $f(\alpha) = \lambda e^{\lambda \alpha}; \ \lambda > 0, \ \alpha > 0$?

2.57) The random variable X has distribution function (*Lomax distribution*, page 93)

$$F(x) = \frac{x}{x+1}, \quad x \geq 0.$$

Check whether there is a subinterval of $[0, \infty)$ on which $F(x)$ is *DFR* or *IFR*.

2.58) Check the aging behavior of systems whose lifetime distributions have

(1) a Frechét distribution with distribution function $F(x) = e^{-(1/x)^2}$, $x > 0$ (sketch its failure rate), and

(2) a power distribution with distribution function $F(x) = 1 - (1/x^2)$, $x \geq 1$. respectively?

2.59) Let $F(x)$ be the distribution function of a nonnegative random variable X with finite mean value μ.

(1) Show that the function $F_s(x)$ defined by

$$F_s(x) = \frac{1}{\mu} \int_0^x (1 - F(t)) \, dt$$

is the distribution function of a nonnegative random variable X_s.

(2) Prove: If X is exponentially distributed with parameter $\lambda = 1/\mu$, then so is X_s and vice versa.

(3) Determine the failure rate $\lambda_s(x)$ of X_s.

2.60) Let X be a random variable with range $\{1, 2, ...\}$ and probability distribution

$$P(X = i) = \left(1 - \frac{1}{n^2}\right) \frac{1}{n^{2(i-1)}}; \quad i = 1, 2, ...$$

Determine the z-transform of X and by means of it $E(X)$, $E(X^2)$, and $Var(X)$.

2.61) Determine the Laplace transform $\hat{f}(s)$ of the density of the Laplace distribution with parameters λ and μ (page 66):

$$f(x) = \frac{1}{2}\lambda e^{-\lambda|x-\mu|}, \quad -\infty < x < +\infty,$$

By means of $\hat{f}(s)$ determine $E(X)$, $E(X^2)$, and $Var(X)$.

CHAPTER 3

Multidimensional Random Variables

The previous chapter essentially dealt with one-dimensional random variables and their probabilistic characterization and properties. Frequently a joint probabilistic analysis of two or more random variables is necessary. For instance, for weather predictions the meteorologist must take into account the interplay of randomly fluctuating parameters as air pressure, temperature, wind force and direction, humidity, et cetera. The operator of a coal power station, in order to be able to properly planning the output of the station, needs to take into account outdoor temperature as well as ash and water content of the coal presently available. These three parameters have a random component and there is a dependency between ash and water content. The information technologist, when analyzing stochastic signals, has jointly to consider their random phases and amplitudes. The forester, who has to estimate the amount of wood in a forest stand, measures both height and stem diameter (at a height of $1.3\,m$) of trees. Even in chapter 2 of this book vectors of random variables occurred without having explicitly hinted to this: When a die is tossed twice, then the outcome is (X_1, X_2). The binomial distribution is derived from a sequence of n binary random variables $(X_1, X_2, ..., X_n)$. More challenging situations will be discussed in Part II of this book: Let, for instance, $X(t)$ be the price of a unit of stock at time t and $0 < t_1 < t_2 < \cdots < t_n$. Then the components of the n-dimensional vector $(X(t_1), X(t_2), ..., X(t_n))$ are the random stock prices at time points t_i. There is an obvious dependency between the $X(t_i)$ so that for the prediction of the stock price development in time the random variables $X(t_i)$ should not be analyzed separately of each other. The same refers to other *time series* as registering temperatures, population sizes, et cetera, at increasing time points.

3.1 TWO-DIMENSIONAL RANDOM VARIABLES

3.1.1 Discrete Components

Let X and Y be two random variables, which are combined to a *random vector* (X, Y). This vector is also called a *two-dimensional random variable* or a *bivariate random variable*. In this section, X and Y are assumed to be discrete random variables with respective ranges $R_X = \{x_0, x_1, ...\}$ and $R_Y = \{y_0, y_1, ...\}$. Then the range of (X, Y) is the set of two-dimensional vectors

$$R_{XY} = \{(x, y), x \in R_X, y \in R_Y\}.$$

The (deterministic) vector (x, y) is called a *realization* of (X, Y).

For instance, if two dice are thrown simultaneously and the outcomes are X and Y, respectively, then the range of (X, Y) is

$$R_{XY} = \{(i,j);\ i,j = 1, 2, ..., 6\}.$$

If X and Y are the random number of traffic accidents occurring a year in the two neighboring towns Atown and Betown, respectively, then

$$R_X = \{0, 1, ...\} \text{ and } R_Y = \{0, 1, ...\},$$

and the range of (X, Y) is $R_{XY} = \{(i,j),\ i,j = 0, 1, 2, ...\}$. It makes sense to consider X and Y together, since weather, seasonal factors, vacation periods, and other conditions induce a dependency between X and Y.

Joint probability distribution Let

$$\{p_i = P(X = x_i;\ i = 0, 1, ...\} \text{ and } \{q_j = P(Y = x_j;\ j = 0, 1, ...\}$$

be the probability distributions of X and Y, respectively. Furthermore, let

$$r_{ij} = P(X = x_i \cap Y = y_j) \text{ for all } (x_i, y_j) \in R_{XY} \tag{3.1}$$

be the probabilities for the joint occurrence of the random events '$X = x_i$' and '$Y = y_j$.' The set of probabilities

$$\{r_{ij};\ i,j = 0, 1, ...\} \tag{3.2}$$

is the *joint* or *two-dimensional probability distribution* of the random vector (X, Y).

From the definition of the r_{ij},

$$p_i = \Sigma_{j=0}^{\infty} r_{ij}, \quad q_j = \Sigma_{i=0}^{\infty} r_{ij}. \tag{3.3}$$

Marginal Distributions The probability distribution $\{p_i,\ i = 0, 1, ...\}$ of X and the probability distribution $\{q_i,\ i = 0, 1, ...\}$ of Y are called the *marginal distributions* of (X, Y). The marginal distributions of (X, Y) do not contain the full information on the joint probability distribution of (X, Y) if there is a dependency between X and Y. However, if X and Y are independent, then the joint probability distribution of (X, Y) and its marginal distributions are equivalent in this regard.

Definition 3.1 (*independence*) Two discrete random variables X and Y are (*statistically*) *independent* if

$$r_{ij} = p_i q_j,\ i,j = 0, 1, ... \qquad \bullet$$

If X and Y are independent, then the value, which X has assumed, has no influence on the value, which Y has assumed and vice versa. This is the situation when throwing two dice simultaneously, or when X denotes the number of shark attacks at humans occurring at the shores of South Africa in 2025 and Y the ones at the shores of Hawaii in 2030. The *mean value* of the product XY is

$$E(XY) = \Sigma_{i=1}^{\infty} \Sigma_{j=1}^{\infty} r_{ij} x_i x_j. \tag{3.4}$$

For independent X and Y, the mean value of XY becomes

$$E(XY) = \sum_{i=1}^{\infty} \sum_{j=1}^{\infty} p_i q_j x_i x_j = (\sum_{i=1}^{\infty} p_i x_i)(\sum_{j=1}^{\infty} q_j y_j)$$

so that
$$E(XY) = E(X) \cdot E(Y). \tag{3.5}$$

Conditional Probability Distribution By formula (1.22), the conditional probabilities of $X = x_i$ given $Y = y_j$ and $Y = y_j$ given $X = x_i$, respectively, are

$$P(X = x_i | Y = y_j) = \frac{r_{ij}}{q_j}, \qquad P(Y = y_j | X = x_i) = \frac{r_{ij}}{p_i}.$$

The sets

$$\left\{ \frac{r_{ij}}{q_j};\ i = 0, 1, ... \right\} \quad \text{and} \quad \left\{ \frac{r_{ij}}{p_i};\ j = 0, 1, ... \right\}$$

are the *conditional probability distributions* of X given $Y = y_j$ and of Y given $X = x_i$, respectively. The corresponding conditional mean values are

$$E(X | Y = y_j) = \sum_{i=0}^{\infty} x_i \frac{r_{ij}}{q_j}, \qquad E(Y | X = x_i) = \sum_{j=0}^{\infty} y_j \frac{r_{ij}}{p_i}.$$

If X and Y are independent, then the conditions have no influence on the respective mean values, since $r_{ij}/q_j = p_i$ and $r_{ij}/p_i = q_j$ (see formula 2.7):

$$E(X | Y = y_j) = E(X), \quad E(Y | X = x_i) = E(Y); \quad i, j = 0, 1, ...$$

The conditional mean value $E(X|Y)$ of X given Y is a random variable, since the condition is random. The range of $E(X|Y)$ is

$$\{ E(X|Y = y_0),\ E(X|Y = y_1),\ ... \},$$

and the mean value of $E(X|Y)$ is $E(X)$, since

$$E(E(X|Y)) = \sum_{j=0}^{\infty} E(X|Y = y_j) P(Y = y_j) = \sum_{j=0}^{\infty} \sum_{i=0}^{\infty} x_i \frac{r_{ij}}{q_j} q_j$$

$$= \sum_{i=0}^{\infty} x_i \sum_{j=0}^{\infty} r_{ij} = \sum_{i=0}^{\infty} x_i p_i = E(X).$$

Because the roles of X and Y can be exchanged,

$$E(E(X|Y)) = E(X) \quad \text{and} \quad E(E(Y|X)) = E(Y). \tag{3.6}$$

Example 3.1 Two dice are thrown. The outcomes are X_1 and X_2, respectively. Let

$$X = \max(X_1, X_2) \quad \text{and} \quad Y = \text{'total number of even figures in } (X_1, X_2).\text{'}$$

The ranges of X and Y are $R_X = \{1, 2, 3, 4, 5, 6\}$ and $R_Y = \{0, 1, 2\}$. Since X_1 and X_2 are independent,

$$P(X_1 = i, X_2 = j) = P(X_1 = i) \cdot P(X_2 = j) = \frac{1}{6} \cdot \frac{1}{6} = \frac{1}{36}.$$

By (3.6), the q_j and the p_i are the corresponding row and column sums in Table 3.1.

X	1	2	3	4	5	6	q_j
Y							
0	1/36	0	3/36	0	5/36	0	9/36
1	0	2/36	2/36	4/36	4/36	6/36	18/36
2	0	1/36	0	3/36	0	5/36	9/36
p_i	1/36	3/36	5/36	7/36	9/36	11/36	1

Table 3.1 Joint distribution and marginal distribution for example 3.1

The mean values of X and Y are

$$E(X) = \tfrac{1}{36}(1 + 2 \cdot 3 + 3 \cdot 5 + 4 \cdot 7 + 5 \cdot 9 + 6 \cdot 11) \approx 4.472,$$

$$E(Y) = \tfrac{1}{36}(0 \cdot 9 + 1 \cdot 18 + 2 \cdot 9) = 1.$$

X and Y are not independent of each other: If $X = 1$, then $Y = 1$. If $X = 2$, then Y can only be 1 or 2 and so on. Hence, it makes sense to determine the conditional distributions, e.g.

$$\left\{ P(X = i | Y = j) = \frac{r_{ij}}{q_j}; \ i = 1, 2, ..., 6 \right\}; \ j = 0, 1, 2.$$

$j = 0: \ \left\{ \tfrac{1}{9}, 0, \tfrac{3}{9}, 0, \tfrac{5}{9}, 0 \right\}, \quad E(X|Y = 0) = \tfrac{35}{9} \approx 3.889.$

$j = 1: \ \left\{ 0, \tfrac{1}{9}, \tfrac{1}{9}, \tfrac{2}{9}, \tfrac{2}{9}, \tfrac{3}{9} \right\}, \quad E(X|Y = 1) = \tfrac{41}{9} \approx 4.556.$

$j = 2: \ \left\{ 0, \tfrac{1}{9}, 0, \tfrac{3}{9}, 0, \tfrac{5}{9} \right\}, \quad E(X|Y = 2) = \tfrac{44}{9} \approx 4.889.$ ◻

3.1.2 Continuous Components

3.1.2.1 Probability Distribution

Let X and Y be continuous, real-valued random variables with distribution functions

$$F_X(x) = P(X \leq x), \ F_Y(y) = P(Y \leq y)$$

and ranges R_X, R_Y, respectively. As with discrete random variables X and Y, (X, Y) is called a *random vector*, a *two-dimensional random variable,* or a *bivariate random variable.* Analogously to the distribution function of a (one-dimensional) random variable, there is a function, which contains the complete probabilistic information on (X, Y). This is the *joint distribution function* $F_{X,Y}(x, y)$ of X and Y defined by

$$F_{X,Y}(x, y) = P(X \leq x, \ Y \leq y), \quad x \in R_X, y \in R_Y,$$

where '$X \leq x, \ Y \leq y$' = '$X \leq x \cap Y \leq y$.' (For discrete random variables X and Y the joint distribution function is defined in the same way.) To discuss the properties of the joint distribu- tion function, it can be assumed without loss of generality that $R_X = R_Y = (-\infty, +\infty)$.

$F_{X,Y}(x,y)$ has the following properties:

(1) $\qquad F_{X,Y}(-\infty,y) = F_{X,Y}(x,-\infty) = 0, \ F_{X,Y}(+\infty,+\infty) = 1.$

(2) $\qquad 0 \le F_{X,Y}(x,y) \le 1.$

(3) $\qquad F_{X,Y}(x,+\infty) = F_X(x), \quad F_{X,Y}(+\infty,y) = F_Y(y).$

(4) For $x_1 \le x_2$ and $y_1 \le y_2$,

$$F_{X,Y}(x_1,y_1) \le F_{X,Y}(x_2,y_1) \le F_{X,Y}(x_2,y_2),$$
$$F_{X,Y}(x_1,y_1) \le F_{X,Y}(x_1,y_2) \le F_{X,Y}(x_2,y_2).$$

Thus, $F_{X,Y}(x,y)$ is nondecreasing in every argument.

(5) $\qquad P(X > x, Y \le y) = F_Y(y) - F_{X,Y}(x,y).$

(6) $\qquad P(X \le x, Y > y) = F_X(x) - F_{X,Y}(x,y).$

(7) $\qquad P(X > x, Y > y) = 1 - F_Y(y) - F_X(x) + F_{X,Y}(x,y).$

A generalization of the formula (2.44) to random vectors (X,Y) is

$$P(a < X \le b, c < Y \le d) = [F_{X,Y}(b,d) - F_{X,Y}(b,c)] - [F_{X,Y}(a,d) - F_{X,Y}(a,c)]. \quad (3.7)$$

Any function $F(x,y)$, which has properties (1) and (4) and is continuous on the left in x and y is the joint distribution function of a random vector (X, Y) if, in addition, the right-hand side of (3.7) is nonnegative for all a, b and c, d with $a < b$ and $c < d$ (see exercise 3.17). Properties (5) – (7) are implications of properties (1) and (4). For instance, to prove (5), the random event '$X > x$, $Y \le y$' is equivalently represented as' $Y \le y$' \ '$X \le x$, $Y \le y$'. Hence, by formula (1.14),

$$P(X > x, Y \le y) = P(Y \le y) - P(X \le x, Y \le y) = F_Y(y) - F_{X,Y}(x,y).$$

Property (6) follows from (5) by changing the roles of X and Y. Property (7) is a special case of formula (3.7) (see exercise (3.16) for a proof of formula (3.7)).

Note Properties (1) to (7) also are true for random vectors with discrete components.

The probability distribution functions of X and Y are the *marginal distribution functions* of the two-dimensional random variable (X, Y), and the pair (F_X, F_Y) is the *marginal distribution* of (X, Y).

Joint Probability Density Assuming its existence, the partial derivative of $F_{X,Y}(x,y)$ with respect to x and y,

$$f_{X,Y}(x,y) = \frac{\partial F_{X,Y}(x,y)}{\partial x \, \partial y}, \qquad (3.8)$$

is called the *joint (probability) density* of (X, Y). Equivalently, the joint density can be defined as a function $f_{X,Y}(x,y)$ satisfying

$$F_{X,Y}(x,y) = \int_{-\infty}^{x} \int_{-\infty}^{y} f_{X,Y}(u,v) \, du \, dv, \quad -\infty < x, y < +\infty. \qquad (3.9)$$

Every joint (probability) density has the two properties

$$f_{X,Y}(x,y) \geq 0, \quad \int_{-\infty}^{+\infty} \int_{-\infty}^{+\infty} f_{X,Y}(x,y)\,dx\,dy = 1. \tag{3.10}$$

Conversely, any function of two variables x and y satisfying these two conditions can be considered the joint density of a random vector (X, Y). From property (3) of the previous page and formula (3.9) one obtains the *marginal densities* of (X, Y) in terms of the joint density:

$$f_X(x) = \int_{-\infty}^{+\infty} f_{X,Y}(x,y)\,dy, \quad f_Y(y) = \int_{-\infty}^{+\infty} f_{X,Y}(x,y)\,dx. \tag{3.11}$$

Analogously to discrete random variables, the marginal distribution $\{F_X, F_Y\}$ or, in terms of the densities, $\{f_X(x), f_Y(y)\}$, does not contain the full information on the joint probability distribution of (X, Y) as given by $F_{X,Y}(x,y)$ if there is a (statistical) dependency between X and Y. If X and Y are independent, then $F_{X,Y}(x,y)$ and its marginal distribution $\{F_X, F_Y\}$ are equivalent in this regard:

Definition 3.2 (independence) Two random variables X and Y are *independent* if

$$F_{X,Y}(x,y) = F_X(x) \cdot F_Y(y). \qquad \bullet$$

Remark For discrete random variables this definition of independence is equivalent to the one given by definition 3.1. Representations of the distribution functions of discrete random variables are given at page 43.

In terms of the densities, X and Y are independent if and only if

$$f_{X,Y}(x,y) = f_X(x) \cdot f_Y(y). \tag{3.12}$$

The mean value of XY is

$$E(XY) = \iint_{-\infty}^{+\infty} xy f(x,y)\,dx\,dy. \tag{3.13}$$

As with discrete random variables (formula 3.5), for independent random variables:

$$E(XY) = E(X) \cdot E(Y). \tag{3.14}$$

Although in many applications the independence assumption is not justified, analytical results can frequently only be derived under this assumption. A reason for this situation is, apart from mathematical challenges, the inherent difficulties the analyst faces when trying to quantify statistical dependency.

Let $R_{\Delta x \Delta y}$ be a rectangle with sufficiently small side lengths Δx and Δy. Then the random vector (X, Y) assumes a realization from this rectangle approximately with probability

$$P((X, Y) \in R_{\Delta x \Delta y}) \approx f_{X,Y}(x,y)\,\Delta x \Delta y.$$

More generally, if B is an area in the plane, then the probability that the vector (X, Y) assumes a realization from B is given by the surface integral

$$P((X, Y) \in B) = \iint_B f_{X,Y}(x,y)\,dx\,dy. \tag{3.15}$$

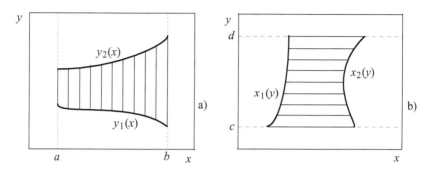

Figure 3.1 Normal regions with regard to the x-axis a) and the y-axis b)

For a *normal region* with regard to the x-axis

$$B = \{a \le x \le b, y_1(x) \le y \le y_2(x)\}$$

(Figure 3.1a), the surface integral (3.15) can be calculated by the double integral

$$P((X, Y) \in B) = \int_a^b \left(\int_{y_1(x)}^{y_2(x)} f_{X,Y}(x,y)\, dy \right) dx. \tag{3.16}$$

For a *normal region* with regard to the y-axis

$$B = \{x_1(x) \le x \le x_2(x), c \le y \le d\}$$

(Figure 3.1b), the surface integral (3.15) can be calculated by the double integral

$$P((X, Y) \in B) = \int_c^d \left(\int_{x_1(y)}^{x_2(y)} f_{X,Y}(x,y)\, dx \right) dy.$$

Double integrals can frequently be more efficiently calculated by transition from the Cartesian coordinates x and y to *curvilinear coordinates* u and v:

$$u = u(x,y), \quad v = v(x,y) \quad \text{or} \quad x = x(u,v), \quad y = y(u,v).$$

Then the normal region B with regard to e.g. the x-axis is transformed to a region B':

$$B' = \{a' \le u \le b', v_1(u) \le v \le v_2(u)\},$$

and the double integral (3.16) becomes

$$\iint_B f_{X,Y}(x,y)\, dx\, dy = \int_{a'}^{b'} \left(\int_{v_1(u)}^{v_2(u)} f_{X,Y}(x(u,v),y(u,v)) \left| \frac{\partial(x,y)}{\partial(u,v)} \right| dv \right) du, \tag{3.17}$$

where

$$\left| \frac{\partial(x,y)}{\partial(u,v)} \right| = \begin{vmatrix} \dfrac{\partial x}{\partial u} & \dfrac{\partial y}{\partial u} \\ \dfrac{\partial x}{\partial v} & \dfrac{\partial y}{\partial v} \end{vmatrix}$$

is the *functional determinant* of the transformation.

If $B = [a < X \le b, \, c < Y \le d]$, then (3.16) becomes

$$P((X, Y) \in B) = \int_a^b \left(\int_c^d f_{X,Y}(x,y) \, dy \right) dx.$$

This integral easily implies formula (3.7).

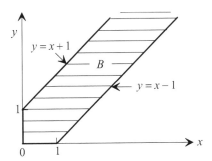

Figure 3.2 Integration region for example 3.2

Example 3.2 The joint probability density of the random vector (X, Y) is

$$f_{X,Y}(x,y) = e^{-(x+y)}; \quad x \ge 0, \; y \ge 0.$$

(1) The corresponding marginal densities are

$$f_X(x) = \int_0^\infty e^{-(x+y)} \, dy = e^{-x}, \quad f_Y(y) = \int_0^\infty e^{-(x+y)} dx = e^{-y}; \; x, y \ge 0.$$

Thus, X and Y are both exponentially distributed with parameter $\lambda = 1$. Moreover, since $e^{-(x+y)} = e^{-x} \cdot e^{-y}$, X and Y are independent.

(2) Let $B = \{|Y - X| \le 1\}$. The region B is hatched in Figure 3.2. The lower bound for B is $y = 0$ if $0 \le x \le 1$ and $y = x - 1$ if $1 \le x$. The upper bound is $y = x + 1$ if $x \ge 0$. Therefore, the outer integral of formula (3.16) has to be split with regard to the x-intervals $[0, 1]$ and $[1, \infty)$:

$$P(|Y - X| \le 1) = \int_0^1 \left(\int_0^{x+1} e^{-(x+y)} \, dy \right) dx + \int_1^\infty \left(\int_{x-1}^{x+1} e^{-(x+y)} \, dy \right) dx$$

$$= \int_0^1 e^{-x} \left[1 - e^{-(x+1)} \right] dx + \int_1^\infty e^{-x} \left[e^{-(x-1)} - e^{-(x+1)} \right] dx$$

$$= 1 - 1/e.$$

Hence, $P(|Y - X| \le 1) \approx 0.632$. $\qquad\qquad\qquad\qquad\qquad\qquad\qquad\qquad\square$

Example 3.3 Let

$$f_{X,Y}(x,y) = \tfrac{1}{2} xy, \quad 0 \le x \le y \le 2.$$

(1) Show that $f_{X,Y}(x,y)$ is a joint probability density.

(2) Determine the probability $P(X^2 > Y)$.

(3) Are X and Y independent?

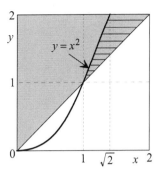

Figure 3.3 Possible (shaded) and favorable (hatched) region for (X, Y) (example 3.3)

(1) It needs to be shown that the conditions (3.10) are fulfilled. $f(x, y)$ is obviously nonnegative. Further,

$$\int_0^2 \left(\int_x^2 \frac{1}{2} x y \, dy \right) dx = \int_0^2 \left(\int_x^2 \frac{1}{2} x y \, dy \right) dx$$

$$= \frac{1}{2} \int_0^2 (2x - x^3/2) \, dx = \frac{1}{2} \left[x^2 - x^4/8 \right]_0^2 = 1.$$

(2) In Figure 3.3 the possible set of realizations of (X, Y) is shaded, and the region B for which $Y^2 > X$ is hatched. The upper bound of B is given by the parabola $y = x^2$ between $x = 1$ and $x = \sqrt{2}$ and the straight line $y = 2$ between $x = \sqrt{2}$ and $x = 2$. The lower bound of B is the straight line $y = x$ between $x = 1$ and $x = 2$. Hence, the desired probability is

$$P(X^2 > Y) = \int_1^{\sqrt{2}} \left(\int_x^{x^2} \frac{1}{2} x y \, dy \right) dx + \int_{\sqrt{2}}^2 \left(\int_x^2 \frac{1}{2} x y \, dy \right) dx$$

$$= \frac{1}{4} \int_1^{\sqrt{2}} \left(x^5 - x^3 \right) dx + \frac{1}{4} \int_{\sqrt{2}}^2 \left(4x - x^3 \right) dx$$

$$= \frac{1}{4} \left(\frac{8}{6} - 1 - \frac{1}{6} + \frac{1}{4} + 8 - 4 - 4 + 1 \right).$$

Thus, $P(X^2 > Y) \approx 0.354$.

(3) The marginal densities $f_X(x)$ and $f_Y(y)$ are

$$f_X(x) = \int_x^2 \frac{1}{2} x y \, dy = \frac{1}{2} x \left[\frac{y^2}{2} \right]_x^2 = \frac{1}{4} \left(4x - x^3 \right), \quad 0 \le x \le 2.$$

$$f_Y(y) = \int_0^y \frac{1}{2} x y \, dy = \frac{1}{2} y \left[\frac{x^2}{2} \right]_0^y = \frac{1}{4} y^3, \quad 0 \le y \le 2.$$

Since

$$f_{X,Y}(x, y) \ne f_X(x) \cdot f_Y(y),$$

X and Y are not independent.   □

Two-Dimensional Uniform Distribution The random vector (X, Y) has a uniform distribution in a finite region B of the (x, y)-plane with positive area $\mu(B)$ if

$$f(x, y) = \frac{1}{\mu(B)}, \quad (x, y) \in B.$$

Outside B the joint density $f(x, y)$ is 0. The conditions (3.10) are fulfilled since

$$\iint_B f(x, y)\, dx\, dy = \iint_B \frac{1}{\mu(B)}\, dx\, dy = \frac{1}{\mu(B)} \iint_B dx\, dy = 1.$$

For any $A \subseteq B$ the probability that (X, Y) assumes a value from A is

$$P((X, Y) \in A) = \frac{\mu(A)}{\mu(B)}.$$

Remark The uniform distribution of a random vector in a plane is identical to the geometric distribution introduced in section 1.3.2 (formula (1.8)) if Ω is a finite subset of a plane.

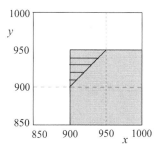

Figure 3.4 Possible and favorable region for example 3.4

Example 3.4 Let X be the daily power production of a power station, and let Y be the daily demand of the consumer. The random vector (X, Y) has a uniform distribution over the region

$$B = \{900 \le x \le 1000,\ 850 \le y \le 950\}.$$

What is the probability that the demand exceeds the supply?

The possible realizations of the random vector are in the shaded region (region B) of Figure 3.4. Its area is 10 000. Hence, the joint density of (X, Y) is

$$f_{X,Y}(x, y) = \frac{1}{10\,000}, \quad (x, y) \in B.$$

The subregion of B, where $Y > X$, is the hatched part of B. Its lower bound is the straight line $y = x$. Hence, the desired probability is

$$P(Y > X) = \int_{900}^{950} \int_x^{950} \frac{1}{10\,000}\, dy\, dx = \frac{1}{10\,000} \int_{900}^{950} (950 - x)\, dx,$$

which works out to be $P(Y > X) = 0.125$.

Of course, no integration is required to arrive at this result, since the area of the hatched part is a half of the area of a square with side length 50. ☐

Theorem 3.1 (1) If X and Y are independent and in the respective intervals $[a,b]$ and $[c,d]$ uniformly distributed, then the random vector (X,Y) has a uniform distribution on the rectangle

$$B = \{a \leq x \leq b, c \leq y \leq d\}.$$

(2) Conversely, if (X,Y) has a uniform distribution on the rectangle B, then the random variables X and Y are independent and uniformly distributed in the intervals $[a,b]$ and $[c,d]$, respectively.

Proof (1) If X is uniformly distributed in $[a,b]$ and Y in $[c,d]$, then

$$F_X(x) = \frac{x-a}{b-a}, \quad a \leq x \leq b,$$

$$F_Y(y) = \frac{y-c}{d-c}, \quad c \leq y \leq d.$$

Hence, by definition 3.2, the joint distribution function of (X,Y) is

$$F_{X,Y}(x,y) = \frac{(x-a)(y-c)}{(b-a)(d-c)}, \quad (x,y) \in B.$$

The corresponding joint density is

$$f_{X,Y}(x,y) = \frac{\partial F(x,y)}{\partial x\, \partial y} = \frac{1}{(b-a)(d-c)}, \quad (x,y) \in B.$$

$f(x,y)$ is the joint density of a random vector (X,Y), which is uniformly distributed on the rectangle B.

(2) If (X,Y) is uniformly distributed in the rectangle B, then its corresponding marginal densities are

$$f_X(x) = \int_c^d f_{X,Y}(x,y)\, dy = \int_c^d \frac{1}{(b-a)(d-c)}\, dy = \frac{1}{b-a}, \quad a \leq x \leq b,$$

$$f_Y(y) = \int_a^b f_{X,Y}(x,y)\, dx = \int_a^b \frac{1}{(b-a)(d-c)}\, dx = \frac{1}{d-c}, \quad c \leq y \leq d,$$

so that $f_{X,Y}(x,y) = f_X(x) \cdot f_Y(y)$. Hence, X and Y are independent and uniformly distributed in the intervals $[a,b]$ and $[c,d]$, respectively. ∎

3.1.2.2 Conditional Probability Distribution

Given a random vector (X,Y), the conditional distribution function of Y given $X = x$ and the corresponding *conditional density of Y given $X = x$* are denoted as

$$F_Y(y|x) = P(Y \leq y | X = x), \quad f_Y(y|x) = dF_Y(y|x)/dy.$$

For continuous random variables, the event '$X = x$' has probability 0 so that the definition of the conditional probability by formula (1.22) cannot directly be applied to deriving $F_Y(y|x)$. Hence, consider for a $\Delta x > 0$ the conditional probability

$$P(Y \le y | x \le X \le x + \Delta x) = \frac{P(Y \le y \cap x \le X \le x + \Delta x)}{P(x \le X \le x + \Delta x)}$$

$$= \frac{\int_{-\infty}^{y} \frac{1}{\Delta x} \left(\int_{x}^{x+\Delta x} f_{X,Y}(u, v) \, du \right) dv}{\frac{1}{\Delta x} [F_X(x + \Delta x) - F_X(x)]} .$$

If $\Delta x \to 0$, then, assuming $f_X(x) > 0$,

$$F_Y(y|x) = \frac{1}{f_X(x)} \int_{-\infty}^{y} f_{X,Y}(x, v) \, dv. \qquad (3.18)$$

Differentiation yields the desired conditional density:

$$f_Y(y|x) = \frac{f_{X,Y}(x, y)}{f_X(x)} . \qquad (3.19)$$

By (3.12), if X and Y are independent, then

$$f_Y(y|x) = f_Y(y).$$

The *conditional mean value of Y given* $X = x$ is

$$E(Y|x) = \int_{-\infty}^{+\infty} y f_Y(y|x) \, dy. \qquad (3.20)$$

The function $m_Y(x) = E(Y|x)$ is called *regression function* of Y with regard to x. It quantifies the average dependency of Y from X. For instance, if X is the body weight and Y the height of a randomly chosen member from a population of adults, then $m_Y(x)$ is the average height of a member of this population with body weight x. Or: the difference $m_Y(x + \Delta x) - m_Y(x)$ is the mean increase in body height if the body weight increases from x to $x + \Delta x$.

The *conditional mean value of Y given X* is

$$E(Y|X) = \int_{-\infty}^{+\infty} y f_Y(y|X) \, dy.$$

$E(Y|X)$ is a <u>random variable</u> with property

$$E(E(Y|X)) = E(Y). \qquad (3.21)$$

This is proved as follows:

$$E(E(Y|X)) = \int_{-\infty}^{+\infty} \int_{-\infty}^{+\infty} y f_Y(y|x) \, dy f_X(x) \, dx$$

$$= \int_{-\infty}^{+\infty} \int_{-\infty}^{+\infty} y \frac{f(x,y)}{f_X(x)} \, dy f_X(x) \, dx = \int_{-\infty}^{+\infty} \int_{-\infty}^{+\infty} y f(x,y) \, dy \, dx.$$

Hence, by (3.11),

$$E(E(Y|X)) = \int_{-\infty}^{+\infty} y f_Y(y) \, dy = E(Y).$$

If X and Y are independent, then

$$E(Y|X = x) = E(Y|X) = E(Y). \qquad (3.22)$$

Clearly, the roles of X and Y can be exchanged in the formulas (3.18) to (3.22).

Formula (3.21), applied to the representation (2.62) of the variance (page 67), can be used to derive a *conditional variance formula* for $Var(X)$ (exercise 3.21):

$$Var(X) = E[Var(X|Y)] + Var[E(X|Y)].\qquad(3.23)$$

Example 3.5 The random vector (X, Y) has the joint probability density

$$f_{X,Y}(x,y) = x + y, \quad 0 \le x, y \le 1.$$

$f_{X,Y}(x,y)$ is nonnegative at the unit square. The marginal densities are

$$f_X(x) = \int_0^1 (x+y)\,dy = [xy + y^2/2]_0^1 = x + 1/2, \quad 0 \le x \le 1,$$

$$f_Y(y) = \int_0^1 (x+y)\,dx = [x^2/2 + yx]_0^1 = y + 1/2, \quad 0 \le y \le 1.$$

Since $f_{X,Y}(x,y) \ne f_X(x) \cdot f_Y(y)$, the random variables X and Y are not independent. (Give an intuitive explanation for this.) The mean value of X is

$$E(X) = \int_0^1 x\,(x + 1/2)\,dx = [x^3/3 + x^2/4]_0^1 = \tfrac{7}{12} \approx 0.5833.$$

In view of the symmetry between x and y in $f_{X,Y}(x,y)$,

$$E(Y) = \tfrac{7}{12} \approx 0.5833.$$

By (3.19), the conditional density of Y on condition $X = x$ is

$$f_Y(y|x) = \frac{x+y}{x+1/2} = 2\,\frac{x+y}{2x+1}, \quad 0 \le x, y \le 1.$$

The regression function $m_Y(x) = E(Y|X = x)$ of Y with regard to x is

$$m_Y(x) = 2 \int_0^1 y\,\frac{x+y}{2x+1}\,dy = \frac{2}{2x+1}\int_0^1 [yx + y^2]\,dy$$

$$= \frac{2}{2x+1}\left[\frac{xy^2}{2} + \frac{y^3}{3}\right]_0^1$$

so that

$$m_Y(x) = \frac{2+3x}{3+6x}, \quad 0 \le x \le 1.$$

In particular,

$$m_Y(0) = \tfrac{2}{3} \approx 0.6667, \quad m_Y(1) = \tfrac{5}{9} \approx 0.5556, \quad m_Y(0.5) = \tfrac{7}{12} = E(Y) \approx 0.5833.$$

The relatively small influence of the conditions at the conditional mean values suggests that the dependency between X and Y is not that strong (Figure 3.5). The conditional mean value of Y given X is the random variable

$$E(Y|X) = \frac{2+3X}{3+6X},$$

which has mean value $E(Y) = 7/12$. ☐

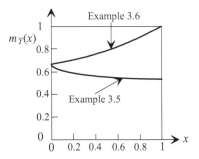

Figure 3.5 Regression functions for examples 3.5 and 3.6

Example 3.6 The random variable Y has probability density

$$f_Y(y) = 3y^2, \quad 0 \le y \le 1.$$

On condition $Y = y$, the random variable X is uniformly distributed in $[0, y]$, $y > 0$.
(1) What is the joint probability density of the random vector (X, Y)?
(2) Determine the conditional mean values $E(Y|X = x)$ and $E(X|Y = y)$.

(1) On condition $Y = y$ with $y > 0$ the density of X is

$$f_X(x|y) = \frac{1}{y}, \quad 0 \le x \le y.$$

Hence, by formula (3.19), the joint density of (X, Y) is

$$f(x,y) = f_X(x|y) \cdot f_Y(y) = \frac{1}{y} \cdot 3y^2 = 3y, \quad 0 \le x \le y \le 1.$$

The (unconditional) density of X one obtains from (3.11):

$$f_X(x) = \int_x^1 3y\, dy = 3 \left[\frac{y^2}{2} \right]_0^1 = 1.5(1 - x^2), \quad 0 \le x \le 1.$$

(2) The regression function $m_Y(x) = E(Y|x)$ of Y with regard to $X = x$ is

$$m_Y(x) = \int_{-\infty}^{+\infty} y \frac{f_{X,Y}(x,y)}{f_X(x)}\, dy = 2\int_x^1 \frac{y^2}{1-x^2}\, dy$$

$$= \frac{2}{3} \frac{1-x^3}{1-x^2}, \quad 0 \le x < 1.$$

The conditional mean value of X given $Y = y$ is

$$E(X|y) = \int_{-\infty}^{+\infty} x f_X(x|y)\, dx = \int_0^y \frac{x}{y}\, dx$$

$$= 0.5\, y, \quad 0 \le x < y. \qquad \square$$

3.1.2.3 Bivariate Normal Distribution

The random vector (X, Y) has a *bivariate (2-dimensional) normal* or a *bivariate (2-dimensional) Gaussian distribution* with parameters

$$\mu_x, \mu_y, \sigma_x, \sigma_y \text{ and } \rho, \quad -\infty < \mu_x, \mu_y < \infty, \quad \sigma_x > 0, \sigma_y > 0, \quad -1 < \rho < 1$$

if it has joint density

$$f_{X,Y}(x,y) = \frac{1}{2\pi\sigma_x\sigma_y\sqrt{1-\rho^2}} \exp\left\{-\frac{1}{2(1-\rho^2)}\left(\frac{(x-\mu_x)^2}{\sigma_x^2} - 2\rho\frac{(x-\mu_x)(y-\mu_y)}{\sigma_x\sigma_y} + \frac{(y-\mu_y)^2}{\sigma_y^2}\right)\right\} \quad (3.24)$$

with $-\infty < x, y < +\infty$. By (3.11), the corresponding marginal densities are seen to be

$$f_X(x) = \frac{1}{\sqrt{2\pi}\,\sigma_x} \exp\left(-\frac{(x-\mu_x)^2}{2\sigma_x^2}\right), \quad -\infty < x < +\infty,$$

$$f_Y(x) = \frac{1}{\sqrt{2\pi}\,\sigma_y} \exp\left(-\frac{(y-\mu_y)^2}{2\sigma_y^2}\right), \quad -\infty < y < +\infty.$$

Hence, if (X, Y) has a bivariate normal distribution with parameters $\mu_x, \sigma_x, \mu_y, \sigma_y$, and ρ, then the random variables X and Y have each a normal distribution with respective parameters μ_x, σ_x and μ_y, σ_y. Since the independence of X and Y is equivalent to

$$f_{X,Y}(x,y) = f_X(x)f_Y(y),$$

X and Y are independent if and only if $\rho = 0$. (In the next section it will be shown that the parameter ρ is the correlation coefficient between X and Y, a measure of the degree of linear statistical dependency between any two random variables.)

The conditional density of Y given $X = x$ is obtained from $f_{X,Y}(x,y)$ and (3.19):

$$f_Y(y|x) = \frac{1}{\sqrt{2\pi}\,\sigma_y\sqrt{1-\rho^2}} \exp\left\{-\frac{1}{2\sigma_y^2(1-\rho^2)}\left(y - \rho\frac{\sigma_y}{\sigma_x}(x-\mu_x) - \mu_y\right)^2\right\}. \quad (3.25)$$

Hence, given $X = x$, the random variable Y has a normal distribution with parameters

$$E(Y|X=x) = \rho\frac{\sigma_y}{\sigma_x}(x-\mu_x) + \mu_y \text{ and } Var(Y|X=x) = \sigma_y^2(1-\rho^2). \quad (3.26)$$

Thus, the regression function

$$m_Y(x) = E(Y|X=x)$$

of Y with regard to $X = x$ for the bivariate normal distribution is a straight line.

Example 3.7 The daily consumptions of tap water X and Y of two neighboring towns have a joint normal distribution with parameters

$$\mu_x = \mu_y = 16\,[10^3\,m^3], \quad \sigma_x = \sigma_y = 2\,[10^3 m^3], \text{ and } \rho = 0.5.$$

The conditional probability density of Y on condition $X = x$ has parameters

$$E(Y|x) = \rho \frac{\sigma_y}{\sigma_x}(x - \mu_x) + \mu_y = 0.5 \cdot \frac{2}{2}(x - 16) = \frac{x}{2} + 8$$

$$Var(Y|x) = \sigma_y^2(1 - \rho^2) = 4(1 - 0.5^2) = 3.$$

Hence,

$$f_Y(y|x) = \frac{1}{\sqrt{2\pi}\,\sqrt{3}}\,\exp\left\{-\frac{1}{2}\left(\frac{y - \frac{x}{2} - 8}{\sqrt{3}}\right)^2\right\}, \quad -\infty < y < +\infty.$$

This is the density of an $N(8 + x/2, 3)$-distributed random variable. Some conditional interval probabilities are:

$$P(14 < Y \le 16 | X = 10) = \Phi\left(\frac{16-13}{\sqrt{3}}\right) - \Phi\left(\frac{14-13}{\sqrt{3}}\right) = 0.958 - 0.718 = 0.240,$$

$$P(14 < Y \le 16 | X = 14) = \Phi\left(\frac{16-15}{\sqrt{3}}\right) - \Phi\left(\frac{14-15}{\sqrt{3}}\right) = 0.718 - 0.282 = 0.436.$$

The corresponding unconditional probability is

$$P(14 < Y \le 16) = \Phi\left(\frac{16-16}{2}\right) - \Phi\left(\frac{14-16}{2}\right) = 0.500 - 0.159 = 0.341. \qquad \square$$

3.1.2.4 Bivariate Exponential Distributions

In this section some joint probability distributions of random vectors (X, Y) with non-negative X and Y are considerered, whose marginal distributions are one-dimensional exponential distributions.

a) A random vector (X, Y) has a *Marshall-Olkin distribution* if its joint distribution function $F_{X,Y}(x,y) = P(X \le x, Y \le y)$ is for $x, y \ge 0$ given by

$$F_{X,Y}(x,y) = 1 - e^{-(\lambda_1+\lambda)x} - e^{-(\lambda_2+\lambda)y} + e^{-\lambda_1 x - \lambda_2 y - \lambda \max(x,y)} \qquad (3.27)$$

with positive parameters λ_1, λ_2, and a nonnegative parameter λ. By property (3) at page 121, the corresponding marginal distribution functions are

$$F_X(x) = 1 - e^{-(\lambda_1+\lambda)x}, \quad F_Y(y) = 1 - e^{-(\lambda_2+\lambda)y}; \quad x, y \ge 0.$$

Using property (7) at page 121 gives the corresponding *joint survival function*

$$\overline{F}_{X,Y}(x,y) = P(X > x, Y > y) = e^{-\lambda_1 x - \lambda_2 y - \lambda \max(x,y)}, \quad x, y \ge 0.$$

The joint density of (X, Y) is

$$f_{X,Y}(x,y) = \begin{cases} \lambda_2(\lambda_1 + \lambda)\,e^{-\lambda_2 y - (\lambda_1+\lambda)x} & \text{if } x > y, \\ \lambda_1(\lambda_2 + \lambda)\,e^{-\lambda_1 x - (\lambda_2+\lambda)y} & \text{if } x \le y. \end{cases}$$

This distribution has the following physical background: A system, which starts operating at time point $t = 0$, consists of two subsystems S_1 and S_2. They are subject to three types of shocks: A shock of type i occurs at time T_i and immediately destroys subsystem S_i, $i = 1, 2$. A shock of type 3 occurs at time T and immediately destroys both subsystems. The subsystems cannot fail for other reasons. The arrival times of the shocks T_1, T_2, and T are asssumed to be independent, exponentially with parameters λ_1, λ_2, and λ distributed random variables. Hence, the respective lifetimes X and Y of the subsystems S_1 and S_2 are

$$X = \min(T_1, T) \text{ and } Y = \min(T_2, T).$$

Thus, the lifetimes of the subsystems are clearly dependent, and their joint survival probability is given by $\overline{F}_{X,Y}(x,y)$.

b) A random vector (X, Y) has a *Gumbel distribution* with positive parameters λ_1, λ_2 and parameter λ, $0 \leq \lambda \leq 1$, if its joint distribution function is given by

$$F_{X,Y}(x,y) = 1 - e^{-\lambda_1 x} + e^{-\lambda_2 y} - e^{-\lambda_1 x - \lambda_2 y - \lambda xy}, \quad x, y \geq 0. \tag{3.28}$$

The corresponding marginal distribution functions are

$$F_X(x) = 1 - e^{-\lambda_1 x}, \quad F_Y(y) = 1 - e^{-\lambda_2 y}, \quad x, y \geq 0,$$

so that the corresponding joint survival probability is

$$\overline{F}_{X,Y}(x,y) = P(X > x, Y > y) = e^{-\lambda_1 x - \lambda_2 y - \lambda xy}, \quad x, y \geq 0,$$

c) Another useful bivariate distribution of a random vector (X, Y) with exponential marginal distributions is given for $x \geq 0$ and $y \geq 0$ by the joint distribution function

$$F_{X,Y}(x,y) = P(X \leq x, Y \leq y) = 1 - e^{-\lambda_1 x} - e^{-\lambda_2 y} - [e^{+\lambda_1 x} + e^{+\lambda_2 y} - 1]^{-1},$$

$\lambda_1, \lambda_2 > 0$. The corresponding marginal distribution functions are the same as the ones of the Gumbel distribution. Again by property (7) at page 121, the joint survival probability is

$$\overline{F}_{X,Y}(x,y) = P(X > x, Y > y) = [e^{+\lambda_1 x} + e^{+\lambda_2 y} - 1]^{-1}; \quad \lambda_1, \lambda_2 > 0, \ x, y \geq 0.$$

3.1.3 Linear Regression and Correlation Analysis

For a given random vector (X, Y) the aim of this section is to approximate Y by a linear function $\tilde{Y}$ of X:

$$\tilde{Y} = aX + b. \tag{3.29}$$

Such an approximation can be expected to yield good results if the regression function $m_Y(x)$ of Y with regard to x is at least approximately a straight line:

$$m_Y(x) = E(Y|X = x) \approx \alpha x + \beta. \tag{3.30}$$

Whether this assumption is realistic in a practical situation, one can empirically check by a scatter diagram of a sample: Let, for instance, X be the speed of a car and Y the corresponding braking time to a full stop. n measurements of both speed and corres-

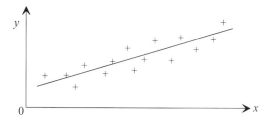

Figure 3.6 Scatter diagram for a linear regression function

ponding braking time had been done. The result is a sample of structure

$$\{(x_i, y_i), \ i = 1, 2, ..., n\}.$$

If the scatter diagram of this sample looks principally like the one in Figure 3.6, then assumption (3.29) is justified.

As criterion for the optimum fit of $\tilde{Y}$ to Y serves the mean squared deviation:

$$Q(a,b) = E[(Y - \tilde{Y}]^2 = E[Y - (aX + b)]^2. \tag{3.31}$$

The parameters a and b have to be determined such that $Q(a,b)$ assumes its absolute minimum. The necessary conditions are

$$\frac{\partial Q(a,b)}{\partial a} = 0, \quad \frac{\partial Q(a,b)}{\partial b} = 0. \tag{3.32}$$

By multiplying out the brackets in (3.31), $Q(a,b)$ is seen to be

$$Q(a,b) = E(Y^2) - 2a\,E(XY) - 2b\,E(Y) + a^2 E(X^2) + 2ab\,E(X) + b^2 \tag{3.33}$$

so that the necessary conditions (3.32) become

$$\frac{\partial Q(a,b)}{\partial a} = -2\,E(XY) + 2aE(X^2) + 2b\,E(X) = 0,$$

$$\frac{\partial Q(a,b)}{\partial b} = -2\,E(Y) + 2a\,E(X) + 2b.$$

The unique solution $(a,b) = (\alpha, \beta)$ is

$$\alpha = \frac{E(XY) - E(X)E(Y)}{Var(X)}, \tag{3.34}$$

$$\beta = E(Y) - \alpha\,E(X). \tag{3.35}$$

Since $\dfrac{\partial^2 Q(a,b)}{\partial a^2} = 2E(X^2), \quad \dfrac{\partial^2 Q(a,b)}{\partial b^2} = 2, \quad$ and $\quad \dfrac{\partial^2 Q(a,b)}{\partial a\,\partial b} = 2\,E(X),$

the sufficient condition for an absolute minimum at $(a,b) = (\alpha, \beta)$ is fulfilled:

$$\frac{\partial^2 Q(a,b)}{\partial a^2} \cdot \frac{\partial^2 Q(a,b)}{\partial b^2} - \left(\frac{\partial^2 Q(a,b)}{\partial a\,\partial b}\right)^2 = 4\left(E(X^2) - [E(X)]^2\right) = 4\,Var(X) > 0.$$

With $\sigma_X^2 = Var(X)$ and $\sigma_Y^2 = Var(Y)$, the smallest possible mean square deviation of Y from $\tilde{Y}$ is obtained from (3.33) by substituting there a and b with α and β:

$$Q(\alpha, \beta) = (\sigma_Y - \alpha \sigma_X)^2. \tag{3.36}$$

$Q(\alpha, \beta)$ is the *residual variance*. The smaller $Q(\alpha, \beta)$, the better is the fit of $\tilde{Y}$ to Y.

Definition 3.3 The straight line

$$\tilde{y} = \alpha x + \beta$$

is called *regression line*. The parameters α and β are the *regression coefficients*. ●

Best Estimate If the regression function $m_Y(x)$ is not linear, then the 'random regression line' $\tilde{Y}(\alpha, \beta) = \alpha X + \beta$ is not the best estimate for Y with regard to the mean squared deviation. Without proof, the following key result is given:

The best estimate for Y is $m_Y(X) = E(Y|X)$, i.e. for all real-valued functions $g(x)$,

$$E(Y - E(Y|X))^2 \leq E(Y - g(X))^2.$$

Only if the regression function $m_Y(x) = E(Y|x)$ is linear, $\tilde{Y}(\alpha, \beta) = \alpha X + \beta$ is the best estimate for Y with regard to the mean-squared deviation. In view of (3.26), this proves an important property of the bivariate normal distribution:

> If (X, Y) has a bivariate normal distribution, then the regression line
>
> $$\tilde{Y}(\alpha, \beta) = \alpha X + \beta$$
>
> is the best possible estimation for Y with respect to the mean-squared deviation.

Covariance The *covariance* between two random variables X and Y is defined as

$$Cov(X, Y) = E([X - E(X)] \cdot [Y - E(Y)]). \tag{3.37}$$

By multiplying out the brackets, one obtains an equivalent formula for the covariance:

$$Cov(X, Y) = E(XY) - E(X) \cdot E(Y). \tag{3.38}$$

The covariance has properties

$$Cov(X, X) = Var(X), \tag{3.39}$$

and $$Cov(X + Y, Z) = Cov(X, Z) + Cov(Y, Z). \tag{3.40}$$

From (3.14) and (3.38):

> If two random variables are independent, then their covariance is 0.

For this reason, the covariance serves as a measure for the degree of statistical dependence between two random variables. Generally one can expect that with increasing absolute value $|Cov(X, Y)|$ the degree of statistical dependence is increasing. But there are examples (given later) which prove that $Cov(X, Y) = 0$ not necessarily implies the independence of X and Y.

In view of being a measure for the dependence of two random variables, it is not surprising that the covariance between X and Y is a factor of α (see (3.34)). If X and Y are independent, then $Cov(X, Y) = 0$. In this case the regression line has slope $\alpha = 0$, i.e., it is a parallel to the x-axis, which gives no indication of a possible dependency between X and Y.

Unfortunately, the covariance does not allow to compare the degree of dependency between two different pairs of random variables, since it principally can assume any real value from $-\infty$ to $+\infty$.

Example 3.8 The random vector (X, Y) has the joint density

$$f_{X,Y}(x, y) = \tfrac{1}{2}xy, \quad 0 \le x \le y \le 2.$$

The marginal distributions are known from example 3.3:

$$f_X(x) = \tfrac{1}{4}(4x - x^3), \ 0 \le x \le 2; \quad f_Y(y) = \tfrac{1}{4}y^3, \ 0 \le y \le 2.$$

X and Y are defined in such a way that they cannot be independent. The corresponding mean values and variances are

$$E(X) = 16/15, \quad Var(X) = 132/675,$$

$$E(Y) = 8/5, \qquad Var(Y) = 8/75.$$

By (3.13),

$$E(XY) = \int_0^2 \int_x^2 xy \tfrac{1}{2}xy \, dy dx = \tfrac{1}{2}\int_0^2 x^2 \left(\int_x^2 y^2 dy \right) dx$$

$$= \tfrac{1}{6}\int_0^2 x^2(8 - x^3) dx = 16/19.$$

With these parameters, the regression coefficients can be calculated:

$$\alpha = \frac{\dfrac{16}{9} - \dfrac{16}{15} \cdot \dfrac{8}{5}}{\dfrac{132}{675}} = 0.36364,$$

$$\beta = \tfrac{8}{5} - \alpha \cdot \tfrac{16}{15} = 1.21212,$$

which gives the regression line

$$\tilde{y} = 0.36364 x + 1.21212 .$$

Thus, an increase of X by one unit approximately implies on average an increase of Y by 0.36364 units. The covariance between X and Y is 0.07111.

In view of the restriction for the joint density to the region $0 \le x \le y \le 2$, one would expect that the regression line assumes at value $x = 2$ the value 2 as well. But this is not the case since $\tilde{y}(2) = 1.93$. This is because the regression function $m_Y(x)$ is not a straight line so that the regression line is only an approximation to $m_Y(x)$. The exact

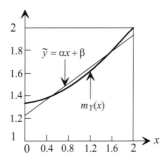

Figure 3.7 Regression function and regression line for example 3.8

average relationship between X and Y is given by the regression function:

$$m_Y(x) = E(Y|X=x) = \int_x^2 y \frac{f_{X,Y}(x,y)}{f_X(x)}\, dy$$

$$= \int_x^2 y \frac{\frac{1}{2}xy}{\frac{1}{4}(4x-x^3)}\, dy = \frac{2}{4-x^2} \int_x^2 y^2\, dy$$

$$= \frac{2}{3} \cdot \frac{8-x^3}{4-x^2}, \quad 0 \le x < 2.$$

Figure 3.7 shows that the largest differences between the regression function and the regression line are at the left- and at the right-hand side of the x-interval $[0, 2]$. □

Correlation Coefficient The *correlation coefficient* $\rho = \rho(X, Y)$ between two random variables X and Y with standard deviations σ_X and σ_Y is defined as the ratio

$$\rho(X, Y) = \frac{E[(X-E(X)) \cdot (Y-E(Y))]}{\sigma_X \sigma_Y} = \frac{E(XY) - E(X) \cdot E(Y)}{\sigma_X \sigma_Y}. \qquad (3.41)$$

The random variables X and Y are *uncorrelated* if $\rho(X, Y) = 0$, they are *positively correlated* if $\rho(X, Y) > 0$, and *negatively correlated* if $\rho(X, Y) < 0$.

The correlation coefficient can be written as the mean value of the product of the standardizations of X and Y:

$$\rho(X, Y) = E\left\{ \left(\frac{(X-E(X))}{\sigma_X} \right) \cdot \left(\frac{(Y-E(Y))}{\sigma_Y} \right) \right\}. \qquad (3.42)$$

There is the following relationship to the covariance between X and Y:

$$\rho(X, Y) = \frac{Cov(X, Y)}{\sigma_X \sigma_Y}. \qquad (3.43)$$

Hence, X and Y are uncorrelated if and only if $Cov(X, Y) = 0$. If X and Y are independent, then X and Y are uncorrelated. But the converse need not be true (see examples 3.11 and 3.12).

The Marshall-Olkin distribution and the Gumbel distribution (pages 132 and 133) are examples for the equivalence of X and Y being independent and uncorrelated:

If (X, Y) has the Marshall-Olkin distribution (3.27), then the correlation coefficient between X and Y is (exercise 3.18)

$$\rho(X, Y) = \frac{\lambda}{\lambda_1 + \lambda_2 + \lambda}.$$

$\rho(X, Y) = 0$ if and only if $\lambda = 0$. X and Y are independent if and only if $\lambda = 0$.

If (X, Y) has the Gumbel distribution (3.28) with $\lambda_1 = \lambda_2 = 1$, then the correlation coefficient between X and Y is (without proof)

$$\rho(X, Y) = \int_0^\infty \frac{e^{-y}}{1 + \lambda y} dy - 1.$$

If $\lambda = 0$, then $\rho(X, Y) = 0$ and X and Y are independent, and, vice versa, if X and Y are independent or $\rho(X, Y) = 0$, then $\lambda = 0$.

With the correlation coefficient, the regression coefficients α and β can be written as (compare to (3.26))

$$\alpha = \rho \frac{\sigma_Y}{\sigma_X}, \quad \beta = E(Y) - \rho \frac{\sigma_Y}{\sigma_X} E(X), \tag{3.44}$$

and another representation of the regression line is

$$\frac{\tilde{y} - E(Y)}{\sigma_Y} = \rho \frac{x - E(X)}{\sigma_X}.$$

Therefore, when X and Y are positively (negatively) correlated, then an increase (decrease) in X will on average lead to an increase (decrease) in Y. If X and Y are uncorrelated, the regression line does not depend on x at all. Nevertheless, even in this case there may be a dependency between X and Y, since X can have influence on the variability of Y. Figure 3.8 illustrates this situation: If $\rho = 0$, the regression line is a parallel to the x-axis, namely $\tilde{y} \equiv E(Y)$. With increasing x the fluctuations of the realizations of Y become larger and larger, but in such a way that $E(Y)$ remains constant.

Figure 3.8 Scatter diagram for (X, Y) indicating a dependence

Theorem 3.2 The correlation coefficient $\rho(X, Y)$ has the following properties:
(1) If X and Y are independent, then $\rho(X, Y) = 0$.
(2) If X and Y are linearly dependent, then $\rho(X, Y) = \pm 1$.
(3) For any random variables X and Y: $-1 \le \rho(X, Y) \le +1$.

Proof (1) The assertion follows from $Cov(X, Y) = \sigma_X \sigma_Y \rho(X, Y)$ and (3.38).
(2) Let $Y = aX + b$ for any a and b. Then, from (2.54) and (2.61),

$$E(Y) = a E(X) + b, \quad \sigma_Y^2 = a^2 Var(X).$$

Now, from (3.42),

$$\rho(X, Y) = E\left\{ \left(\frac{(X - E(X))}{\sigma_X}\right) \cdot \left(\frac{a(X - E(X))}{|a|\sigma_X}\right) \right\} = E\left(\frac{a(X - E(X))^2}{|a|\sigma_X^2} \right)$$

$$= \frac{a}{|a|} \cdot \frac{\sigma_X^2}{\sigma_X^2} = \frac{a}{|a|} = \begin{cases} +1 \text{ if } a > 0 \\ -1 \text{ if } a < 0 \end{cases}.$$

(3) Using (3.43), the residual variance (3.36) can be written in the form

$$Q(\alpha, \beta) = \sigma_Y^2 (1 - \rho^2).$$

Since a quadratic deviation can never be negative and σ_Y^2 is positive anyway, the factor $1 - \rho^2$ must be positive. But $1 - \rho^2 > 0$ is equivalent to $-1 \le \rho \le +1$. ∎

According to this theorem, the correlation coefficient can be interpreted as the covariance standardized to the interval $[-1, +1]$. In case of independence the correlation coefficient is 0; for linear (deterministic) dependence this coefficient assumes one of its extreme values -1 or +1. Thus, unlike the covariance, the correlation coefficient allows for comparing the (linear) dependencies between different pairs of random variables. However, the following examples show that even in case of (nonlinear) functional dependence the correlation coefficient can be so close to 1 that the difference is negligibly small, whereas, on the other hand, the correlation coefficient can be 0 for non-linear functional dependence.

Example 3.9 The bending strength Y of a steel rod of a given length is given by the equation $Y = cX^2$, where X is the diameter of the rod and the parameter c is a material constant. X is a random variable, which has a uniform distribution in the interval $[3.92\,cm,\ 4.08\,cm]$. The input parameters for $\rho(X, Y)$ are

$$E(X) = 4,$$

$$Var(X) = \frac{1}{0.16} \int_{3.92}^{4.08} x^2 dx - 16 = \frac{1}{0.48} \left[x^3 \right]_{3.92}^{4.08} - 16 = 0.0021333,$$

$$E(Y) = \frac{c}{0.16} \int_{3.92}^{4.08} x^2 dx = 16.0021333 \cdot c,$$

$$Var(Y) = \frac{c^2}{0.16} \int_{3.92}^{4.08} x^4 dx - [c\,E(Y)]^2 = 0.1365380 \cdot c^2,$$

and

$$E(XY) = \frac{c}{0.16} \int_{3.92}^{4.08} x^3 dx = 64.0256000 \cdot c.$$

Hence, the correlation coefficient between X and Y is

$$\rho(X, Y) = \frac{64.0256 \cdot c - 4 \cdot 16.0021333 \cdot c}{0.0461877 \cdot 0.3695105 \cdot c} = 0.9999976.$$

Although there is no linear functional relationship between X and Y, their correlation coefficient is practically 1. (The extreme degree of numerical accuracy is required to make sure that the calculated correlation coefficient does not exceed 1.)   □

Example 3.10 Let $Y = \sin X$, where X has a uniform distribution in the interval $[0, \pi]$, i.e., it has density $f_X(x) = 1/\pi$, $0 \le x \le \pi$. The input parameters for $Cov(X, Y)$ are

$$E(X) = \pi/2,$$

$$E(Y) = \frac{1}{\pi} \int_0^\pi \sin x \, dx = \frac{1}{\pi} [-\cos x]_0^\pi = 2/\pi.$$

$$E(XY) = \frac{1}{\pi} \int_0^\pi x \sin x \, dx = \frac{1}{\pi} [\sin x - x \cos x]_0^\pi = 1.$$

Hence, $Cov(X, Y) = 0$ so that $\rho(X, Y) = 0$ as well. Despite X and Y being functionally related, they are uncorrelated. (Give an intuitive explanation for this.)   □

As mentioned before in section 3.1.2.3, if the random vector (X, Y) has a bivariate normal distribution, then the random variables X and Y are independent if and only if they are uncorrelated. There are bivariate distributions, which do not have this property, i.e., dependent random variables can be uncorrelated. This will be demonstrated by the following two examples.

Example 3.11 The random vector (X, Y) has the joint probability density

$$f_{X,Y}(x,y) = \frac{x^2 + y^2}{4\pi} \exp\left\{ \left(-\frac{x^2 + y^2}{2} \right) \right\}, \quad -\infty < x, y < +\infty.$$

Next the marginal densities of $f_{X,Y}(x,y)$ have to be determined:

$$f_X(x) = \int_{-\infty}^{+\infty} \frac{x^2 + y^2}{4\pi} \exp\left\{ \left(-\frac{x^2 + y^2}{2} \right) \right\} dy$$

$$= \frac{e^{-x^2/2}}{2\sqrt{2\pi}} \left(x^2 \int_{-\infty}^{+\infty} \frac{1}{\sqrt{2\pi}} e^{-y^2/2} dy + \int_{-\infty}^{+\infty} y^2 \frac{1}{\sqrt{2\pi}} e^{-y^2/2} dy \right).$$

The integrand of the first integral is the density of an $N(0, 1)$-distribution; the second integral is the variance of an $N(0, 1)$-random variable. Both integrals are equal to 1 so that

$$f_X(x) = \frac{1}{2\sqrt{2\pi}}(x^2 + 1)e^{-x^2/2}, \quad -\infty < x, y < +\infty.$$

Since $f_{X,Y}(x,y)$ is symmetric in x and y,

$$f_Y(y) = \frac{1}{2\sqrt{2\pi}}(y^2 + 1)e^{-y^2/2}, \quad -\infty < x, y < +\infty.$$

Obviously, $f_{X,Y}(x,y) \neq f_X(x) \cdot f_Y(y)$ so that X and Y are not independent.

The mean value of XY is

$$E(XY) = \int_{-\infty}^{+\infty} xy \frac{x^2 + y^2}{4\pi} \exp\left\{\left(-\frac{x^2 + y^2}{2}\right)\right\} dx\, dy$$

$$= \frac{1}{4\pi}\left(\int_{-\infty}^{+\infty} x^3 e^{-x^2/2} dx\right)\left(\int_{-\infty}^{+\infty} y^3 e^{-y^2/2} dy\right).$$

Both integrals in the last line are 0, since their integrands are odd functions with regard to the origin. But $E(X)$ and $E(Y)$ are 0 as well, since $f_X(x)$ and $f_Y(y)$ are symmetric functions with regard to the origin. Hence, $E(XY) = E(X) \cdot E(Y)$. Thus, X and Y are uncorrelated, but not independent. □

Regression line and correlation coefficient are defined for discrete random variables as well. The next example gives a discrete analogue to the previous one.

Example 3.12 Let X and Y be two discrete random variables with ranges
$$R_X = \{-2, -1, +1, +2\} \text{ and } R_Y = \{-1, 0, +1\}.$$

Their joint distribution is given by Table 3.2:

X $\backslash$ Y	-2	-1	+1	+2	q_j
-1	1/16	1/8	1/8	1/16	6/16
0	1/16	1/16	1/16	1/16	4/16
+1	1/16	1/8	1/8	1/16	6/16
p_i	3/16	5/16	5/16	3/16	1

Table 3.2 Joint and marginal distribution for Example 3.12

From Table 3.2: The input parameters into the covariance between X and Y are

$$E(X) = \frac{1}{16}[3 \cdot (-2) + 5 \cdot (-1) + 5 \cdot (+1) + 3 \cdot (+2)] = 0,$$

$$E(Y) = \frac{1}{16}[6 \cdot (-1) + 4 \cdot 0 + 6 \cdot (+1)] = 0,$$

$$E(XY) = \frac{1}{16}[(-2)(-1) + 2 \cdot (-1)(-1) + 2 \cdot (+1)(-1) + (+2)(-1)]$$

$$+ \frac{1}{16}[(-2) \cdot 0 + 2 \cdot (-1) \cdot 0 + 2 \cdot (+1) \cdot 0 + (+2) \cdot 0]$$

$$+ \frac{1}{16}[(-2)(+1) + 2 \cdot (-1)(+1) + 2 \cdot (+1)(+1) + (+2)(+1)] = 0.$$

Hence, $Cov(X, Y) = \rho(X, Y) = 0$ so that X and Y are uncorrelated.

On the other hand,

$$P(X = 2, \ Y = -1) = \frac{1}{16} \neq P(X = 2) \cdot P(Y = -1) = \frac{3}{16} \cdot \frac{6}{16} = \frac{9}{128}$$

so that X and Y are not independent.   □

In applications it is usually assumed that the random vector (X, Y) has a bivariate normal distribution. Reasons for this are the following ones

1) The regression line $\tilde{y} = \alpha x + \beta$ coincides with the regression function

$$m_Y(x) = E(Y|X = x).$$

Hence, $\tilde{Y} = \alpha X + \beta$ is the best estimate for Y with regard to the mean squared deviation of Y from $\tilde{Y}$.

2) X and Y are independent if and only if X and Y are uncorrelated.

3) Applicability of statistical procedures.

Statistical Approach to Linear Regression The approach to the linear regression analysis adopted so far in this section is based on assuming that the joint distribution of the random vector (X, Y) is known, including the numerical parameters involved. The statistical approach is to estimate the numerical parameters based on a sample $\{(x_i, y_i); \ i = 1, 2, ..., n\}$. This sample is obtained by repeating the random experiment with outcome (X, Y) independently and under identical conditions n times and registering the realizations (x_i, y_i). The principle of minimizing the mean squared deviation (3.31) is now applied to minimizing the arithmetic mean of the squared deviations of the observed values y_i from the ones given by the regression line $\tilde{y} = \alpha x + \beta$, whose coefficients α and β are to be estimated:

$$Q(\alpha, \beta) = \frac{1}{n} \sum_{i=1}^{n} (y_i - \tilde{y}_i)^2 = \frac{1}{n} \sum_{i=1}^{n} (y_i - \alpha x_i - \beta)^2 \to \min. \qquad (3.45)$$

This method of parameter estimation is called *the method of least squares*. Differentiating (3.45) with respect to α and β yields necessary and in this case also sufficient conditions for the best least square estimates of α and β (of course, the factor $1/n$ can be ignored):

$$\sum_{i=1}^{n} x_i y_i - \alpha \sum_{i=1}^{n} x_i^2 - n\bar{x}\bar{y} + \alpha n\bar{x}^2 = 0,$$

$$\beta = \bar{y} - \alpha\bar{x}.$$

The unique solution is

$$\hat{\alpha} = \frac{\sum\limits_{i=1}^{n} x_i y_i - n\bar{x}\bar{y}}{\sum\limits_{i=1}^{n} x_i^2 - n\bar{x}^2} = \frac{\sum\limits_{i=1}^{n}(x_i - \bar{x})(y_i - \bar{y})}{\sum\limits_{i=1}^{n}(x_i - \bar{x})^2}, \tag{3.46}$$

$$\hat{\beta} = \bar{y} - \hat{\alpha}\bar{x},$$

where $\bar{x}$ and $\bar{y}$ are the arithmetic means

$$\bar{x} = \frac{1}{n}\sum_{i=1}^{n} x_i, \quad \bar{y} = \frac{1}{n}\sum_{i=1}^{n} y_i.$$

$\hat{\alpha}$ and $\hat{\beta}$ are (*point*) *estimates* of the unknown regression coefficients α and β. With the additional notation

$$s_X^2 = \frac{1}{n-1}\sum_{i=1}^{n}(x_i - \bar{x})^2, \quad s_Y^2 = \frac{1}{n-1}\sum_{i=1}^{n}(y_i - \bar{y})^2,$$

$$s_{XY} = \frac{1}{n-1}\sum_{i=1}^{n}(x_i - \bar{x})(y_i - \bar{y}) = \frac{1}{n-1}\left(\sum_{i=1}^{n} x_i y_i - n\bar{x}\bar{y}\right),$$

the *empirical regression coefficients* $\hat{\alpha}$ and $\hat{\beta}$ can be rewritten as

$$\hat{\alpha} = \frac{s_{XY}}{s_X^2} = r \cdot \frac{s_Y}{s_X}, \quad \hat{\beta} = \bar{y} - r \cdot \frac{s_Y}{s_X}\bar{x}, \tag{3.47}$$

where s_{XY}, the *empirical* or *sample covariance,* is an estimate for the (theoretical) covariance $Cov(X, Y)$ between X and Y, and

$$r = r(X, Y) = \frac{s_{XY}}{s_X \cdot s_Y}, \tag{3.48}$$

the *empirical* or *sample correlation coefficient,* is an estimate for the (theoretical) correlation coefficient $\rho = \rho(X, Y)$ between X and Y. With this notation and interpretation the analogies between (3.43) and (3.47) as well as (3.41) and (3.48) are obvious.

It is interesting that the same estimates of the regression coefficients would have been obtained if all mean values in (3.34) are replaced with the corresponding arithmetic means. (Note that variances are mean values as well.) The fact that in s_X^2, s_Y^2, and s_{XY} the factor $1/(n-1)$ appears instead of $1/n$ is motivated by theorem 4.2 (page 188).

Example 3.13 In a virgin forest stand of yellowwoods (*Podocarpus latifolius*) in the Soutpansberg, South Africa, 12 trees had been randomly selected and had their stem diameters (1.3 m above ground) and heights measured. Table 3.3 shows the results:

Tree number		1	2	3	4	5	6	7	8	9	10	11	12
Stem diameter [*cm*]	x_i	44	62	50	84	38	95	76	104	35	99	57	78
Height [*m*]	y_i	32	48	38	56	31	62	57	73	28	76	41	49

Table 3.3 Stem diameters and the corresponding tree heights

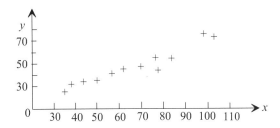

Figure 3.9 Scatter diagram for example 3.13

Then,

$$\bar{x} = 68.50, \quad \bar{y} = 49.25, \quad s_x = 24.21, \quad s_y = 16.03, \quad s_{X,Y} = 378.14.$$

This gives the empirical correlation coefficient as

$$r = \frac{s_{XY}}{s_X \cdot s_Y} = \frac{378.14}{24.21 \cdot 16.03} = 0.974.$$

Hence, there is a strong linear connection between stem diameter and tree height. This numerical result is in concordance with Figure 3.9. The empirical regression line, therefore, adequately quantifies the average relationship between stem diameter and tree height:

$$\tilde{y} = \hat{\alpha} x + \hat{\beta} = 0.645 x + 5.068.$$

Hence, the average increase of the height of a yellowwood is $0.645\,m$ if the stem diameter increases by $1cm$. $\square$

3.2 *n*-DIMENSIONAL RANDOM VARIABLES

Let $X_1, X_2, ..., X_n, \ n \geq 2$, be continuous random variables with distribution functions

$$F_{X_1}(x_1), F_{X_2}(x_2), \cdots, F_{X_n}(x_n) \tag{3.49}$$

and probabiliy densities

$$f_{X_1}(x_1), f_{X_2}(x_2), \cdots, f_{X_n}(x_n). \tag{3.50}$$

In what follows, let

$$\mathbf{X} = (X_1, X_2, ..., X_n).$$

The *joint distribution function* of the random vector $\mathbf{X}$ is defined as

$$F_{\mathbf{X}}(x_1, x_2, ..., x_n) = P(X_1 \leq x_1, X_2 \leq x_2, ..., X_n \leq x_n). \tag{3.51}$$

The *marginal distribution functions* $F_{X_i}(x_i)$ are obtained from $F_{\mathbf{X}}(x_1, x_2, ..., x_n)$ by

$$F_{X_i}(x_i) = F(\infty, ..., \infty, x_i, \infty, ..., \infty); \quad i = 1, 2, ..., n.$$

Basic properties of the joint distribution function are:

1) $F_\mathbf{X}(x_1, x_2, \cdots, x_n) = 0$ if one or more of the x_i are equal to $-\infty$.

2) $F_\mathbf{X}(+\infty, +\infty, \cdots, +\infty) = 1$,

3) $F_\mathbf{X}(x_1, x_2, ..., x_n)$ is nondecreasing in each $x_1, x_2, ..., x_n$.

Apart from the marginal distribution functions, $F_\mathbf{X}(x_1, x_2, ..., x_n)$ yields the joint distributions of all subvectors of $\mathbf{X}$. Let, for instance,

$$\{X_i, X_j\} \subset \{X_1, X_2, ..., X_n\}; \ i < j, \ n > 2.$$

Then the joint distribution function $F_{X_i, X_j}(x_i, x_j)$ of the random vector (X_i, X_j) is

$$F_{X_i, X_j}(x_i, x_j) = F_\mathbf{X}(\infty, \cdots \infty, x_i, \infty, \cdots, \infty, x_{j+1}, \infty, \cdots, \infty).$$

In this way, the joint distribution functions of all subvectors

$$\{X_{i_1}, X_{i_2}, ..., X_{i_k}\} \subset \{X_1, X_2, ..., X_n\}, \ k < n,$$

can be obtained. For instance, the joint distribution function of $(X_1, X_2, ..., X_k)$ is

$$F_{X_1, X_2, ..., X_k}(x_1, x_2, ..., x_k) = F_\mathbf{X}(x_1, x_2 ..., x_k, \infty, \infty, ..., \infty), \quad k < n.$$

The *joint probability density* of $\mathbf{X}$ is the nth mixed partial derivative of the joint distribution function with respect to the $x_1, x_2, ..., x_n$:

$$f_\mathbf{X}(x_1, x_2, ..., x_n) = \frac{\partial^n F_\mathbf{X}(x_1, x_2, ..., x_n)}{\partial x_1 \partial x_2 \cdots \partial x_n}. \tag{3.52}$$

The characteristic properties of the two-dimensional densities can be extended in a straightforward way to the n-dimensional densities. In particular, properties (3.11) are special cases of

$$f_\mathbf{X}(x_1, x_2, ..., x_n) \geq 0, \quad \int_{-\infty}^{+\infty} \cdots \int_{-\infty}^{+\infty} f_\mathbf{X}(x_1, x_2, ..., x_n)\, dx_1 dx_2 \cdots dx_n = 1, \tag{3.53}$$

and the marginal densities are for all $i = 1, 2, ..., n$,

$$f_{X_i}(x_i) = \int_{-\infty}^{+\infty} \cdots \int_{-\infty}^{+\infty} f_\mathbf{X}(x_1, x_2, \cdots, x_n)\, dx_1 \cdots dx_{i-1}\, dx_{i+1} \cdots dx_n. \tag{3.54}$$

Definition 3.4 (*independence*) The random variables $X_1, X_2, ..., X_n$ are (*completely*) *independent* if and only if

$$F_\mathbf{X}(x_1, x_2, ..., x_n) = F_{X_1}(x_1) \cdot F_{X_2}(x_2) \cdot \ \cdots \ \cdot F_{X_n}(x_n). \qquad \bullet$$

For the practical relevance of this definition, see comment after formula (3.14), page 122. In terms of the densities, the $X_1, X_2, ..., X_n$ are independent if and only if

$$f_\mathbf{X}(x_1, x_2, ..., x_n) = f_{X_1}(x_1) \cdot f_{X_2}(x_2) \cdot \cdots \cdot f_{X_n}(x_n). \tag{3.55}$$

Definition 3.4 also includes discrete random variables. However, for discrete random variables X_i (complete) independence can be equivalently defined by

$$P(X_1 = x_1, X_2 = x_2, \cdots, X_n = x_n) = P(X_1 = x_1) \cdot P(X_2 = x_2) \cdots \cdot P(X_n = x_n) \quad (3.56)$$

for all $x_i \in R_{X_i}$; $i = 1, 2, ..., n$.

The intuitive meaning of independence is that the values, which any of the X_i have assumed, has no influence on the values, which the remaining X_j have taken on.

If the X_i are independent, the set of the marginal distributions

$$\{F_{X_1}(x_1), F_{X_2}(x_2), ..., F_{X_n}(x_n)\}$$

contains the same amount of information on the probability distribution of the random vector $\mathbf{X}$ as the joint probability distribution function.

If the $X_1, X_2, ..., X_n$ are independent, then every subset $\{X_{i_1}, X_{i_2}, ..., X_{i_k}\}$ of the set $\{X_1, X_2, ..., X_n\}$ is independent as well. In particular, all possible pairs of random variables (X_i, X_j), $i \neq j$, are independent (*pairwise independence* of the $X_1, X_2, ..., X_n$). As the following example shows, pairwise independence of the $X_1, X_2, ..., X_n$ does not necessarily imply their complete independence.

Example 3.14 Let $A_1, A_2,$ and A_3 be pairwise independent random events and $X_1, X_2,$ and X_3 their respective indicator variables:

$$X_i = \begin{cases} 1 & \text{if } A_i \text{ occurs,} \\ 0 & \text{otherwise,} \end{cases} \quad i = 1, 2, 3.$$

By (3.56), complete independence of the $X_1, X_2,$ and X_3 would imply that

$$P(X_1 = 1, X_2 = 1, X_3 = 1) = P(X_1 = x_1) \cdot P(X_2 = x_2) \cdot P(X_3 = x_3),$$

or equivalently that

$$P(A_1 \cap A_2 \cap A_3) = P(A_1) \cdot P(A_2) \cdot P(A_3).$$

However, we know from example 1.20 that the pairwise independence of random events $A_1, A_2,$ and A_3 does necessarily imply their complete independence. $\square$

The joint density of (X_i, X_j), $i < j$, is

$$f_{X_i, X_j}(x_i, x_j) = \int_{-\infty}^{+\infty} \cdots \int_{-\infty}^{+\infty} f_{\mathbf{X}}(x_1, x_2, ..., x_n) \, dx_1 \cdots dx_{i-1} dx_{i+1} \cdots dx_{j-1} dx_{j+1} \cdots dx_n,$$

whereas the joint density of $(X_1, X_2, ..., X_k)$, $k < n$, is

$$f_{X_1, X_2, ..., X_k}(x_1, x_2, ..., x_k) = \int_{-\infty}^{+\infty} \cdots \int_{-\infty}^{+\infty} f_{\mathbf{X}}(x_1, x_2, ..., x_k, x_{k+1} \cdots x_n) \, dx_{k+1} \cdots dx_n.$$

Conditional densities can be obtained analogously to the two-dimensional case: For instance, the conditional density of $(X_1, X_2, ..., X_n)$ given $X_i = x_i$ is

$$f_{X_1, ..., X_{i-1}, X_{i+1}, ..., X_n}(x_1, ..., x_{i-1}, x_{i+1}, ..., x_n | x_i) = \frac{f_{\mathbf{X}}(x_1, x_2, ..., x_n)}{f_{X_i}(x_i)}, \quad (3.57)$$

and the conditional density of $(X_1, X_2, ..., X_n)$ given $(X_1 = x_1, X_2 = x_2, ..., X = x_k)$ is

$$f_{X_{k+1},X_{k+2},\dots,X_n}(x_{k+1},x_{k+2},\dots,x_n|x_1,x_2,\dots,x_k) = \frac{f_{\mathbf{X}}(x_1,x_2,\dots,x_n)}{f_{X_1,X_2,\dots,X_k}(x_1,x_2,\dots,x_k)}. \tag{3.58}$$

for $k < n$. Let $y = h(x_1,x_2,\dots,x_n)$ be a real-valued function of n variables. Then the mean value of the random variable $Y = h(X_1,X_2,\dots,X_n)$ is defined as

$$E(Y) = \int_{-\infty}^{+\infty}\cdots\int_{-\infty}^{+\infty} h(x_1,x_2,\dots,x_n)f_{\mathbf{X}}(x_1,x_2,\dots,x_n)\,dx_1 dx_2\cdots dx_n. \tag{3.59}$$

In particular, the mean value of the product $Y = X_1 X_2\cdots X_n$ is

$$E(X_1 X_2\cdots X_n) = \int_{-\infty}^{+\infty}\cdots\int_{-\infty}^{+\infty} x_1 x_2\cdots x_n f_{\mathbf{X}}(x_1,x_2,\dots,x_n)\,dx_1 dx_2\cdots dx_n.$$

Due to (3.55), for independent X_i this n-dimensional integral simplifies to the product of n one-dimensional integrals:

$$E(X_1 X_2\cdots X_n) = E(X_1)E(X_2)\cdots E(X_n). \tag{3.60}$$

The mean value of the product of independent random variables is equal to the product of the mean values of these random variables.

The *conditional mean value* of $Y = h(X_1,\dots,X_n)$ on condition $X_1 = x_1,\dots,X_k = x_k$ is

$$E(Y|x_1,x_2,\cdots,x_k) = \tag{3.61}$$

$$= \int_{-\infty}^{+\infty}\int_{-\infty}^{+\infty}\cdots\int_{-\infty}^{+\infty} h(x_1,x_2,\dots,x_n)\frac{f_{\mathbf{X}}(x_1,x_2,\dots,x_n)}{f_{X_1,X_2,\dots,X_k}(x_1,x_2,\dots,x_k)}\,dx_{k+1}dx_{k+2}\cdots dx_n.$$

Replacing in (3.61) the realizations $x_1,x_2,\dots,x_k$ with the corresponding random variables $X_1,X_2,\dots,X_k$ yields the <u>random mean value</u> of Y on condition $X_1,X_2,\dots,X_k$:

$$E(Y|X_1,X_2,\dots,X_k) = \left(\int_{-\infty}^{+\infty}\cdots\int_{-\infty}^{+\infty} h(X_1,X_2,\cdots,X_k,x_{k+1},\cdots x_n)\right. \tag{3.62}$$

$$\left. \times\frac{f_{\mathbf{X}}(X_1,X_2,\cdots,X_k,x_{k+1},\cdots x_n)}{f_{X_1,X_2,\dots,X_k}(X_1,X_2,\dots,X_k)}\,dx_{k+1}dx_{k+2}\cdots dx_n\right).$$

The mean value of this random variable (with respect to all $X_1,X_2,\dots,X_k$) is

$$E_{X_1,X_2,\dots,X_k}(E(Y|X_1,X_2,\dots,X_k)) = E(Y). \tag{3.63}$$

For instance, the mean value of $E(Y|X_1,X_2,\dots,X_k)$ with respect to the random variables $X_1,X_2,\dots,X_{k-1}$ is the random variable:

$$E_{X_1,X_2,\dots,X_{k-1}}(E(Y|X_1,X_2,\dots,X_k)) = E(Y|X_k). \tag{3.64}$$

Now it is obvious how to obtain the conditional mean values $E(Y|x_{i_1},x_{i_2},\cdots,x_{i_k})$ and $E(Y|X_{i_1},X_{i_2},\cdots,X_{i_k})$ with regard to any subsets of

$$\{x_1,x_2,\dots,x_n\} \text{ and } \{X_1,X_2,\dots,X_n\},$$

respectively.

Let $c_{ij} = Cov(X_i, X_j)$ be the covariance between X_i and X_j; $i, j = 1, 2, ..., n$. It is useful to unite the c_{ij} in the *covariance matrix* $\mathbf{C}$:

$$\mathbf{C} = ((c_{ij})) ; \quad i, j = 1, 2, ..., n.$$

The main diagonal of $\mathbf{C}$ consists of the variances of the X_i:

$$c_{ii} = Var(X_i); \quad i = 1, 2, ..., n.$$

n-Dimensional Normal Distribution Let $\mathbf{X} = (X_1, X_2, \cdots, X_n)$ be an n-dimensional random vector with $\mu_i = E(X_i)$ for $i = 1, 2, ..., n$, and covariance matrix $\mathbf{C} = ((c_{ij}))$. Furthermore, let $|\mathbf{C}|$ and $\mathbf{C}^{-1}$ be the positive determinant and the inverse of $\mathbf{C}$, respectively, as well as

$$\mu = (\mu_1, \mu_2, \cdots, \mu_n), \text{ and } \mathbf{x} = (x_1, x_2, \cdots, x_n).$$

$(X_1, X_2, \cdots, X_n)$ has an *n-dimensionally normal* (or *Gaussian*) *distribution* if it has joint density

$$f_{\mathbf{X}}(\mathbf{x}) = \frac{1}{\sqrt{(2\pi)^n |\mathbf{C}|}} \exp\left(-\frac{1}{2}(\mathbf{x} - \mu)\mathbf{C}^{-1}(\mathbf{x} - \mu)^T\right), \tag{3.65}$$

where $(\mathbf{x} - \mu)^T$ is the transpose of the vector

$$\mathbf{x} - \mu = (x_1 - \mu_1, x_2 - \mu_2, \cdots, x_n - \mu_n).$$

By doing the matrix-vector-multiplication in (3.65), $f_{\mathbf{X}}(\mathbf{x})$ becomes

$$f_{\mathbf{X}}(\mathbf{x}) = \frac{1}{\sqrt{(2\pi)^n |\mathbf{C}|}} \exp\left(-\frac{1}{2|\mathbf{C}|} \sum_{i=1}^{n} \sum_{j=1}^{n} C_{ij}(x_i - \mu_i)(x_j - \mu_j)\right), \tag{3.66}$$

where C_{ij} is the cofactor of c_{ij}.

For $n = 2$, $x_1 = x$, and $x_2 = y$, (3.66) specializes to the density of the bivariate normal distribution (3.24). Generalizing from the bivariate special case, it can be shown that the random variables X_i have an $N(\mu_i, \sigma_i^2)$-distribution with $\sigma_i^2 = c_{ii}$, $i = 1, 2, ..., n$, if $\mathbf{X}$ has an n-dimensional normal distribution, i.e., the marginal distributions of $\mathbf{X}$ are the one-dimensional normal distributions

$$N(\mu_i, \sigma_i^2); \quad i = 1, 2, ..., n.$$

If the X_i are uncorrelated, then $\mathbf{C} = ((c_{ij}))$ is a diagonal matrix with $c_{ij} = 0$ for $i \neq j$ so that the joint density $f_{\mathbf{X}}(x_1, x_2, ..., x_n)$ assumes the product form (3.55):

$$f_{\mathbf{X}}(x_1, x_2, \cdots, x_n) = \prod_{i=1}^{n} \left[\frac{1}{\sqrt{2\pi} \, \sigma_i} \exp\left(-\frac{1}{2}\left(\frac{x_i - \mu_i}{\sigma_i}\right)^2\right) \right]. \tag{3.67}$$

Hence, the $X_1, X_2, ..., X_n$ are independent if and only if they are uncorrelated.

Theorem 3.3 The random vector $(X_1, X_2, ..., X_n)$ have an n-dimensionally normal distribution. If the random variables $Y_1, Y_2, ..., Y_m$ are linear combinations of the X_i, i.e., if there exist constants a_{ij} so that

$$Y_i = \sum_{j=1}^{n} a_{ij} X_j; \quad i = 1, 2, ..., m,$$

then $(Y_1, Y_2, ..., Y_m)$ has an m-dimensional normal distribution (without proof). ∎

The following two n-dimensional distributions are generalizations of the bivariate distributions (3.27) and (3.28), respectively.

n-Dimensional Marshall-Olkin Distribution The random vector $\mathbf{X} = (X_1, X_2, ..., X_n)$ has an *n-dimensional Marshall-Olkin distribution* with positive parameters $\lambda_1, \lambda_2, ...,$ and λ_n and with nonnegative parameter λ if it has the joint survival probability

$$\overline{F}_{\mathbf{X}}(x_1, x_2, ..., x_n) = P(X_1 > x_1, X_2 > x_2, ..., X_n > x_n)$$
$$= e^{-\lambda_1 x_1 - \lambda_2 x_2 - \cdots - \lambda_n x_n - \lambda \max(x_1, x_2, ..., x_n)}, \quad x_i \geq 0, i = 1, 2, ..., n.$$

n-Dimensional Gumbel Distribution The random vector $\mathbf{X} = (X_1, X_2, ..., X_n)$ has an *n-dimensional Gumbel distribution* with positive parameters $\lambda_1, \lambda_2, ..., \lambda_n$ and with parameter λ, $0 \leq \lambda \leq 1$, if it has the joint survival probability

$$\overline{F}_{\mathbf{X}}(x_1, x_2, ..., x_n) = P(X_1 > x_1, X_2 > x_2, ..., X_n > x_n)$$
$$= e^{-\lambda_1 x_1 - \lambda_2 x_2 - \cdots - \lambda_n x_n - \lambda x_1 x_2 \cdots x_n)}, \quad x_i \geq 0, i = 1, 2, ..., n.$$

3.3 EXERCISES

3.1) Two dice are thrown. Their respective random outcomes are X_1 and X_2. Let $X = \max(X_1, X_2)$ and Y be the number of even components of (X_1, X_2). X and Y have the respective ranges $R_X = \{1, 2, 3, 4, 5, 6\}$ and $R_Y = \{0, 1, 2\}$.

(1) Determine the joint probability distribution of the random vector (X, Y) and the corresponding marginal distributions. Are X and Y independent?

(2) Determine $E(X)$, $E(Y)$, and $E(XY)$.

3.2) Every day a car dealer sells X cars of type 1 and Y cars of type 2. The following table shows the joint distribution $\{r_{ij} = P(X = i, Y = j); i, j = 0, 1, 3\}$ of (X, Y).

	Y	0	1	2
X	0	0.1	0.1	0
	1	0.1	0.3	0.1
	2	0	0.2	0.1

(1) Determine the marginal distributions of (X, Y).

(2) Are X and Y independent?

(3) Determine the conditional mean values $E(X|Y = i)$, $i = 0, 1, 2$.

3.3) Let B be the upper half of the circle $x^2 + y^2 = 1$. The random vector (X, Y) is uniformly distributed over B.

(1) Determine the joint density of (X, Y).

(2) Determine the marginal distribution densities.

(3) Are X and Y independent? Is theorem 3.1 applicable to answer this question?

3.4) Let the random vector (X, Y) have a uniform distribution over a circle with radius $r = 2$.

Determine the distribution function of the point (X, Y) from the center of this circle.

3.5) Tessa and Vanessa have agreed to meet at a café between 16 and 17 o'clock. The arrival times of Tessa and Vanessa are X and Y, respectively. The random vector (X, Y) is assumed to have a uniform distribution over the square

$$B = \{(x, y); \ 16 \le x \le 17, \ 16 \le y \le 17\}.$$

Who comes first will wait for 40 minutes and then leave.

What is the probability that Tessa and Vanessa will miss each other?

3.6) Determine the mean length of a chord, which is randomly chosen in a circle with radius r. Consider separately the following ways how to randomly choose a chord:

(1) For symmetry reasons, the direction of the chord can be fixed in advance. Draw the diameter of the circle, which is perpendicular to this direction. The midpoints of the chords are uniformly distributed over the whole length of the diameter.

(2) For symmetry reasons, one end point of the chord can be fixed at the periphery of the circle. The direction of a chord is uniformly distributed over the interval in $[0, \pi]$.

(3) How do you explain the different results obtained under (1) an (2)?

3.7) Matching bolts and nuts have the diameters X and Y, respectively. The random vector (X, Y) has a uniform distribution in a circle with radius $1mm$ and midpoint $(30mm, 30mm)$. Determine the probabilities

(1) $P(Y > X)$, and (2) $P(Y \le X < 29)$.

3.8) The random vector (X, Y) is defined as follows: X is uniformly distributed in the interval $[0, 10]$. On condition $X = x$, the random variable Y is uniformly distributed in the interval $[0, x]$. Determine

(1) $f_{X,Y}(x, y)$, $f_X(x|y)$, and $f_Y(y|x)$,

(2) $E(Y)$, $E(Y|X = 5)$, (3) $P(5 \le Y < 10)$.

3.9) Let
$$f_{X,Y}(x,y) = cx^2 y, \ 0 \le x, y \le 1,$$
be the joint probability density of the random vector (X, Y).
(1) Determine the constant c and the marginal densities.
(2) Are X and Y independent?

3.10) The random vector (X, Y) has the joint probability density
$$f_{X,Y}(x,y) = \tfrac{1}{2} e^{-x}, \ 0 \le x, \ 0 \le y \le 2.$$
(1) Determine the marginal densities and the mean values $E(X)$ and $E(Y)$.
(2) Determine the conditional densities $f_X(x|y)$ and $f_Y(y|x)$. Are X and Y independent?

3.11) Let
$$f(x,y) = \tfrac{1}{2} \sin(x+y), \ 0 \le x, y \le \tfrac{\pi}{2},$$
be the joint probability density of the random vector (X, Y).
(1) Determine the marginal densities.
(2) Are X and Y independent?
(3) Determine the conditional mean value $E(Y|X=x)$.
(4) Compare the numerical values $E(Y|X=0)$ and $E(Y|X=\pi/2)$ to $E(Y)$. Are the results in line with your anwer to (2)?

3.12) The temperatures X and Y, measured daily at the same time at two different locations, have the joint density
$$f_{X,Y}(x,y) = \frac{xy}{3} \exp\left[-\frac{1}{2}\left(x^2 + \frac{y^3}{3}\right) \right], \ 0 \le x, y \le \infty.$$
Determine the probabilities
$$P(X > Y) \text{ and } P(X < Y \le 3X).$$

3.13) A large population of rats had been fed with individually varying mixtures of wholegrain wheat and puffed wheat to see whether the composition of the food has any influence on the lifetimes of the rats. Let Y be the lifetime of a rat and X the corresponding ratio of wholegrain it had in its food. An evaluation of (real life) data justifies the assumption that the random vector (X, Y) has a bivariate normal distribution with parameters (in *months*)
$$\mu_x = 0.50, \ \sigma_x^2 = 0.028, \ \mu_y = 6.0, \ \sigma_y^2 = 3.61, \text{ and } \rho = 0.92.$$
With these parameters, X and Y are unlikely to assume negative values.

(1) Determine the regression function $m_Y(x)$, $0 \le x \le 1$, and the corresponding residual variance.

(2) Determine the probability $P(Y \ge 8, X \le 0.6)$.

You may use software you are familiar with to numerically calculate this probability. Otherwise, only produce the double integral.)

3.14) In a forest stand, the stem diameter X (measured $1.3\,m$ above ground) and the corresponding tree height Y have a bivariate normal distribution with joint density

$$f_{X,Y}(x,y) = \frac{1}{0.48\pi}\exp\left\{-\frac{25}{18}\left(\frac{(x-0.3)^2}{\sigma_x^2} - 2\rho\frac{(x-0.3)(y-30)}{0.4} + \frac{(y-30)^2}{25}\right)\right\}.$$

Remark With this joint density, negative values of X and Y are extremely unlikely.

Determine

(1) the correlation coefficient $\rho = \rho(X, Y)$, and

(2) the regression line $\tilde{y} = \alpha x + \beta$.

3.15) The prices per unit X and Y of two related stocks have a bivariate normal distribution with parameters

$$\mu_X = 24, \ \sigma_X^2 = 49, \ \mu_Y = 36, \ \sigma_Y^2 = 144, \ \text{and} \ \rho = 0.8.$$

(1) Determine the probabilities

$$P(|Y-X| \le 10) \ \text{and} \ P(|Y-X| > 15).$$

You may make use of software you are familiar with to numerically calculate these probabilities. Otherwise only produce the respective double integrals.

(2) Determine the regression function $m_Y(x)$ and corresponding residual variance.

3.16) (X, Y) has the joint distribution function $F_{X,Y}(x,y)$. Show that

$$P(a < X \le b, c < Y \le d) = [F_{X,Y}(b,d) - F_{X,Y}(b,c)] - [F_{X,Y}(a,d) - F_{X,Y}(a,c)]$$

for $a < b$ and $c < d$. (This is formula (3.7), page 121.) For illustration, see the Figure:

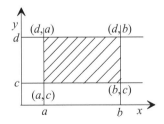

The area integral of the joint probability density $f_{X,Y}(x,y)$ over the hatched area gives the desired probability.

3.17) Let a function of two variables x and y be given by

$$F(x,y) = \begin{cases} 0 & \text{for } x+y \le 0, \\ 1 & \text{for } x+y > 0. \end{cases}$$

Show that $F(x,y)$ does not fulfill the condition

$$[F(b,d) - F(b,c)] - [F(a,d) - F(a,c)] \ge 0$$

for all $a, b, c,$ and d with $a < b$ and $c < d$. Hence, although $F(x,y)$ is continuous on the left in x and y and nondecreasing in x and y, it cannot be the joint distribution function of a random vector (X,Y).

3.18) The vector (X, Y) has the joint distribution function $F_{X,Y}(x,y)$. Show that

$$P(X > x, Y > y) = 1 - F_Y(y) - F_X(x) + F_{X,Y}(x,y).$$

3.19) The random vector (X, Y) has the joint distribution function (*Marshall-Olkin distribution*, page 132) with parameters $\lambda_1 > 0$, $\lambda_2 > 0$, and $\lambda \ge 0$

$$F_{X,Y}(x,y) = 1 - e^{-\lambda_1 x} - e^{-\lambda_2 y} - e^{-\lambda_1 x - \lambda_2 y - \lambda \max(x,y)}.$$

Show that the correlation coefficient between X and Y is given by

$$\rho(X, Y) == \frac{\lambda}{\lambda_1 + \lambda_2 + \lambda}.$$

3.20) At time $t = 0$, a parallel system S consisting of two elements e_1 and e_2 starts operating. Their lifetimes X_1 and X_2 are dependent with joint survival function

$$\overline{F}(x_1, x_2) = P(X_1 > x_1, X_2 > x_2) = \frac{1}{e^{+0.1x_1} + e^{+0.2x_2} - 1}, \quad x_1, x_2 \ge 0.$$

(1) What are the distribution functions of X_1 and X_2?

(2) What is the probability that the system survives the interval $[0, 10]$?

Note By definition, a parallel system is fully operating at a time point t if at least one of its elements is still operating at time t, i.e., a parallel system fails at that time point when the last of its operating elements fails. See also example 4.16, page 176.

3.21) Prove the *conditional variance formula*

$$Var(X) = E[Var(X|Y)] + Var[E(X|Y)].$$

Hint Make use of formulas (2.62) and (3.21).

3.22) The random edge length X of a cube has a uniform distribution in the interval $[4.8, 5.2]$. Determine the correlation coefficient $\rho = \rho(X, Y)$, where $Y = X^3$ is the volume of the cube.

3.23) The edge length X of a equilateral triangle is uniformly distributed in the interval $[9.9, 10.1]$. Determine the correlation coefficient between X and the area Y of the triangle.

3.24) The random vector (X, Y) has the joint density
$$f_{X,Y}(x,y) = 8xy, \;\; 0 < y \le x \le 1.$$

Determine

(1) the correlation coefficient $\rho(X, Y)$,

(2) the regression line $\tilde{y} = \alpha x + \beta$ of Y with regard to X,

(3) the regression function $y = m_Y(x)$.

3.25) The random variables U and V are uncorrelated and have mean value 0. Their variances are 4 and 9, respectively.

Determine the correlation coefficient $\rho(X, Y)$ between the random variables
$$X = 2U + 3V \text{ and } Y = U - 2V.$$

3.26) The random variable Z is uniformly distributed in the interval $[0, 2\pi]$.

Check whether the random variables $X = \sin Z$ and $Y = \cos Z$ are uncorrelated.

CHAPTER 4

Functions of Random Variables

4.1 FUNCTIONS OF ONE RANDOM VARIABLE

4.1.1 Probability Distribution

Functions of a random variable have already played important roles in the previous two chapters. For instance, the nth moment of a random variable X is the mean value of the random variable $Y = X^n$, the variance of X is the mean value of the random variable $Y = (X - E(X))^2$, a standard random variable S is defined by

$$S = \frac{X - E(X)}{\sqrt{Var(X)}},$$

and the Laplace transform of the density of X is defined as the mean value of the random variable $Y = e^{-sX}$. In each case, a function $y = h(x)$ is given, which assigns a value y to each realization x of X. Since it is random, which value X assumes, it is also random which value $h(x)$ takes on. In this way, a new random variable is generated, which is denoted as $Y = h(X)$. Hence, the focus is not in the first place on the values assumed by X, but on the values assumed by Y. The situation is quite analogous to the one which occurred when making the transition from the outcomes ω, $\omega \in \Omega$, of the underlying random experiment to the corresponding values of a random variable $X = X(\omega)$ (section 2.1). Theoretically, one could straightly assign to every elementary event ω the value $y = h(X(\omega))$ instead of making a detour via X, as the probability distribution of Y is fully determined by the one of X:

$$P(Y \in A) = P(X \in h^{-1}(A)),$$

where h^{-1} is the inverse function of h. A motivation for making this detour is given by an example: The area of a circle with diameter D has to be determined. In view of a random measurement error Δ, the true diameter D is not known so that one has to work with an estimate for D, namely with the random variable $X = D + \Delta$. This gives instead of the true area of the circle $\mathbf{A} = h(D) = \frac{\pi}{4} D^2$ only a random estimate of $\mathbf{A}$:

$$Y = h(X) = \frac{\pi}{4} X^2.$$

The aim is to obtain from the probability distribution of X, assumed to be known, the desired probability distribution of Y. Another situation: A random signal X is emitted by a source (the *useful signal*), which arrives at the receiver as $Y = \sin X$. The receiver knows that this coding takes place, and he has information on the probability distribution of Y. Based on this knowledge, the receiver needs to extract information on the probability distribution of the useful signal.

a) Strictly increasing $h(x)$ Let X be a continuous random variable with distribution function $F_X(x) = P(X \le x)$ and with range

$$R_X = [a, b], \quad -\infty \le a < b \le +\infty.$$

$h(x)$ is assumed to be a differentiable and strictly increasing function on R_X. Hence, to every x_0 there exists exactly one y_0 so that $y_0 = h(x_0)$ and vice versa. This implies the existence of the inverse function $h(x)$, which will be denoted as

$$x = x(y) = h^{-1}(y).$$

Its defining property is $h^{-1}(h(x)) = x$ for all $x \in R_X$. The domain of definition of h^{-1} is given by

$$R_Y = \{y, \; y = h(x), \; x \in R_X\}.$$

R_Y is also the range of the random variable $Y = h(X)$.

To derive the distribution function of Y note that the random event "$h(X) \le y_0$" occurs if and only if the random event "$X \le h^{-1}(y_0) = x_0$" occurs. Therefore, for all $y \in R_Y$, the distribution function of Y can be obtained from F_X:

$$F_Y(y) = P(Y \le y) = P(h(X) \le y) = P(X \le h^{-1}(y)) = F_X(h^{-1}(y)), \quad y \in R_Y.$$

Using the chain rule, differentiation of $F_Y(y)$ with regard to y yields the probability density $f_Y(y)$ of Y:

$$f_Y(y) = \frac{dF_Y(y)}{dy} = f_X(h^{-1}(y)) \cdot \frac{dh^{-1}(y)}{dy} = f_X(x(y)) \cdot \frac{dx}{dy}.$$

b) Strictly decreasing $h(x)$ Under otherwise the same assumptions and notations as under a), let $h(x)$ be a strictly decreasing function in R_X. In this case, the random event "$h(X) \le y_0$" occurs if and only if the random event "$X > h^{-1}(y_0) = x_0$" occurs. Hence, for all $y \in R_Y$,

$$F_Y(y) = P(Y \le y) = P(h(X) \le y) = P(X > h^{-1}(y)) = 1 - F_X(h^{-1}(y)), \quad y \in R_Y.$$

Differentiation of $F_Y(y)$ with regard to y yields the corresponding density:

$$f_Y(y) = \frac{dF_Y(y)}{dy} = -f_X(h^{-1}(y)) \cdot \frac{dh^{-1}(y)}{dy} = -f_X(x(y)) \cdot \frac{dx}{dy} = f_X(x(y)) \cdot \left(-\frac{dx}{dy}\right).$$

Summarizing If $y = h(x)$ is strictly increasing, the distribution function of $Y = h(X)$ is

$$F_Y(y) = F_X(h^{-1}(y)), \quad y \in R_Y. \tag{4.1}$$

If $y = h(x)$ strictly decreasing, then

$$F_Y(y) = 1 - F_X(h^{-1}(y)), \quad y \in R_Y. \tag{4.2}$$

In both cases, the probability density of $Y = h(X)$ is

$$f_Y(y) = f_X(h^{-1}(y)) \cdot \left|\frac{dh^{-1}(y)}{dy}\right| = f_X(x(y)) \cdot \left|\frac{dx}{dy}\right|. \tag{4.3}$$

In the important special case of a linear transformation $h(x) = ax + b$, the inverse function of $h(x)$ is $h^{-1}(y) = (y - b)/a$ so that the results (4.1) to (4.3) specialize to

$$F_Y(y) = F_X\left(\tfrac{y-b}{a}\right) \qquad \text{for } a > 0,$$

$$F_Y(y) = 1 - F_X\left(\tfrac{y-b}{a}\right) \quad \text{for } a < 0, \tag{4.4}$$

$$f_Y(y) = \left|\tfrac{1}{a}\right| f_X\left(\tfrac{y-b}{a}\right) \quad \text{for } a \neq 0.$$

As pointed out before, in this case

$$E(Y) = c\,E(X) + d, \quad Var(Y) = a^2 Var(X). \tag{4.5}$$

Example 4.1 The distribution density of the random variable X is

$$f_X(x) = 1/x^2, \ \ x \geq 1.$$

Integration yields the distribution function of the *shifted Lomax distribution*

$$F_X(x) = \tfrac{x-1}{x}, \ \ x \geq 1.$$

Distribution function and density of the random variable $Y = e^{-X}$ has to be determined. The function $h(x) = e^{-x}$ transforms the range $R_X = [1, \infty)$ of X to the range

$$R_Y = (0, 1/e]$$

of $Y = e^{-X}$. Since $h(x)$ is strictly decreasing and $x(y) = h^{-1}(y) = -\ln y$, equations (4.2) and (4.3) yield

$$F_Y(y) = -\frac{1}{\ln y} \ \text{ and } \ f_Y(y) = \frac{1}{y\,(\ln y)^2}, \ \ 0 < y \leq \tfrac{1}{e}. \qquad \square$$

Example 4.2 X has an exponential distribution with parameter $\lambda = 1$:

$$f_X(x) = e^{-x}, \ \ x \geq 0.$$

The density of $Y = 3 - X^3$ has to be determined. Since $y = h(x) = 3 - x^3$, the range of $Y = h(X)$ is $R_Y = (-\infty, 3)$. Moreover,

$$x(y) = h^{-1}(y) = (3 - y)^{1/3} \ \text{ and } \ \frac{dx}{dy} = -\tfrac{1}{3}(3 - y)^{-2/3}, \ \ y \in R_Y.$$

With these relations, equation (4.3) yields

$$f_Y(y) = \frac{e^{-(3-y)^{1/3}}}{3\,(3 - y)^{2/3}}, \ \ -\infty < y < 3. \qquad \square$$

Example 4.3 A body with mass m moves along a straight line with a random velocity X, which is uniformly distributed in the interval $[0.8, 1.2]$. What is the probability density of the body's kinetic energy $Y = \tfrac{1}{2}mX^2$, and what is its mean kinetic energy? X has density

$$f_X(x) = \tfrac{1}{0.4} = 2.5, \ \ 0.8 \leq x \leq 1.2.$$

By the transformation $h(x) = \frac{1}{2}mx^2$, the range $R_X = [0.8, 1.2]$ of X is transformed to the range $R_Y = [0.32\,m, 0.72\,m]$ of Y. Since

$$x(y) = h^{-1}(y) = \sqrt{\frac{2y}{m}}, \quad \frac{dx}{dy} = \sqrt{\frac{1}{2my}}, \quad y \in R_Y,$$

and $f_X(x)$ is constant in R_X, equation (4.3) yields

$$f_Y(y) = \frac{2.5}{\sqrt{2m}} \cdot \frac{1}{\sqrt{y}}, \quad 0.32\,m \le y \le 0.72\,m.$$

The mean kinetic energy of the body with mass m is

$$E(Y) = \int_{R_Y} y f_Y(y)\,dy = \frac{2.5}{\sqrt{2m}} \int_{0.32\,m}^{0.72\,m} y \cdot \frac{1}{\sqrt{y}}\,dy$$

$$= \frac{2.5}{\sqrt{2m}} \left[\frac{2}{3} y^{3/2} \right]_{0.32\,m}^{0.72\,m} = \frac{5}{3\sqrt{2m}} \left[(0.72\,m)^{3/2} - (0.32\,m)^{3/2} \right]$$

so that $E(Y) = 0.50\underline{6}\,m$. $\qquad\qquad\qquad\qquad\qquad\qquad\qquad\qquad$ $\square$

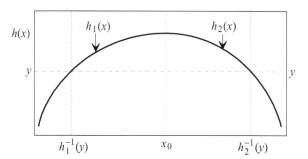

Figure 4.1 Nonmonotone $h(x)$

Nonmonotone $h(x)$ Equations analogous to (4.1) to (4.2) can also be established for nonmonotone functions $h(x)$.

As a special case, let us assume that $y = h(x)$ assumes an absolute maximum at $x = x_0$ (Figure 4.1). More exactly, let

$$h(x) = \begin{cases} h_1(x) & \text{for } x \le x_0, \\ h_2(x) & \text{for } x > x_0, \end{cases}$$

where $h_1(x)$ and $h_2(x)$ are strictly increasing and strictly decreasing, respectively, in their respective domains of definition. Then the random event "$Y \le y$" with $Y = h(X)$ can be written in the following form:

$$\text{"}Y \le y\text{"} = \text{"}h_1(x) \le y\text{"} \cup \text{"}h_2(x) \le y\text{"}$$

(Figure 4.1). Hence,

$$F_Y(y) = P(h(X) \le y) = P(h_1(X) \le y) + P(h_2(X) \le y)$$
$$= P(X \le h_1^{-1}(y)) + P(X > h_2^{-1}(y)).$$

Thus, $F_Y(y)$ can be represented as

$$F_Y(y) = F_X(h_1^{-1}(y)) + 1 - F_X(h_2^{-1}(y)), \quad y \in R_Y. \tag{4.6}$$

Differentiating $F_Y(y)$ and letting $x_1 = h_1^{-1}(y)$ and $x_2 = h_2^{-1}(y)$ yields the probability density of Y:

$$f_Y(y) = f_X(x_1(y)) \left| \frac{dx_1}{dy} \right| + f_X(x_2(y)) \left| \frac{dx_2}{dy} \right|, \quad y \in R_Y. \tag{4.7}$$

This representation of $f_Y(y)$ is also valid if $h(x)$ assumes at $x = x_0$ an absolute minimum.

Example 4.4 A lawn sprinkler moves the direction of its nozzle from horizontal to perpendicular, i.e., within the angular area from 0 to $\pi/2$, with constant angular velocity. Possible rotation movements of the nozzle do not play any role in what follows. It has to be checked, whether in this way the lawn, assumed to be a horizontal plane, is evenly irrigated, i.e., every part of the lawn receives on average the same amount of water per unit time.

(x, z)-coordinates are introduced in that plane, in which the trajectory of a water drop is embedded. The nozzle is supposed to be in the origin $(0,0)$ of this plane. It is known from physics that a drop of water, which leaves the nozzle at time 0 with velocity s and angle α to the lawn, is at time t at location (air resistance being negelected)

$$x = st\cos\alpha, \quad z = st\sin\alpha - \tfrac{1}{2}gt^2,$$

where t is such that $z \ge 0$, and g denotes the *gravitational constant*:

$$g = 6.6726 \cdot 10^{-11} m^3 kg^{-1} s^{-2}.$$

As soon as z becomes 0, the drop of water lands. This happens at time

$$t_L = 2\tfrac{s}{g}\sin\alpha.$$

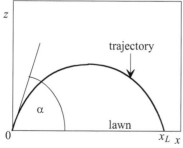 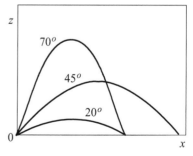

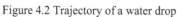

Figure 4.2 Trajectory of a water drop   Figure 4.3 Trajectories of several drops

The corresponding x-coordinate is (Figure 4.2)

$$x_L = a \sin 2\alpha \quad \text{with} \quad a = s^2/g,$$

since $\sin 2\alpha = 2 \sin \alpha \cos \alpha$. From this results the well-known fact that under the assumptions stated, a drop of water, just as any other particle, flies farthest if the start angle is 45^0 (Figure 4.3). Since the nozzle moves with constant angular velocity, the start angle of a drop of water leaving the nozzle at a random time point is a random variable $\hat{\alpha}$ with density

$$f_{\hat{\alpha}}(\alpha) = \frac{2}{\pi}, \quad 0 \le \alpha \le \frac{\pi}{2}, \tag{4.8}$$

i.e. $\hat{\alpha}$ is uniformly distributed in the interval $[0, \pi/2]$. The lawn, under the irrigation policy adopted, will be evenly irrigated if and only if the random landing point

$$X = a \sin 2\hat{\alpha}$$

with range $R_X = [0, a]$ has a uniform distribution in the interval $[0, \pi/2]$ as well. This seems to be unlikely, and the probabilistic analysis will confirm this suspicion.

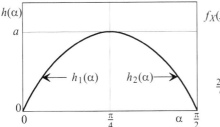
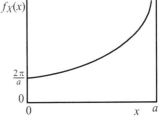

Figure 4.4 Graph of $h(\alpha) = a \sin 2\alpha$ Figure 4.5 'Irrigation density'

The function $x = h(\alpha) = a \sin 2\alpha$, $0 \le \alpha \le \pi/2$, assumes its absolute maximum a at the location $\alpha = \pi/4$ (Figure 4.4). The function $x = h(\alpha) = h_1(\alpha)$ is strictly increasing in $[0, \frac{\pi}{4}]$, and $x = h(\alpha) = h_2(\alpha)$ is strictly decreasing in the interval $[\frac{\pi}{4}, \frac{\pi}{2}]$. In view of this, for all $0 \le x \le a$,

$$\alpha_1 = h_1^{-1}(x) = \frac{1}{2} \arcsin \frac{x}{a},$$

$$\alpha_2 = h_2^{-1}(x) = \frac{\pi}{2} - \frac{1}{2} \arcsin \frac{x}{a}.$$

Differentiation with regard to x yields

$$\left| \frac{d\alpha_1}{dx} \right| = \left| \frac{d\alpha_2}{dx} \right| = \frac{1}{2a\sqrt{1 - (x/a)^2}}.$$

Now (4.7) and (4.8) yield

$$f_X(x) = \frac{2}{\pi} \frac{1}{2a\sqrt{1 - (x/a)^2}} + \frac{2}{\pi} \frac{1}{2a\sqrt{1 - (x/a)^2}}$$

so that the final result is

$$f_X(x) = \frac{2}{\pi a \sqrt{1 - (x/a)^2}}, \quad 0 \le x \le a.$$

This density tends to ∞ if $x \to a$ (Figure 4.5). Therefore, the outer area to be irrigated will get more water than the area next to the nozzle. A 'fair' irrigation can only be achieved with varying angular speed of the nozzle. (Note that in order to be in line with the adequate (x, z)-system of coordinates used in this example, the roles of the variables x and y in formulas (4.3) and (4.7) have been taken over by α and x, respectively.)   □

The derivation of the density $f_Y(y)$ for $Y = h(X)$ (formulas (4.3) and (4.7)) was done in two basic steps:

1) The distribution function $F_Y(y)$ is expressed in terms of F_X.

2) The distribution function $F_Y(y)$ is differentiated.

For nonmonotonic functions $y = h(x)$ it is frequently more convenient, instead of meticulously following (4.7), to do these two steps individually, tailored to the respective problem. This will be illustrated by the following example.

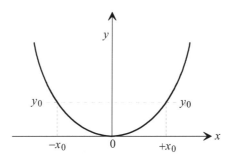

Figure 4.6 Parabola $y = x^2$

Example 4.5 X has both distribution function and density $F_X(x)$ and $f_X(x)$ in the range $R_X = (-\infty, +\infty)$. The density of $Y = X^2$ is to be determined.

The parabola $y = x^2$ assumes its absolute minimum at $x = 0$ so that it is clearly not a monotonic function. The random event '$Y \le y_0$' happens if and only if (Figure 4.6)

$$- \sqrt{y_0} \le X \le + \sqrt{y_0} \, .$$

Hence, $F_Y(y) = P(-\sqrt{y} \le X \le + \sqrt{y})$ so that, by equation (2.5), page 42,

$$F_Y(y) = F_X(\sqrt{y}) - F_X(-\sqrt{y}).$$

Differentiation yields

$$f_Y(y) = \frac{1}{2\sqrt{y}} \left[f_X(\sqrt{y}) - f_X(-\sqrt{y}) \right], \quad 0 \le y < \infty.$$

In particular, for an $N(0, 1)$-distributed random variable X, the density of $Y = X^2$ is

$$f_Y(y) = \frac{1}{2\sqrt{y}}\left[\frac{1}{\sqrt{2\pi}}e^{-y/2} + \frac{1}{\sqrt{2\pi}}e^{-y/2}\right] = \frac{1}{\sqrt{2\pi y}}e^{-y/2}, \quad 0 < y < \infty.$$

This is the density of a χ^2-*distribution* (*chi-square distribution*) *with one degree of freedom*. $\square$

Note A random variable X has a *chi-square distribution* with *n degrees of freedom* (or, equivalently, with *parameter n*) if it has density

$$f_X(x) = \frac{1}{2^{n/2}\Gamma(n/2)}x^{\frac{n}{2}-1}e^{-x/2}, \quad 0 < x < \infty, \quad n = 1, 2, ..., \tag{4.9}$$

where the Gamma function $\Gamma(\cdot)$ is defined by formula (2.75), page 75.

Mean Value of Y According to formula (2.51), the mean value $E(Y)$ of a random vari- able Y with density $f_Y(y)$ is

$$E(Y) = \int_{R_Y} y f_Y(y)\, dy.$$

If Y has structure $Y = h(X)$ with a strictly monotone function $y = h(x)$, then, by (4.3),

$$E(Y) = \int_{R_Y} y f_X(h^{-1}(y))\left|\frac{dx}{dy}\right| dy.$$

Substituting $y = h(x)$ and $x = h^{-1}(y)$, respectively, yields

$$E(Y) = \int_{R_x} h(x) f_X(x)\, dx. \tag{4.10}$$

Hence, knowledge of $f_Y(y)$ is not necessary for obtaining $E(Y)$. We already made use of this in chapters 2 and 3 when determining moments, variance, and other parameters.

Continuation of Example 4.3 The mean kinetic energy $E(Y)$ of the body has to be calculated by formula (4.10). Since the density of X is

$$f_X(x) = \frac{1}{0.4} = 2.5, \quad 0.8 \le x \le 1.2,$$

the mean kinetic energy is

$$E(Y) = E\left(\tfrac{1}{2}mX^2\right) = \tfrac{1}{2}mE(X^2) = \tfrac{1}{2}m\int_{0.8}^{1.2}x^2\, 2.5\, dx$$

$$= 1.25\, m\left[\frac{x^3}{3}\right]_{0.8}^{1.2} = \frac{1.25}{3}m\left[1.2^3 - 0.8^3\right] = 0.50\underline{6}\, m. \quad \square$$

Continuation of Example 4.4 The mean x-coordinate of the random landing point $X = a\sin 2\hat{\alpha}$ of a drop of water will be calculated by formula (4.10): Since the density of $\hat{\alpha}$ is given by (4.8),

$$E(X) = a\int_0^{\pi/2}(\sin 2\alpha)\tfrac{2}{\pi}\, d\alpha = \frac{2a}{\pi}\left[-\tfrac{1}{2}\cos 2\alpha\right]_0^{\pi/2} = 2a/\pi \approx 0.6366. \quad \square$$

4.1.2 Random Numbers

Computers, even scientific calculators, are equipped with software for the generation of *random numbers*, i.e., a computer can randomly pick numbers from the interval $[0,1]$. More exactly, a computer can *generate* or *simulate* arbitrarily frequently and independently of each other realizations of a random variable X, which has a uniform distribution in the interval $[0,1]$. The result of n successive, independent simulations is a set of numbers

$$\{x_1, x_2, ..., x_n\}, x_i \in [0, 1]. \tag{4.11}$$

This set is called a *sequence of random numbers* or, more precisely, a *sequence of random numbers generated from a* $[0,1]$-*uniform distribution*. In applications, however, one will only in rare cases directly need random numbers simulated from a uniform distribution. Hence the following problem needs to be solved:

Problem Let X have a uniform distribution in the interval $[0,1]$. Does there exist a function $y = h(x)$, $0 \le x \le 1$, with property that the random variable $Y = h(X)$ has a desired distribution function $F(y)$?

By asuumption, the distribution function of X is

$$F_X(x) = \begin{cases} 0 & \text{for} & x < 1, \\ x & \text{for} & 0 \le x < 1, \\ 1 & \text{for} & x > 1. \end{cases} \tag{4.12}$$

The function, which solves the problem, is simply $h = F^{-1}$, where F^{-1} is the inverse function of F, i.e. $F^{-1}(F(y)) = y$ for all $y \in R_Y$. This can be seen as follows:

For $Y = F^{-1}(X)$, taking into account (4.12),

$$P(Y \le y) = P(F^{-1}(X) \le y) = P(X \le F(y)) = F_X(F(y)) = F(y).$$

Thus, $Y = F^{-1}(X)$ has indeed the desired distribution function $F_Y(y) = F(y)$. This result is summarized in the following theorem (compare to formula (4.1)):

Theorem 4.1 Let X be a uniformly in $[0, 1]$ distributed random variable with distribution function $F_X(x)$, and $F(y)$ be a strictly monotone, but otherwise arbitrary distribufunction. Then the random variable $Y = F^{-1}(X)$ has distribution function

$$F_Y(y) = F(y).$$

Vice versa, if X is a random variable with distribution function $F_X(x)$, then $Y = F_X(X)$ has a uniform distribution in $[0, 1]$.   ■

Now it is obvious, how to generate from the sequence of random numbers (4.11), simulated from a $[0, 1]$-uniform distribution, a sequence of random numbers, which is simulated from a probability distribution given by $F_Y(y)$:

$$\{y_1, y_2, ..., y_n\} \text{ with } y_i = F^{-1}(x_i), \ i = 1, 2, ..., n. \tag{4.13}$$

The set of numbers (4.13) will be called simply a *sequence of random numbers from a probability distribution given by* $F_Y(y)$. If, for instance, $F_Y(y)$ is the distribution function of a Weibull distributed random variable, then (4.13) is called a *sequence of Weibull distributed random numbers*; analogously, there are *sequences of normally distributed random numbers* and so on.

Of course, these numbers are not random at all, but are realizations of a random variable Y with distribution function $F_Y(y)$. More precisely: The sequence (4.13) of real numbers $y_1, y_2, ..., y_n$ is generated by the outcomes of n independent repetitions of a random experiment with random outcome Y.

In the literature, the terminology 'to simulate a sequence of random numbers from a given distribution' is used equivalently to 'simulate a random variable with a given probability distribution', e.g., to 'simulate an exponenially distributed random variable' or to 'simulate a normally distributed random variable'.

Example 4.6 Based on a random variable X, which has a uniform distribution in the interval $[0, 1]$, a random variable Y is to be generated, which has an exponential distribution with parameter λ :

$$F(y) = P(Y \le y) = 1 - e^{-\lambda y}, \ \ y \ge 0.$$

First, the equation $x = 1 - e^{-\lambda y}$ has to be solved for y:

$$y = F^{-1}(x) = -\frac{1}{\lambda} \ln(1 - x), \ \ 0 \le x < 1.$$

Hence, the random variable

$$Y = F^{-1}(X) = -\frac{1}{\lambda} \ln(1 - X)$$

has an exponential distribution with parameter λ. Thus, if the sequence (4.11) of uniformly in $[0, 1]$-distributed random numbers is given, the corresponding sequence of exponentially with parameter λ distributed random numbers is

$$\{y_1, y_2, ..., y_n\},$$

where $y_i = F^{-1}(x_i) = -\frac{1}{\lambda} \ln(1 - x_i), \ \ i = 1, 2, ..., n.$ $\qquad\qquad$ □

It is not always possible to find an explicit formula for the inverse function F^{-1} of F. For instance, if $F(y)$ is the distribution function of a normal distribution with parameters μ and σ^2, then the equation

$$x = F(y) = \frac{1}{\sqrt{2\pi} \, \sigma} \int_{-\infty}^{y} e^{-\frac{(u - \mu)^2}{2\sigma^2}} \, du$$

cannot explicitly solved for y. However, given the x_i, the numerical calculation of the corresponding y_i, i.e., the numerical calculation of a sequence of normally distributed random numbers, is no problem at all.

Generalization Let Y and Z be two random variables with strictly monotone distribution functions $F_Y(y)$ and $F_Z(z)$, respectively. Is there a function $z = h(y)$ so that

$$Z = h(Y)?$$

This function can be derived by twofold application of theorem 4.1: According to this theorem, the random variable $X = F_Y(Y)$ has a uniform distribution in $[0,1]$. Hence, again by this theorem, the random variable $F_Z^{-1}(X)$ has distribution function F_Y so that the desired function $z = h(y)$ is

$$z = F_Z^{-1}(F_Y(y)).$$

Thus, if $Z = F_Z^{-1}(F_Y(Y))$, then Y has distribution function F_Y, and Z has distribution function $F_Z(z)$.

Example 4.7 Let Y and Z be two random variables with distribution functions

$$F_Y(y) = 1 - e^{-y}, \ y \geq 0, \ \text{and} \ F_Z(z) = \sqrt{z}, \ 0 \leq z \leq 1.$$

For which function $z = h(y)$ is $Z = h(Y)$?

The random variable

$$X = F_Y(Y) = 1 - e^{-Y}$$

with realizations x, $0 \leq x \leq 1$, is uniformly distributed in $[0,1]$. Moreover,

$$F_Z^{-1}(x) = x^2.$$

Hence, the desired function is

$$z = h(y) = (1 - e^{-y})^2 , \ y \geq 0,$$

so that there is the following relationship between Y and Z:

$$Z = \left(1 - e^{-Y}\right)^2. \qquad\qquad \square$$

Discrete Random Variables Sequences of random numbers of type (4.11), simulated from a uniform distribution in $[0, 1]$, can also be used to simulate sequences of random numbers from discrete random variables.

For instance, if Y is a random variable with range $R_Y = \{-3, -1, +1, +3\}$ and probability distribution

$$\{P(Y = -3) = 0.2, \ P(Y = -1) = 0.1, \ P(Y = +1) = 0.4, \ P(Y = +3) = 0.3\},$$

then sequences of random numbers from this probability distribution can be simulated from a random variable X, which has a uniform distribution in $[0,1]$, as follows:

$$Y = \begin{cases} -3 & \text{for } 0.0 \leq X \leq 0.2, \\ -1 & \text{for } 0.2 < X \leq 0.3, \\ +1 & \text{for } 0.3 < X \leq 0.7, \\ +3 & \text{for } 0.7 < X \leq 1.0. \end{cases}$$

This representation of Y is not unique, since the assignment of subintervals of $[0,1]$ to the values of Y only requires that the length of subintervals correspond to the respective probabilities. So, another, equivalent representation of Y would be, e.g.,

$$Y = \begin{cases} -3 & \text{for } 0.8 \leq X \leq 0.2, \\ -1 & \text{for } 0.7 < X \leq 0.8, \\ +1 & \text{for } 0.0 \leq X \leq 0.4, \\ +3 & \text{for } 0.4 < X \leq 0.7. \end{cases}$$

The method of simulating sequences of random numbers from a given distribution based on sequences of uniformly in $[0,1]$-distributed random numbers is, for obvious reasons, called the *inverse transformation method.* There are a couple of other simulation techniques for generating sequences of random numbers, e,.g. the *failure* or *hazard rate method* and the *rejection method.* They do, however, not fit in the framework of section 4.1.

One question still needs to be answered: How are sequences of random numbers from a $[0,1]$-uniform distribution generated?

It can be done manually by repeating a Laplace random experiment (page 12) with outcomes 0,1,...,9 several times. For instance, 10 balls, with respective numbers 0, 1, ...,9 attached to them, are put into a bowl. A ball is randomly selected. Its number i_1 is the first decimal. The ball is returned to the bowl. After shaking it, a second ball is randomly drawn from the bowl; its number i_2 is the second decimal, and so on. When having done this m-times, the number

$$0.i_1 i_2 \cdots i_m$$

has been generated. After having repeated this procedure n times, a sequence of n in $[0,1]$ uniformly distributed random numbers has been simulated. Or, by repeating the Laplace experiment 'flipping a coin' with outcomes '1' (head) or '0' (tail) m times, one obtains a binary number with m digits. Decades ago, researchers would obtain $[0,1]$-uniformly distributed sequences of random numbers from voluminous *tables of random numbers.*

Note In what follows, the attribute '$[0,1]$-uniform(ly)' will be omitted.

But how are nowadays sequences of random numbers generated by a computer? The answer is quite surprising: Usually by deterministic algorithms. From the numerical point of view, these algorithms are most efficient. But they only yield *sequences of pseudo-random numbers.* Extensive statistical tests, however, have established that sequences of pseudo-random numbers, when properly generated, have the same statistical properties as sequences of (genuine) random numbers, i.e., sequences of pseudo-random numbers and sequences random numbers cannot be distinguished from each other.

There are three basic properties, which any sequences of (pseudo-) random numbers $x_1, x_2, ..., x_n$ for sufficiently large n must fulfill:

1) The $x_1, x_2, ..., x_n$ are in $[0,1]$ uniformly distributed in the sense that every subinterval of $[0,1]$ of the same length contains about the same number of x_i.

2) Within the sequence $x_1, x_2, ..., x_n$ no dependencies can be found. In particular, the structure of any subsequence (denoted as ss) of $x_1, x_2, ..., x_n$ does not contain any information on any other subsequence of $x_1, x_2, ..., x_n$, which is disjoint to ss.

3) The sequence $x_1, x_2, ..., x_n$ is not periodic, i.e., there is no positive integer p with property that there exists an element x_p of this sequence with $x_p = x_1$ and after x_p the numbers develop in the same way as from the start, i.e.,

$$x_1, x_2, \cdots, x_p = x_1, \; x_{p+1} = x_2, \; x_{p+2} = x_3, ..., x_{2p} = x_1, \cdots$$

In this case, the sequence $x_1, x_2, ..., x_n$ would consist of identical subsequences of length p (only the last one is likely to be shorter).

Congruence Method This method is probably mostly used by *random number generators* (of computers) to produce sequences of pseudo-random numbers.

Starting with a nonnegative integer z_1 (the *seed*) a sequence of pseudo-random numbers $x_1, x_2, ...$ with

$$x_i = z_i/m, \quad i = 1, 2, ... \tag{4.14}$$

is generated as follows:

$$z_{i+1} = (az_i + b) \bmod m, \quad i = 1, 2, ... \tag{4.15}$$

with integers a, b, and m, which in this order are called *factor*, *increment*, and *module*, $a > 0$, $b \geq 0$, $m > 0$.

Note The relation $z = y \bmod m$ (read: z is equal to y modulo m) between three numbers z, y, and m means that z is the remainder, which is left after the division of y by m.

Each of the figures z_i generated by (4.15) is an element of the set $\{0, 1, ..., m-1\}$. Thus, the sequence $\{z_i\}$ must have a finite period p with $p \leq m$, Therefore, the algorithm has to start with an m as large as possible or necessary, respectively, so that with re- gard to the respective application a sufficiently large sequence of random numbers has been generated before the sequence reaches length p. The specialized literature gives recommendations how to select the parameters a, b, and z_1 to make sure that the generated sequences of pseudo-random numbers have the properties 1 to 3 listed above.

If $b = 0$, then the algorithm is called the *multiplicative congruence method*, and for $b > 0$ it is called the *linear congruence method*.

Example 4.8 Let $a = 21$, $b = 53$, $m = 256$, and $z_1 = 101$. The corresponding recursive equations (4.15) are

$$z_{i+1} = (21 z_i + 53) \bmod 256, \quad i = 1, 2, ... \tag{4.16}$$

The first seven equations are

$$z_2 = (21 \cdot 101 + 53) \bmod 256 = 2174 \bmod 256 = 126,$$
$$z_3 = (21 \cdot 126 + 53) \bmod 256 = 2699 \bmod 256 = 139,$$
$$z_4 = (21 \cdot 139 + 53) \bmod 256 = 2972 \bmod 256 = 156,$$
$$z_5 = (21 \cdot 156 + 53) \bmod 256 = 3329 \bmod 256 = 1,$$
$$z_6 = (21 \cdot 1 + 53) \bmod 256 = 74 \bmod 256 = 74,$$
$$z_7 = (21 \cdot 74 + 53) \bmod 256 = 1607 \bmod 256 = 71,$$
$$z_8 = (21 \cdot 71 + 53) \bmod 256 = 1544 \bmod 256 = 8.$$

The corresponding first eight numbers in the sequence of pseudo-random numbers calculated by $x_i = z_i/256$ are

$$x_1 = 0.39453; \quad x_2 = 0.49219; \quad x_3 = 0.54297; \quad x_4 = 0.60938;$$
$$x_5 = 0.00391; \quad x_6 = 0.28906; \quad x_7 = 0.27734; \quad x_8 = 0.03125.$$

Of course, with a sequence of eight pseudo-random numbers one cannot confirm that the sequence generated by (4.16) and (4.14) satisfies the three basic properties above. This example and the following one can only explain the calculation steps.   □

Mid-Square Method From a $2k$-figure integer z_i one generates the subsequent figure z_{i+1} by identifying it with the middle $2k$ figures of z_i^2. If z_i^2 has less than $2k$ figures, then the missing ones will be replaced with 0 at the front of z_i^2. The figure z_i yields the decimals of the pseudo-random number x_i after the point. The specialized literature gives hints how to select z_1 and k so that the generated sequence of pseudo-random numbers $x_1, x_2, ..., x_n$ fulfills the basic properties 1 to 3 listed above.

Example 4.9 Let $k = 2$ and $z_1 = 4567$. The first 7 numbers of the corresponding sequences $\{z_i\}$ and $\{x_i\}$ are

$$z_1 = 4567 \qquad z_1^2 = 20857489 \qquad x_1 = 0.4567$$
$$z_2 = 8574 \qquad z_2^2 = 73513476 \qquad x_2 = 0.8574$$
$$z_3 = 5134 \qquad z_3^2 = 26357956 \qquad x_3 = 0.5134$$
$$z_4 = 3579 \qquad z_4^2 = 12809241 \qquad x_4 = 0.3579$$
$$z_5 = 8092 \qquad z_5^2 = 65480464 \qquad x_5 = 0.8092$$
$$z_6 = 4804 \qquad z_6^2 = 23078416 \qquad x_6 = 0.4804$$
$$z_7 = 0784 \qquad z_7^2 = 00614656 \qquad x_7 = 0.0784$$

It is obvious that after sufficiently many steps one must return to an x_i already obtained before. This is because the total number of 4-figure integers is 10000. Hence, with regard to this example, the generated sequence $x_1, x_2, ...$ of pseudo-random numbers must have a period p not exceeding 10 000.   □

The generation of random numbers is the basis for computer-aided modelling (simulation) of complex stochastic systems in industry, economy, military, science, humanity, or other areas in order to determine properties or relevant parameters of these systems. Such properties/parameters are, for instance, productivity, stability, availability, safety, efficiency criteria, mean values, variances, state probabilities, ... By computer-aided simulation, systems can be qualitatively and quantitatively evaluated, which in view of their complexity or lack of input data and other information cannot be analyzed by only using analytical methods. Simulation considerably reduces costly and time consuming experiments, which otherwise have to be carried out under real-life conditions. The application of computer-aided simulation is facilitated by special software packages.

4.2 FUNCTIONS OF SEVERAL RANDOM VARIABLES

4.2.1 Introduction

A rectangle with side lengths a and b has the area $\mathbf{A} = ab$. In view of random measurement errors one has only the random side lengths X and Y, which give for $\mathbf{A}$ the random estimate $\hat{\mathbf{A}} = XY$. If this rectangle is the base of a cylinder with random height Z, then a random estimate of its volume is $\mathbf{V}$ is $\hat{\mathbf{V}} = \hat{\mathbf{A}}Z = XYZ$.

If instead of the exact values of voltage V and resistance R in view of random fluctuations only the random values $\hat{V}$ and $\hat{R}$ are given and if the conditions for Ohm's law are fulfilled, then instead of the exact value of the corresponding amperage $I = V/R$, one has only the random estimate $\hat{I} = \hat{V}/\hat{R}$.

Has an investor per year the random profits (losses) from shares, bonds, and funds X, Y, and Z, respectively, then her/his annual total profit (loss) will be $P = X + Y + Z$.

If the signal $\sin Y$ with random Y has been sent and will have its its amplitude $(= 1)$ randomly distorted to X during transmission, then the receiver obtains the message

$$X \sin Y.$$

Consists a system of two subsystems with respective random lifetimes X and Y and fails it as soon as the first subsystem fails, then its lifetime is $\min(X, Y)$. If this system only fails if when both subsystems are down, then its lifetime is $\max(X, Y)$. These are examples for functions of two or more random variables which motivate the subject of the rest of this chapter.

The following sections 4.2.2 to 4.2.6 essentially deal with functions of two random variables $Z = h(X, Y)$. If the generalization to functions of an arbitrary number of random variables $Z = h(X_1, X_2, ..., X_n)$ is straightforward, then the corresponding results will be given. This is usually only then the case when the $X_1, X_2, ..., X_n$ are independent.

4.2.2 Mean Value

The random vector (X, Y) have the joint density $f_{X,Y}(x,y)$ and range $R_{X,Y}$ given by the normal region with regard to the x-axis

$$R_{X,Y} = \{(x,y); \ a \le x \le b, \ y_1(x) \le y \le y_2(x)\}$$

(Figure 3.1, page 123). Let $z = h(x,y)$ be a function on $R_{X,Y}$ and $Z = h(X, Y)$. Then, by formula (3.59), the mean value of Z, provided its existence, is defined as

$$E(Z) = \int_a^b \int_{y_1(x)}^{y_2(x)} h(x,y) f_{X,Y}(x,y) \, dy dx. \tag{4.17}$$

Since outside of $R_{X,Y}$ the joint density is 0, it is not wrong to write this mean value as

$$E(Z) = \int_{-\infty}^{+\infty} \int_{-\infty}^{+\infty} h(x,y) f_{X,Y}(x,y) \, dy dx.$$

For the calculation of $E(Z)$ this formula may not help very much, since in each case the bounds prescribed by $R_{X,Y}$ have to be inserted.

If the random variables X and Y are discrete with respective ranges $R_X = \{x_0, x_1, ...\}$, $R_Y = \{y_0, y_1, ...\}$, and joint distribution

$$\{r_{ij} = P(X = x_i, Y = y_j; \ i, j = 0, 1, ...\},$$

then
$$E(Z) = \Sigma_{i=0}^{\infty} \Sigma_{j=0}^{\infty} h(x_i, y_j) r_{ij}. \tag{4.18}$$

Example 4.10 The random vector (X, Y) has a uniform distribution in the rectangle $R_{X,Y} = \{0 \le x \le \pi, \ 0 \le y \le 1\}$. The mean value of the random variable $Z = X \sin(XY)$ has to be calculated.

Since a rectangle is a normal region, formula (4.17) is directly applicable with $f_{X,Y}(x,y) = 1/\pi$ for all $(x,y) \in R_{X,Y}$ and $h(x,y) = x \sin(xy)$:

$$E(Z) = \int_0^\pi \int_0^1 x \sin(xy) \frac{1}{\pi} \, dy dx = \frac{1}{\pi} \int_0^\pi x \left(\int_0^1 x \sin(xy) \, dy \right) dx$$

$$= \frac{1}{\pi} \int_0^\pi x \left(\left[-\frac{\cos(xy)}{x} \right]_0^1 \right) dx = \frac{1}{\pi} \int_0^\pi x(1 - \cos x) \, dx = \frac{1}{\pi} [x - \sin x]_0^\pi.$$

Hence, $E(Z) = 1$.   □

Example 4.11 A target, which is positioned in the origin $(0,0)$ of the (x,y)-coordinate system is subject to permanent artillery fire. The random x-coordinate X and the random y-coordinate Y of the impact marks of the shells are independent and identical as $N(0, \sigma^2)$-distributed random variables. (The assumption $E(X) = E(Y) = 0$ means that there are no systematic deviations from the target.)

Let Z be the random distance of an impact mark to the target (origin). The aim is to determine the probability distribution of Z and $E(Z)$.

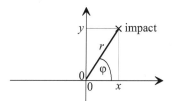

Figure 4.7 Impact mark and polar coordinates

By (2.81) and (3.13), the joint probability density of (X, Y) is

$$f_{X,Y}(x,y) = \frac{1}{\sqrt{2\pi}\,\sigma}\,e^{-\frac{x^2}{2\sigma^2}} \cdot \frac{1}{\sqrt{2\pi}\,\sigma}\,e^{-\frac{y^2}{2\sigma^2}} = \frac{1}{2\pi\sigma^2}\,e^{-\frac{x^2+y^2}{2\sigma^2}}, \quad -\infty < x, y < +\infty.$$

Since the distance of the impact mark to the target is $Z = \sqrt{X^2 + Y^2}$, the distribution function of Z is principally given by

$$F_Z(z) = P(Z \le z) = \iint\limits_{\{(x,y),\ \sqrt{x^2+y^2}\ \le z\}} \frac{1}{2\pi\sigma^2}\,e^{-\frac{x^2+y^2}{2\sigma^2}}\,dx\,dy. \tag{4.19}$$

To facilitate the evaluation of this double integral, a transition is made to polar coordinates (special curvilinear coordinates, page 123) according to Figure 4.7:

$$x = r\cos\varphi, \quad y = r\sin\varphi \text{ or } r = \sqrt{x^2+y^2}, \quad \varphi = \arctan\frac{y}{x}$$

with
$$\frac{\partial x}{\partial r} = \cos\varphi, \ \frac{\partial x}{\partial \varphi} = -r\sin\varphi, \ \frac{\partial y}{\partial r} = \sin\varphi, \ \frac{\partial y}{\partial \varphi} = r\cos\varphi.$$

The corresponding functional determinant is (page 123)

$$\left| \frac{\partial(x,y)}{\partial(r,\varphi)} \right| = \left| \begin{array}{cc} \frac{\partial x}{\partial r} & \frac{\partial y}{\partial r} \\ \frac{\partial x}{\partial \varphi} & \frac{\partial y}{\partial \varphi} \end{array} \right| = \left| \begin{array}{cc} \cos\varphi & \sin\varphi \\ -r\sin\varphi & r\cos\varphi \end{array} \right| = r(\cos\varphi)^2 + r(\sin\varphi)^2 = r.$$

Integrating over the full circle $\{(x,y),\ \sqrt{x^2+y^2} \le z\}$ in (4.19) is, in polar coordinates equivalent to integrating over the area $[0 \le r \le z,\ 0 \le \varphi \le 2\pi]$. By (3.17), page 123, the integral (4.19) reduces to

$$F_Z(z) = \int_0^z \int_0^{2\pi} \frac{1}{2\pi\sigma^2}\,e^{-\frac{r^2}{2\sigma^2}}\,r\,d\varphi\,dr = \frac{1}{\sigma^2}\int_0^z r\,e^{-\frac{r^2}{2\sigma^2}}\,dr = 1 - e^{-\frac{z^2}{2\sigma^2}}, \quad z \ge 0.$$

This is a Weibull-distribution with parameters $\beta = 2$ and $\theta = \sqrt{2}\,\sigma$, i.e., the random variable Z is Rayleigh-distributed. Hence, by formula (2.78), its mean value is

$$E(Z) = \sqrt{2}\,\sigma\,\Gamma(1.5) \approx 1.2533\,\sigma. \qquad \square$$

4.2.3 Product of Two Random Variables

Let (X, Y) be a random vector with joint probability density $f_{X,Y}(x,y)$, and

$$Z = XY.$$

The distribution function of Z is given by

$$F_Z(z) = \iint\limits_{\{(x,y); \, xy \le z\}} f_{X,Y}(x,y)\, dx\, dy$$

with (see Figure 4.8)

$$\{(x,y); \, xy \le z\} = \{-\infty < x \le 0, \tfrac{z}{x} \le y < \infty\} \cup \{0 \le x < \infty, \, -\infty < y \le \tfrac{z}{x}\}.$$

Hence,

$$F_Z(z) = \int_{-\infty}^{0} \int_{z/x}^{+\infty} f_{X,Y}(x,y)\, dy\, dx + \int_{0}^{+\infty} \int_{-\infty}^{z/x} f_{X,Y}(x,y)\, dy\, dx.$$

Differentiation with regard to z yields the probability density of Z:

$$f_Z(z) = \int_{-\infty}^{0} \left(-\tfrac{1}{x}\right) f_{X,Y}(x, \tfrac{z}{x})\, dx + \int_{0}^{\infty} \tfrac{1}{x} f_{X,Y}(x, \tfrac{z}{x})\, dx.$$

This representation can be simplified to

$$f_Z(z) = \int_{-\infty}^{+\infty} \left|\tfrac{1}{x}\right| f_{X,Y}(x, \tfrac{z}{x})\, dx, \quad z \in (-\infty, +\infty). \tag{4.20}$$

For nonnegative X and Y,

$$F_Z(z) = \int_{0}^{+\infty} \int_{0}^{z/x} f_{X,Y}(x,y)\, dy\, dx, \quad z \ge 0,$$

$$f_Z(z) = \int_{0}^{+\infty} \tfrac{1}{x} f_{X,Y}(x, \tfrac{z}{x})\, dx, \qquad z \ge 0. \tag{4.21}$$

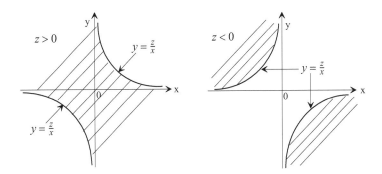

Figure 4.8 Derivation of the distribution function of a product

Example 4.12 The random vector (X, Y) has the joint density

$$f_{X,Y}(x,y) = 6x^2 y, \quad 0 \le x, y \le 1.$$

Since both X and Y are nonnegative, formula (4.21) can be applied to determine the density of $Z = XY$: Since $z/x \leq 1$,

$$f_Z(z) = \int_z^1 \frac{1}{x} (6x^2 \cdot \frac{z}{x}) \, dx = 6z(1-z), \quad 0 \leq z \leq 1.$$

The calculation of the mean value of Z yields

$$E(Z) = \int_0^1 z \, [6z(1-z)] \, dz = 6 \left[\frac{z^3}{3} - \frac{z^4}{4} \right]_0^1 = \frac{1}{2}.$$

The marginal distribution densities of (X, Y) are

$$f_X(x) = 3x^3, \quad 0 \leq x \leq 1, \quad \text{and} \quad f_Y(y) = 2y, \quad 0 \leq y \leq 1.$$

Hence, $f_{X,Y}(x,y) = f_X(x) \cdot f_Y(y)$ so that X and Y are independent. $\qquad \square$

4.2.4 Ratio of Two Random Variables

Let (X, Y) be a random vector with joint probability density $f_{X,Y}(x,y)$, and

$$Z = \frac{Y}{X}.$$

The distribution function of Z is given by

$$F_Z(z) = \iint\limits_{\{(x,y); \frac{y}{x} \leq z\}} f_{X,Y}(x,y) dx \, dy$$

with (Figure 4.9)

$$\left\{ (x,y); \frac{y}{x} \leq z \right\} = \{-\infty < x \leq 0, \, zx \leq y < \infty\} \cup \{0 \leq x < \infty, \, -\infty < y \leq zx\}.$$

Hence

$$F_Z(z) = \int_{-\infty}^0 \int_{zx}^{+\infty} f_{X,Y}(x,y) \, dy \, dx + \int_0^{+\infty} \int_{-\infty}^{zx} f_{X,Y}(x,y) \, dy \, dx.$$

Differentiation with regard to z yields the probability density of Z:

$$f_Z(z) = \int_{-\infty}^{+\infty} |x| \, f_{X,Y}(x, zx) \, dx. \qquad (4.22)$$

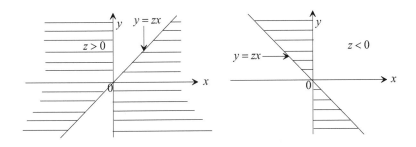

Figure 4.9 Derivation of the distribution function of a ratio

In case of nonnegative X and Y,

$$F_Z(z) = \int_0^{+\infty} \int_0^{zx} f_{X,Y}(x,y)\, dy\, dx, \quad z \geq 0,$$

$$f_Z(z) = \int_0^{+\infty} x f_{X,Y}(x,zx)\, dx, \quad z \geq 0. \tag{4.23}$$

Example 4.13 The random vector (X, Y) has the joint density

$$f_{X,Y}(x,y) = \lambda\mu\, e^{-(\lambda x + \mu y)}, \quad x \geq 0,\, y \geq 0;\ \lambda > 0,\, \mu > 0. \tag{4.24}$$

The structure of this joint density implies that X and Y are independent and have exponential distributions with parameters λ and μ, respectively. Hence, the density of the ratio $Z = Y/X$ is

$$f_Z(z) = \int_0^\infty x \lambda\mu\, e^{-(\lambda + \mu z)x}\, dx, \quad z \geq 0.$$

A slight transformation yields

$$f_Z(z) = \frac{\lambda\mu}{\lambda + \mu z} \int_0^\infty x(\lambda + \mu z) e^{-(\lambda + \mu z)x} dx, \quad z \geq 0.$$

The integral is the mean value of an exponentially distributed random variable with parameter $\lambda + \mu z$. Therefore,

$$f_Z(z) = \frac{\lambda\mu}{(\lambda + \mu z)^2}, \quad z \geq 0, \tag{4.25}$$

$$F_Z(z) = 1 - \frac{\lambda}{\lambda + \mu z}, \quad z \geq 0.$$

This is the *Lomax distribution* (page 93). □

Example 4.14 A system has the random lifetime (= time to failure) X. After a failure it is replaced with a new system. It takes Y time units to replace a failed system. Thus, within a (lifetime-replacement) cycle, the random fraction during which the system is operating, is

$$A = \frac{X}{X+Y}.$$

A is called the *availability* of the system (in a cycle). Determining the distribution function of A can be reduced to determining the distribution function of the ratio $Z = Y/X$ since

$$F_A(t) = P(A \leq t) = P\left(\frac{X}{X+Y} \leq t\right) = 1 - P\left(\frac{Y}{X} < \frac{1-t}{t}\right).$$

Hence,

$$F_A(t) = 1 - F_Z\left(\frac{1-t}{t}\right), \quad 0 < t \leq 1.$$

Differentiation with respect to t yields the probability density of A:

$$f_A(t) = \frac{1}{t^2} f_Z\left(\frac{1-t}{t}\right), \quad 0 < t \leq 1.$$

Specifically, if the joint density of (X, Y) is given by (4.24) then $f_Z(z)$ is given by (4.25) so that we again get a Lomax distribution:

$$f_A(t) = \frac{\lambda\mu}{[(\lambda-\mu)t+\mu]^2}, \quad F_A(t) = \frac{\lambda t}{(\lambda-\mu)t+\mu}, \quad 0 \le t \le 1.$$

For $\lambda \ne \mu$, the mean value of A is (easily obtained by formula (2.52), page 64)

$$E(A) = \frac{\mu}{\mu-\lambda}\left[1 + \frac{\lambda}{\mu-\lambda}\right]\ln\frac{\lambda}{\mu}.$$

In particular, let $\lambda/\mu = 1/4$. Then the probability that the system availability assumes a value between 0.7 and 0.9 is

$$P(0.7 \le A \le 0.9) = F_A(0.9) - F_A(0.7) = \frac{0.9}{4-3\cdot0.9} - \frac{0.7}{4-3\cdot0.7} = 0.324.$$

In view of $E(X) = 1/\lambda$ and $E(Y) = 1/\mu$ the assumption $\lambda/\mu = 1/4$ implies that the mean lifetime of the system is on average four times larger than its mean replacement time. Hence, one would expect that the mean availability of the system is 0.75. But the true value is slightly lower: $E(A) \approx 0.717$.

If $\lambda = \mu$, then A is uniformly distributed over $[0, 1]$. In this case, $E(A) = 1/2$. $\square$

4.2.5 Maximum of Random Variables

Let (X, Y) be a random vector with joint density $f_{X,Y}(x,y)$ and

$$Z = \max(X, Y).$$

The random event '$Z \le z$' occurs if and only if both X and Y assume values which do not exceed z. Hence (Figure 4.10),

$$F_Z(z) = P(Z \le z) = P(X \le z, Y \le z) = \int_{-\infty}^{z}\int_{-\infty}^{z} f_{X,Y}(x,y)\,dxdy.$$

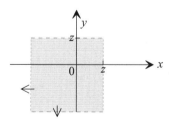

Figure 4.10 Integration region for the maximum

Example 4.15 The random vector (X, Y) has a Marshall-Olkin distribution with joint distribution function given by (3.27): For $\lambda_1 > 0$, $\lambda_2 > 0$, $\lambda > 0$, and $x, y \ge 0$,

$$F_{X,Y}(x,y) = 1 - e^{-(\lambda_1+\lambda)x} - e^{-(\lambda_2+\lambda)y} + e^{-\lambda_1 x - \lambda_2 y - \lambda\max(x,y)}.$$

so that

$$P(Z > z) = 1 - F_Z(z) = 1 - F_{X,Y}(z, z) = e^{-(\lambda_1 + \lambda)z} + e^{-(\lambda_2 + \lambda)z} - e^{-(\lambda_1 + \lambda_2 + \lambda)z}.$$

Hence, by formula (2.52), page 64, the mean value of $Z = \max(X, Y)$ is

$$E(Z) = \frac{1}{\lambda_1 + \lambda} + \frac{1}{\lambda_2 + \lambda} - \frac{1}{\lambda_1 + \lambda_2 + \lambda}. \qquad (4.26)$$

As a practical application, if a system consists of two subsystems with respective life-times X and Y, and the systems fails when both subsystems have failed, then its mean lifetime is given by (4.26). In particular, in case of independent, identically distribut-ed lifetimes X and Y (i.e., $\lambda = 0$, $\lambda_1 = \lambda_2$):

$$E(Z) = \frac{1.5}{\lambda_1}.$$

In this case, a 'spare' system increases the mean system life by the factor 1.5.   □

Now the random variables $X_1, X_2, ..., X_n$ are assumed to be independent with distribu-tion functions $F_{X_i}(z) = P(X_i \le z)$, $i = 1, 2, ..., n$. Let

$$Z = \max\{X_1, X_2, ..., X_n\}. \qquad (4.27)$$

Since the random event "$Z \le z$" occurs if and only if

$$'X_1 \le z, X_2 \le z, ..., X_n \le z',$$

and the events '$X_i \le z$' are independent, the distribution function of Z is

$$F_Z(z) = F_{X_1}(z) \cdot F_{X_2}(z) \cdots F_{X_n}(z). \qquad (4.28)$$

Example 4.16 A system consists of n subsystems $s_1, s_2, ..., s_n$. All of them start oper-ating at time point $t = 0$ and fail independently of each other. The system operates as long as at least one of its subsystems is operating. Thus, $n - 1$ out of the n subsystems are virtually spare systems. Hence, if X_i denotes the lifetime of subsystem s_i, then the lifetime Z of the system is given by (4.27) and has distribution function (4.28). In engineering reliability, systems like that are called *parallel systems*. Its failure behav-ior is illustrated by Figure 4.11. Each of the n edges in the graph with parallel edges depicted there symbolizes a subsystem. The system works if and only if there is at least one 'operating edge', which connects *entrance node* **en** and *exit node* **ex**.

As a special case, let us assume that the lifetimes X_i are identically exponentially dis-tributed with parameter λ :

$$F_{X_i}(x) = 1 - e^{-\lambda x}, \lambda > 0, i = 1, 2, ..., n.$$

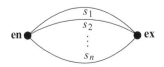

Figure 4.11 Illustration of parallel system

Then the system lifetime has distribution function $F_Z(z) = (1 - e^{-\lambda z})^n$, $z \geq 0$, so that the mean system lifetime is

$$E(Z) = \int_0^\infty \left[1 - (1 - e^{-\lambda z})^n \right] dz.$$

The substitution $x = 1 - e^{-\lambda z}$ yields

$$E(Z) = \frac{1}{\lambda} \int_0^1 \frac{1 - x^n}{1 - x} dx = \frac{1}{\lambda} \int_0^1 \left[1 + x + \cdots + x^{n-1} \right] dx.$$

Hence,
$$E(Z) = \frac{1}{\lambda} \left[1 + \frac{1}{2} + \cdots + \frac{1}{n} \right].$$

Because of the divergence of the harmonic series $\sum_{i=1}^\infty 1/i$, an arbitrary large mean system lifetime can be achieved by installing a sufficient number of subsystems. $\square$

4.2.6 Minimum of Random Variables

Let the random vector (X, Y) have the joint density $f_{X,Y}(x,y)$, and let $Z = \min(X, Y)$ have distribution function $F_Z(z) = P(Z \leq z)$. Then, by integrating over the hatched area in Figure 4.12,

$$F_Z(z) = \iint_{\{(x,y);\ x \leq z,\ y \leq z\}} f_{X,Y}(x,y) \, dxdy = \int_{-\infty}^z \int_{-\infty}^z f_{X,Y}(x,y) \, dxdy.$$

Integrating over the non-hatched area yields

$$\overline{F}_Z(z) = P(Z > z) = P(X > z, Y > z) = \int_z^\infty \int_z^\infty f_{X,Y}(x,y) \, dxdy.$$

For independent X and Y,

$$\overline{F}_Z(z) = \overline{F}_X(z) \cdot \overline{F}_Y(z).$$

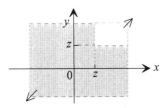

Figure 4.12 Integration region for the minimum

Example 4.17 A system consists of two subsystems with respective lifetimes X and Y. The system fails as soon as the first subsystem fails. Then $Z = \min(X, Y)$ is the mean lifetime of the system. Let, for instance, the random vector (X, Y) have the Gumbel-distribution (3.28) with parameters $\lambda_1 = \lambda_2 = 1$ and parameter λ, $0 \leq \lambda \leq 1$. Then,

$$\overline{F}_Z(z) = P(Z > z) = e^{-2z - \lambda z^2}, \quad z \geq 0,$$

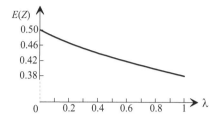

Figure 4.13 Decrease of the mean lifetime for $\lambda \to 1$

and, by formula (2.52), the mean lifetime is

$$E(Z) = \int_0^\infty e^{-(2z+\lambda z^2)}dz.$$

Figure 4.13 shows the graph of the mean lifetime depending on λ. With increasing dependence between X and Y ($\lambda \to 1$), the mean lifetime decreases almost linearly from 0.5 (independence) to about 0.38. (The correlation coefficient between X and Y is given at page 138.)   □

Now let $X_1, X_2, ..., X_n$ be independent random variables and

$$Z = \min\{X_1, X_2, ..., X_n\}.$$

Then, $P(Z > x) = P(X_1 > z, X_2 > z, ..., X_n > z)$ so that

$$\overline{F}_Z(z) = P(Z > z) = \overline{F}_{X_1}(z) \cdot \overline{F}_{X_2}(z) \cdots \overline{F}_{X_n}(z). \tag{4.29}$$

Thus, the distribution function of the minimum of n independent random variables is

$$F_Z(z) = P(Z \le z) = 1 - \overline{F}_{X_1}(z) \cdot \overline{F}_{X_2}(z) \cdots \overline{F}_{X_n}(z). \tag{4.30}$$

Generalizing example 4.17, if a system, consisting of n independently operating subsystems $s_1, s_2, ..., s_n$, starts operating at time $z = 0$ and fails as soon as one of its subsystems fails, then its survival function is given by (4.29). In Figure 4.14, if the chain between entrance node **en** and exit node **ex** of the graph is interrupted by a failed subsystem, then the system as a whole fails. In reliability engineering, systems like this are called *series systems*. If, in particular, the lifetimes of the subsystems are identically exponentially distributed with parameter λ, then $\overline{F}_Z(z) = e^{-n\lambda z}$, $z \ge 0$, and the corresponding mean system lifetime is $E(Z) = 1/\lambda n$. Every installation of another subsystem decreases both the survival probablity and the mean lifetime of a series system. For instance, if one subsystem survives the interval $[0,1]$ with probability $e^{-\lambda} = 0.99$, then 100 of such subsystems in series survive this interval only with probability $0.99^{100} \approx 0.37$. Therefore, in technological designs, combinations of parallel and series systems are preferred.   □

Figure 4.14 Illustration of a series system

4.3 SUMS OF RANDOM VARIABLES

4.3.1 Sums of Discrete Random Variables

Mean Value of a Sum The random vector (X, Y) with discrete components X and Y has the joint distribution

$$\{r_{ij} = P(X = x_i \cap Y = y_j;\ i, j = 0, 1, ...\},$$

and the marginal distributions

$$p_i = P(X = x_i) = \textstyle\sum_{j=0}^{\infty} r_{ij},$$

$$q_j = P(Y = y_j) = \textstyle\sum_{i=0}^{\infty} r_{ij}.$$

Then the mean value of the sum $Z = X + Y$ is

$$E(Z) = \textstyle\sum_{i=0}^{\infty} \sum_{j=0}^{\infty} (x_i + y_j) r_{ij}$$

$$= \textstyle\sum_{i=0}^{\infty} x_i \sum_{j=0}^{\infty} r_{ij} + \sum_{i=0}^{\infty} y_j \sum_{j=0}^{\infty} r_{ij}$$

$$= \textstyle\sum_{i=0}^{\infty} x_i p_i + \sum_{j=0}^{\infty} y_j q_j.$$

Thus,

$$E(X + Y) = E(X) + E(Y). \tag{4.31}$$

By induction, for any discrete random variables $X_1, X_2, ..., X_n$,

$$E(X_1 + X_2 + \cdots + X_n) = E(X_1) + E(X_2) + \cdots + E(X_n). \tag{4.32}$$

Distribution of a Sum Let X and Y be <u>independent</u> random variables with common range $\mathbf{R} = \{0, 1, ...\}$ and probability distributions

$$\{p_i = P(X = i;\ i = 0, 1, ...\} \text{ and } \{q_j = P(Y = j;\ j = 0, 1, ...\}.$$

Then,

$$P(Z = k) = P(X + Y = k) = \textstyle\sum_{i=0}^{k} P(X = i) P(Y = k - i).$$

Letting $r_k = P(Z = k)$ yields for all $k = 0, 1, ...$

$$r_k = p_0 q_k + p_1 q_{k-1} + \cdots + p_k q_0.$$

Thus, according to formula (2.114) at page 98, the discrete probability distribution $\{r_k;\ k = 0, 1, ...\}$ is the *convolution* of the probability distributions of X and Y. The z-transforms of X and Y are defined by (2.110):

$$M_X(z) = \textstyle\sum_{i=0}^{\infty} p_i z^i,$$

$$M_Y(z) = \textstyle\sum_{i=0}^{\infty} q_i z^i.$$

By (2.116),

$$M_Z(z) = M_X(z)\, M_Y(z). \tag{4.33}$$

> *The z-transform $M_Z(z)$ of the sum $Z = X + Y$ of two independent discrete random variables X and Y with common range $\mathbf{R} = \{0, 1, ...\}$ is equal to the product of the z-transforms of X and Y.*

By induction, if $Z = X_1 + X_2 + \cdots + X_n$ with independent X_i, then

$$M_Z(z) = M_{X_1}(z) \, M_{X_2}(z) \cdots M_{X_n}(z). \tag{4.34}$$

Example 4.18 Let $Z = X_1 + X_2 + \cdots + X_n$ be a sum of independent random variables, where X_i has a Poisson distribution with parameter λ_i; $i = 1, 2, ..., n$, i.e.,

$$P(X_i = k) = \frac{\lambda_i^k}{k!} e^{-\lambda_i}, \quad k = 0, 1, ...$$

The z-transform of X_i is (page 91)

$$M_{X_i}(z) = e^{\lambda_i (z-1)}. \tag{4.35}$$

From (4.34),

$$M_Z(z) = e^{(\lambda_1 + \lambda_2 + \cdots + \lambda_n)(z-1)}.$$

The functional structure of $M_Z(z)$ is the same as the one of $M_{X_i}(z)$. Thus, the sum of independent, Poisson distributed random variables has a Poisson distribution, the parameter of which is the sum of the parameters of the Poisson distributions of these random variables. (This way of reasoning is only possible, because, as pointed out in section 2.5, to every probability distribution there belongs exactly one z-transform and vice versa.) □

Example 4.19 Let $Z = X_1 + X_2 + \cdots + X_n$ be a sum of independent random variables, where X_i has a binomial distribution with parameters n_i and p_i, $i = 1, 2, ..., n$, i.e.,

$$P(X_i = k) = \binom{n_i}{k} p_i^k (1 - p_i)^{n_i - k}, \quad k = 0, 1, ..., n_i.$$

Then (page 98), the z-transform of X_i is

$$M_{X_i}(z) = [p_i z + 1 - p_i]^{n_i}.$$

Hence, the z-transform of the sum is

$$M_Z(z) = \prod_{i=1}^{n} [p_i z + 1 - p_i]^{n_i}.$$

Under the additional assumption that

$$p_i = p, \quad i = 1, 2, ..., n,$$

this representation of the z-transform of Z simplifies to

$$M_Z(z) = [pz + 1 - p]^{n_1 + n_2 + \cdots + n_n}.$$

Comparing this $M_Z(z)$ with $M_{X_i}(z)$ shows that in case of $p_i = p$ the sum Z has again a binomial distribution, but with parameters p and $n_1 + n_2 + \cdots + n_n$. □

4.3.2 Sums of Continuous Random Variables

4.3.2.1 Sum of Two Random Variables

Distribution The random vector (X, Y) have the joint density $f_{X,Y}(x,y)$. Based on this information, the distribution function $F_Z(z) = P(Z \le z)$ of the sum $Z = X + Y$ has to be determined.

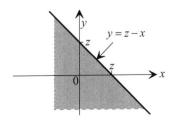

Figure 4.15 Integration region for the sum

Figure 4.15 illustrates the situation: Those realizations (x,y) of (X, Y), which satisfy the condition $x + x \le z$ or $y \le z - x$, respectively, are in the hatched area. If the vector (X, Y) assumes such a realization, then the random event '$X + Y \le z$' occurs. Hence, $F_Z(z)$ is given by the double integral

$$F_Z(z) = \int_{-\infty}^{+\infty} \int_{-\infty}^{z-x} f_{X,Y}(x,y)\,dy dx.$$

Differentiation with regard to z yields the density of Z:

$$f_Z(z) = \frac{d}{dz} \int_{-\infty}^{+\infty} \int_{-\infty}^{z-x} f_{X,Y}(x,y)\,dy dx = \int_{-\infty}^{+\infty} \frac{d}{dz} \int_{-\infty}^{z-x} f_{X,Y}(x,y)\,dy dx$$

so that

$$f_Z(z) = \int_{-\infty}^{+\infty} f_{X,Y}(x, z - x)\,dx. \tag{4.36}$$

If X and Y are nonnegative, then $f_{X,Y}(x,y)$ is 0 for $x < 0$ and/or $y < 0$. In this case, only such x and $z - x$ can contribute to the integral in (4.36), which satisfy $x \ge 0$ and $z - x \ge 0$. Hence,

$$f_Z(z) = \int_0^z f_{X,Y}(x, z - x)\,dx. \tag{4.37}$$

If X and Y are independent, then $f_{X,Y}(x,y) = f_X(x) \cdot f_Y(y)$ so that in this case formulas (4.36) and (4.37) become

$$f_Z(z) = \int_{-\infty}^{+\infty} f_X(x) f_Y(z - x)\,dx, \tag{4.38}$$

$$f_Z(z) = \int_0^z f_X(x) f_Y(z - x)\,dx. \tag{4.39}$$

These integrals are the *convolutions* of f_X and f_Y (formulas (2.125) and (2.126)).

> *The density of the sum of two independent random variables X and Y is the convolution of the densities of X and Y.*

By formula (2.127), the Laplace transform of the density of the sum of two independent random variables X and Y is equal to the product of their Laplace transforms:

$$\hat{f}_Z(s) = \hat{f}_X(s) \cdot \hat{f}_Y(s). \tag{4.40}$$

The distribution function of Z for independent X and Y one simply gets by integrating the density $f_Z(z)$ given by (4.38) and (4.39), respectively. A heuristic approach is the following one: On condition $Y = y$ the distribution function of $Z = X + Y$ is

$$F_Z(Z \le z | Y = y) = P(X + y \le z) = P(X \le z - y) = F_X(z - y).$$

Since $dF_Y(y) = f_Y(y) dy$ is the 'probability' of the event '$Y = y$' (see comment after formula (2.50), page 61),

$$F_Z(z) = \int_{-\infty}^{+\infty} F_X(z - y) f_Y(y) \, dy, \tag{4.41}$$

or

$$F_Z(z) = \int_{-\infty}^{+\infty} F_X(z - y) \, dF_Y(y). \tag{4.42}$$

For nonnegative X and Y the formulas (4.41) and (4.42) become

$$F_Z(z) = \int_0^z F_X(z - y) f_Y(y) \, dy, \tag{4.43}$$

$$F_Z(z) = \int_0^z F_X(z - y) \, dF_Y(y). \tag{4.44}$$

In the terminology used so far, the intergral in (4.41) is the convolution of the functions F_X and f_Y. The integral (4.42), however, is called the *convolution of the <u>distribution functions</u>* F_X and F_Y. Of course, the roles of X and Y can be exchanged in formulas (4.36) to (4.44) since $X + Y = Y + X$.

Example 4.20 It is assumed that the random vector (X, Y) has a uniform distribution over the square $[0 \le x \le T, 0 \le y \le T]$, i.e.

$$f_{X,Y}(x,y) = \begin{cases} 1/T^2, & 0 \le x, y \le T \\ 0, & \text{otherwise} \end{cases}.$$

By theorem 3.1, this assumption implies that X and Y are independent and in the interval $[0, T]$ uniformly distributed random variables. Hence, formula (4.39) is applicable for determining the density of $Z = X + Y$:

$$f_Z(z) = \int_0^z f_{X,Y}(x, z - x) \, dx = \begin{cases} \int_0^z \frac{1}{T^2} dx, & 0 \le z \le T \\ \int_{z-T}^T \frac{1}{T^2} dx, & T < z \le 2T \end{cases}.$$

Therefore,

$$f_Z(z) = \begin{cases} \dfrac{z}{T^2}, & 0 \le z \le T \\ \dfrac{1}{T^2}(2T - z), & T < z \le 2T \end{cases}.$$

Figure 4.16 shows the graph of $f_Z(z)$. It motivates the name *triangular distribution*. But it is also called *Simpson distribution*. The corresponding distribution function is

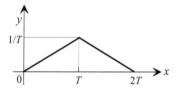

Figure 4.16 Density of the triangular distribution

$$F_Z(z) = \int_0^z f_Z(u)\,du = \begin{cases} \frac{1}{2}\left(\frac{z}{T}\right)^2, & 0 \le z \le T \\ \frac{z}{T}\left(2 - \frac{z}{2T}\right) - 1, & T < z \le 2T \end{cases}.$$

The symmetry of the density with regard to $x = T$ implies that $E(Z) = T$. Hence,

$$E(Z) = E(X) + E(Y). \qquad \square$$

Example 4.21 Let the random vector (X, Y) have the joint density

$$f_{X,Y}(x,y) = \lambda\mu\,e^{-(\lambda x + \mu y)}, \quad x \ge 0, y \ge 0; \ \lambda > 0, \ \mu > 0.$$

From example 4.13 we know that X and Y are independent and have exponential distributions with parameters λ and μ, respectively. Hence, formula (4.39) is applicable to determine the density of the sum $Z = X + Y$:

$$f_Z(z) = \int_0^z \lambda e^{-\lambda x}\mu e^{-\mu(z-x)}\,dx = \lambda\mu\,e^{-\mu z}\int_0^z e^{-(\lambda-\mu)x}\,dx.$$

Two cases have to be considered separately:

1) $\lambda = \mu$: $\qquad\qquad\qquad\qquad f_Z(z) = \lambda^2 z e^{-\lambda z}, \ z \ge 0.$

This is an Erlang distribution with parameters λ and $n = 2$ (page 75).

2) $\lambda \ne \mu$: $\qquad\qquad\qquad f_Z(z) = \frac{\lambda\mu}{\lambda-\mu}\left[e^{-\mu z} - e^{-\lambda z}\right], \ z \ge 0.$

The mean value of $Z = X + Y$ is $(\lambda \ne \mu)$

$$E(Z) = \int_0^\infty z f_Z(z)\,dz = \frac{\lambda\mu}{\lambda-\mu}\left[\int_0^\infty z e^{-\mu z}\,dz - \int_0^\infty z e^{-\lambda z}\,dz\right]$$

$$= \frac{1}{\lambda} + \frac{1}{\mu} = E(X) + E(Y). \qquad \square$$

Mean Value of a Sum In the previous two examples, the mean value of a sum proved to be equal to the sum of the mean values of the terms. This is generally true, whether X and Y are independent or not (but $E(X)$ and $E(Y)$ must be finite):

$$E(X + Y) = \int_{-\infty}^{+\infty}\int_{-\infty}^{+\infty}(x + y)f_{X,Y}(x,y)\,dydx$$

$$= \int_{-\infty}^{+\infty} x \int_{-\infty}^{+\infty} f_{X,Y}(x,y)\,dydx + \int_{-\infty}^{+\infty} y \int_{-\infty}^{+\infty} f_{X,Y}(x,y)\,dxdy$$

$$= \int_{-\infty}^{+\infty} x\left(\int_{-\infty}^{+\infty} f_{X,Y}(x,y)\,dy\right)dx + \int_{-\infty}^{+\infty} y\left(\int_{-\infty}^{+\infty} f_{X,Y}(x,y)\,dx\right)dy.$$

Now, by using properties (3.11) of the joint density,

$$E(X+Y) = \int_{-\infty}^{+\infty} x\, f_X(x)\, dx + \int_{-\infty}^{+\infty} y f_Y(y)\, dy = E(X) + E(Y). \qquad (4.45)$$

The mean value of the sum of two random variables is equal to the sum of their mean values.

Variance of a Sum To present the variance of the sum $Z = X + Y$ in a convenient way, we need again the concept of the covariance between X and Y as defined by (3.37) or (3.38) (page 135):

$$Cov(X, Y) = E([X - E(X)] \cdot [Y - E(Y)]).$$

By definition (2.60) of the variance,

$$Var(Z) = E(Z - E(Z))^2 = E(X + Y - E(X) - E(Y))^2$$
$$= E([X - E(X)] + [Y - E(Y)])^2$$
$$= E(X - E(X))^2 + 2E([Y - E(Y)]\, E([Y - E(Y)]) + E(Y - E(Y))^2.$$

Hence, the variance of the sum is

$$Var(X + Y) = Var(X) + 2Cov(X, Y) + Var(Y). \qquad (4.46)$$

If X and Y are independent, then $Cov(X, Y) = 0$. In this case,

$$Var(X + Y) = Var(X) + Var(Y). \qquad (4.47)$$

The variance of the sum of two independent random variables is equal to the sum of their variances.

Bivariate Normal Distribution Let the random vector (X, Y) have a bivariate normal distribution with parameters

$$\mu_x,\ \mu_y,\ \sigma_x,\ \sigma_y,\ \text{and } \rho;\quad -\infty < \mu_x, \mu_y < \infty,\ \ \sigma_x > 0,\ \sigma_y > 0,\ \ -1 < \rho < 1.$$

Then (X, Y) has the joint density (page 131)

$$f_{X,Y}(x,y) = \frac{1}{2\pi\sigma_x\sigma_y\sqrt{1-\rho^2}}\, \exp\left\{ -\frac{1}{2(1-\rho^2)}\left(\frac{(x-\mu_x)^2}{\sigma_x^2} - 2\rho\frac{(x-\mu_x)(y-\mu_y)}{\sigma_x\sigma_y} + \frac{(y-\mu_y)^2}{\sigma_y^2} \right) \right\}.$$

To determine the density $f_Z(z)$ of $Z = X + Y$, formula (4.36) has to be applied. Letting

$$u = x - \mu_x \text{ and } v = z - \mu_x - \mu_y$$

yields $f_Z(z)$ in the form

$$f_Z(z) = \frac{1}{2\pi\sigma_x\sigma_y\sqrt{1-\rho^2}}\, \int_{-\infty}^{+\infty} \exp\left\{ -\frac{1}{2(1-\rho^2)}\left(\frac{u^2}{\sigma_x^2} - 2\rho\frac{u(v-u)}{\sigma_x\sigma_y} + \frac{(v-u)^2}{\sigma_y^2} \right) \right\} du.$$

The following transformation in the integrand of this formula requires some routine effort, but will prove to be advantageous:

$$\frac{u^2}{\sigma_x^2} - 2\rho\frac{u(v-u)}{\sigma_x\sigma_y} + \frac{(v-u)^2}{\sigma_y^2} = \frac{\sigma_x^2 + 2\rho\sigma_x\sigma_y + \sigma_y^2}{\sigma_x^2\sigma_y^2}u^2 - 2\frac{\sigma_x + \rho\sigma_y}{\sigma_x\sigma_y^2}uv + \frac{1}{\sigma_y^2}v^2$$

$$= \left(\frac{\sqrt{\sigma_x^2 + 2\rho\sigma_x\sigma_y + \sigma_y^2}}{\sigma_x\sigma_y}u - \frac{\sigma_x + \rho\sigma_y}{\sigma_y\sqrt{\sigma_x^2 + 2\rho\sigma_x\sigma_y + \sigma_y^2}}v\right)^2 + \frac{1-\rho^2}{\sigma_x^2 + 2\rho\sigma_x\sigma_y + \sigma_y^2}v^2.$$

Now this expression is inserted into the integrand and after having done this the following substitution is done:

$$t = \frac{1}{\sqrt{1-\rho^2}}\left(\frac{\sqrt{\sigma_x^2 + 2\rho\sigma_x\sigma_y + \sigma_y^2}}{\sigma_x\sigma_y}u - \frac{\sigma_x + \rho\sigma_y}{\sigma_y\sqrt{\sigma_x^2 + 2\rho\sigma_x\sigma_y + \sigma_y^2}}v\right).$$

These transformations result in the following form for $f_Z(z)$:

$$f_Z(z) = \frac{1}{2\pi\sqrt{\sigma_x^2 + 2\rho\sigma_x\sigma_y + \sigma_y^2}}\exp\left(-\frac{v^2}{2(\sigma_x^2 + 2\rho\sigma_x\sigma_y + \sigma_y^2)}\right)\int_{-\infty}^{+\infty}e^{-t^2/2}dt.$$

Since $\int_{-\infty}^{+\infty}e^{-t^2/2}dt = \sqrt{2\pi}$, the final result is

$$f_Z(z) = \frac{1}{\sqrt{2\pi(\sigma_x^2 + 2\rho\sigma_x\sigma_y + \sigma_y^2)}}\exp\left(-\frac{(z-\mu_x-\mu_y)^2}{2(\sigma_x^2 + 2\rho\sigma_x\sigma_y + \sigma_y^2)}\right), \quad -\infty < z < \infty. \quad (4.48)$$

Comparing $f_Z(z)$ with the density (2.81) of the one-dimensional normal distribution verifies the following corollary from (4.48):

> *If the random vector (X,Y) has a two-dimensional normal distribution with parameters*
>
> $\mu_x, \mu_y, \sigma_x, \sigma_y,$ and $\rho;$ $-\infty < \mu_x, \mu_y < \infty,$ $\sigma_x > 0, \sigma_y > 0,$ $-1 < \rho < 1,$
>
> *then the sum $Z = X + Y$ has a one-dimensional normal distribution with parameters*
>
> $$E(Z) = \mu_x + \mu_y \text{ and } Var(Z) = \sigma_x^2 + 2\rho\sigma_x\sigma_y + \sigma_y^2. \quad (4.49)$$

The Laplace transform of any $N(\mu, \sigma^2)$ distributed random variable is, by formula (2.129), page 102,

$$\hat{f}(s) = e^{-\mu s + \frac{1}{2}\sigma^2 s^2}.$$

If X and Y are independent, then the Laplace transform of Z is the product of the Laplace transforms of X and Y:

$$\hat{f}_Z(s) = e^{-\mu_x s + \frac{1}{2}\sigma_x^2 s^2} \cdot e^{-\mu_y s + \frac{1}{2}\sigma_y^2 s^2} = e^{-(\mu_x + \mu_y)s + \frac{1}{2}(\sigma_x^2 + \sigma_y^2)s^2}.$$

This proves once more that the sum $Z = X + Y$ of two independent, normally distributed random variables X and Y is normally distributed with parameters

$$E(Z) = \mu_x + \mu_y \text{ and } Var(Z) = \sigma_x^2 + \sigma_y^2, \text{ i.e. } Z = N(\mu_x + \mu_y, \sigma_x^2 + \sigma_y^2). \quad (4.50)$$

Example 4.22 Let X and Y be the annual profits Bobo makes from her investments in equities and bonds, respectively. She has analyzed her profits over a couple of years, and knows that the random vector (X, Y) has a bivariate normal distribution with parameters (in $, influence of inflation eliminated)

$$\mu_x = 2160, \; \mu_y = 3420, \; \sigma_x = 1830, \; \sigma_y = 2840, \; \text{and } \rho = -0.28.$$

(1) What probability distribution has Bobo's total profit $Z = X + Y$?

(2) What is the probability that her total 'profit' is actually negative?

(1) According to (4.46), Z has a normal distribution with parameters

$$\mu_z = 5580, \; \sigma_z^2 = \sigma_x^2 + 2\rho\sigma_x\sigma_y + \sigma_y^2 = 8\,504\,068$$

so that $\sigma_z \approx 2916$.

(2) $P(Z < 0) = P\left(\dfrac{Z - 5580}{2916} < -\dfrac{5580}{2916}\right) \approx \Phi(-1.91) \approx 0.028.$ $\qquad\qquad\square$

Continuation of Example 3.7 (page 131) The daily consumptions of tap water X and Y of two neighboring towns have a bivariate normal distribution with parameters

$$\mu_x = \mu_y = 16\,[10^3\,m^3], \; \sigma_x = \sigma_y = 2\,[10^3 m^3], \; \text{and } \rho = 0.5.$$

What is the probability that the total daily tap water consumption $Z = X + Y$ of the two towns exceeds the amount of $36\,[10^3\,m^3]$, which is the maximal amount manageable by the municipality?

Z has a normal distribution with parameters

$$\mu_z = 32\,[10^3\,m^3] \; \text{and} \; \sigma_z^2 = \sigma_x^2 + 2\rho\sigma_x\sigma_y + \sigma_y^2 = 12\,[10^6\,m^6]$$

so that $\sigma_z \approx 3.464$. Hence,

$$P(Z > 36) = P\left(\dfrac{Z - 32}{3.464} > \dfrac{36 - 32}{3.464}\right) \approx \Phi(-1.155) \approx 0.124. \qquad\square$$

4.3.2.2 Sum of $n \geq 2$ Random Variables

In this section, $X_i; \; i = 1, 2, ..., n;$ are random variables with respective distribution functions, densities, mean values, and variances

$$F_i(x_i), \; f_i(x_i), \; \mu_i = E(X_i), \; \text{and } \sigma_i^2 = Var(X_i); \; i = 1, 2, ..., n.$$

The joint density of $\mathbf{X} = (X_1, X_2, ..., X_n)$ is denoted as $f_{\mathbf{X}}(x_1, x_2, ..., x_n)$. All mean values and variances are assumed to be finite. The covariance between X_i and X_j is according to (3.37) defined as

$$Cov(X_i, X_j) = E([X_i - E(X_i)][X_j - E(X_j)]).$$

The sum of the X_i is again denoted as $Z = X_1 + X_2 + \cdots + X_n$, and its distribution function and density as $F_Z(z)$ and $f_Z(z)$.

Mean Value of a Sum

$$E(Z) = E(X_1 + X_2 + \cdots + X_n) = E(X_1) + E(X_2) + \cdots + E(X_n). \qquad (4.51)$$

| *The mean value of the sum of n (discrete or continuous) random variables is equal to the sum of the mean values of these random variables.*

This can be proved analogously to formula (4.45) by making use of the relationship (3.54) between $f_{\mathbf{X}}$ and the f_{X_i} or simply by induction starting with formula (4.45):

If, for instance, the mean value $E(X_1 + X_2 + X_3)$ has to be determined, let

$$X = X_1 + X_2 \quad \text{and} \quad Y = X_3$$

and apply (4.45) as follows:

$$E(X_1 + X_2 + X_3) = E(X) + E(Y)$$
$$= E(X_1 + X_2) + E(X_3)$$
$$= E(X_1) + E(X_2) + E(X_3).$$

Variance of a Sum The variance of the sum $Z = \sum_{i=1}^{n} X_i$ of n random variables X_i results from its representation as

$$Var(Z) = E(Z - E(Z))^2 = E([X_1 - E(X_1)] + [X_2 - E(X_2)] + \cdots + [X_n - E(X_n)])^2.$$

Since

$$Cov(X_i, X_i) = Var(X_i) \quad \text{and} \quad Cov(X_i, X_j) = Cov(X_j, X_i),$$

the generalization of formula (4.46) is

$$Var\left(\sum_{i=1}^{n} X_i\right) = \sum_{i=1}^{n} Var(X_i) + 2 \sum_{i,j=1; i<j}^{n} Cov(X_i, X_j). \qquad (4.52)$$

Thus, for uncorrelated X_i,

$$Var(X_1 + X_2 + \cdots + X_n) = Var(X_1) + Var(X_2) + \cdots + Var(X_n). \qquad (4.53)$$

| *The variance of a sum of uncorrelated random variables is equal to the sum of the variances of these random variables.*

Let $\alpha_1, \alpha_2, \cdots, \alpha_n$ be any sequence of finite real numbers. Then, by (2.54) and (2.61),

$$E\left(\sum_{i=1}^{n} \alpha_i X_i\right) = \sum_{i=1}^{n} \alpha_i E(X_i), \qquad (4.54)$$

$$Var\left(\sum_{i=1}^{n} \alpha_i X_i\right) = \sum_{i=1}^{n} \alpha_i^2 Var(X_i) + 2 \sum_{i,j=1, i<j}^{n} \alpha_i \alpha_j Cov(X_i, X_j). \qquad (4.55)$$

If the X_i are uncorrelated, the latter formula simplifies to

$$Var\left(\sum_{i=1}^{n} \alpha_i X_i\right) = \sum_{i=1}^{n} \alpha_i^2 Var(X_i). \qquad (4.56)$$

Now let us interpret a sequence $\{X_1, X_2, ..., X_n\}$ of independent, identically as X distributed random variables as a *random sample* taken from X, i.e., a random experiment with outcome X is repeated n times. Mean value and variance of X and, hence, of all the X_i are $E(X) = \mu$ and $Var(X) = \sigma^2$. Then formulas (4.54) and (4.56) simplify to

$$E\left(\sum_{i=1}^{n} X_i\right) = n\mu, \quad Var\left(\sum_{i=1}^{n} X_i\right) = n\sigma^2. \tag{4.57}$$

Under the same assumptions, application of (4.54) and (4.56) to the arithmetic mean

$$\overline{X} = \tfrac{1}{n} \sum_{i=1}^{n} X_i$$

yields with $\alpha_i = 1/n$

$$E(\overline{X}) = \mu \text{ and } Var(\overline{X}) = \frac{\sigma^2}{n}. \tag{4.58}$$

Note Formulas (4.51) to (4.58) hold both for discrete and continuous random variables.

Definition 4.1 A function $\hat{\theta} = \hat{\theta}(X_1, X_2, ..., X_n)$ of a sample $\{X_1, X_2, ..., X_n\}$ taken from a random variable X is called an *unbiased estimator* of a parameter θ of X if

$$E(\hat{\theta}) = \theta. \qquad \bullet$$

Parameters can, e.g., be $\theta = \mu = E(X)$, $\theta = \sigma^2 = Var(X)$, or $\theta = \beta$ in case of the beta or Weibull distribution. The left formula of (4.58) shows that $\hat{\theta} = \overline{X}$ is an *unbiased estimator* of $\theta = \mu = E(X)$. Verbally, when estimating the mean value of X by $\overline{X}$, only random deviations of $\overline{X}$ from $\mu = E(X)$ can be observed, no systematic ones. In addition, the right formula in (4.58) shows that with increasing number of measurements the accuracy of $\overline{X}$ as estimator for μ improves since $Var(\overline{X})$ tends to 0 if $n \to \infty$.

After having done the n repetitions of the random experiment, a sequence of real numbers $\{x_1, x_2, ..., x_n\}$ has been obtained, i.e., $X_i = x_i$; $i = 1, 2, ..., n$. This sequence gives *empirical estimators* for μ and σ^2:

$$\overline{x} = \tfrac{1}{n} \sum_{i=1}^{n} x_i, \quad s^2 = \tfrac{1}{n-1} \sum_{i=1}^{n} (x_i - \overline{x})^2.$$

Now, as announced after formula (3.48), page 143, we are in a position to justify the factor $\frac{1}{n-1}$ in the formula for s^2.

Theorem 4.2 Let $\{X_1, X_2, ..., X_n\}$ be a random sample from a random variable X with $0 < \sigma^2 = Var(X) < \infty$. Then the random sample function

$$S^2 = \tfrac{1}{n-1} \sum_{i=1}^{n} (X_i - \overline{X})^2$$

is an unbiased estimator of $\sigma^2 = Var(X)$.

Proof We have to prove $E(S^2) = \sigma^2$. For this reason, S^2 is written in the form

$$S^2 = \tfrac{1}{n-1} \sum_{i=1}^{n} X_i^2 - \tfrac{n}{n-1} \overline{X}^2. \tag{4.59}$$

In what follows, use will be made of the independence of the X_k and their identical distribution as X:

$$E(X_i \cdot X_j) = E(X_i) \cdot E(X_j) = [E(X)]^2 \text{ for } i \neq j.$$

Then

$$E\left(\sum_{i=1}^{n} X_i^2\right) = n E(X^2) \tag{4.60}$$

so that only the second moment of $\overline{X}$ has to be determined:

$$E(\overline{X}^2) = \frac{1}{n^2} E\left(\sum_{i=1}^{n} X_i\right)^2 = \frac{1}{n^2} E\left(\sum_{i,j=1}^{n} X_i X_j\right)$$

$$= \frac{1}{n^2} E\left(\sum_{i=1}^{n} X_i^2\right) + \frac{1}{n^2} E\left(\sum_{\substack{i,j=1 \\ i \neq j}}^{n} X_i X_j\right)$$

$$= \frac{1}{n} E(X^2) + \frac{n-1}{n} E(X^2).$$

Substituting this result and (4.60) into (4.59) gives

$$E(S^2) = \sigma^2. \qquad \blacksquare$$

Distribution of a Sum The density of the sum $Z = X_1 + X_2 + \cdots + X_n$ of n independent, continuous random variables X_i is obtained by repeated application of (4.36), page 181. To do this in an efficient way, next the convolution symbol '*' will be introduced: For any two integrable functions f and g, their convolution is denoted as

$$f * g(z) = \int_{-\infty}^{+\infty} f(z-x)g(x)\, dx = \int_{-\infty}^{+\infty} g(z-x)f(x)\, dx = g * f(z). \tag{4.61}$$

Thus, the *convolution product* is *commutative*, i.e.

$$f * g(z) = g * f(z),$$

just as the product of two real numbers: $a \cdot b = b \cdot a$.

The convolution of the densities $f_{X_1}, f_{X_2}, ..., f_{X_n}$ is obtained by repeated application of (4.61): Firstly, $f_{X_1} * f_{X_2}$ is calculated. Then the convolution of f_{X_3} with $f_{X_1} * f_{X_2}$ is determined to obtain $f_{X_1} * f_{X_2} * f_{X_3}$ and so on. The final result is the probability density of Z:

$$f_Z(z) = f_{X_1} * f_{X_2} * \cdots * f_{X_n}(z). \tag{4.62}$$

In particular, if the X_i are identically distributed with density f, then f_Z is the *n-fold convolution* of f with itself or, equivalently, the nth *convolution power* $f^{*(n)}(z)$ of f. $f^{*(n)}(z)$ can be recursively obtained as follows:

$$f^{*(i)}(z) = \int_{-\infty}^{+\infty} f^{*(i-1)}(z-x)f(x)\,dx, \tag{4.63}$$

$i = 2, 3, ..., n$; $f^{*(1)}(x) \equiv f(x)$.

For nonnegative random variables X_i, this formula becomes

$$f^{*(i)}(z) = \int_0^z f^{*(i-1)}(z-x)f(x)\,dx, \quad z \geq 0. \tag{4.64}$$

From (4.40), by induction: The Laplace transform of the density f_Z of the sum of n independent random variables $Z = X_1 + X_2 + \cdots + X_n$ is equal to the product of the Laplace transforms of these random variables:

$$L(f_Z) = L(f_{X_1})L(f_{X_2}) \cdots L(f_{X_n}). \tag{4.65}$$

The convolution of the distribution functions F_{X_1} and F_{X_2} is defined by (4.42) as

$$F_{X_1} * F_{X_2}(z) = \int_{-\infty}^{+\infty} F_{X_1}(z-y)\,dF_{X_2}(y). \tag{4.66}$$

The repeated application of (4.66) yields the distribution function of the sum Z of the n independent random variables $X_1, X_2, ..., X_n$ in the form

$$F_Z(z) = F_{X_1} * F_{X_2} * \cdots * F_{X_n}(z). \tag{4.67}$$

In particular, if the X_i are independent and identically distributed with distribution function F, then $F_Z(z)$ is equal to the nth *convolution power* of F:

$$F_Z(z) = F^{*(n)}(z). \tag{4.68}$$

$F_Z(z)$ can be recursively obtained from

$$F^{*(i)}(z) = \int_{-\infty}^{+\infty} F^{*(i-1)}(z-x)\,dF(x); \tag{4.69}$$

$n = 2, 3, ...;$ $F^{*(0)}(x) \equiv 1$, $F^{*(1)}(x) \equiv F(x)$.

If the X_i are nonnegative, then (4.69) becomes

$$F^{*(i)}(z) = \int_0^z F^{*(i-1)}(z-x)\,dF(x). \tag{4.70}$$

The convolution powers of any order n can explicitly be given for the Erlang distribution and for the normal distribution.

Erlang Distribution Let the random variables X_1 and X_2 be independent and exponentially distributed with parameters λ_1 and λ_2:

$$f_{X_i}(x) = \lambda_i e^{-\lambda_i x},$$

$$F_{X_i}(x) = 1 - e^{-\lambda_i x}; \quad x \geq 0, \; i = 1, 2.$$

Formula (4.37) yields the density of $Z = X_1 + X_2$:

$$f_Z(z) = \int_0^z \lambda_2 e^{-\lambda_2(z-x)} \lambda_1 e^{-\lambda_1 x}\,dx$$

$$= \lambda_1 \lambda_2 e^{-\lambda_2 z} \int_0^z e^{-(\lambda_1 - \lambda_2)x}\,dx.$$

At this stage, two cases have to be treated separately:

1) $\lambda_1 = \lambda_2 = \lambda$: $\qquad\qquad\qquad f_Z(z) = \lambda^2 z e^{-\lambda z}, \quad z \geq 0.$ $\qquad\qquad$ (4.71)

This is the density of an Erlang distribution with parameters $n = 2$ and λ (page 75).

2) $\lambda_1 \neq \lambda_2$:

$$f_Z(z) = \frac{\lambda_1 \lambda_2}{\lambda_1 - \lambda_2} \left(e^{-\lambda_2 z} - e^{-\lambda_1 z} \right), \quad z \geq 0.$$

Now let $X_1, X_2, ..., X_n$ be independent, identically distributed exponential random variables with density $f(x) = \lambda e^{-\lambda x}; \ x \geq 0$. The Laplace transform of f is (page 101)

$$\hat{f}(s) = \frac{\lambda}{s + \lambda}.$$

Hence, by (4.65), the Laplace transform of the density of $Z = X_1 + X_2 + \cdots + X_n$ is

$$\hat{f}_Z(s) = \left(\frac{\lambda}{s + \lambda} \right)^n.$$

The pre-image of this Laplace transform is

$$f_Z(z) = \lambda \frac{(\lambda z)^{n-1}}{(n-1)!} e^{-\lambda z}, \quad z \geq 0,$$

(Verify this by calculating the Laplace transform of $f_Z(z)$.) This is the density of an Erlang distribution with parameters n and λ. Hence, the density of an Erlang distribution with parameters n and λ is the n th convolution power of the density of an exponential distribution $f(x) = \lambda e^{-\lambda x}$, which is an Erlang distribution with the parameters $n = 1$ and λ.

Normal Distribution Let X_1 and X_2 be two independent, normally distributed random variables: $X_1 = N(\mu_1, \sigma_1^2)$, $X_2 = N(\mu_2, \sigma_2^2)$. Then we know from formula (4.50) that $Z = X_1 + X_2$ is normally distributed with parameters $\mu_1 + \mu_2$ and $\sigma_1^2 + \sigma_2^2$:

$$Z = N(\mu_1 + \mu_2, \sigma_1^2 + \sigma_2^2).$$

By induction: the sum of n independent random variables $X_i = N(\mu_i, \sigma_i^2)$,

$$Z = X_1 + X_2 + \cdots + X_n,$$

is normally distributed with parameters

$$E(Z) = \mu_1 + \mu_2 + \cdots + \mu_n \quad \text{and} \quad Var(Z) = \sigma_1^2 + \sigma_2^2 + \cdots + \sigma_n^2,$$

or, more concise,

$$Z = N\left(\Sigma_{i=1}^n \, \mu_i, \ \Sigma_{i=1}^n \, \sigma_i^2 \right). \qquad\qquad (4.72)$$

In terms of the density,

$$f_Z(z) = \frac{1}{\sqrt{2\pi \left(\Sigma_{i=1}^n \sigma_i^2 \right)}} \exp\left(-\frac{\left(z - \Sigma_{i=1}^n \mu_i \right)^2}{2 \left(\Sigma_{i=1}^n \sigma_i^2 \right)} \right), \quad -\infty < z < +\infty.$$

In terms of the convolution,

$$f_Z(z) = f_{X_1} * f_{X_2} * \cdots * f_{X_n}(z).$$

If the X_i are identically distributed as $X = N(\mu, \sigma^2)$, then each X_i has density

$$f_X(x) = \frac{1}{\sqrt{2\pi}\,\sigma} e^{-\frac{(x-\mu)^2}{2\sigma^2}}, \quad -\infty < x < +\infty,$$

and f_Z is the nth convolution power of f_X:

$$f_Z(z) = f^{*(n)}(z) = \frac{1}{\sqrt{2\pi n}\,\sigma} e^{-\frac{(x-n\mu)^2}{2n\sigma^2}}, \quad -\infty < x < +\infty.$$

Example 4.23 (1) The daily power consumption X and Y of two customers has a bi-variate normal distribution with parameters

$$\mu_x = 200, \ \mu_y = 300, \ \sigma_x = 26, \ \sigma_y = 32 \,[\text{in } 10^3 kWh], \ \text{and } \rho = 0.6.$$

Calculate a) the probability that the daily total consumption $Z = X + Y$ of the two customers is between 450 and 550, and

b) the probability of the same event as under a), but on condition that X and Y are independent.

(2) Determine the probability that the daily total consumption of 10 independent customers, each of them has a daily consumption of X as given under (1), is between 1950 and 2050.

(1) a) By (4.49), the daily total consumption of the two customers has mean value

$$E(Z) = 200 + 300 = 500$$

and variance/standard deviation

$$Var(Z) = \sigma_x^2 + 2\rho\sigma_x\sigma_y + \sigma_y^2 = 26^2 + 2 \cdot 0.6 \cdot 26 \cdot 32 + 32^2 = 2698.4$$

so that

$$\sqrt{Var(Z)} = 51.95.$$

The desired probability is

$$P(450 \le Z \le 550) = \Phi\left(\frac{550-500}{51.95}\right) - \Phi\left(\frac{450-500}{51.95}\right)$$
$$= \Phi(0.92) - \Phi(-0.92) = 2\Phi(0.92) - 1$$
$$= 0.664.$$

b) Since X and Y are independent, $\rho = 0$. Hence,

$$Var(Z) = \sigma_x^2 + \sigma_y^2 = 26^2 + 32^2 = 1700 \ \text{and} \ \sqrt{Var(Z)} = 41.23.$$

Therefore, the desired probability is obtained as follows:

$$P(450 \le Z \le 550) = \Phi\left(\frac{550 - 500}{41.23}\right) - \Phi\left(\frac{450 - 500}{41.23}\right)$$

$$= \Phi(1.213) - \Phi(-1.213) = 2\Phi(1.213) - 1$$

$$= 0.774.$$

(2) According (4.72), the daily total consumption of 10 independent customers has a normal distribution with parameters

$$E(Z) = 10 \cdot 200 = 2000, \quad Var(Z) = 10 \cdot 26^2 = 6760, \quad \sqrt{Var(Z)} = 82.22.$$

Therefore, the desired probability is

$$P(1950 \le Z \le 2050) = \Phi\left(\frac{2050 - 2000}{82.22}\right) - \Phi\left(\frac{1950 - 2000}{82.22}\right)$$

$$= \Phi(0.608) - \Phi(-0.608) = 2\Phi(0.608) - 1$$

$$= 0.456. \qquad \square$$

Example 4.24 A bulk goods freighter has to be loaded with at least $2000\,t$ of iron ore. The ore arrives by goods wagons, whose load weights $X_1, X_2, \cdots$ are independent and have an $N(50, 64)$-distribution.

How many wagons are needed to make sure that the freighter can be loaded with the required minimum load with a probability of at least 0.99?

Let $Z_n = X_1 + X_2 + \cdots + X_n$. n has to be determined as the smallest integer with property $P(Z_n \ge 2000) \ge 0.99$. This relation is equivalent to

$$P(Z_n < 2000) \le 0.01. \qquad (4.73)$$

By (4.72), $Z_n = N(50n, 64n)$. The corresponding standardization is

$$Y_n = N(0, 1) = \frac{Z_n - 50n}{8\sqrt{n}}.$$

Hence, (4.73) can be written in the equivalent form

$$P(Z_n < 2000) = P\left(Y_n < \frac{2000 - 50n}{8\sqrt{n}}\right) = \Phi\left(\frac{2000 - 50n}{8\sqrt{n}}\right) \le 0.01.$$

The 0.01-percentile of the standard normal distribution is -2.32, i.e.,

$$\Phi(-2.32) = 0.01.$$

Hence, relation (4.73) is equivalent to

$$\frac{2000 - 50n}{8\sqrt{n}} \le -2.32 \quad \text{or} \quad \frac{50n - 2000}{8\sqrt{n}} \ge 2.32.$$

By squaring and some simple algebra these relations are seen to be equivalent to

$$(n - 40.069)^2 \ge 5.5 \quad \text{or} \quad n \ge 42.41.$$

Hence, at least 43 waggons are needed. $\qquad \square$

4.3.3 Sums of a Random Number of Random Variables

Frequently, sums of a random number of random variables have to be investigated. For instance, the total claim size an insurance company is confronted with a year is the sum of a random number of random individual claim sizes. The total repair cost a machine causes a year is the sum of random number of random repair costs, the increase of a population a year is determined by the random number of individuals producing children and the random number of children produced by an individual, etc.

Wald's Identities Let $\{X_1, X_2, ...\}$ be a sequence of independent random variables, which are identically distributed as X with $E(X) < \infty$. Let further N be a positive, integer-valued random variable, which is independent of all $X_1, X_2, ...$ Then mean value and variance of the sum $Z = X_1 + X_2 + \cdots + X_N$ are given by *Wald's identities*:

$$E(Z) = E(X) \cdot E(N), \tag{4.74}$$

$$Var(Z) = Var(X) E(N) + [E(X)]^2 Var(N). \tag{4.75}$$

The proof of these relations is easily done by conditioning:

$$E(Z) = \sum_{n=1}^{\infty} E(X_1 + X_2 + \cdots + X_N | N = n) P(N = n)$$

$$= \sum_{n=1}^{\infty} E(X_1 + X_2 + \cdots + X_n) P(N = n) = \sum_{n=1}^{\infty} E(nX) P(N = n)$$

$$= E(X) \sum_{n=1}^{\infty} n P(N = n) = E(X) \cdot E(N).$$

This proves (4.74). To verify (4.75), the second moment of Z is determined:

$$E(Z^2) = \sum_{n=1}^{\infty} E(Z^2 | N = n) P(N = n)$$

$$= \sum_{n=1}^{\infty} E([X_1 + X_2 + \cdots + X_n]^2) P(N = n).$$

By making use of formula (2.62), page 67,

$$E(Z^2) = \sum_{n=1}^{\infty} \{ Var(X_1 + X_2 + \cdots + X_n) + [E(X_1 + X_2 + \cdots + X_n)]^2 \} P(N = n)$$

$$= \sum_{n=1}^{\infty} \{ n\, Var(X) + n^2 [E(X)]^2 \} P(N = n)$$

$$= Var(X) \sum_{n=1}^{\infty} n P(N = n) + [E(X)]^2 \sum_{n=1}^{\infty} n^2 P(N = n)$$

$$= Var(X) E(N) + [E(X)]^2 E(N^2).$$

Hence,

$$Var(Z) = E(Z^2) - [E(Z)]^2$$

$$= Var(X) E(N) + [E(X)]^2 E(N^2) - [E(X)]^2 [E(N)]^2$$

$$= Var(X) E(N) + [E(X)]^2 Var(N).$$

This is the identity (4.75).

Wald's identities (4.74) and (4.75) remain valid if the assumption that N is independent of all X_i is somewhat weakened by introducing the concept of a stopping time.

Definition 4.2 (*stopping time*) A positive, integer-valued random variable N is said to be a *stopping time* for the sequence of independent random variables $\{X_1, X_2, ...\}$ if the occurrence of the random event '$N = n$' is completely determined by the finite sequence $X_1, X_2, ..., X_n$, and, therefore, independent of all $X_{n+1}, X_{n+2}, ..., n \geq 1.$ ●

Note A random event A is said to be independent of a random variable X if the indicator variable of A is independent of X (see also example 3.14, page 146).

Sometimes, a stopping time defined in this way is called a *Markov time*, and only a finite Markov time is called a stopping time. (A random variable Y is said to be *finite* if $P(Y < \infty) = 1$. In this case, $E(Y) < \infty$.)

The notation 'stopping time' can be motivated as follows: The $X_1, X_2, ...$ are observed one after the other. As soon as the event '$N = n$' occurs, the observation is stopped, i.e., the $X_{n+1}, X_{n+2}, ...$ will not be observed.

Theorem 4.3 Let $\{X_1, X_2, ...\}$ be a sequence of random variables, which are identically distributed as X with $E(X) < \infty$, and let N be a finite stopping time for this sequence. Then

$$E(Z) = E(X) \cdot E(N). \tag{4.76}$$

Proof Let binary random variables Y_i be defined as follows:

$$Y_i = \begin{cases} 1 & \text{if } N \geq i \\ 0 & \text{if } N < i \end{cases}, \quad i = 1, 2, ...$$

The event '$Y_i = 1$' occurs if and only if no stopping has been done after the observation of the $i - 1$ random variables $X_1, X_2, ..., X_{i-1}$. Since N is a stopping time, Y_i is independent of the $X_i, X_{i+1}, ...$. Moreover,

$$E(Y_i) = P(N \geq i) \text{ and } E(X_i Y_i) = E(X_i) E(Y_i)$$

so that

$$E(\textstyle\sum_{i=1}^{N} X_i) = E(\textstyle\sum_{i=1}^{\infty} X_i Y_i)$$

$$= \textstyle\sum_{i=1}^{\infty} E(X_i) E(Y_i) = E(X) \textstyle\sum_{i=1}^{\infty} E(Y_i)$$

$$= E(X) \textstyle\sum_{i=1}^{\infty} P(N \geq i).$$

Now formula (2.9) at page 46 implies (4.76). ■

Example 4.25 a) Let $X_i = 1$ if $i\,th$ flipping a fair coin yields 'head' and $X_i = 0$ if the outcome is 'tail'. The X_i are independent and identically distributed as

$$X = \begin{cases} 1 & \text{if } head \text{ occurs,} \\ -1 & \text{if } tail \text{ occurs.} \end{cases}$$

Then, a finite stopping time for the sequence $X_1, X_2, ...$ is

$$N = \min \{n; \ X_1 + X_2 + \cdots + X_n = 10\}. \tag{4.77}$$

Since $E(X) = 1/2$,

$$E(X_1 + X_2 + \cdots + X_N) = \tfrac{1}{2} \cdot E(N).$$

According to the definition of N,

$$X_1 + X_2 + \cdots + X_N = 10$$

so that $E(N) = 20$.

b) Let $X_i = 1$ if the i th flipping a fair coin yields 'head' and $X_i = -1$ otherwise. Then N given by (4.77) is again a finite stopping time for $X_1, X_2, \dots$. A formal application of Wald's equation yields

$$E(X_1 + X_2 + \cdots + X_N) = E(X) \cdot E(N).$$

The left hand side of this equation is equal to 10. The right hand side contain the factor $E(X) = 0$. Therefore, Wald's equation (4.76) is not applicable. ☐

4.4 EXERCISES

4.1 In a game reserve, the random position (X, Y) of a leopard has a uniform distribution in a semicircle with radius $r = 10 \, km$ (figure). Determine $E(X)$ and $E(Y)$.

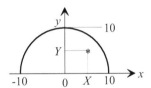

Illustration to Exercise 4.1

4.2) From a circle with radius $R = 9$ and center $(0,0)$ a point is randomly selected.

(1) Determine the mean value of the distance of this point to the nearest point at the periphery of the circle.

(2) Determine the mean value of the geometric mean of the random variables X and Y, i.e. $E(\sqrt{XY})$.

4.3) X and Y are independent, exponentially with parameter $\lambda = 1$ distributed random variables. Determine

(1) $E(X - Y)$,

(2) $E(|X - Y|)$, and

(3) distribution function and density of $Z = X - Y$.

4.4) X and Y are independent random variables with
$E(X) = E(Y) = 5$, $Var(X) = VarY) = 9$, and let $U = 2X + 3Y$ and $V = 3X - 2Y$.
Determine $E(U)$, $E(V)$, $Var(U)$, $Var(V)$, $Cov(U, V)$, and $\rho(U, V)$.

4.5) X and Y are independent, in the interval $[0,1]$ uniformly distributed random variables. Determine the densities of
(1) $Z = \min(X, Y)$, and (2) $Z = XY$.

4.6) X and Y are independent and $N(0, 1)$-distributed. Determine the density $f_Z(z)$ of
$$Z = X/Y.$$
Which type of probability distributions does $f_Z(z)$ belong to?

4.7) X and Y are independent and identically Cauchy distributed with parameters $\lambda = 1$ and $\mu = 0$, i.e. they have densities (page 74)
$$f_X(x) = \frac{1}{\pi}\frac{1}{1+x^2}, \quad f_Y(y) = \frac{1}{\pi}\frac{1}{1+y^2}, \quad -\infty < x, y < +\infty.$$
Verify that the sum $Z = X + Y$ has a Cauchy distribution as well.

4.8) The joint density of the random vector (X, Y) is
$$f(x,y) = 6x^2y, \quad 0 \le x, y \le 1.$$
Determine the distribution density of the product $Z = XY$.

4.9) The random vector (X, Y) has the joint density
$$f_{X,Y}(x,y) = 2e^{-(x+y)} \text{ for } 0 \le x \le y < \infty.$$
Determine the densities of $Z = \max(X, Y)$ and $Z = \min(X, Y)$.

4.10) The resistance values X, Y, and Z of 3 resistors connected in series are assumed to be independent, normally distributed random variables with respective mean values 200, 300, and 500 $[\Omega]$, and standard deviations 5, 10, and 20 $[\Omega]$.
(1) What is the probability that the total resistance exceeds 1020 $[\Omega]$?
(2) Determine that interval $[1000 - \varepsilon, 1000 + \varepsilon]$ to which the total resistance belongs with probability 0.95.

4.11) A supermarket employs 24 shopassistants. 20 of them achieve an average daily turnover of \$ 8000, whereas 4 achieve an average daily turnover of \$ 10 000. The corresponding standard deviations are \$ 2400 and \$ 3000, respectively. The daily turnovers of all shopassistants are independent and have a normal distribution. Let Z be the daily total turnover of all shop-assistants.
(1) Determine $E(Z)$ and $Var(Z)$.
(2) What is the probability that the daily total turnover Z is greater than \$ 190 000?

4.12) A helicopter is allowed to carry at most 8 persons given that their total weight does not exceed 620kg. The weights of the passengers are independent, identically normally distributed random variables with mean value 76kg and variance 324kg^2.

(1) What are the probabilities of exceeding the permissible load with 7 and 8 passengers, respectively?

(2) What would the maximum total permissible load have to be to ensure that with probability 0.99 the helicopter will be allowed to fly 8 passengers?

4.13) Let X be the height of the woman and Y be the height of the man in married couples in a certain geographical region. By analyzing a sufficiently large sample, a statistician found that the random vector (X, Y) has a joint normal distribution with parameters

$$E(X) = 168\,cm, \quad Var(X) = 64\,cm^2, \quad E(Y) = 175\,cm, \quad Var(Y) = 100\,cm^2, \quad \rho = 0.86.$$

(1) Determine the probability $P(X > Y)$ that in married couples in this area a wife is taller than her spouse.

(2) Determine the same probability on condition that there is no correlation between X and Y, and interprete the result in comparison to (1).

Hint If you do not want to use a statistical software package, make use of the fact that the desired probability has structure $P(X > Y) = P(X + (-Y) > 0)$ and apply formula (4.48), page 185.

4.14) Let A and B be independent random variables, identically distributed with the probability density (Laplace distribution, page 66)

$$f(x) = \frac{1}{2}e^{-|x|}, \quad -\infty < x < +\infty.$$

Determine the probability density of the sum $X = A + B$.

4.15) Let A and B be independent random variables, identically distributed with the probability density

$$f(x) = \frac{2}{\pi}\frac{1}{e^x + e^{-x}}, \quad -\infty < x < +\infty.$$

Determine the probability density of the sum $X = A + B$.

Remark In terms of the *cosinus hyberbolicus*, this density can be rewritten as

$$f(x) = \frac{1}{\pi}\frac{1}{\cosh x}, \quad -\infty < x < +\infty.$$

CHAPTER 5

Inequalities and Limit Theorems

5.1 INEQUALITIES

5.1.1 Inequalities for Probabilities

Inequalities in probability theory are useful tools for estimating probabilities and moments of random variables if their exact calculation is only possible with extremely high effort or is even impossible in view of incomplete information on the underlying probability distribution. In what follows, all occurring mean values and variances are assumed to be finite.

Inequality of Chebyshev (also called *Bienaymé-Chebyshev inequality*) For any random variable X with mean value $\mu = E(X)$, variance $\sigma^2 = Var(X)$, and for any $\varepsilon > 0$,

$$P(|X - \mu| \geq \varepsilon) \leq \frac{\sigma^2}{\varepsilon^2}. \tag{5.1}$$

To prove (5.1), assume for convenience that X has density $f(x)$. Then,

$$\sigma^2 = \int_{-\infty}^{+\infty} (x - \mu)^2 f(x)\, dx \geq \int_{\{x,\, |x-\mu| \geq \varepsilon\}} (x - \mu)^2 f(x)\, dx$$

$$\geq \int_{\{x,\, |x-\mu| \geq \varepsilon\}} \varepsilon^2 f(x)\, dx = \varepsilon^2 P(|X - \mu| \geq \varepsilon).$$

This proves the *two-sided Chebyshev inequality* (5.1). The following *one-sided Chebyshev inequality* is proved analogously:

$$P(X - \mu \geq \varepsilon) \leq \frac{\sigma^2}{\sigma^2 + \varepsilon^2}.$$

Corollary By letting $\varepsilon = n\sigma$, one gets from formula (5.1) $n\sigma$-*rules*:

$$P(|X - \mu| \geq n\sigma) \leq 1/n^2 \ \text{ or } \ P(|X - \mu| < n\sigma) > 1 - 1/n^2. \tag{5.2}$$

Example 5.1 The height X of trees in a forest stand has mean value $\mu = 20\,m$ and standard deviation $\sigma = 2\,m$. To obtain an upper limit of the probability that the height of a tree differs at least $4\,m$ from μ, Chebyshev's inequality (5.1) is applied:

$$P(|X - 20| \geq 4) \leq 4/16 = 0.250.$$

For the sake of comparison, assume that the height of trees in this forest stand has a normal distribution. Then the exact probability that the height of a tree differs at least $4\,m$ from μ is

$$P(|X-20| \geq 4) = P(X-20 \geq 4) + P(X-20 \leq -4) = 2\,\Phi(-2) = 0.046 \,.$$

In this case Chebyshev's inequality gives a rather rough upper bound. On the other hand, this inequality requires little input. □

Example 5.2 Let $X_1, X_2, ..., X_n$ be the outcomes of n Bernoulli trials (pages 49, 51), with $p = 1/6$, i.e.

$$X_i = \begin{cases} 1 \text{ with probability } 1/6, \\ 0 \text{ with probability } 5/6, \end{cases} \text{ and } X = \Sigma_{i=1}^{n} X_i.$$

X can be interpreted as the number of the occurrences of "6" when tossing a fair die n times. By making use of the Chebyshev inequality, the smallest integer $n = n_0$ with property

$$P\left(\left| \frac{X}{n} - \frac{1}{6} \right| \geq 0.01 \right) \leq 0.05 \quad \text{for all } n \geq n_0$$

has to be found. Note that X/n is the relative frequency of the occurrence of "6" when tossing the die n times. Since X has a binomial distribution with

$$\mu = E(X) = np = n/6 \text{ and } Var(X) = np(1-p) = 5n/36.$$

X/n has mean 1/6 and variance $\sigma^2 = Var(X/n) = \frac{1}{n^2} Var(X) = \frac{5}{36 \cdot n}$. This implies

$$P\left(\left| \frac{X}{n} - \frac{1}{6} \right| \geq 0.01 \right) \leq \frac{5}{(0.01)^2 \cdot 36 \cdot n} \leq 0.05.$$

Hence, $\dfrac{5}{(0.01)^2 \cdot 36 \cdot 0.05} \leq n$ so that $n_0 \geq 27778$. □

Inequalities of Gauss Let X be a continuous random variable with $\mu = E(X)$ and unimodal density with mode x_m. Then the *Gauss inequalities* are

$$P(|X-\mu| \geq \varepsilon) \leq \frac{4}{9} \frac{\sigma^2 + (\mu - x_m)^2}{(\varepsilon - |\mu - x_m|)^2}, \quad \varepsilon > 0. \tag{5.3}$$

$$P(|X-x_m| \geq \varepsilon) \leq \frac{4}{9\varepsilon^2} [\sigma^2 + (\mu - x_m)^2], \quad \varepsilon > 0. \tag{5.4}$$

(5.3) is also called *Camp-Meidell inequality*.

For $\mu = x_m$, in particular for symmetric densities with symmetry center μ, the inequalities (5.3) and (5.4) are identical. In this case one obtains an improvement of the Chebyshev inequality (but under the additional assumptions of the Gauss inequalities):

$$P(|X-\mu| \geq \varepsilon) \leq (2\sigma/3\varepsilon)^2 \,. \tag{5.5}$$

Corollary By letting $\varepsilon = n\sigma$ and assuming unimodality with $\mu = x_m$, one gets from formula (5.3) or (5.4) $n\sigma$-*rules*:

$$P(|X-\mu| \geq n\sigma) \leq \frac{4}{9n^2} \text{ or } P(|X-\mu| \leq n\sigma) \geq 1 - \frac{4}{9n^2}; \quad n = 1, 2, ... \tag{5.6}$$

Table 5.1 compares the lower bounds for the probabilities $P(|X - \mu| \leq n\sigma)$, which are given by the $n\sigma$-rules (5.2) and (5.6), respectively, with the exact probabilities of the events $'|X - \mu| \leq n\sigma'$, $n = 1, 2, ..., 5$, if X has a normal distribution $N(\mu, \sigma^2)$.

$P(|X-\mu| \leq n\sigma)$	$n = 1$	$n = 2$	$n = 3$	$n = 4$	$n = 5$
Chebyshev inequality	> 0	> 0.750	> 0.889	>0.938	> 0.960
Gauss inequality	> 0.556	> 0.889	> 0.951	> 0.972	> 0.982
Normal distribution	$= 0.683$	$= 0.955$	$= 0.997$	> 0.999	> 0.999

Table 5.1 Lower bounds (5.2) and (5.6) and exact values for normal distribution

Inequalities of Markov Type Let $y = h(x)$ be a nonnegative, strictly increasing function on $[0, \infty)$. Then, for any $\varepsilon > 0$, the *general Markov inequality* is

$$P(|X| \geq \varepsilon) \leq \frac{E(h|X|))}{h(\varepsilon)}. \tag{5.7}$$

(5.7) is proved as follows:

$$E(h(|X|)) = \int_{-\infty}^{+\infty} h(|y|) f(y)\, dy$$

$$\geq \int_{+\varepsilon}^{+\infty} h(|y|) f(y)\, dy + \int_{-\infty}^{-\varepsilon} h(|y|) f(y)\, dy$$

$$\geq h(|\varepsilon|) \int_{+\varepsilon}^{+\infty} f(y)\, dy + h(|\varepsilon|) \int_{-\infty}^{-\varepsilon} f(y)\, dy$$

$$= h(\varepsilon) P(|X| \geq \varepsilon),$$

which is equivalent to (5.7). Letting $h(x) = x^a$, $a > 0$, inequality (5.7) yields *Markov's inequality* as such:

$$P(|X| \geq \varepsilon) \leq \frac{E(|X|^a)}{\varepsilon^a}. \tag{5.8}$$

From (5.8) Chebyshev's inequality is obtained by letting $a = 2$ and replacing X with $X - \mu$.

If $h(x) = e^{bx}$, $b > 0$, Markov's inequality (5.7) yields an *exponential inequality*:

$$P(|X| \geq \varepsilon) \leq e^{-b\varepsilon} E\left(e^{b|X|}\right). \tag{5.9}$$

Markov's inequality (5.8) and the exponential inequality (5.9) are usually superior to Chebyshev's inequality, since, given X and ε, their right-hand sides can be minimized with respect to a and b. On the other hand, to determine the mean values in formulas (5.8) and (5.9), the probability distribution of X needs to be known. But in this case the exact value of the desired probability $P(|X| \geq \varepsilon)$ can be calculated anyway. Hence, application of (5.8) and (5.9) makes sense only if the expected values involved are known from whatsoever source (expert opinions) or they are estimated based on a sample taken from X, i.e., the random experiment with output X is independently repeated n times to get a sequence of values of X: $x_1, x_2, ..., x_n$. For instance, the mean value $m = E(|X|^a)$ occurring in (5.8) would have to be estimated by the arithmetic mean of the $|x_i|^n$:

$$\hat{m} = \tfrac{1}{n} \sum_{i=1}^{n} |x_i|^a .$$

If the variance σ^2 in (5.1) is unknown, it also has to be estimated from a sample $\{x_1, x_2, ..., x_n\}$. The estimator is

$$s^2 = \tfrac{1}{n-1} \sum_{i=1}^{n} (x_i - \bar{x})^2 \text{ with } \bar{x} = \tfrac{1}{n} \sum_{i=1}^{n} x_i .$$

Continuation of Example 5.1 Let us check whether the upper bound of Chebyshev's inequality (5.1) can by improved by (5.8) if X has a normal distribution with mean μ and standard deviation $\sigma = 2$.

For $a = 1$, the mean value $E(|X - \mu|^a)$ becomes (see page 79),

$$E(|X - \mu|) = \sqrt{\tfrac{2}{\pi}}\, \sigma \approx 0.798 \cdot 2 = 1.596.$$

Hence, (5.8) yields

$$P(|X - \mu| \ge 4) \le \tfrac{E(|X-\mu|)}{4} = \tfrac{1.596}{4} = 0.399.$$

This is a worse result than the one given by Chebyshev's inequality ($a = 2$).

Now let $a = 4$. Then (see page 83, note that $X - \mu$ has mean value 0)

$$E(|X - \mu|^4) = \mu_4 = E((X - \mu)^4) = 3\sigma^4.$$

Hence, (5.8) yields

$$P(|X - \mu| \ge 4) \le \tfrac{E(|X-\mu|^4)}{\varepsilon^4} = \tfrac{3 \cdot 2^4}{4^4} = \tfrac{48}{256} = 0.1875.$$

This is a substantial improvement of the bound given by Chebyshev's inequality. $\square$

5.1.2 Inequalities for Moments

Inequalities of Chebyshev Let functions $g(x)$ and $h(x)$ be either both nonincreasing or both nondecreasing. Then,

$$E[g(X)]\, E[h(X)] \le E[g(X)h(X)]. \tag{5.10}$$

If g is nonincreasing and h nondecreasing or vice versa, then

$$E[g(X)]\, E[h(X)] \ge E[g(X)h(X)].$$

As an important special case, let

$$g(x) = x^r \text{ and } h(x) = x^s; \quad r, s \ge 0.$$

Then, from (5.10),

$$E(|X^r|)\, E(|X^s|) \le E(|X^{r+s}|).$$

Inequality of Schwarz

$$[E(|XY|)]^2 \leq E(|X|^2)\, E(|Y|^2). \tag{5.11}$$

Hölder's Inequality Let r and s be positive numbers satisfying $\frac{1}{r} + \frac{1}{s} = 1$. Then

$$E(|XY|) \leq [E(|X|^r)]^{1/r}[E(|Y|^s)]^{1/s}. \tag{5.12}$$

For $r = s = 2$, Hölder's inequality implies the inequality of Schwarz.

Inequality of Minkovski (Triangle Inequality) For $r \geq 1$,

$$[E(|X+Y|^r)]^{1/r} \leq [E(|X|^r)]^{1/r} + [E(|Y|^r)]^{1/r}. \tag{5.13}$$

Inequality of Jensen Let $h(x)$ be a convex (concave) function. Then, for any X,

$$h(E(X)) \underset{(\geq)}{\overset{\leq}{}} E(h(X)). \tag{5.14}$$

In particular, if X is nonnegative and $h(x) = x^a$ (convex for $a \geq 1$ and $a \leq 0$, concave for $0 \leq a \leq 1$), $h(x) = e^x$ (convex), and $h(x) = \ln x$ (concave), the respective inequalities of Jensen are

$$[E(X)]^a \leq E(X^a) \quad \text{for } a > 1 \text{ or } a < 0,$$

$$[E(X)]^a \geq E(X^a) \quad \text{for } 0 < a < 1,$$

$$e^{E(X)} \leq E\!\left(e^X\right), \tag{5.15}$$

$$\ln E(X) \geq E(\ln X).$$

Example 5.3 To get an impression on the sharpness of the inequalities of Schwarz and Minkowski, let us consider a random vector (X, Y) with joint density

$$f_{X,Y}(x,y) = x + y, \ 0 \leq x, y \leq 1,$$

and marginal densities (see example 3.5, page 129)

$$f_X(x) = x + 1/2, \quad f_Y(y) = y + 1/2; \quad 0 \leq x, y \leq 1.$$

Schwarz inequality: The second moment of X is

$$E(X^2) = \int_0^1 x^2(x + 1/2)\, dx = 5/12.$$

For symmetry reasons, $E(Y^2) = 5/12$ as well. Thus, (5.11) yields

$$[E(XY)]^2 \leq 0.174$$

so that the upper bound for $E(XY)$ is 0.417. For the sake of comparison, the exact value of $E(XY)$ is

$$E(XY) = \int_0^1 \int_0^1 xy\,(x+y)\, dx\, dy = 2 \int_0^1 \int_0^1 x^2 y \, dx\, dy = 0.33\underline{3}.$$

Minkovsky inequality: For $r = 1$, inequality (5.13) is trivial (left- and right-hand side are equal). Let $r = 2$. Then (5.13) becomes

$$\sqrt{E(X+Y)^2} \le \sqrt{E(X^2)} + \sqrt{E(Y)^2}.$$

Since $E(X^2) = E(Y^2) = 5/12$, an upper bound for $\sqrt{E(X+Y)^2}$ is 1.291:

$$\sqrt{E(X+Y)^2} \le 1.291.$$

For the sake of comparison:

$$E(X+Y)^2 = \int_0^1\int_0^1 (x^2 + 2xy + y^2)(x+y)\,dx\,dy$$

$$= \int_0^1\int_0^1 (x^3 + 3x^2y + 3xy^2 + y^3)\,dx\,dy$$

$$= \int_0^1 \left(\tfrac{1}{4} + y + \tfrac{3}{2}y^2 + y^3\right)dy = \tfrac{1}{4} + \tfrac{1}{2} + \tfrac{1}{2} + \tfrac{1}{4} = \tfrac{3}{2}.$$

Hence, $\sqrt{E(X+Y)^2} = 1.225$. $\qquad\qquad\qquad\qquad\qquad\qquad\qquad\qquad\square$

5.2 LIMIT THEOREMS

5.2.1 Convergence Criteria for Sequences of Random Variables

There are three large classes of limit theorems in probability theory: 1) The laws of the large numbers, 2) the central limit theorem and its numerous modifications, and 3) the local limit theorems. The laws of the large numbers are essentially statements on the convergence behavior of arithmetic means of random variables. They constitute the theoretical foundation of statistical methods for the estimation of parameters of probability distributions based on samples. They also have applications in simulation procedures for the numerical solutions of stochastic and even deterministic problems. The central limit theorem justifies the application of the normal distribution as distribution of random variables, which are known to arise by the additive superposition of numerous random influences. Local limit theorems investigate the convergence of probability densities of continuous random variables and the convergence of the probabilities $P(X = x_i)$ of discrete random variables X.

Limit theorems in probability theory are subject to certain convergence criteria for sequences of random variables, which next have to be introduced (even if in a more or less heuristic way).

1) Convergence in Probability A sequence of random variables $\{X_1, X_2, ...\}$ converges *in probability* towards a random variable X if for all $\varepsilon > 0$,

$$\lim_{i\to\infty} P(|X_i - X| > \varepsilon) = 0. \tag{5.16}$$

2) Convergence in Mean A sequence of random variables $\{X_1, X_2, ...\}$ with property

$$E(|X_i|) < \infty; \quad i = 1, 2, ...$$

converges *in mean* towards a random variable X if

$$\lim_{n \to \infty} E(|X_i - X|) = 0 \quad \text{and} \quad E(|X|) < \infty. \tag{5.17}$$

3) Mean Square Convergence A sequence of random variables $\{X_1, X_2, ...\}$ with

$$E(|X_i|^2) < \infty; \quad i = 1, 2, ...,$$

converges *in mean square* or *in square mean* towards a random variable X if

$$\lim_{n \to \infty} E(|X_i - X|^2) = 0 \quad \text{and} \quad E(|X|^2) < \infty. \tag{5.18}$$

4) Convergence with Probability 1 A sequence of random variables $\{X_1, X_2, ...\}$ converges *with probability* 1 or *almost sure* towards a random variable X if

$$P(\lim_{i \to \infty} X_i = X) = 1.$$

5) Convergence in Distribution Let the random variables X_i have the distribution functions $F_{X_i}(x)$; $i = 1, 2, ...$ Then $\{X_1, X_2, ...\}$ converges towards a random variable X with distribution function $F_X(x)$ *in distribution* if, for all points of continuity x of $F_X(x)$,

$$\lim_{i \to \infty} F_{X_i}(x) = \lim_{i \to \infty} P(X_i \leq x) = P(X \leq x) = F_X(x).$$

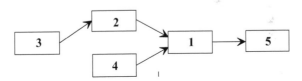

Figure 5.1 Relations between the convergence criteria 1-5.

Figure 5.1 shows the implications between the convergence critria. The integers refer to the respective convergence criteria listed above.

Under additional assumptions, the opposite implications may be true as well (in what follows, the notation $X_n \overset{k}{\to} X$ refers to the convergence criterion **k** above):

1) If $X_n \overset{5}{\to} c$ is true with a finite constant c, then $X_n \overset{1}{\to} c$, i.e., in case of a constant limit, convergence in probability and convergence in distribution are equivalent.

2) If $X_n \overset{1}{\to} X$ is true, then there exists a subsequence $\{X_{n_1}, X_{n_2}, ...\}$ of the given sequence $\{X_1, X_2, ...\}$ so that $X_{n_i} \overset{4}{\to} X$ for $i \to \infty$.

5.2.2 Laws of Large Numbers

5.2.2.1 Weak Laws of Large Numbers

There are *weak* and *strong laws of large numbers*. They essentially deal with the convergence behavior of arithmetic means $\overline{X}_n$ for $n \to \infty$, where

$$\overline{X}_n = \frac{1}{n} \sum_{i=1}^{n} X_i.$$

Definition 5.1 A sequence of random variables $\{X_1, X_2, ...\}$ satisfies the *weak law of large numbers* if there exists a sequence of real numbers $\{a_1, a_2, ...\}$ so that the sequence $\{\overline{X}_1 - a_1, \overline{X}_2 - a_2, ...\}$ converges in probability towards 0.   ●

A direct consequence of the Chebyshev's inequality (5.1) is the following version of the weak law of large numbers.

Theorem 5.1 Let $\{X_1, X_2, ...\}$ be a sequence of independent, identically distributed random variables with finite mean μ and variance σ^2. Then the sequence of arithmetic means $\{\overline{X}_1, \overline{X}_2, ...\}$ converges in probability towards μ, i.e., for all $\varepsilon > 0$,

$$\lim_{n \to \infty} P\left(\left| \overline{X}_n - \mu \right| > \varepsilon \right) = 0.$$

Proof In view of $Var(\overline{X}_n) = \sigma^2/n$, Chebyshev's inequality (5.1) yields

$$P\left(\left| \overline{X}_n - \mu \right| > \varepsilon \right) \leq \frac{\sigma^2}{n\varepsilon^2}. \tag{5.19}$$

Letting $n \to \infty$ proves the theorem.   ■

Bernoulli's Weak Law of the Large Numbers The first version of the weak law of the large numbers can be found in *Bernoulli* (1713), the first textbook on probability theory. Jacob Bernoulli considered the limit behavior of the sequence $\{X_1, X_2, ...\}$, where the X_i are the indicator variables for the occurrence of a random event A in a series of n independent trials:

$$X_i = \begin{cases} 1 & \text{if } A \text{ occurs,} \\ 0 & \text{otherwise.} \end{cases} \quad i = 1, 2, ...$$

The sum $Z_n = \sum_{i=1}^{n} X_i$ is the number of occurrences of the random event A in this series, and the arithmetic mean

$$\hat{p}_n(A) = \overline{X}_n = \frac{1}{n} \sum_{i=1}^{n} X_i$$

is the relative frequency of the occurrence of event A in a series of n trials. From section 2.2.2, page 51, we know that Z_n has a binomial distribution with parameters n and $p = P(A)$ so that

$$E(Z_n) = np \quad \text{and} \quad Var(Z_n) = np(1-p).$$

Therefore, the relative frequency $\hat{p}_n(A)$ has mean value

$$E(\hat{p}_n(A)) = \frac{1}{n} \sum_{i=1}^{n} E(X_i) = \frac{1}{n} (n P(A)) = P(A) = p$$

and variance

$$Var(\hat{p}_n(A)) = \frac{p(1-p)}{n}.$$

Now, applying (5.1) to the sequence $\{\hat{p}_1(A), \hat{p}_2(A), ...\}$ yields for all $\varepsilon > 0$,

$$P(|\hat{p}_n(A) - P(A)| > \varepsilon) \le \frac{p(1-p)}{n\varepsilon^2} \to 0 \text{ as } n \to \infty.$$

This proves *Bernoulli's weak law of the large numbers*:

> *The relative frequency $\hat{p}_n(A)$ of the occurrence of the random event A in a series of n independent trials converges to $p = P(A)$ in probability as $n \to \infty$:*
>
> $$\lim_{n\to\infty} \hat{p}_n(A) = P(A).$$

Two more variants of the weak law of the large numbers will be added.

Theorem 5.2 (Chebyshev) Let $\{X_1, X_2, ...\}$ be a sequence of (not necessarily independent) random variables X_i with finite means $\mu_i = E(X_i)$; $i = 1, 2, ...$. On condition

$$\lim_{i\to\infty} Var(X_i) = 0,$$

the sequence $\{X_1 - \mu_1, X_2 - \mu_2, ...\}$ converges in probability towards 0. ■

The following theorem does not need assumptions on variances. Instead, the pairwise (not the complete, page 145) independence of the sequence $\{X_1, X_2, ...\}$ is required, i.e., X_i and X_j are independent for $i \ne j$.

Theorem 5.3 (Chintchin) Let $\{X_1, X_2, ...\}$ be a sequence of pairwise independent, identically distributed random variables with finite mean μ. Then the corresponding sequence of arithmetic means $\{\bar{X}_1, \bar{X}_2, ...\}$ converges in probability towards μ. ■

5.2.2.2 Strong Laws of Large Numbers

These laws of the large numbers are called *strong*, since the almost sure convergence implies the convergence in probability (Figure 5.1). Thus, almost sure convergence is a stronger property than convergence in probability.

Definition 5.2 A sequence of random variables $\{X_1, X_2, ...\}$ satisfies the *strong law of the large numbers* if there is a sequence of real numbers $\{a_1, a_2, ...\}$ so that the sequence $\{\bar{X}_1 - a_1, \bar{X}_2 - a_2, ...\}$ converges with probability 1 towards 0:

$$P(\lim_{n\to\infty} (X_i - a_i) = 0) = 1.$$
●

If a sequence of random variables satisfies the strong law of the large numbers with a sequence of real numbers $\{a_1, a_2, ...\}$, then it satisfies the weak law of the large numbers with the same sequence of real numbers. The converse is generally not true. Here two versions of the strong law of the large numbers are given.

Theorem 5.4 (*Kolmogorov*) Let $\{X_1, X_2, ...\}$ be a sequence of independent, identically distributed random variables with finite mean μ. Then the sequence of arithmetic means $\{\overline{X}_1, \overline{X}_2, ...\}$ converges with probability 1 towards μ.   ∎

Theorems 5.4 implies that the sequence of relative frequencies $\{\hat{p}_1(A), \hat{p}_2(A), ...\}$ also converges towards $p = P(A)$ with probability 1. Thus, Bernoulli's law of the large numbers is both weak and strong. The following theorem abandons the assumption of identically distributed random variables.

Theorem 5.5 (*Kolmogorov*) Let $\{X_1, X_2, ...\}$ be a sequence of independent random variables with parameters $\mu_i = E(X_i)$ and $\sigma_i^2 = Var(X_i)$; $i = 1, 2, ...$ On condition

$$\sum_{i=1}^{\infty}(\sigma_i/i)^2 < \infty,$$

the sequence $\{Y_1, Y_2, ...\}$ with

$$Y_n = \overline{X}_n - \frac{1}{n}\sum_{i=1}^{n}\mu_i$$

converges with probability 1 towards 0.   ∎

5.2.3 Central Limit Theorem

The central limit theorem provides theoretical reasons for the significant role of the normal distribution in probability theory and its applications. Intuitively, it states that a random variable, which arises from additive superposition of many random influences with none of them being dominant, has approximately a normal distribution. The simplest version of the central limit theorem is the following one:

Theorem 5.6 (*Lindeberg* and *Lèvy*) Let $Z_n = X_1 + X_2 + \cdots + X_n$ be the sum of n independent, identically distributed random variables X_i with finite mean $E(X_i) = \mu$ and finite variance $Var(X_i) = \sigma^2$, and let S_n be the standardization of Z_n, i.e.

$$S_n = \frac{Z_n - n\mu}{\sigma\sqrt{n}}.$$

Then, $$\lim_{n\to\infty} P(S_n \leq x) = \frac{1}{\sqrt{2\pi}} \int_{-\infty}^{x} e^{-u^2/2} du = \Phi(x),$$

where $\Phi(x)$ is the distribution function of the standard normal distribution $N(0, 1)$.   ∎

Corollary Under the conditions of theorem 5.6, Z_n has for sufficiently large n appro- ximately a normal distribution with mean value $n\mu$ and variance $n\sigma^2$:

$$Z_n \approx N(n\mu, n\sigma^2). \qquad (5.20)$$

Thus, Z_n is *asymptotically normally distributed* as $n \to \infty$. The fact that Z_n has mean value $n\mu$ and variance $n\sigma^2$ follows from (4.57), page 188.

As a rule of thumb, (5.20) gives satisfactory results if $n \geq 20$. Sometimes even $n \geq 10$ is sufficient. The following theorem shows that the assumptions of theorem 5.6 can be partially weakened.

Theorem 5.7 (*Lindeberg and Feller*) Let $Z_n = X_1 + X_2 + \cdots + X_n$ be the sum of in-dependent random variables X_i with densities $f_{X_i}(x)$, finite means $\mu_i = E(X_i)$, and finite variances $\sigma_i^2 = Var(X_i)$. Let further S_n be the standardization of Z_n :

$$S_n = \frac{Z_n - E(Z_n)}{\sqrt{Var(Z_n)}} = \frac{Z_n - \sum_{i=1}^{n} \mu_i}{\sqrt{\sum_{i=1}^{n} \sigma_i^2}}.$$

Then the limit relation

$$\lim_{n \to \infty} P(S_n \leq x) = \Phi(x) = \frac{1}{\sqrt{2\pi}} \int_{-\infty}^{x} e^{-u^2/2} du \qquad (5.21)$$

is uniformly true for all x and $Var(Z_n)$ has the properties

$$\lim_{n \to \infty} \sqrt{Var(Z_n)} \to \infty \quad \text{and} \quad \lim_{n \to \infty} \max_{i=1,2,\ldots,n} \left(\frac{\sigma_i}{\sqrt{Var(Z_n)}} \right) \to 0 \qquad (5.22)$$

if and only if the *Lindeberg condition*

$$\lim_{n \to \infty} \frac{1}{Var(Z_n)} \sum_{i=1}^{n} \int_{\{x, |x-\mu_i| > \varepsilon \sqrt{Var(Z_n)}\}} (x - \mu_i)^2 f_{X_i}(x) dx = 0$$

is fulfilled for all $\varepsilon > 0$. ∎

The properties (5.22) imply that no term X_i in the sum dominates the rest and that for $n \to \infty$ the contributions of the X_i to the sum uniformly tend to 0. Under the assumptions of theorem 5.6, the X_i a priori have this property.

Example 5.4 Weekdays a car dealer sells on average one car (of a certain make) per $\mu = 2.4$ days with a standard deviation of $\sigma = 1.6$.

1) What is the probability that the dealer sells at least 35 cars a quarter (75 weekdays)? Let $X_i; i = 1, 2, \ldots, X_0 = 0$ be the time span between selling the $(i-1)$th and the ith car. Then $Z_n = X_1 + X_2 + \cdots + X_n$ is the time point, at which the nth car is sold (sel-ling times assumed to be negligibly small). Hence, the probability $P(Z_{35} \leq 75)$ has to be determined.

If the X_i are assumed to be independent, then

$$E(Z_{35}) = 35 \cdot 2.4 = 84 \text{ and } Var(Z_{35}) = 35 \cdot 1.6^2 = 89.6.$$

In view of (5.20), Z_{35} has approximately an $N(84, 89.6)$-distribution. Hence,

$$P(Z_{35} \leq 75) \approx \Phi\left(\frac{75 - 84}{9.466}\right) = \Phi(-0.95) = 0.171.$$

2) How many cars n_{min} the dealer does have to stock at least at the beginning of a quarter to make sure that every customer can immediately buy a car with a probability of not smaller than 0.95?

$n = n_{min}$ is the smallest n with property that

$$P(Z_{n+1} > 75) \geq 0.95.$$

Equivalently, n_{min} is the smallest n with property

$$P(Z_{n+1} \leq 75) \leq 0.05 \text{ or } \Phi\left(\frac{75 - 2.4\,(n+1)}{1.6\sqrt{n+1}}\right) \leq 0.05.$$

Since the 0.05-percentile of an $N(0, 1)$-distribution is $x_{0.05} = -1.64$, the latter inequality is equivalent to

$$\frac{75 - 2.4\,(n + 1)}{1.6\sqrt{n + 1}} \leq -1.64 \text{ or } (n - 30.85)^2 \geq 37.7.$$

Hence, $n_{min} = 37$. $\qquad\qquad\qquad\qquad\qquad\qquad\qquad\qquad\qquad\qquad\qquad$ $\square$

Normal Approximation to the Binomial Distribution Any binomially with parameters n and p distributed random variable Z_n can be represented as the sum of n independent (0,1)-random variables of structure

$$X_i = \begin{cases} 1 & \text{with probability } p, \\ 0 & \text{with probability } 1 - p, \end{cases} \quad 0 \leq p \leq 1.$$

Thus, $Z_n = X_1 + X_2 + \cdots + X_n$ so that the assumptions of central limit theorem 5.6 are fulfilled with $\mu = p$ and $\sigma^2 = np(1 - p)$:

$$E(Z_n) = np, \quad Var(Z_n) = np\,(1 - p). \tag{5.23}$$

A corollary of theorem 5.6 is

Theorem 5.8 (*Central limit theorem of Moivre-Laplace*) If the random variable X has a binomial distribution with parameters n and p, then, for all x,

$$\lim_{n \to \infty} P\left(\frac{Z_n - np}{\sqrt{np(1-p)}} \leq x\right) = \frac{1}{\sqrt{2\pi}} \int_{-\infty}^{x} e^{-u^2/2} du. \qquad\blacksquare$$

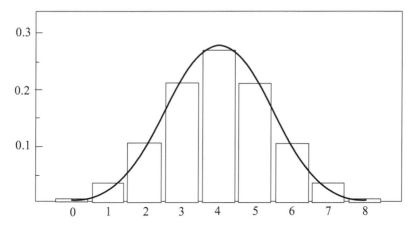

Figure 5.2 Approximation of the normal distribution to the binomial distribution

As a special case of formula (5.20), Z_n has approximately a normal distribution:

$$Z_n \approx N(np, np(1-p)).$$

Thus,

$$P(i_1 \le Z_n \le i_2) \approx \Phi\left(\frac{i_2 + \frac{1}{2} - np}{\sqrt{np(1-p)}}\right) - \Phi\left(\frac{i_1 - \frac{1}{2} - np}{\sqrt{np(1-p)}}\right); \quad 0 \le i_1 \le i_2 \le n.$$

$$(5.24)$$

$$P(Z_n = i) = \binom{n}{i} p^i (1-p)^{n-i} \approx \Phi\left(\frac{i + \frac{1}{2} - np}{\sqrt{np(1-p)}}\right) - \Phi\left(\frac{i - \frac{1}{2} - np}{\sqrt{np(1-p)}}\right), \quad 0 \le i \le n.$$

The term $\pm 1/2$ is called *continuity correction*. It improves the accuracy of the approximation, since a discrete distribution is approximated by a continuous one. Because the distribution function of Z_n has only jumps at integers i, there is

$$F_{Z_n}(i) = F_{Z_n}(i + \tfrac{1}{2}), \quad i = 0, 1, ..., n.$$

The approximation formulas (5.24) are the better the larger n is and the closer p is to 1/2. Because the normal distribution is used to approximate the distribution of a non-negative random variable, the condition

$$E(Z_n) \ge 3\sqrt{Var(Z_n)} \tag{5.25}$$

should be satisfied (see page 79, there written as $\mu \ge 3\sigma$) to make sure the approximation yields satisfactory results. In view of (5.23), this condition is equivalent to

$$n > 9\frac{1-p}{p}. \tag{5.26}$$

Thus, for $p = 1/2$, only 10 summands may be sufficient to get good approximations, whereas for $p = 0.1$ the number n required is at least 82. In practice the following rules of thumb will usually do:

$$E(Z_n) = np > 35 \text{ and/or } Var(Z_n) = np(1-p) > 10.$$

Continuation of Example 2.5 (page 52) From a large delivery of calculators a sample of size $n = 100$ is taken. The delivery will be accepted if there are at most four defective calculators in the sample. The average rate of defective calculators from the producer is known to be 2%.

1) What is the probability P_{risk} that the delivery will be rejected (producer's risk)?

2) What is the probability C_{risk} to accept the delivery although it contains 7% defective calculators (consumer's risk)?

1) The underlying binomial distribution has parameters $n = 100$ and $p = 0.02$:

$$p_i = P(Z_{100} = i) = \binom{100}{i} (0.02)^i (0.98)^{100-i}, \quad i = 0, 1, ..., 100.$$

The random number Z_{100} of defective calculators in the sample has mean value and standard deviation

$$E(Z_{100}) = 2 \text{ and } \sqrt{Var(Z_{100})} = \sqrt{100 \cdot 0.02 \cdot 0.98} = 1.4.$$

This gives for the exact value

$$P_{risk} = 1 - p_0 - p_1 - p_2 - p_3 - p_4 = 0.051$$

the approximative value

$$P_{risk} \approx P(Z_n \geq 5) \approx P\left(\frac{Z_n - 2}{1.4} \geq \frac{5-2-0.5}{1.4}\right)$$

$$= 1 - \Phi(1.786) \approx 1 - 0.962$$

$$= 0.038.$$

This approximative value is not satisfactory since p is too small. Condition (5.26) is far from being fulfilled.

2) In this case, $p = 0.07$ so that

$$E(Z_{100}) = 7 \text{ and } \sqrt{Var(Z_{100})} = 2.551.$$

This gives for C_{risk} the approximative value

$$C_{risk} = P(Z_{100} \leq 4) = P\left(\frac{Z_n - 7}{2.551} \leq \frac{4-7+0.5}{2.551}\right) = \Phi(-1.176)$$

$$= 0.164.$$

The exact value is 0.163.

Taking into account the continuity correction proved essential both for calculating the approximative values of P_{risk} and C_{risk}.   □

Normal Approximation to the Poisson Distribution From example 4.18 (page 180) or from Theorem 7.7 (page 285) we know that the sum of independent, Poisson distributed random variables has a Poisson distribution, the parameter of which is the sum of the parameters of the Poisson distributions of these random variables. This implies that every Poisson with parameter λ distributed random variable X can be represented as a sum Z_n of n independent, identically Poisson with parameter λ/n distributed random variables X_i:

$$X = Z_n = X_1 + X_2 + \cdots + X_n, \quad n = 1, 2, ..., \tag{5.27}$$

with

$$P(X_i = k) = \frac{(\lambda/n)^k}{k!} e^{-(\lambda/n)}; \quad k = 0, 1, ...,$$

and

$$E(X_i) = Var(X_i) = \frac{\lambda}{n}; \quad i = 1, 2, ..., n.$$

Random variables X (or, equivalently, their probability distributions), which can be represented for any integer $n > 1$ as the sum of n independent, identically distributed random variables, are called *infinitely divisible*. Other probability distributions, which have this property, are the normal, the Cauchy, and the gamma distribution.

X as given by the sum (5.27) is Poisson distributed with parameters

$$E(X) = \lambda \text{ and } Var(X) = \lambda.$$

Since the sum representation (5.27) satisfies the assumptions of the central limit theorem 5.6, X has approximately the normal distribution

$$X \approx N(\lambda, \lambda), \quad F_X(x) \approx \Phi\left(\frac{x - \lambda}{\sqrt{\lambda}}\right)$$

so that, using the continuity correction $1/2$ as in case of the normal approximation to the binomial distribution,

$$P(i_1 \leq X \leq i_2) \approx \Phi\left(\frac{i_2 + \frac{1}{2} - \lambda}{\sqrt{\lambda}}\right) - \Phi\left(\frac{i_1 - \frac{1}{2} - \lambda}{\sqrt{\lambda}}\right),$$

$$\tag{5.28}$$

$$P(X = i) \approx \Phi\left(\frac{i + \frac{1}{2} - \lambda}{\sqrt{\lambda}}\right) - \Phi\left(\frac{i - \frac{1}{2} - \lambda}{\sqrt{\lambda}}\right).$$

Since the distribution of a nonnegative random variable is approximated by the normal distribution, analogously to (5.25), the assumption

$$E(X) = \lambda > 3 \sqrt{Var(X)} = 3\sqrt{\lambda}$$

has to be made. Hence, the normal approximation to the Poisson distribution can only be expected to yield good results if $\lambda > 9$.

Continuation of Example 2.8 (page 56). Let X be the random number of staff of a company being on sick leave a day. Long-term observations have shown that X has a Poisson distribution with parameter $\lambda = E(X) = 10$.

What is the probability that the number of staff being on sick leave a day is 9, 10, or 11? The normal approximation to this probability is

$$P(9 \leq X \leq 11) \approx \Phi\left(\frac{11 + \frac{1}{2} - 10}{\sqrt{10}}\right) - \Phi\left(\frac{9 - \frac{1}{2} - 10}{\sqrt{10}}\right)$$

$$= \Phi(0.474) - \Phi(-474) = 2\,\Phi(0.474) - 1$$

$$= 0.364.$$

This value almost coincides with the exact one, which is 0.3639. Again, making use of the continuity correction is crucial for obtaining a good result. The approximation for p_{10}, for instance, is

$$p_{10} = \frac{10^{10}}{10!!}e^{-10} \approx \Phi\left(\frac{10 + \frac{1}{2} - 10}{\sqrt{10}}\right) - \Phi\left(\frac{10 - \frac{1}{2} - 10}{\sqrt{10}}\right) = 2\Phi(0.158) - 1$$

$$= 0.1255.$$

The exact value is 0.1251.   □

5.2.3 Local Limit Theorems

The central limit theorems investigate the convergence of distribution functions of sums of random variables towards a limit distribution function. The *local limit theorems* consider the convergence of probabilities $P(Z = x_i)$ towards a limit probability if Z is the sum of discrete random variables, or they deal with the convergence behavior of the densities of sums of continuous random variables. This section presents three theorems of this type without proof.

Theorem 5.9 (*Local limit theorem of Moivre-Laplace*) The random variable X have a binomial distribution with parameters n and p:

$$P(X = i) = b(i; n, p) = \binom{n}{i} p^i (1 - p)^{n-i}; \quad i = 0, 1, ..., n.$$

Then,

$$\lim_{n \to \infty} \left\{ \sqrt{np(1-p)}\, b(i; n, p) - \frac{1}{\sqrt{2\pi}} \exp\left[-\frac{1}{2}\left(\frac{i - np}{\sqrt{np(1-p)}}\right)^2\right] \right\} = 0.$$

The convergence is uniform with regard to $i = 0, 1, ..., n$.   ■

Theorem 5.9 implies that for sufficiently large n an acceptable approximation for the probability $b(i;n,p)$ is

$$b(i;n,p) \approx \frac{1}{\sqrt{2\pi} \sqrt{np(1-p)}} \exp\left[-\frac{1}{2}\left(\frac{i-np}{\sqrt{np(1-p)}}\right)^2\right]. \tag{5.29}$$

Theorem 5.10 (*Poisson approximation to the binomial distribution*) If the parameters n and p of the binomial distribution tend to ∞ and 0, respectively, in such a way that their product np stays constant λ, $\lambda > 0$, then

$$\lim_{\substack{n\to\infty \\ p\to 0 \\ np=\lambda}} b(i;n,p) = \frac{\lambda^i}{i!} e^{-\lambda}; \quad i = 0,1,\dots .$$

Proof From the definition of the binomial coefficient $\binom{n}{i}$ (see formula (1.5)),

$$\frac{b(i;n,p)}{b(i-1;n,p)} = \frac{n-1+1}{i} \cdot \frac{p}{1-p} = \frac{np}{i(1-p)} = \left(1-\frac{1}{i}\right)\left(\frac{P}{1-P}\right). \tag{5.30}$$

After having taken the limit, the $b(i;n,p)$ can no longer depend on n and p, but are only functions of i and λ, which are denoted as $h(i,\lambda)$. From (5.30),

$$\lim_{\substack{n\to\infty \\ p\to 0 \\ np=\lambda}} \left\{\frac{b(i;n,p)}{b(i-1;n,p)}\right\} = \frac{h(i,\lambda)}{h(i-1,\lambda)} = \frac{\lambda}{i}, \quad i = 1,2,\dots$$

Therefore, the limit probabilities of the binomial distribution satisfy

$$h(i,\lambda) = \frac{\lambda}{i} h(i-1,\lambda) : \quad i = 1,2,\dots$$

For $i = 1$ and $i = 2$, this functional equation becomes

$$h(1,\lambda) = \lambda\, h(0,\lambda) \quad \text{and} \quad h(2,\lambda) = \frac{\lambda}{2} h(1,\lambda) = \frac{\lambda^2}{2!} h(0,\lambda).$$

Induction yields

$$h(i,\lambda) = \frac{\lambda^i}{i!} h(0,\lambda).$$

The normalizing condition (2.6) at page 43 gives the still unknown constant $h(0,\lambda)$:

$$\sum_{i=0}^{\infty} h(i;\lambda) = h(0,\lambda) \sum_{i=0}^{\infty} \frac{\lambda^i}{i!} = h(0,\lambda) e^{\lambda} = 1$$

so that $h(0,\lambda) = e^{-\lambda}$. This completes the proof of the theorem:

$$h(i,\lambda) = \frac{\lambda^i}{i!} e^{-\lambda}; \quad i = 0,1,\dots . \qquad \blacksquare$$

Note: The result of this theorem is formula (2.40) at page 57.

Example 5.5 Let X have a binomial distribution with parameters $n = 12$ and $p = 0.4$. For the exact probability

$$p_4 = \binom{12}{4} (0.4)^4 (0.6)^8 = 0.2128$$

the local limit theorem (5.29) yields the appoximative value

$$p_4 \approx \frac{1}{\sqrt{2\pi} \sqrt{12 \times 0.4 \times 0.6}} \exp\left[-\frac{1}{2} \left(\frac{4 - 12 \times 0.4}{\sqrt{12 \times 0.4 \times 0.6}} \right)^2 \right] = 0.2104,$$

whereas the central limit theorem (5.24) provides the approximative value

$$p_4 \approx \Phi\left(\frac{4 + \frac{1}{2} - 12 \times 0.4}{\sqrt{12 \times 0.4 \times 0.6}} \right) - \Phi\left(\frac{4 - \frac{1}{2} - 12 \times 0.4}{\sqrt{12 \times 0.4 \times 0.6}} \right)$$

$$= \Phi(-0.17680) - \Phi(-0.7660) = 0.2149.$$

The Poisson approximation with $np = 4.8$ gives the worst result:

$$p_4 \approx \frac{4.8^4}{4!} e^{-4.8} = 0.1820. \qquad \square$$

To formulate the next local limit theorem for sums of discrete random variables, the following definition is needed:

Definition 5.3 A discrete random variable X, which for given real numbers a and b with $b > 0$, can only take on values of the form

$$x_k = a + kb; \quad k = 0, \pm 1, \pm 2, ..., \qquad (5.31)$$

is called *lattice distributed*. The corresponding probability distribution of X is called a *lattice distribution*. The largest constant b, which allows the representation of all realizations of X by (5.31), is called the *lattice constant* of X or its probability distribution. Specifically, a lattice distribution with $a = 0$ is an *arithmetic distribution*. ●

Lattice distributed random variables obviously include all integer-valued random variables as geometrically, binomially, and Poisson distributed random variables.

Theorem 5.11 (*Gnedenko*) Let $\{X_1, X_2, ...\}$ be a sequence of independent, identically lattice distributed random variables with values (5.31), finite mean value μ, finite, positive variance σ^2, and

$$P_n(m) = P(X_1 + X_2 + \cdots + X_n = na + mb); \quad m = 0, \pm 1, \pm 2, ...$$

Then the following limit relation is true uniformly in m if and only if b is the lattice constant of the $X_1, X_2, ...$:

$$\lim_{n \to \infty} \left\{ \frac{\sigma \sqrt{n}}{b} P_n(m) - \frac{1}{\sqrt{2\pi}} \exp\left[-\frac{1}{2} \left(\frac{an + mb - \mu n}{\sigma \sqrt{n}} \right)^2 \right] \right\} = 0. \qquad \blacksquare$$

Finally, a local limit theorem is given which deals with the convergence of the density of sums of random variables.

Theorm 5.12 (*Gnedenko*) Let $\{X_1, X_2, ...\}$ be a sequence of independent, identically distributed, continuous random variables with bounded density, mean value $\mu = 0$, and positive, finite variance σ^2. If $f_n(x)$ denotes the density of

$$\frac{1}{\sigma\sqrt{n}} \sum_{i=1}^{n} X_i,$$

then $f_n(x)$ converges uniformly in x to the density of the standard normal distribution:

$$\lim_{n\to\infty} f_n(x) = \varphi(x) = \frac{1}{\sqrt{2\pi}} e^{-x^2/2}, \quad -\infty < x < +\infty. \qquad \blacksquare$$

5.3 EXERCISES

5.1) On average, 6% of the citizens of a large town suffer from severe hypertension. Let X be the number of people in a sample of n randomly selected citizens from this town which suffer from this disease.

(1) By making use of Chebyshev's inequality find the smallest positive integer n_{min} with property

$$P(\left|\tfrac{1}{n}X - 0.06\right| \ge 0.01) \le 0.05 \quad \text{for all } n \text{ with } n \ge n_{min}.$$

(2) Find a positive integer n_{min} satisfying this relationship by using theorem 5.6.

5.2) The measurement error X of a measuring device has mean value $E(X) = 0$ and variance $Var(X) = 0.16$. The random outcomes of n independent measurements are $X_1, X_2, ..., X_n$, i.e., the X_i are independent, identically as X distributed random variables.

(1) By the Chebyshev's inequality, determine the smallest integer $n = n_{min}$ with property that the arithmetic mean of n measurements differs from $E(X) = 0$ by less than 0.1 with a probability of at least 0.99.

(2) On the additional assumption that X is continuous with unimodal density and mode $x_m = 0$, solve (1) by applying the Gauss inequality (5.4).

(3) Solve (1) on condition that $X = N(0, 0.16)$.

5.3) A manufacturer of TV sets knows from past experience that 4% of his products do not pass the final quality check.

(1) What is the probability that in the total monthly production of 2000 sets between 60 and 100 sets do not pass the final quality check?

(2) How many sets have at least to be produced a month to make sure that at least 2000 sets pass the final quality check with probability 0.9?

5.4) The daily demand for a certain medication in a country is given by a random variable X with mean value 28 packets per day and with a variance of 64. The daily demands are independent of each other and distributed as X.

(1) What amount of packets should be ordered for a year with 365 days so that the total annual demand does not exceed the supply with probability 0.99?

(2) Let X_i be the demand at day $i = 1, 2, \ldots$, and

$$\bar{X}_n = \frac{1}{n} \sum_{i=1}^{n} X_i.$$

Determine the smallest integer $n = n_{min}$ so that the probability of the occurrence of the event

$$\left| \bar{X}_n - 28 \right| \geq 1$$

does not exceed 0.05.

5.5) According to the order, the rated nominal capacitance of condensers in a large delivery should be $300 \, \mu F$. Their actual rated nominal capacitances are, however, random variables X with

$$E(X) = 300 \text{ and } Var(X) = 144.$$

(1) By means of Chebyshev's inequality determine an upper bound for the probability of the event A that X differs from the rated nominal capacitance by more than 5%.

(2) Under the additional assumption that X is a continuous random variable with unimodal density and mode $x_m = 300$, determine an upper bound for the probability of the event that X differs from the rated nominal capacitance by more than 5%.

(3) Determine the exact probability of the event that X differs from the rated nominal capacitance by more than 5% on condition that

$$X = N(300, 144).$$

(4) A delivery contains 600 condensers. Their capacitances are independent and identically distributed as X. The distribution of X has the same properties as stated under (2). By means of a Gauss inequality give a lower bound for the probability that the arithmetic mean of the capacitances of the condensers in the delivery differs from $E(X) = 300$ by less than 2.

5.6) A digital transmission channel distorts on average 1 out of 10 000 bits during transmission. The bits are transmitted independently of each other.

(1) Give the <u>exact</u> <u>formula</u> for the probability of the random event A that amongst 10^6 sent bits there are at least 80 bits distorted.

(2) Determine the probability of A by approximation of the normal distribution to the binomial distribution.

5.7) Solve the problem of example 2.4 (page 51) by making use of the normal approximation to the binomial distribution and compare with the exact result.

5.8) Solve the problem of example 2.6 (page 54) by making use of the normal approximation to the hypergeometric distribution and compare with the exact result.

5.9) The random number of asbestos particles per $1mm^3$ in the dust of an industrial area is Poisson distributed with parameter $\lambda = 8$.

What is the probability that in $1cm^3$ of dust there are

(1) at least 10 000 asbestos particles, and

(2) between 8000 and 12 000 asbestos particles (including the bounds)?

5.10) The number of e-mails, which daily arrive at a large company, is Poisson distributed with parameter

$$\lambda = 22\,400.$$

What is the probability that daily between between 22 300 and 22 500 e-mails arrive?

5.11) In $1kg$ of a tapping of cast iron melt there are on average 1.2 impurities.

What is the probability that in a $1000kg$ tapping there are at least 1240 impurities?

The spacial distribution of the impurities in a tapping is assumed to be Poisson.

5.12) After six weeks, 24 seedlings, which had been planted at the same time, reach the random heights $X_1, X_2, ..., X_{24}$, which are independent, identically exponentially distributed as X with mean value $\mu = 32cm$.

Based on the Chebyshev inequality determine

(1) an upper bound for the probability $P(|\overline{X}_{24} - 32| \geq 6)$ that the arithmetic mean

$$\overline{X}_{24} = \frac{1}{24} \sum_{i=1}^{24} X_i$$

differs from μ by more than $6\,cm$,

(2) a lower bound for the probability $P(|\overline{X}_{24} - 32| \leq 6)$ that the deviation of $\overline{X}_{24}$ from μ does not exceed $6cm$.

5.13) Under otherwise the same notation and assumptions as in exercise 5.12, only 6 seedlings had been planted. Let

$$\overline{X}_6 = \frac{1}{6} \sum_{i=1}^{6} X_i.$$

(1) By the one-sided Chebyshev inequality determine an upper bound for the probability $P(\overline{X}_6 \geq 38)$.

(2) Determine the exact value of the probability $P(\overline{X}_6 \geq 38)$.

(3) By the Gauss inequality (5.3) determine a lower bound for the probability

$$P(|\overline{X}_6 - 32| < 6).$$

5.14) The continuous random variable X is uniformly distributed on $[0, 2]$.

(1) Draw the graph of the function

$$p(\varepsilon) = P(|X - 1| \geq \varepsilon)$$

in dependence of ε, $0 \leq \varepsilon \leq 1$.

(2) Compare this graph with the upper bound for the probability

$$P(|X - 1| \geq \varepsilon)$$

given by the Chebyshev inequality, $0 \leq \varepsilon \leq 1$.

(3) Try to improve the Chebyshev upper bound for

$$P(|X - 1| \geq \varepsilon)$$

by the Markov upper bound (5.8) for $a = 3$ and $a = 4$.

PART II

Stochastic Processes

CHAPTER 6

Basics of Stochastic Processes

6.1 MOTIVATION AND TERMINOLOGY

A random variable X is the outcome of a random experiment under fixed conditions. A change of these conditions will influence the outcome of the experiment, i.e. the probability distribution of X will change. Varying conditions can be taken into account by considering random variables which depend on a deterministic parameter t: $X = X(t)$. This approach leads to more general random experiments than the ones defined in section 1.1. To illustrate such generalized random experiments, two simple examples will be considered.

Example 6.1 a) At a fixed geographical point, the temperature is measured every day at 12:00. Let x_i be the temperature measured on the ith day of a year. The value of x_i will vary from year to year and, hence, it can be considered a realization of a random variable X_i. Thus, X_i is the (random) temperature measured on the ith day of a year at 12:00. Apart from random fluctuations of the temperature, the X_i also depend on a deterministic parameter, namely on the time, or, more precisely, on the day of the year. However, if one is only interested in the temperatures X_1, X_2, X_3 on the first 3 days (or any other 3 consecutive days) of the year, then these temperatures are at least approximately identically distributed. Nevertheless, indexing the daily temperatures is necessary, because modeling the obviously existing statistical dependence between the daily temperatures requires knowledge of the joint probability distribution of the random vector (X_1, X_2, X_3). This situation and the problems connected with it motivate the introduction of the generalized random experiment *daily measurement of the temperature at a given geographical point at 12:00 during a year.* The random outcomes of this generalized random experiment are sequences of random variables $\{X_1, X_2, ..., X_{365}\}$ with the X_i being generally neither independent nor identically distributed. If on the ith day temperature x_i has been measured, then the vector $(x_1, x_2, ..., x_{365})$ can be interpreted as a function $x = x(t)$, defined at discrete time points t, $t \in [1, 2, ..., 365]$: $x(t) = x_i$ for $t = i$. Vector $(x_1, x_2, ..., x_{365})$ is a realization of the random vector $(X_1, X_2, ..., X_{365})$.

b) If a sensor graphically records the temperature over the year, then the outcome of the measurement is a continuous function of time t: $x = x(t)$, $0 \le t \le 1$, where $x(t)$ is realization of the random temperature $X(t)$ at time t at a fixed geographical location. Hence it makes sense to introduce the generalized random experiment *continuous measurement of the temperature during a year at a given geographical location.* It will be denoted as $\{X(t), \ 0 \le t \le 1\}$.

A complete probabilistic characterization of this generalized random experiment requires knowledge of the joint probability distributions of all possible random vectors

$$(X(t_1), X(t_2), ..., X(t_n)); \quad 0 \le t_1 < t_2 < \cdots < t_n \le 1; \ n = 1, 2, ...$$

This knowledge allows for statistically modelling the dependence between the $X(t_i)$ in any sequence of random variables $X(t_1), X(t_2), ..., X(t_n)$. It is quite obvious that for small time differences $t_{i+1} - t_i$ there is a strong statistical dependence between $X(t_i)$ and $X(t_{i+1})$. But there may also be a dependence between $X(t_i)$ and $X(t_k)$ for large time differences $t_k - t_i$ due to the inertia of weather patterns over an area. □

Example 6.2 The deterministic parameter, which influences the outcome of a random experiment, needs not be time. For instance, if at a fixed time point and a fixed observation point the temperature is measured along a vertical of length L to the earth's surface, then one obtains a function $x = x(h)$, $0 \le h \le L$, which obviously depends on the distance h of the measurement point to the earth's surface. But if the experiment is repeated in the following years under the same conditions (same time, location, and measurement procedure), then, in view of the occurrence of nonpredictable influences, different functions $x = x(h)$ will be obtained. Hence, the temperature at distance h is a random variable $X(h)$ and the generalized random experiment *measuring the temperature along a vertical of length L*, denoted as $\{X(h), 0 \le h \le L\}$, has outcomes, which are real functions of h: $x = x(h)$, $0 \le h \le L$.

In this situation, it also makes sense to consider the temperature in dependence of both h and the time point of observation t: $x = x(h, t)$; $0 \le h \le L$, $t \ge 0$. Then the observation x depends on a vector of deterministic parameters:

$$x = x(\theta), \ \theta = (h, t).$$

In this case, the outcomes of the corresponding generalized random experiment are surfaces in the (h, t, x)-space. However, this book only considers one-dimensional parameter spaces.

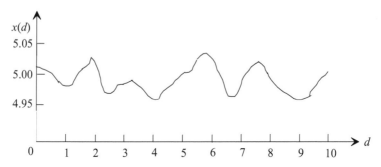

Figure 6.1 Random variation of the diameter of a nylon rope

An already 'classical' example for illustrating the fact that the parameter need not be time is due to *Cramer, Leadbetter* (1967): A machine is supposed to continuously produce ropes of length $10\,m$ with a given nominal diameter of $5\,mm$. Despite maintaining constant production conditions, minor variations of the rope diameter can technologically not be avoided. Thus, when measuring the actual diameter x of a single rope at a distance d from the origin, one gets a function $x = x(d)$ with $0 \le d \le 10$. This function will randomly vary from rope to rope. This suggests the introduction of the generalized random experiment *continuous measurement of the rope diameter in dependence on the distance d from the origin*. If $X(d)$ denotes the diameter of a randomly selected rope at a distance d from the origin, then it makes sense to introduce the corresponding generalized random experiment

$$\{X(d),\ 0 \le d \le 10\}$$

with outcomes $x = x(d)$, $0 \le d \le 10$ (Figure 6.1). $\square$

In contrast to the random experiments considered in chapter 1, the outcomes of which are real numbers, the outcomes of the generalized random experiments, dealt with in examples 2.1 and 2.2, are real functions. Hence, in the literature such generalized random experiments are frequently called *random functions*. However, the terminology *stochastic processes* is more common and will be used throughout the book. In order to characterize the concept of a stochastic process more precisely, further notation is required: Let the random variable of interest X depend on a parameter t, which assumes values from a set $\mathbf{T}$: $X = X(t)$, $t \in \mathbf{T}$. To simplify the terminology and in view of the overwhelming majority of applications, in this book the parameter t is interpreted as time. Thus, $X(t)$ is the random variable X at time t and $\mathbf{T}$ denotes the whole observation time span. Further, let $\mathbf{Z}$ denote the set of all values the random variables $X(t)$ can assume for all $t \in \mathbf{T}$.

Stochastic Process A family of random variables $\{X(t),\, t \in \mathbf{T}\}$ is called a *stochastic process with parameter space* $\mathbf{T}$ *and state space* $\mathbf{Z}$.

If $\mathbf{T}$ is a finite or countably infinite set, then $\{X(t),\, t \in \mathbf{T}\}$ is called a *stochastic process in discrete time* or a *discrete-time stochastic process.* Such processes can be written as a sequences of random variables $\{X_1, X_2, ...\}$ (example 6.1 a). On the other hand, every sequence of random variables can be thought of a stochastic process in discrete time. If $\mathbf{T}$ is an interval, then $\{X(t),\, t \in \mathbf{T}\}$ is a *stochastic process in continuous time* or a *continuous-time stochastic process.* A stochastic process $\{X(t),\, t \in \mathbf{T}\}$ is said to be *discrete* if its state space $\mathbf{Z}$ is a finite or a countably infinite set, and a stochastic process $\{X(t),\, t \in \mathbf{T}\}$ is said to be *continuous* if $\mathbf{Z}$ is an interval. Thus, there are discrete stochastic processes in discrete time, discrete stochastic processes in continuous time, continuous stochastic processes in discrete time, and continuous stochastic processes in continuous time. Throughout this book the state space $\mathbf{Z}$ is usually assumed to be a subset of the real axis.

If the stochastic process $\{X(t),\, t \in \mathbf{T}\}$ is observed over the whole time period $\mathbf{T}$, i.e. the values of $X(t)$ are registered for all $t \in \mathbf{T}$, then one obtains a real function

$$x = x(t),\ t \in \mathbf{T}.$$

Such a function is called a *sample path*, a *trajectory*, or a *realization* of the stochastic process. In this book the concept *sample path* is used. The sample paths of a stochastic process in discrete time are, therefore, sequences of real numbers, whereas the sample paths of stochastic processes in continuous time can be any functions of time. The sample paths of a discrete stochastic process in continuous time are piecewise constant functions (step functions). The set of all sample paths of a stochastic process with parameter space $\mathbf{T}$ is, therefore, a subset of all functions over the domain $\mathbf{T}$.

In engineering, science, and economics there are many time-dependent random phenomena which can be modeled by stochastic processes: In an electrical circuit it is not possible to keep the voltage strictly constant. Random fluctuations of the voltage are for instance caused by *thermal noise.* If $v(t)$ denotes the voltage measured at time point t, then $v = v(t)$ is a sample path of a stochastic process $\{V(t),\, t \ge 0\}$ where $V(t)$ is the random voltage at time t (Figure 6.2). Producers of radar and satellite supported communication systems have to take into account a phenomenon called *fading.*

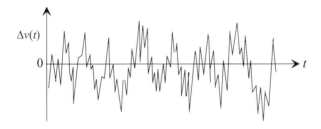

Figure 6.2 Voltage fluctuations caused by random noise

This is characterized by random fluctuations in the energy of received signals caused by the dispersion of radio waves as a result of inhomogeinities in the atmosphere and by *meteorological* and *industrial noise*. Both meteorological and industrial noise create electrical discharges in the atmosphere which occur at random time points with randomly varying intensity. 'Classic' applications of stochastic processes in economics are modeling the fluctuations of share prices, rendits, and prices of commodities over time. In operations research, stochastic processes describe the development in time of the 'states' of queueing, inventory, and reliability systems. In statistical quality control, they model the fluctuation of quality criteria over time. In medicine, the development in time of 'quality parameters' of health as blood pressure and cholesterol level as well as the spread of epidemics are typical examples of stochastic processes.

Important impulses for the development and application of stochastic processes came from biology: stochastic models for population dynamics from cell to mammal level, competition models (predator-prey), capture-recapture models, growth processes, and many more.

6.2 CHARACTERISTICS AND EXAMPLES

From the mathematical point of view, the given heuristic explanation of a stochastic process needs to be supplemented. Let $F_t(x)$ be the distribution function of $X(t)$:

$$F_t(x) = P(X(t) \leq x), \ t \in \mathbf{T}.$$

The family of the one-dimensional distribution functions

$$\{F_t(x), t \in \mathbf{T}\}$$

is the *one- dimensional probability distribution* of $\{X(t), t \in \mathbf{T}\}$. In view of the statistical dependence, which generally exists between the $X(t_1), X(t_2), ..., X(t_n)$ for any $t_1, t_2, ..., t_n$, the family of the one-dimensional distribution functions $\{F_t(x), t \in \mathbf{T}\}$ does not completely characterize a stochastic process (see examples 6.1 and 6.2).

A stochastic process $\{X(t), t \in \mathbf{T}\}$ is only then completely characterized if for all positive integers $n = 1, 2, ...,$ for all n-tuples $\{t_1, t_2, ..., t_n\}$ with $t_i \in \mathbf{T}$, and for all vectors $\{x_1, x_2, ..., x_n\}$ with $x_i \in \mathbf{Z}$, the joint distribution function of the random vector $(X(t_1), X(t_2), ..., X(t_n))$ is known:

$$F_{t_1, t_2, ..., t_n}(x_1, x_2, ..., x_n) = P(X(t_1) \leq x_1, X(t_2) \leq x_2, ..., X(t_n) \leq x_n). \tag{6.1}$$

The set of all these joint distribution functions defines the *probability distribution* of the stochastic process. For a discrete stochastic process, it is generally simpler to characterize its probability distribution by the probabilities

$$P(X(t_1) \in A_1, X(t_2) \in A_2, ..., X(t_n) \in A_n)$$

for all $t_1, t_2, ..., t_n$ with $t_i \in \mathbf{T}$ and $A_i \subseteq \mathbf{Z}; \ i = 1, 2, ..., n; \ n = 1, 2, ...$

Trend Function Assuming the existence of $E(X(t))$ for all $t \in \mathbf{T}$, the *trend* or *trend function* of the stochastic process $\{X(t), t \in \mathbf{T}\}$ is the mean value of $X(t)$ as a function of t:

$$m(t) = E(X(t)), \quad t \in \mathbf{T}. \tag{6.2}$$

Thus, the trend function of a stochastic process describes its average development of the process in time. If the densities $f_t(x) = dF_t(x)/dx$ exist, then

$$m(t) = \int_{-\infty}^{+\infty} x\, f_t(x)\, dx, \quad t \in \mathbf{T}.$$

Covariance Function The *covariance function* of a stochastic process $\{X(t), t \in \mathbf{T}\}$ is the covariance between the random variables $X(s)$ and $X(t)$ as a function of s and t. Hence, in view of (3.37) and (3.38), page 135,

$$C(s,t) = Cov(X(s), X(t)) = E([X(s) - m(s)][X(t) - m(t)]); \quad s, t \in \mathbf{T}, \tag{6.3}$$

or

$$C(s,t) = E(X(s)X(t)) - m(s)m(t); \quad s, t \in \mathbf{T}. \tag{6.4}$$

In particular,

$$C(t,t) = Var(X(t)). \tag{6.5}$$

The covariance function is a symmetric function of s and t:

$$C(s,t) = C(t,s). \tag{6.6}$$

Since the covariance function $C(s,t)$ is a measure for the degree of the statistical dependence between $X(s)$ and $X(t)$, one expects that

$$\lim_{|t-s| \to \infty} C(s,t) = 0. \tag{6.7}$$

Example 6.3 shows that this need not be the case.

Correlation Function The *correlation function* of $\{X(t), t \in \mathbf{T}\}$ is the correlation coefficient $\rho(s,t) = \rho(X(s), X(t))$ between $X(s)$ and $X(t)$ as a function of s and t. According to (3.43),

$$\rho(s,t) = \frac{Cov(X(s), X(t))}{\sqrt{Var(X(s)}\,\sqrt{Var(X(t)}}. \tag{6.8}$$

The covariance function of a stochastic process is also called *autocovariance function* and the correlation function *autocorrelation function*. This terminology avoids mistakes, when dealing with covariances and correlations between $X(s)$ and $Y(t)$ for different stochastic processes $\{X(t), t \in \mathbf{T}\}$ and $\{Y(t), t \in \mathbf{T}\}$. The *cross covariance function* between these two processes is defined as

$$C(s,t) = Cov(X(s), Y(t)) = E([X(s) - m_X(s)][Y(t) - m_Y(t)]); \quad s, t \in \mathbf{T}, \tag{6.9}$$

with $m_X(t) = E(X(t))$ and $m_Y(t) = E(Y(t))$. Correspondingly, the *cross correlation function* between the processes $\{X(t), t \in \mathbf{T}\}$ and $\{Y(t), t \in \mathbf{T}\}$ is

$$\rho(s,t) = \frac{Cov(X(s), Y(t))}{\sqrt{Var(X(s)}\,\sqrt{Var(Y(t)}}. \tag{6.10}$$

As pointed out in section 3.1.3 (page 139), the advantage of the correlation coefficient to the covariance is that it allows for comparing the (linear) dependencies between different pairs of random variables. Being able to compare the dependency between two stochastic processes by their cross-correlation function is important for processes, which are more or less obviously dependent as, for instance, the development in time of air temperature and air moisture or air temperature and CO_2 content of the air.

Semi-variogram The *semi-variogram* or, shortly, *variogram* of a stochastic process $\{X(t),\ t \in \mathbf{T}\}$ is defined as

$$\gamma(s,t) = \frac{1}{2} E[(X(t) - X(s)]^2 \tag{6.11}$$

as a function of s and t; $s, t \in \mathbf{T}$. The variogram is obviously a symmetric function in s and t: $\gamma(s,t) = \gamma(t,s)$.

The concept of a variogram has its origin in geostatistics for describing properties of *random fields*, i.e., stochastic processes, which depend on a multi-dimensionally deterministic parameter $\mathbf{t}$, which refers to a location, but may also include time.

Example 6.3 (*cosine wave with random amplitude*) Let

$$X(t) = A \cos \omega t,$$

where A is a nonnegative random variable with $E(A) < \infty$. The process $\{X(t),\ t \geq 0\}$ can be interpreted as the output of an oscillator which is selected from a set of identical ones. (Random deviations of the amplitudes from a nominal value are technologically unavoidable.) The trend function of this process is

$$m(t) = E(A) \cos \omega t.$$

By (6.4), its covariance function is

$$C(s,t) = E([A \cos \omega s][A \cos \omega t]) - m(s)m(t)$$
$$= [E(A^2) - (E(A))^2](\cos \omega s)(\cos(\omega t)).$$

Hence,

$$C(s,t) = Var(A)(\cos \omega s)(\cos \omega t).$$

Obviously, the process does not have property (6.7). Since there is a functional relationship between $X(s)$ and $X(t)$ for any s and t, $X(s)$ and $X(t)$ cannot tend to become independent for $|t - s| \to \infty$. Actually, the correlation function $\rho(s,t)$ between $X(s)$ and $X(t)$ is equal to 1 for all (s,t). $\square$

The stochastic process considered in example 6.3 has a special feature: For a given value a that the random variable A has assumed, the process develops in a strictly deterministic way. That means, by only observing a sample path of such a process over an arbitrarily small time interval, one can predict the further development of the

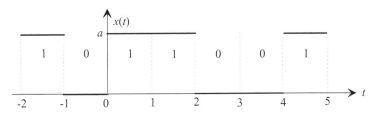

Figure 6.3 Pulse code modulation

sample path with absolute certainty. (The same comment refers to examples 6.6 and 6.7.) More complicated stochastic processes arise when random influences continuously, or at least repeatedly, affect the phenomenon of interest. The following example belongs to this category.

Example 6.4 (*pulse code modulation*) A source generates symbols 0 or 1 independently with respective probabilities p and $1 - p$. The symbol '0' is transmitted by sending nothing during a time interval of length one. The symbol '1' is transmitted by sending a pulse with constant amplitude a during a time unit of length one. The source has started operating in the past. A stochastic signal (sequence of symbols) generated in this way is represented by the stochastic process $\{X(t), t \in (-\infty, +\infty)\}$ with

$$X(t) = \sum_{n=-\infty}^{+\infty} A_n h(t-n), \quad n \le t < n+1, \tag{6.12}$$

where the A_n; $n = 0, \pm 1, \pm 2, ...$; are independent binary random variables defined by

$$A_n = \begin{cases} 0 & \text{with probability} \quad p, \\ a & \text{with probability} \quad 1-p, \end{cases}$$

and $h(t)$ is given by

$$h(t) = \begin{cases} 1 & \text{for} \quad 0 \le t < 1, \\ 0 & \text{elsewhere.} \end{cases}$$

For any t,

$$X(t) = \begin{cases} 0 & \text{with probability} \quad p, \\ a & \text{with probability} \quad 1-p. \end{cases}$$

For example, the section of a sample path $x = x(t)$ plotted in Figure 6.3 is generated by the following partial sequence of a signal:

$$\cdots 1\ 0\ 1\ 1\ 0\ 0\ 1 \cdots.$$

The role of the function $h(t)$ is to keep $X(t)$ at level 0 or 1, respectively, in the intervals $[n, n+1)$. Note that the time point $t = 0$ coincides with the beginning of a new transmission period. The process has a constant trend function:

$$m(t) \equiv a \cdot P(X(t) = a) + 0 \cdot P(X(t) = 0) = a(1-p).$$

For $n \leq s, t < n + 1; \; n = 0, \pm 1, \pm 2, ...,$

$$E(X(s)X(t)) = E(X(s)X(t)|X(s) = a) \cdot P(X(s) = a)$$
$$+ \; E(X(s)X(t)|X(s) = 0) \cdot P(X(s) = 0)$$
$$= a^2(1 - p).$$

Therefore,

$$Cov(X(s), X(t)) = a^2(1 - p) - a^2(1 - p)^2 = a^2 p(1 - p) \text{ for } n \leq s, t < n + 1.$$

If $m \leq s < m + 1$ and $n \leq t < n + 1$ with $m \neq n$, then $X(s)$ and $X(t)$ are independent random variables. Hence, the covariance function of $\{X(t), \; t \in (-\infty, +\infty)\}$ is

$$C(s, t) = \begin{cases} a^2 p(1 - p) & \text{for } n \leq s, t < n + 1; \;\; n = 0, \pm 1, \pm 2, ... \\ 0 & \text{elsewhere} \end{cases}.$$

Although the stochastic process analyzed in this example has a rather simple structure, it is of considerable importance in physics, electrical engineering, and communication; for more information, see e.g. *Gardner* (1989). A modification of the pulse code modulation process is considered in example 6.8. As the following example shows, the pulse code modulation is a special shot noise process. □

Example 6.5 (*shot noise process*) At time points T_n, pulses of random intensity A_n are induced. The sequences $\{T_1, T_2, ...\}$ and $\{A_1, A_2, ...\}$ are assumed to be discrete-time stochastic processes with properties

1) With probability 1, $T_1 < T_2 < \cdots$ and $\lim\limits_{n \to \infty} T_n = \infty$,

2) $E(A_n) < \infty; \; n = 1, 2, ...$.

In communication theory, the sequence $\{(T_n, A_n); \; n = 1, 2, ...\}$ is called a *pulse process*. (In section 7.1, it will be called a *marked point process*.) The function $h(t)$, the *response* of a system to a pulse, has properties

$$h(t) = 0 \text{ for } t < 0 \quad \text{and} \quad \lim_{t \to \infty} h(t) = 0. \tag{6.13}$$

The stochastic process $\{X(t), \; t \in (-\infty, +\infty)\}$ defined by

$$X(t) = \sum_{n=1}^{\infty} A_n h(t - T_n) \tag{6.14}$$

is called a *shot noise process* or just *shot noise*. It quantifies the additive superposition of the responses of a system to pulses. The factors A_n are sometimes called *amplitudes* of the shot noise process. In many applications, the A_n are independent, identically distributed random variables, or, as in example 6.4, even constant.

If the sequences of the T_n and A_n are doubly infinite,

$$\{T_n; \; n = 0, \pm 1, \pm 2, ...\} \quad \text{and} \quad \{A_n; \; n = 0, \pm 1, \pm 2, ...\},$$

then the shot noise process $\{X(t), \; t \in (-\infty, +\infty)\}$ is defined as

$$X(t) = \sum_{n=-\infty}^{n=+\infty} A_n \, h(t - T_n). \tag{6.15}$$

A well-known physical phenomenon, which can be modeled by a shot noise process, is the fluctuation of the anode current in vacuum tubes (*tube noise*). This fluctuation is caused by random current impulses, which are initiated by emissions of electrons from the anode at random time points (*Schottky effect*); see *Schottky* (1918). The term *shot noise* has its origin in the fact that the effect of firing small shot at a metal slab can be modeled by a stochastic process of structure (6.15). More examples of shot noise processes are discussed in chapter 7, where special assumptions on the underlying pulse process are made. □

6.3 CLASSIFICATION OF STOCHASTIC PROCESSES

Stochastic processes are classified with regard to properties which reflect, e.g., their dependence on time, the statistical dependence of their developments over disjoint time intervals, and the influence of the history or the current state of a stochastic process on its future evolvement. In the context of example 6.1: Has the date any influence on the daily temperature at 12:00? (That need not be the case if the measurement point is near to the equator.) Or, has the sample path of the temperature in January any influence on the temperature curve in February? For reliably predicting tomorrow's temperature at 12:00, is it sufficient to know the present temperature or would knowledge of the temperature curve during the past two days allow a more accurate prediction? What influence has time on trend or covariance function?

Special importance have those stochastic processes for which the joint distribution functions (6.1) only depend on the distances between t_i and t_{i+1}, i.e., only the relative positions of $t_1, t_2, ..., t_n$ to each other have an impact on the joint distribution of the random variables $X(t_1), X(t_2), ..., X(t_n)$.

Strong Stationarity A stochastic process $\{X(t), t \in \mathbf{T}\}$ is said to be *strongly stationary* or *strictly stationary* if for all $n = 1, 2, ...$, for any real τ, for all n-tuples

$$(t_1, t_2, ..., t_n) \quad \text{with } t_i \in \mathbf{T} \text{ and } t_i + \tau \in \mathbf{T}; \ i = 1, 2, ..., n;$$

and for all n-tuples $(x_1, x_2, ..., x_n)$, the joint distribution function of the random vector $(X(t_1), X(t_2), ..., X(t_n))$ has property

$$F_{t_1, t_2, ..., t_n}(x_1, x_2, ..., x_n) = F_{t_1+\tau, t_2+\tau, ..., t_n+\tau}(x_1, x_2, ..., x_n). \tag{6.16}$$

That means, the probability distribution of a strongly stationary stochastic process is invariant against absolute time shifts. In particular, by letting $n = 1$ and $t = t_1$, property (6.16) implies that $F_t(x) = F_{t+\tau}(x)$ for all τ with arbitrary but fixed t and x. That means $F_t(x)$ actually does not depend on t. Hence, for strongly stationary processes there exists a distribution function $F(x)$, which does not depend on t, so that

$$F_t(x) = F(x) \quad \text{for all } t \in T \text{ and } x \in \mathbf{Z}. \tag{6.17}$$

Hence, trend and variance function of $\{X(t), t \in \mathbf{T}\}$ do not depend on t either:

$$m(t) = E(X(t)) \equiv m, \quad Var(X(t)) \equiv \sigma^2 \tag{6.18}$$

(given that the parameters m and σ^2 exist). The trend function of a strongly stationary process is, therefore, a parallel to the time axis, and the fluctuations of its sample paths around the trend function experience no systematic changes with increasing t.

What influence has the strong stationarity of a stochastic process on its covariance function?

To answer this question, the special values $n = 2$, $t_1 = 0$, $t_2 = t - s$, and $\tau = s$ are substituted in (6.16). This yields for all $s < t$,

$$F_{0, t-s}(x_1, x_2) = F_{s, t}(x_1, x_2),$$

i.e. the joint distribution function of the random vector (X_s, X_t), and, therefore, the mean value of the product $X_s X_t$, depend only on the difference $\tau = t - s$, and not on the absolute values of s and t. Hence, by formulas (6.4) and (6.18), $C(s, t)$ must have the same property:

$$C(s, t) = C(s, s + \tau) = C(0, \tau) = C(\tau).$$

Thus, the covariance function of strongly stationary processes depends only on one variable:

$$C(\tau) = Cov(X(s), X(s + \tau)) \text{ for all } s \in \mathbf{T}. \tag{6.19}$$

Since the covariance function $C(s, t)$ of any stochastic process is symmetric in the variables s and t, the covariance function of a strongly stationary process is a symmetric function with symmetry center $\tau = 0$, i.e. $C(\tau) = C(-\tau)$ or, equivalently,

$$C(\tau) = C(|\tau|). \tag{6.20}$$

In practical situations it is generally not possible to determine the probability distributions of all possible random vectors $\{X(t_1), X(t_2), \cdots, X(t_n)\}$ in order to check whether a stochastic process is strongly stationary or not. But the user of stochastic processes is frequently satisfied with the validity of properties (6.18) and (6.19). Hence, based on these two properties, another concept of stationarity had been introduced. It is, however, only defined for second-order processes:

Second-Order Process A stochastic process $\{X(t), t \in \mathbf{T}\}$ is called a *second-order process* if

$$E(X^2(t)) < \infty \text{ for all } t \in \mathbf{T}. \tag{6.21}$$

The existence of the second moments of $X(t)$ as required by assumption (6.21) implies the existence of the covariance function $C(s, t)$ for all s and t, and, therefore, the existence of the variances $Var(X(t))$ and mean values $E(X(t))$ for all $t \in \mathbf{T}$ (see inequality of Schwarz (5.11), page 195). (In deriving (6.20) we have implicitly assumed the existence of the second moments $E(X^2(t))$ without referring to it.)

Weak Stationarity A stochastic process $\{X(t), t \in \mathbf{T}\}$ is said to be *weakly station-ary* if it is a second order process and has properties (6.18) and (6.19):

1) $m(t) = m$ for all $t \in \mathbf{T}$.

2) $C(\tau) = Cov(X(s), X(s + \tau))$ for all $s \in \mathbf{T}$.

From (6.18) with $t = 0$:

$$Var(X(0)) = C(0) = \sigma^2. \qquad (6.22)$$

The covariance function $C(\tau)$ of weakly stationary process has two characteristic pro-perties (without proof):

1) $|C(\tau)| \le \sigma^2$ for all τ,

2) $C(\tau)$ is *positive semi-definite*, i.e. for all n, all real numbers $a_1, a_2, ..., a_n$, and for all $t_1, t_2, ..., t_n$; $t_i \in \mathbf{T}$,

$$\sum_{i=1}^{n} \sum_{j=1}^{n} a_i a_j C(t_i - t_j) \ge 0.$$

A strongly stationary process is not necessarily weakly stationary, since there are strongly stationary processes, which are not second order processes. But, if a second order process is strongly stationary, then, as shown above, it is also weakly stationary. Weakly stationary processes are also called *wide-sense stationary*, *covariance statio-nary*, or *second-order stationary*.

Further important properties of stochastic processes are based on properties of their increments:

The *increment* of a stochastic process $\{X(t), t \in \mathbf{T}\}$ with respect to the interval $[t_1, t_2)$ is the difference $X(t_2) - X(t_1)$.

Hence, the variogram $\gamma(s, t)$ as defined by (6.11) is a half of the second moment of the increment $X(t) - X(s)$.

Homogeneous Increments A stochastic process $\{X(t), t \in \mathbf{T}\}$ is said to have *homo-geneous* or *stationary increments* if for arbitrary, but fixed $t_1, t_2 \in \mathbf{T}$ the increment $X(t_2 + \tau) - X(t_1 + \tau)$ has the same probability distribution for all values of τ with pro-perty $t_1 + \tau \in \mathbf{T}$, $t_2 + \tau \in \mathbf{T}$.

An equivalent definition of processes with homogeneous increments is:

The stochastic process $\{X(t), t \in \mathbf{T}\}$ has homogeneous increments if the probability distribution of the increments $X(t + \tau) - X(t)$ does not depend on t for any fixed τ; $t, t + \tau \in \mathbf{T}$.

Thus, the development in time of a stochastic process with homogeneous increments in any interval of the same length is governed by the same probability distribution. This motivates the term *stationary increments*.

A stochastic process with homogeneous (stationary) increments need not be station-ary in any sense.

Taking into account (6.22), the variogram of a stochastic process with homogeneous increments has a simple structure:

$$\gamma(s, s + \tau) = \frac{1}{2} E[(X(s) - X(s + \tau))^2]$$

$$= \frac{1}{2} E[((X(s) - m) - (X(s + \tau) - m))^2]$$

$$= \frac{1}{2} E[(X(s) - m)^2 - 2(X(s) - m)(X(s + \tau) - m)) + (X(s + \tau) - m)^2]$$

$$= \frac{1}{2}\sigma^2 - C(\tau) + \frac{1}{2}\sigma^2$$

so that

$$\gamma(\tau) = \sigma^2 - C(\tau).$$

Therefore, in case of a process with homogeneous increments, the variogram does yield additional information on the process compared to the covariance function.

Independent Increments A stochastic process $\{X(t), t \in \mathbf{T}\}$ has *independent increments* if for all $n = 2, 3, \ldots$ and for all n-tuples $(t_1, t_2, \ldots, t_n)$ with $t_1 < t_2 < \cdots < t_n$, $t_i \in \mathbf{T}$, the increments

$$X(t_2) - X(t_1), \ X(t_3) - X(t_2), \ \cdots, \ X(t_n) - X(t_{n-1})$$

are independent random variables.

The meaning of this concept is that the development of the process in an interval **I** has no influence on the development of the process on intervals, which are disjoint to **I**. Thus, when the price of a share is governed by a process with independent increments and there was sharp increase in year n, then this information is worthless with regard to predicting the development of the share price in year $n+1$.

Gaussian Process A stochastic process $\{X(t), t \in \mathbf{T}\}$ is a *Gaussian process* if the random vectors $(X(t_1), X(t_2), \ldots, X(t_n))$ have a joint normal (Gaussian) distribution for all n-tuples $(t_1, t_2, \ldots, t_n)$ with $t_i \in \mathbf{T}$ and $t_1 < t_2 < \cdots < t_n$; $n = 1, 2, \ldots$.

Gaussian processes have an important property:

| *A Gaussian process is strongly stationary if and only if it is weakly stationary.*

Gaussian processes will play an important role in Chapter 11.

Markov Process A stochastic process $\{X(t), t \in \mathbf{T}\}$ has the *Markov(ian) property* if for all $(n + 1)$-tuples $(t_1, t_2, \ldots, t_{n+1})$ with $t_i \in \mathbf{T}$ and $t_1 < t_2 < \cdots < t_{n+1}$, and for any $A_i \subseteq \mathbf{Z}$; $i = 1, 2, \ldots, n + 1$;

$$P(X(t_{n+1}) \in A_{n+1} | X(t_n) \in A_n, X(t_{n-1}) \in A_{n-1}, \ldots, X(t_1) \in A_1)$$

$$= P(X(t_{n+1}) \in A_{n+1} | X(t_n) \in A_n). \tag{6.23}$$

The Markov property can be interpreted as follows: If t_{n+1} is a time point in the future, t_n the present time poin,t and, correspondingly, $t_1, t_2, ..., t_{n-1}$ are time points in the past, then the future development of a process having the Markov property does not depend on its evolvement in the past, but only on its present state. Stochastic processes having the Markov property are called *Markov processes*.

A Markov process with finite or countably infinite parameter space $\mathbf{T}$ is called a *discrete-time Markov process*. Otherwise it is called a *continuous-time Markov process*. Markov processes with finite or countably infinite state spaces $\mathbf{Z}$ are called *Markov chains*. Thus, a discrete-time Markov chain has both a discrete state space and a discrete parameter space. Deviations from this terminology can be found in the literature.

Markov processes play an important role in all sorts of applications, mainly for four reasons: 1) Many practical phenomena can be modeled by Markov processes. 2) The input necessary for their practical application is generally more easy to provide than the necessary input for other classes of stochastic processes. 3) Computer algorithms are available for numerical evaluations. 4) Stochastic processes $\{X(t), t \in \mathbf{T}\}$ with independent increments and parameter space $\mathbf{T} = [0, \infty)$ always have the Markov property. The practical importance of Markov processes is illustrated by numerous examples in chapters 8 and 9.

Theorem 6.1 A Markov process is strongly stationary if and only if its one-dimensional probability distribution does not depend on time, i.e., if there exists a distribution function $F(x)$ with

$$F_t(x) = P(X(t) \le x) = F(x) \quad \text{for all } t \in \mathbf{T}. \qquad \blacksquare$$

Thus, condition (6.17), which is necessary for any a stochastic process to be strongly stationary, is necessary and sufficient for a Markov process to be strongly stationary.

Mean-Square Continuous A second order process $\{X(t), t \in \mathbf{T}\}$ is said to be *mean-square continuous at point $t = t_0 \in \mathbf{T}$* if

$$\lim_{h \to 0} E([X(t_0 + h) - X(t_0)]^2) = 0. \qquad (6.24)$$

The process $\{X(t), t \in \mathbf{T}\}$ is said to be *mean-square continuous in the region $\mathbf{T}_0$*, $\mathbf{T}_0 \subseteq \mathbf{T}$, if it is mean-square continuous at all points $t \in \mathbf{T}_0$.

According to section 5.2.1 (page 205), the convergence used in (6.24) is called *convergence in mean square*. There is a simple criterion for a second order stochastic process to be mean-square continuous at t_0:

A second order process $\{X(t), t \in \mathbf{T}\}$ is mean-square continuous at t_0 if and only if its covariance function $C(s, t)$ is continuous at $(s, t) = (t_0, t_0)$.

As a corollary from this statement:

A weakly stationary process $\{X(t), t \in (-\infty, +\infty)\}$ is mean-square continuous in $(-\infty, +\infty)$ if and only if it is mean-square continuous at time point $t = 0$.

The following two examples make use of two formulas from trigonometry:

$$\cos\alpha\,\cos\beta = \tfrac{1}{2}[\cos(\beta-\alpha)+\cos(\alpha+\beta)],$$

$$\cos(\beta-\alpha) = \cos\alpha\,\cos\beta + \sin\alpha\,\sin\beta.$$

Example 6.6 (*cosine wave with random amplitude and random phase*) In modifying example 6.3, let

$$X(t) = A\,\cos(\omega t + \Phi),$$

where A is a nonnegative random variable with finite mean value and finite variance. The random parameter Φ is assumed to be uniformly distributed over $[0, 2\pi]$ and independent of A. The stochastic process $\{X(t),\ t \in (-\infty, +\infty)\}$ can be thought of as the output of an oscillator, selected from a set of oscillators of the same kind, which have been turned on at different times (see, e.g., *Helstrom* (1989)). Since

$$E(\cos(\omega t + \Phi)) = \frac{1}{2\pi}\int_0^{2\pi}\cos(\omega t + \varphi)\,d\varphi = \frac{1}{2\pi}[\sin(\omega t + \varphi)]_0^{2\pi} = 0,$$

the trend function of this process is identically zero:

$$m(t) \equiv 0.$$

Its covariance function is

$$C(s,t) = E\{[A\cos(\omega s + \Phi)][A\cos(\omega t + \Phi)]\}$$

$$= E(A^2)\frac{1}{2\pi}\int_0^{2\pi}\cos(\omega s + \varphi)\,\cos(\omega t + \varphi)\,d\varphi$$

$$= E(A^2)\frac{1}{2\pi}\int_0^{2\pi}\tfrac{1}{2}\{\cos\omega(t-s) + \cos[\omega(s+t) + 2\varphi]\}\,d\varphi.$$

The first integrand is a constant with respect to integration. Since the integral of the second term is zero, $C(s,t)$ depends only on the difference $\tau = t - s$:

$$C(\tau) = \tfrac{1}{2}E(A^2)\,\cos w\tau.$$

Thus, the process is weakly stationary.   □

Example 6.7 Let the stochastic process $\{X(t),\ t \in (-\infty, +\infty)\}$ be is defined by

$$X(t) = A\,\cos\omega t + B\,\sin\omega t,$$

where A and B are two uncorrelated random variables satisfying

$$E(A) = E(B) = 0 \quad \text{and} \quad Var(A) = Var(B) = \sigma^2 < \infty.$$

Since $Var(X(t)) = \sigma^2 < \infty$ for all t, $\{X(t),\ t \in (-\infty, +\infty)\}$ is a second order process. Its trend function is identically zero: $m(t) \equiv 0$. Thus,

$$C(s,t) = E(X(s)X(t)).$$

For A and B being uncorrelated, $E(AB) = E(A)E(B)$. Hence,

$$C(s,t) = E(A^2\cos \omega s \cos \omega t + B^2 \sin \omega s \sin \omega t)$$
$$+ E(AB \cos \omega s \sin \omega t + AB \sin \omega s \cos \omega t)$$
$$= \sigma^2 (\cos \omega s \cos \omega t + \sin \omega s \sin \omega t)$$
$$+ E(AB)(\cos \omega s \sin \omega t + \sin \omega s \cos \omega t)$$
$$= \sigma^2 \cos \omega (t - s).$$

Thus, the covariance function depends only on the difference $\tau = t - s$:

$$C(\tau) = \sigma^2 \cos \omega \tau$$

so that the process $\{X(t), t \in (-\infty, +\infty)\}$ is weakly stationary.   □

Example 6.8 (*randomly delayed pulse code modulation*) Based on the stochastic process $\{X(t), t \in (-\infty, +\infty)\}$ defined in example 6.4, the stochastic process

$$\{Y(t), t \in (-\infty, +\infty)\} \text{ with } Y(t) = X(t - Z)$$

is introduced, where Z is uniformly distributed over $[0, 1]$. When shifting the sample paths of the process $\{X(t), t \in (-\infty, +\infty)\}$ Z time units to the right, one obtains the corresponding sample paths of the process $\{Y(t), t \in (-\infty, +\infty)\}$. For instance, shifting the section of the sample path, shown in Figure 6.3, $Z = z$ time units to the right yields the corresponding section of the sample path of the process $\{Y(t), t \in (-\infty, +\infty)\}$ depicted in Figure 6.4.

The trend function of the process $\{Y(t), t \in (-\infty, +\infty)\}$ is

$$m(t) \equiv a(1 - p).$$

To determine the covariance function, let $B = B(s,t)$ denote the random event that $X(s)$ and $X(t)$ are separated by a switching point $n + Z$; $n = 0, \pm 1, \pm 2, \dots$. Then

$$P(B) = |t - s|, \quad P(\bar{B}) = 1 - |t - s|.$$

The random variables $X(s)$ and $X(t)$ are independent if $|t - s| > 1$ and/or B occurs. Therefore,

$$C(s,t) = 0 \text{ if } |t - s| > 1 \text{ and/or } B \text{ occurs.}$$

If $|t - s| \le 1$, $X(s)$ and $X(t)$ are only then independent if B occurs. Hence, the covariance function of $\{Y(t), t \in (-\infty, +\infty)\}$ given $|t - s| \le 1$ can be obtained as follows:

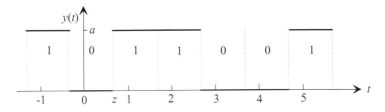

Figure 6.4 Randomly delayed pulse code modulation

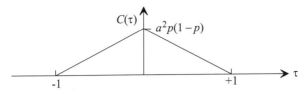

Figure 6.5 Covariance function of the randomly delayed pulse code modulation

$$C(s,t) = E(X(s)X(t)|B)P(B) + E(X(s)X(t)|\bar{B})P(\bar{B}) - m(s)m(t)$$
$$= E(X(s))E(X(t))P(B) + E([X(s)]^2)P(\bar{B}) - m(s)m(t)$$
$$= [a(1-p)]^2|t-s| + a^2(1-p)(1-|t-s|) - [a(1-p)]^2.$$

Finally, with $\tau = t - s$, the covariance function becomes

$$C(\tau) = \begin{cases} a^2 p(1-p)(1-|\tau|) & \text{for } |\tau| \le 1 \\ 0 & \text{elsewhere} \end{cases}.$$

The process $\{Y(t), t \in (-\infty, +\infty)\}$ is weakly stationary. Analogously to the transition from example 6.3 to example 6.6, stationarity is achieved by introducing a uniformly distributed phase shift in the pulse code modulation of example 6.4. ◻

6.4 TIME SERIES IN DISCRETE TIME

6.4.1 Introduction

All examples in sections 6.2 and 6.3 dealt with stochastic processes in continuous time. In this section, examples for discrete-time processes are considered, which are typical in time-series analysis. The material introduced in the previous sections is extended and supplemented with time-series specific terminology and techniques.

A *time series* is a realization (trajectory, sample path) of a stochastic process in discrete time $\{X(t_1), X(t_2), ...\}$. The time (parameter) space $\mathbf{T}$ of this process is finite, i.e. $\mathbf{T} = \{t_1, t_2, ..., t_n\}$, or only a finite piece of a trajectory of a stochastic process with unbounded time space $\mathbf{T} = \{t_1, t_2, ...\}$ has been observed. Thus, a time series is simply a sequence of real numbers

$$x_1, x_2, ..., x_n$$

with property that the underlying stochastic process has assumed value x_i at time t_i:

$$X(t_i) = x_i = x(t_i); \quad i = 1, 2, ..., n.$$

Frequently it is assumed that the $t_1, t_2, ..., t_n$ are equidistant, i.e.,

$$t_i = i\Delta t; \quad i = 1, 2, ..., n.$$

If the underlying stochastic process $\{X(t), t \in \mathbf{T}\}$ is a process in continuous time, it also can give rise to a time series in discrete time, simply by scanning the state of the process at discrete (possibly equidistant) time points. As with stochastic processes, the parameter 'time' in time series need not be the time. Time series occur in all areas, where the development of economical, physical, technological, biological, etc. phenomena is controlled by stochastic processes. Hence, with regard to application of time series, it can be referred to the introduction of this chapter. Figures 6.1 and 6.2 are actually time series plots. When analyzing time series, the emphasis is on numerical aspects how to extract as much as possible information from the time series with regard to trend, seasonal, and random influences as well as prediction and to a lesser extent on theoretical implications regarding the underlying stochastic process.

In elementary time series analysis, the underlying stochastic process $\{X(t), t \in \mathbf{T}\}$ is assumed to have a special structure: $X(t)$ is given by the additive superposition of three components:

$$X(t) = T(t) + S(t) + R(t), \tag{6.25}$$

where $T(t)$ is the *trend* of the time series and $S(t)$ is a *seasonal component*. Both $T(t)$ and $S(t)$ are deterministic functions of t, whereas $R(t)$ is a random variable, which, in what follows, is assumed to have mean value $E(R(t)) = 0$ for all t. The seasonal component captures periodic fluctuations of the observations as they commonly arise when observing e.g. meterological parameters as temperature and rainfall against the time. This means that a single observation of the process $\{X(t), t \in \mathbf{T}\}$ made at time t has structure

$$x(t) = T(t) + S(t) + r(t), \tag{6.26}$$

where $r(t)$ is a realization of the random variable $R(t)$.

As a numerical example for a time series, Table 6.1 shows the average of the daily maximum temperatures per month in Johannesburg over a time period of 24 months (in ^0C) and Figure 6.6 the corresponding time series plot. The effect of a seasonal component is clearly visible.

It may make sense to add other deterministic components to the model (6.25), for instance, a component which takes into account short-time cyclic fluctuations of the observations, e.g. systematic fluctuations of the temperature during a day or long-time cyclic changes in the electromagnetic radiation of the sun due to the 33-year period of sunspot fluctuations. It depends on what information is wanted. If the averages of the daily maximum temperatures are of interest, then the fluctuations of the temperature during a day are not relevant. If the oxygen content in the water of a river is measured against the time, then two additional components in (6.25), namely the water temperature and the speed of the running water, should be included. This short section is based on the model (6.25) for the structure of a time series.

The reader will have noticed that the term *trend* has slightly different meanings in stochastic processes and in time series analysis:

a) The trend of a stochastic process $\{X(t), t \in \mathbf{T}\}$ is the mean value $m(t) = E(X(t))$ as a function of time. Hence, a stochastic process of structure (6.25) has trend function

Month i	1	2	3	4	5	6	7	8	9	10	11	12
x_i	26.3	25.6	24.3	22.1	19.1	16.5	16.4	19.8	22.8	25.0	25.3	26.1
Month i	13	14	15	16	17	18	19	20	21	22	23	24
x_i	27.4	26.3	24.8	22.4	18.6	16.7	15.9	20.2	23.4	24.2	25.9	27.0

Table 6.1 Monthly average maximal temperature in Johannesburg

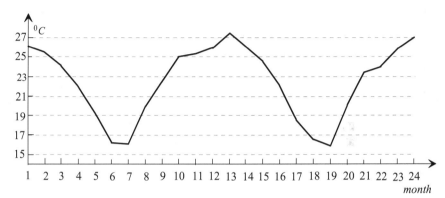

Figure 6.6 Time plot to Table 6.1

$$m(t) = T(t) + S(t),$$

since, by assumption, $E(R(t)) \equiv 0$.

b) In time series analysis, the trend $T(t)$ gives information on the average development of the observations <u>in the longrun</u>. More exactly, the trend of a time series can
principally be obtained by excluding all possible sources of variations of the observations (deterministic and random ones in model (6.25)). Later numerical methods are
proposed how to do this.

Note If $T(t)$ is a parallel to the t-axis, then the time series analysts say '*the time series has no
trend*'. This terminology should not be extended to the trend functions $m(t)$ of stochastic processes. A constant trend function is after all a trend function as well.

6.4.2 Smoothing of Time Series

Smoothing techniques are simple and efficient methods to partially or completely
'level out' deterministic and/or random fluctuations within observed time series, and
in doing this they provide information on the trend $T(t)$ of a time series. The idea
behind smoothing is a technique, which is well-established in the theory of linear
systems, and which is denoted there as *filtration*. Its basis is a *linear filter*, which
transforms a given time series $\{x_i\} = \{x_0, x_1, ..., x_n\}$ of length $n + 1$ into a sequence

$$\{y_i\} = \{y_a, y_{a+1}, ..., y_{n-b}\}$$

of length $n + 1 - a - b$ as follows:

$$y_k = \sum_{i=k-a}^{k+b} w_{i-k} x_i; \quad k = a, a+1, ..., n-b; \quad 0 \le a, b \le n, \tag{6.27}$$

or

$$y_k = w_{-a} x_{k-a} + w_{-a+1} x_{k-a+1} + \cdots + w_b x_{k+b}; \quad k = a, a+1, ..., n-b.$$

The parameter w_i are the *weights* assigned to the respective observations x_j, whereas the interval $[-a, b]$ determines the *bandwidth* of the filter. The weights will usually be positive, but can also be negative. They must satisfy the *normalizing condition*

$$\sum_{i=-a}^{i=b} w_i = 1. \tag{6.28}$$

To illustrate the filter, let $a = b = 2$. Then (6.27) becomes

$$y_k = w_{-2} x_{k-2} + w_{-1} x_{k-1} + w_0 x_k + w_1 x_{k+1} + w_2 x_{k+2}.$$

Thus, y_k is calculated as the sum of those weighted values, which the time series $\{x_i\}$ assumes at time points $k-2$, $k-1$, k, $k+1$, and $k+2$. It is obvious that in this way a 'smoother' sequence than $\{x_i\}$ is generated, i.e. $\{y_i\}$ will exhibit fewer fluctuations, and its fluctuations will have on average smaller amplitudes than $\{x_i\}$. Depending on the aim of smoothing, bandwidth and weights have to be chosen accordingly. If the aim is to level out periods of seasonal influence in order, e.g., to get information on the trend of $\{x_i\}$, then a large bandwidth must be applied. The weights w_i should generally be chosen in such a way that the influence of the x_i on the value of y_k decreases with increasing timely distance $|t_k - t_i|$ of x_i to y_k.

Moving Averages A simple special case of (6.27) is to assume $a = b$ and

$$w_i = \begin{cases} \dfrac{1}{2b+1} & \text{for } i = -b, -b+1, \cdots, b-1, b, \\ 0 & \text{otherwise.} \end{cases}$$

This case is denoted as $M.A.(2b+1)$. The corresponding bandwidth is $[-b, +b]$ and comprises $2b + 1$ time points.

Special cases: 1) If $b = 1$, then y_k is calculated from three observations ($M.A.(3)$):

$$y_k = \frac{1}{3}[x_{k-1} + x_k + x_{k+1}].$$

2) If $b = 2$, then y_k is calculated from 5 observations ($M.A.(5)$):

$$y_k = \frac{1}{5}[x_{k-2} + x_{k-1} + x_k + x_{k+1} + x_{k+2}].$$

Frequently, the time point k is interpreted as the presence, so that time points smaller than k belong to the past and time points greater than k to the future. Particularly interesting is the case when y_k is calculated from the present value and past values of $\{x_i\}$. This case is given by (6.27) with $b = 0$. For instance, with $a = 2$ and equal weights,

$$y_k = \frac{1}{3}[x_k + x_{k-1} + x_{k-2}].$$

In this case it makes sense to interpret y_k as a prediction of the unknown value x_{k+1}.

Smoothing with the Discrete Epanechnikov Kernel   The *Epanechnikov kernel* is given by bandwidth $[-b, b]$ and weights

$$w_i = \left[1 - \frac{i^2}{(b+1)^2} \right] c \quad \text{for} \quad i = 0, \pm 1, ..., \pm b.$$

The factor c makes sure that condition (6.28) is fulfilled:

$$c = \left[1 + \frac{b(4b+5)}{3(b+1)} \right]^{-1}.$$

For instance, if $b = 2$, then $c \approx 0.257$ and y_k is given by

$$y_k = w_{-2} x_{k-2} + w_{-1} x_{k-1} + w_0 x_k + w_1 x_{k+1} + w_2 x_{k+2}$$

$$= [0.556 x_{k-2} + 0.889 x_{k-1} + x_k + 0.889 x_{k+1} + 0.556 x_{k+2}] c.$$

This filter is convenient for numerical calculations: 1) Its input is fully determined by its bandwidth parameter b, and 2) the weights have the symmetry property $w_{-i} = w_i$. Moreover, the observation x_k has the strongest impact on y_k, and the impact of the x_i on y_k becomes smaller with increasing distance of t_i to t_k. The larger the parameter b, the stronger is the smoothing effect.

Exponential (Geometrical) Smoothing   This type of smoothing uses all the 'past' values and the "present" value of the given time series $\{x_0, x_1, ..., x_n\}$ to calculate y_k from the observations $x_k, x_{k-1}, ..., x_0$ in the following way:

$$y_k = \lambda c(k) x_k + \lambda(1-\lambda)c(k)x_{k-1} + \cdots + \lambda(1-\lambda)^k c(k)x_0, \quad k = 0, 1, ..., n, \quad (6.29)$$

where the parameter λ satisfies $0 < \lambda < 1$. Hence, the weights are

$$w_{-i} = \lambda(1-\lambda)^i c(k) \quad \text{for} \quad i = k, k-1, ..., 1, 0.$$

The bandwidth limitation $a = a(k) = k+1$ depends on k, whereas $b = 0$. The factor $c(k)$ ensures that condition (6.28) is fulfilled (apply formula (2.18) with $x = 1 - \lambda$):

$$c(k) = \frac{1}{1 - (1-\lambda)^{k+1}}. \quad (6.30)$$

Since $c(0) = 1/\lambda$ and $c(1) = 1/\lambda(2-\lambda)$, smoothing starts with $y_0 = x_0$, and

$$y_1 = \frac{1}{2-\lambda} x_1 + \frac{1-\lambda}{2-\lambda} x_0 = \frac{1}{2-\lambda}[x_1 + (1-\lambda)x_0].$$

A strong smoothing of $\{x_i\}$ will be achieved with small values of λ since in this case even the 'more distant' values have a nonnegligible effect on y_k. To achieve the

k	2	4	6	8	10	12	14	16	18	20	22
$\lambda = 0.2$	2.778	1.694	1.355	1.202	1.120	1.074	1.046	1.029	1.018	1.012	1.007
$\lambda = 0.4$	1.563	1.149	1.049	1.017	1.006	1.002	1.001	1.000	1.000	1.000	1.000

Table 6.2 Convergence of $c(k)$ towards 1 with increasing k

desired result, one should try different values of λ. As a rule of thumb, start with a value between 0.1 and 0.3.

Table 6.2 shows that even for fairly small values of λ the factor $c(k)$ tends to 1 rather fast. Therefore, in particular when smoothing large time series (which possibly originated in the 'distant past'), $c(k) = 1$ is frequently assumed to be true right from the beginning, i.e., for all $k = 0, 1, \ldots$. Under this assumption, equation (6.29) can be written in the recursive form

$$y_k = \lambda x_k + (1 - \lambda)y_{k-1}; \quad y_0 = x_0, \ k = 1, 2, \ldots, n. \tag{6.31}$$

Table 6.3 gives some principal guidelines about the choice of λ when smoothing.

Effect of the choice of λ on:	λ large	λ small
Smoothing	little	strong
Weights of distant observations	small	large
Weights of near observations	large	small

Table 6.3 Choice of λ in exponential smoothing

Table 6.4 shows once more the original time series $\{x_i\}$ from Table 6.1, the respective sequences $\{y_i\}$ obtained by M.A.(3), by the Epanechnikov kernel (Ep) with $b = 2$, and by exponential smoothing with $\lambda = 0.6$ and (6.31), starting with $y_1 = x_1$ (Ex 0.6). Figure 6.7 illustrates the results for exponential smoothing and for the Epanechnikov approach. With the parameters selected, the sequences $\{y_i\}$ essentially follow the seasonal (periodic) fluctuations, but cleary, the original time series has been smoothed.

Short-Time Forecasting The recursive equation (6.31) provides an easy and efficient possibility for making short-time predictions: Since y_k only depends on the observations x_i made at time points before or at time k, y_k can be considered an estimate of the value the time series $\{x_i\}$ will assume at time point $k + 1$. If this estimate is denoted as $\hat{x}_{k+1}$, equation (6.31) can be rewritten as

$$\hat{x}_{k+1} = \lambda x_k + (1 - \lambda)\hat{x}_{k-1}; \quad y_0 = x_0, \ k = 1, \ldots, n.$$

This equation contains all the information on the development of the time series up to time point k, and gives an estimate of the value of the next observation at time $k + 1$.

Month i	1	2	3	4	5	6	7	8	9	10	11	12
x_i	26.3	25.6	24.3	22.1	19.1	16.5	16.4	19.8	22.8	25.0	25.3	26.1
M.A.3		25.4	24.0	21.8	19.2	17.3	17.6	19.7	22.5	24.4	25.5	26.3
Ep $b=2$			23.6	22.6	19.5	18.3	18.5	20.0	22.1	24.0	25.4	26.1
Ex 0,6	26.3	25.9	24.9	23.2	20.7	18.2	17.1	18.7	21.2	23.5	24.6	25.5
Month i	13	14	15	16	17	18	19	20	21	22	23	24
x_i	27.4	26.3	24.8	22.4	18.6	16.7	15.9	20.2	23.4	24.2	25.9	27.0
M.A.3	26.6	26.2	24.5	21.9	19.2	17.1	17.6	19.8	22.6	24.5	25.7	
Ep $b=2$	26.2	25.6	24.1	21.8	19.5	18.3	18.5	20.0	22.1	24.2		
Ex 0.6	26.6	26.4	25.4	23.6	20.6	18.3	16.9	18.9	21.6	23.2	24.9	26.2

Table 6.4 Data from Table 6.1 and the effect of smoothing

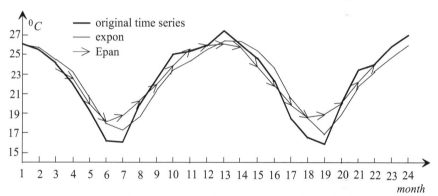

Figure 6.7 Time series plot for Tables 6.1 and 6.4

6.4.3 Trend Estimation

To obtain information on the trend $T(t)$ of a time series by smoothing methods, the bandwidths of the M.A. technique and of the Epanechnikov kernel must be sufficiently large to be able to filter out seasonal (periodic) fluctuations. The time series given by Table 6.1, as with most other meterological and many economical time series, has a period of 12 months. Thus, good smoothing results can be expected with M.A.b with $b \geq 12$. In case of exponential smoothing, the parameter λ needs to be small enough to achieve good smoothing results. All these techniques require sufficiently long time series with respect to the length of the periods of seasonal influences.

Smoothing techniques, however, do not yield the trend as a (continuous) function. But they give an indication which type of continuous function can be used to model the trend best. In many cases, a linear trend function

$$T(t) = \alpha + \beta t \qquad (6.32)$$

will give a satisfactory fit, at least piecewise. Thus, when the original time series $\{x_i\}$ has been smoothed to a time series without seasonal component $\{y_i\}$, then the problem of fitting a linear trend function to $\{y_i\}$ is equivalent to determining the empirical regression line to the values $\{y_i\}$. According to formulas (3.46), page 143, estimates for the coefficients α and β are

$$\hat\alpha = \frac{\sum\limits_{i=1}^{n}(y_i - \bar y)(t_i - \bar t)}{\sum\limits_{i=1}^{n}(t_i - \bar t)^2} = \frac{\sum\limits_{i=1}^{n} y_i t_i - n\bar y \bar t}{\sum\limits_{i=1}^{n} t_i^2 - n\bar t^2}, \quad \hat\beta = \bar y - \hat\alpha \bar t, \qquad (6.33)$$

where the y_i just as the x_i belong to the time points t_i. For estimations of more complicated trend functions, i.e. polynomial ones of higher order than 1, the use of a statistical software package is recommended.

Removing the seasonal influences from a time series of structure (6.26) led to the time series $\{y_i\}$. The next step might be to eliminate the influence of the trend from the time series as well. In many cases this can be achieved, at least approximately, by going over from the time series $\{y_i\}$ to the time series $\{r_i\}$ with

$$r_i = y_i - T(t_i), \quad i = 1, 2, ..., n, \qquad (6.34)$$

where $T(t_i)$ is the value of the trend at time t_i (obtained by smoothing the sequence $\{y_i\}$). Thus, $\{r_i\} = \{r_1, r_2, ..., r_n\}$ is the time series, which arises from the original time series $\{x_i\}$ by eliminating both seasonal influences and trend. Hence, fluctuations within the sequence $\{r_i\}$ are purely due to random influences on the development of a time series. The sequence $\{r_i\}$ is frequently assumed to be the trajectory of a weakly stationary discrete-time stochastic process $\{R(t_1), R(t_2), ..., R(t_n)\}$. The next section deals with some stationary discrete-time stochastic processes $\{R(t), t \in \mathbf{T}\}$, which are quite popular in time series analysis as models for the random component in time series.

Example 6.9 Let us again consider the time series of Table 6.1. This series is too short for long-time predictions of the development of the monthly average maximum temperatures in Johannesburg, but it is suitable as a numerical example. To eliminate the seasonal fluctuations, the $M.A.(13)$ technique is applied. Table 6.5 shows the results. For instance, the values y_7 and y_{18} in the smoothed series $\{y_7, y_7, ..., y_{18}\}$ are

$$y_7 = \frac{1}{13}\sum_{i=1}^{13} x_i = \frac{1}{13}(26.3 + 25.6 + 24.3 + 22.1 + 19.1 + 16.5 + 16.4$$

$$+ 19.8 + 22.8 + 25.0 + 25.3 + 26.1 + 27.4) = 22.8,$$

$$y_{18} = \frac{1}{13}\sum_{i=12}^{24} x_i = \frac{1}{13}(26.1 + 27.4 + 26.3 + 24.8 + 22.4 + 18.6 + 16.7$$

$$+ 15.9 + 20.2 + 23.4 + 24.2 + 25.9 + 27.0) = 23.0.$$

month i	7	8	9	10	11	12	13	14	15	16	17	18
y_i	22.8	22.8	22.8	22.6	22.3	22.2	22.1	22.4	22.7	22.8	22.9	23.0
$T(t_i)$	22.4	22.5	22.5	22.5	22.6	22.6	22.6	22.6	22.7	22.7	22.7	22.9
r_i	0.4	0.3	0.3	0.1	-0.3	-0.4	-0.5	-0.2	0.0	0.1	0.2	0.1

Table 6.5 Results of a time series analysis for the data of Table 6.1

The time points t_i in Table 6.5 refer to the respective month, i.e. $t_i = i$, $i = 7, 8, ..., 18$, so that

$$\bar{y} = \frac{1}{12} \sum_{i=7}^{18} y_i = 22,6 \quad \text{and} \quad \bar{t} = \frac{1}{12} \sum_{i=7}^{18} i = 12.5.$$

Table 6.5 supports the assumption that the trend of the time series $\{x_i\}$ in the interval $[7, 18]$ is a linear one. By (6.33), estimates of its slope and intercept are $\hat{\alpha} = 0.0308$ and $\hat{\beta} = 22.215$. Hence, the linear trend of this time series between $t = 7$ and $t = 18$ is

$$T(t) = 0.0308 \, t + 22.215, \quad 7 \le t \le 18. \tag{6.35}$$

Letting $t = 7, 2, ..., 18$ yields the third row in Table 6.5 and the fourth row contains the effects $r_i = y_i - T(t_i)$ of the 'purely random component' $R(t)$. Figure 6.8 shows the 'smoothed values' y_i and the linear trend (6.35) obtained from these values.   □

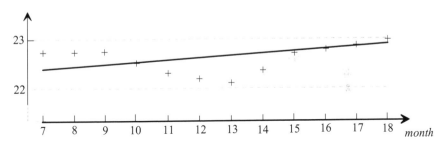

Figure 6.8 Linear trend and M.M.(13)-smoothed values for example 6.9

Some statistical procedures require as input time series which are sample paths of (weakly) stationary stochastic processe (see section 6.4.4). If the time series $\{x_i\}$ has trend $T(t)$, then the underlying stochastic process cannot be stationary. By replacing, however, the original time series $\{x_i\}$ with

$$\{y_i = x_i - T(t_i); \ i = 1, 2, ..., n\},$$

one frequently gets a time series, which is at least approximately the sample path of a discrete-time stationary process. At least, the time series $\{y_i\}$ has no trend.

For getting into theory and applications of time series, the text *Chatfield* (2012) is recommended. Other recent books are e.g. *Madsen* (2008) and *Prado, West* (2010).

6.4.4 Stationary Discrete-Time Stochastic Processes

This section deals with some discrete-time stochastic processes and their stationary representations, which play an important role in time series analysis. They are designed as models for the underlying mathematical structure of stochastic processes, which generate the observed time series, or at least as models for their random components. Knowledge of this structure is particularly essential for the prediction of not yet observed values and for analyzing stochastic signals in communication theory. The models are related to smoothing techniques, but now the x_i are no longer real numbers observed over a time interval, but time-dependent random variables pointed out before, discrete-time stochastic processes are actually sequences of random variables. Hence, in what follows they are written as $\{..., X_{-2}, X_{-1}, X_0, X_1, X_2, ...\}$ if the process started 'in the past', and $\{X_0, X_1, ...\}$ or $\{X_1, X_2, ...\}$ otherwise.

Purely Random Sequence Let $\{..., X_{-2}, X_{-1}, X_0, X_1, X_2, ...\}$ be a sequence of independent random variables, which are identically distributed as X with

$$E(X) = 0 \text{ and } Var(X) = \sigma^2. \tag{6.36}$$

The trend function of this sequence is identically equal to 0:

$$m(t) = 0; \quad t = 0, \pm 1, \pm 2, ...$$

The covariance function of the *purely random sequence* is

$$C(s, t) = \begin{cases} 0 & \text{for } s \neq t, \\ \sigma^2 & \text{for } s = t, \end{cases}$$

or, letting $\tau = t - s$,

$$C(\tau) = \begin{cases} \sigma^2 & \text{for } \tau = 0, \\ 0 & \text{for } \tau \neq 0. \end{cases} \tag{6.37}$$

The purely random sequence is also called *discrete white noise*. If, in addition, the X_i are normally distributed, then $\{..., X_{-2}, X_{-1}, X_0, X_1, X_2, ...\}$ is called a *Gaussian discrete white noise*. The purely random sequence is the most popular discrete-time stochastic process for modelling a random noise, which superimposes an otherwise deterministic time-dependent phenomenon. An example for this is the stochastic process given by (6.25). Its components $S(t)$ and $T(t)$ are deterministic.

Sequence of Moving Averages of Order n. Notation: *M.A.(n)*. Let the random variable Y_t be given by

$$Y_t = \sum_{i=0}^{n} c_i X_{t-i}; \quad t = 0, \pm 1, \pm 2, ... ;$$

where n is a positive integer, $c_0, c_1, ..., c_n$ are finite real numbers, and $\{X_t\}$ is the purely random sequence with parameters (6.36) for all $t = 0, \pm 1, \pm 2, ...$ Thus, the random variable Y_t is constructed from the 'present' X_t and from the n 'preceding' random variables $X_{t-1}, X_{t-2}, ..., X_{t-n}$. This is again the *principle of moving averages* introduced in the previous section fo ther realizations of the X_t. In view of (4.56), page 187,

$$Var(Y_t) = \sigma^2 \sum_{i=0}^{n} c_i^2 < \infty, \quad t = 0, \pm 1, \pm 2, ...,$$

so that $\{Y_t, \ t = 0, \pm 1, \pm 2, ... \}$ is a second-order process. Its trend function is identically equal to 0:

$$m(t) = E(Y_t) = 0 \quad \text{for} \ t = 0, \pm 1, \pm 2, ... $$

For integer-valued s and t,

$$C(s,t) = E(Y_s Y_t) = E\left(\left[\sum_{i=0}^{n} c_i X_{s-i}\right] \cdot \left[\sum_{k=0}^{n} c_k X_{t-k}\right]\right)$$

$$= E\left(\sum_{i=0}^{n} \sum_{k=0}^{n} c_i c_k X_{s-i} X_{t-k}\right).$$

Since $E(X_{s-i} X_{t-k}) = 0$ for $s - i \neq t - k$, the double sum is 0 when $|t - s| > n$. Otherwise there exist i and k so that $s - i = t - k$. In this case $C(s,t)$ becomes

$$C(s,t) = E\left(\sum_{\substack{0 \le i \le n \\ 0 \le |t-s|+i \le n}} c_i c_{|t-s|+i} \, X_{s-i}^2\right)$$

$$= \sigma^2 \sum_{i=0}^{n-|t-s|} c_i c_{|t-s|+i}.$$

Letting $\tau = t - s$, the covariance function $C(s,t) = C(\tau)$ becomes

$$C(\tau) = \begin{cases} \sigma^2 [c_0 c_{|\tau|} + c_1 c_{|\tau|+1} + \cdots + c_{n-|\tau|} c_n] & \text{for} \ \ 0 \le |\tau| \le n \\ 0 & \text{for} \quad \ |\tau| > n \end{cases}. \quad (6.38)$$

Thus, the sequence of moving averages $\{Y_t, \ t = 0, \pm 1, \pm 2, ... \}$ is weakly stationary.

Special case: Let $c_i = \dfrac{1}{n+1}$; $\ i = 0, 1, ..., n$. Then the sequence $M.A.(n)$ becomes

$$Y_t = \frac{1}{n+1} \sum_{i=0}^{n} X_{t-i}; \quad t = 0, \pm 1 \ , \pm 2, ...,$$

and the covariance function (6.38) simplifies to

$$C(\tau) = \begin{cases} \dfrac{\sigma^2}{n+1}\left(1 - \dfrac{|\tau|}{n+1}\right) & \text{for} \ \ 0 \le |\tau| \le n, \\ 0 & \text{for} \ |\tau| > n. \end{cases}$$

Sequence of Moving Averages of Unbounded Order. Notation: *M.A.*(∞). Let

$$Y_t = \sum_{i=0}^{\infty} c_i X_{t-i}; \quad t = 0, \pm 1, \pm 2, \dots, \tag{6.39}$$

where $\{X_t\}$ is the purely random sequence with parameters (6.36), and the c_i are real numbers.

Remark The random sequence $\{Y_t, \; t = 0, \pm 1, \pm 2, \dots\}$ defined in this way is sometimes called a *linear stochastic process.*

To guarantee the convergence of the infinite series (6.39) in mean square, the c_i must satisfy

$$\sum_{i=0}^{\infty} c_i^2 < \infty. \tag{6.40}$$

From (6.38), the covariance of the sequence *M.A.*(∞) is

$$C(\tau) = \sigma^2 \sum_{i=0}^{\infty} c_i c_{|\tau|+i}; \quad \tau = 0, \pm 1, \pm 2, \dots. \tag{6.41}$$

In particular, the variance of Y_t is

$$Var(Y_t) = C(0) = \sigma^2 \sum_{i=0}^{\infty} c_i^2; \quad t = 0, \pm 1, \pm 2, \dots.$$

If the doubly infinite sequence of real numbers

$$\{\dots, c_{-2}, c_{-1}, c_0, c_1, c_2, \dots\}$$

satisfies the condition

$$\sum_{i=-\infty}^{\infty} c_i^2 < \infty,$$

then the doubly infinite series of random variables

$$\{\dots, Y_{-2}, Y_{-1}, Y_0, Y_1, Y_2, \dots\}$$

defined by

$$Y_t = \sum_{i=-\infty}^{\infty} c_i X_{t-i}; \quad t = 0, \pm 1, \pm 2, \dots, \tag{6.42}$$

is also weakly stationary, and it has covariance function

$$C(\tau) = \sigma^2 \sum_{i=-\infty}^{\infty} c_i c_{|\tau|+i}; \quad \tau = 0, \pm 1, \pm 2, \dots$$

and variance

$$Var(Y_t) = \sigma^2 \sum_{i=-\infty}^{\infty} c_i^2; \quad t = 0, \pm 1, \pm 2, \dots.$$

In order to distinguish between the sequences of structure (6.39) and (6.42), they are called *one-* and *two-sided sequences of moving averages*, respectively.

Autoregressive Sequence of Order 1 (Notation: $AR(1)$) Let a and b be finite real numbers with $|a| < 1$. Then a doubly infinite series $\{Y_t\}$ is recursively generated by the equation

$$Y_t = a\,Y_{t-1} + b\,X_t; \quad t = 0, \pm1, \pm2, ..., \tag{6.43}$$

where $\{X_t\}$ is the purely random sequence with parameters (6.36). (Note the analogy to the recursive equation (6.31).) Thus, the 'present' state Y_t depends directly on the preceding one Y_{t-1} and on a random noise term $b\,X_t$ with mean value 0 and variance $b^2\sigma^2$. The n-fold application of (6.43) yields

$$Y_t = a^n Y_{t-n} + b \sum_{i=0}^{n-1} a^i X_{t-i}. \tag{6.44}$$

This formula shows that the influence of a past state Y_{t-n} on the present state Y_t on average decreases as the distance n between Y_{t-n} and Y_t increases. Hence it can be anticipated that the solution of the recurrent equation (6.43) is a stationary process. This stationary solution is obtained by letting n tend to infinity in (6.44): Since there holds $\lim_{n \to \infty} a^n = 0$,

$$Y_t = b \sum_{i=0}^{\infty} a^i X_{t-i}, \quad t = 0, \pm1, \pm2, ... \tag{6.45}$$

The doubly infinite random sequence $\{Y_t; t = 0, \pm1, \pm2, ...\}$ generated in this way is called a *first-order autoregressive sequence* or an *autoregressive sequence of order* 1 (shortly: $AR(1)$). This sequence is a special case of the random sequence defined by (6.38), since letting there $c_i = b a^i$ makes the sequences (6.38) and (6.45) formally identical. Moreover, condition (6.40) is fulfilled:

$$b^2 \sum_{i=0}^{\infty} (a^i)^2 = b^2 \sum_{i=0}^{\infty} a^{2i} = \frac{b^2}{1-a^2} < \infty.$$

Thus, an autoregressive sequence of order 1 is a weakly stationary sequence. Its covariance function is given by formula (6.41) with $c_i = b a^i$:

$$C(\tau) = (b\sigma)^2 \sum_{i=0}^{\infty} a^i a^{|\tau|+i} = (b\sigma)^2 a^{|\tau|} \sum_{i=0}^{\infty} a^{2i}$$

so that

$$C(\tau) = \frac{(b\sigma)^2}{1-a^2} a^{|\tau|}; \quad \tau = 0, \pm1, \pm2, ... $$

Autoregressive Sequence of Order r (Notation: $AR(r)$) In generalization of the recursive equation (6.43), let for a given sequence of real numbers $a_1, a_2, ..., a_r$ with finite a_i and finite integer r random variables Y_t be generated by

$$Y_t = a_1 Y_{t-1} + a_2 Y_{t-2} + \cdots + a_r Y_{t-r} + b\,X_t, \tag{6.46}$$

where $\{X_t\}$ is a purely random sequence with parameters (6.36). The sequence $\{Y_t; t = 0, \pm1, \pm2, ... \}$ is called an *autoregressive sequence of order r*.

It is interesting to investigate whether analogously to the previous example a weakly stationary sequence

$$Y_t = \sum_{i=0}^{\infty} c_i X_{t-i}; \quad t = 0, \pm 1, \pm 2, \dots, \tag{6.47}$$

exists, which is solution of (6.46). Substituting (6.47) into (6.46) yields a linear algebraic system of equations for the unknown parameters c_i :

$$c_0 = b$$

$$c_1 - a_1 c_0 = 0$$

$$c_2 - a_1 c_1 - a_2 c_0 = 0$$

$$\cdots$$

$$c_r - a_1 c_{r-1} - \cdots - a_r c_0 = 0$$

$$c_i - a_1 c_{i-1} - \cdots - a_r c_{i-r} = 0; \quad i = r+1, r+2, \cdots.$$

It can be shown that a nontrivial solution $\{c_0, c_1, \cdots\}$ of this system exists, which satisfies condition (6.40) if the absolute values of the solutions $y_1, y_2, \dots, y_r$ of the algebraic equation

$$y^r - a_1 y^{r-1} - \cdots - a_{r-1} y - a_r = 0 \tag{6.48}$$

are all less than 1, i.e., they are within the unit circle. (Note, this is solely a property of the sequence $a_1, a_2, \dots, a_r$.) In this case, the sequence $\{Y_t; t = 0, \pm 1, \pm 2, \dots\}$ given by (6.47) is a weakly stationary solution of (6.46).

Special Case $r = 2$ Let y_1 and y_2 be the solutions of

$$y^2 - a_1 y - a_2 = 0 \tag{6.49}$$

with $|y_1| < 1$ and $|y_2| < 1$. Then, without proof, the covariance function of the corresponding weakly stationary autoregressive sequence of order 2 is

for $y_1 \neq y_2$

$$C(\tau) = C(0) \frac{(1 - y_1^2) y_2^{|\tau|+1} - (1 - y_2^2) y_1^{|\tau|+1}}{(y_2 - y_1)(1 + y_1 y_2)}; \quad \tau = 0, \pm 1, \pm 2, \dots, \tag{6.50}$$

and for $y_1 = y_2 = y_0$

$$C(\tau) = C(0) \left(1 + \frac{1 - y_0^2}{1 + y_0^2} |\tau| \right) y_0^{|\tau|}; \quad \tau = 0, \pm 1, \pm 2, \dots, \tag{6.51}$$

where the variance $C(0) = Var(Y_t)$ both in (6.50) and (6.51) is

$$C(0) = \frac{1 - a_2}{(1 + a_2)\left[(1 - a_2)^2 - a_1^2\right]} (b\sigma)^2.$$

If the solutions of (6.49) are complex, say,

$$y_1 = y_0 e^{i\omega} \text{ and } y_2 = y_0 e^{-i\omega}$$

with real numbers y_0 and ω, then the covariance function assumes a more convenient form than (6.50):

$$C(\tau) = C(0)\,\alpha y_0^{|\tau|}\sin(\omega|\tau| + \beta); \quad \tau = 0, \pm 1, \pm 2, ...,$$

where

$$\alpha = \frac{1}{\sin\beta} \quad \text{and} \quad \beta = \arctan\left(\frac{1+y_0^2}{1-y_0^2}\tan\omega\right).$$

If $y_1 = y_2 = y_0$, then this representation of $C(\tau)$ is identical to (6.51).

Example 6.10 Consider an autoregressive sequence of order 2 given by

$$Y_t = 0.6Y_{t-1} - 0.05Y_{t-2} + 2X_t; \quad t = 0, \pm 1, \pm 2, ...$$

with $\sigma^2 = Var(X_t) = 1$. It is obvious that the influence of Y_{t-2} on Y_t is small compared to the influence of Y_{t-1} on Y_t. The corresponding algebraic equation (6.49) is

$$y^2 - 0.6y + 0.05 = 0.$$

The solutions are $y_1 = 0.1$ and $y_2 = 0.5$. The absolute values of y_1 and y_2 are smaller than 1 so that the random sequence, generated by (6.46), is weakly stationary. Its covariance is obtained from (6.50):

$$C(\tau) = 7.017\,(0.5)^{|\tau|} - 1.063\,(0.1)^{|\tau|}; \quad \tau = 0, \pm 1, \pm 2, ...$$

As expected, with increasing $|\tau| = |t - s|$, i.e, with increasing timely distance between Y_t and Y_s, the covariance is decreasing. The variance has for all t the value

$$Var(Y_t) = C(0) = 5.954. \qquad \square$$

Autoregressive Mean Average (r, s)-Models. (Notation: **$ARMA(r, s)$**). Let the random sequence $\{Y_t;\, t = 0, \pm 1, \pm 2, ...\}$ be generated by

$$Y_t = +a_1 Y_{t-1} + a_2 Y_{t-2} + \cdots + a_r Y_{t-r} \qquad (6.52)$$
$$+ b_0 X_t + b_1 X_{t-1} + \cdots + b_s X_{t-s},$$

where $\{X_t\}$ is the purely random sequence with parameters (6.36). It can be shown that (6.52) also generates a stationary random sequence $\{Y_t\}$ if the absolute values of the solutions of the algebraic equation (6.48) are less than 1.

The practical work with $ARMA$-models and its special cases is facilitated by the use of statistical software packages. Important problems are: Estimation of the parameters a_i and b_i in (6.46) and (6.52), estimation of trend functions, detection and quantification of possible cyclic, seasonal, and other systematic influences. In particular, reliable predictions are only possible if structure and properties of the random component $\{R(t),\, t \in \mathbf{T}\}$ as stationarity, Markov property, and other properties not taken into account in this short section are known.

6.5 EXERCISES

6.1) A stochastic process $\{X(t),\ t>0\}$ has the one-dimensional distribution

$$\{F_t(x) = P(X(t) \le x) = 1 - e^{-(x/t)^2},\ x \ge 0,\ t>0\}.$$

Is this process weakly stationary?

6.2) The one-dimensional distribution of a stochastic process $\{X(t),\ t>0\}$ is

$$F_t(x) = P(X(t) \le x) = \frac{1}{\sqrt{2\pi t}\ \sigma} \int_{-\infty}^{x} e^{-\frac{(u-\mu t)^2}{2\sigma^2 t}}\ du$$

with $\mu > 0,\ \sigma > 0;\ x \in (-\infty + \infty)$.

Determine its trend function $m(t)$ and, for $\mu = 2$ and $\sigma = 0.5$, sketch the functions

$$y_1(t) = m(t) + \sqrt{Var(X(t))} \quad \text{and} \quad y_2(t) = m(t) - \sqrt{Var(X(t))}\ .$$

6.3) Let $X(t) = A\ \sin(\omega t + \Phi)$, where A and Φ are independent, non-negative random variables with Φ uniformly distributed over $[0, 2\pi]$ and $E(A) < \infty$.

(1) Determine trend, covariance, and correlation function of $\{X(t),\ t \in (-\infty, +\infty)\}$.

(2) Is the stochastic process $\{X(t),\ t \in (-\infty, +\infty)\}$ weakly and/or strongly stationary?

6.4) Let $X(t) = A(t)\ \sin(\omega t + \Phi)$ where $A(t)$ and Φ are independent, non-negative random variables for all t, and let Φ be uniformly distributed over $[0, 2\pi]$.

Verify: If $\{A(t),\ t \in (-\infty, +\infty)\}$ is a weakly stationary process, then the stochastic process $\{X(t),\ t \in (-\infty, +\infty)\}$ is also weakly stationary.

6.5) Let $\{a_1, a_2, ..., a_n\}$ be a sequence of real numbers, and $\{\Phi_1, \Phi_2, ..., \Phi_n\}$ be a sequence of independent random variables, uniformly distributed over $[0, 2\pi]$.

Determine covariance and correlation function of the process $\{X(t),\ t \in (-\infty, +\infty)\}$ given by

$$X(t) = \sum_{i=1}^{n} a_i \sin(\omega t + \Phi_i).$$

6.6)* A modulated signal (pulse code modulation) $\{X(t),\ t \in (-\infty, +\infty)\}$ is given by

$$X(t) = \sum_{-\infty}^{+\infty} A_n\ h(t-n),$$

where the A_n are independent and identically distributed random variables which can only take on values -1 and $+1$ and have mean value 0. Further, let

$$h(t) = \begin{cases} 1 & \text{for } 0 \le t < 1/2 \\ 0 & \text{elsewhere} \end{cases}.$$

(1) Sketch a possible sample path of the stochastic process $\{X(t), t \in (-\infty, +\infty)\}$.

(2) Determine the covariance function of this process.

(3) Let $Y(t) = X(t - Z)$, where the random variable Z has a uniform distribution over $[0, 1]$.

Is $\{Y(t), t \in (-\infty, +\infty)\}$ a weakly stationary process?

6.7) Let $\{X(t), t \in (-\infty, +\infty)\}$ and $\{Y(t), t \in (-\infty, +\infty)\}$ be two independent, weakly stationary stochastic processes, whose trend functions are identically 0 and which have the same covariance function $C(\tau)$.

Verify: The stochastic process $\{Z(t), t \in (-\infty, +\infty)\}$ with

$$Z(t) = X(t) \cos \omega t - Y(t) \sin \omega t$$

is weakly stationary.

6.8) Let $X(t) = \sin \Phi t$, where Φ is uniformly distributed over the interval $[0, 2\pi]$.

Verify: (1) The discrete-time stochastic process $\{X(t); t = 1, 2, ...\}$ is weakly, but not strongly stationary

(2) The continuous-time stochastic process $\{X(t), t \geq 0\}$ is neither weakly nor strongly stationary.

6.9) Let $\{X(t), t \in (-\infty, +\infty)\}$ and $\{Y(t), t \in (-\infty, +\infty)\}$ be two independent stochastic processes with trend and covariance functions

$$m_X(t), \ m_Y(t) \ \text{and} \ C_X(s, t), \ C_Y(s, t),$$

respectively. Further, let

$$U(t) = X(t) + Y(t) \ \text{and} \ V(t) = X(t) - Y(t), \ t \in (-\infty, +\infty).$$

Determine the covariance functions of the stochastic processes $\{U(t), t \in (-\infty, +\infty)\}$ and $\{V(t), t \in (-\infty, +\infty)\}$.

6.10) The following table shows the annual, inflation-adjusted profits of a bank in the years between 2005 to 2015 [in $\$10^6$].

Year	1 (2005)	2	3	4	5	6	7	8	9	10	11
Profit x_i	0.549	1.062	1.023	1.431	2.100	1.809	2.250	3.150	3.636	3.204	4.173

(1) Determine the smoothed values $\{y_i\}$ obtained by applying $M.A.(3)$.

(2) Based on the y_i, determine the trend function (assumed to be a straight line).

(3) Draw the original time series plot, the smoothed version based on the y_i, and the trend function in one and the same Figure.

6.11) The following table shows the production figures x_i of cars of a company over a time period of 12 years (in 10^3).

Year i	1	2	3	4	5	6	7	8	9	10	11	12
x_i	3.08	3.40	4.00	5.24	7.56	10.68	13.72	18.36	23.20	28.36	34.68	40.44

(1) Draw a time series plot. Is the underlying trend function linear?

(2) Smooth the time series $\{x_i\}$ by the Epanechnikov kernel with bandwidth $[-2,+2]$.

(3) Smooth the time series $\{x_i\}$ by exponential smoothing with parameter $\lambda = 0.6$ and predict the output for year 13 by the recursive equation (6.31).

6.12) Let $Y_t = 0.8Y_{t-1} + X_t$; $t = 0,\pm1,\pm2,...$,   where $\{X_t; t = 0,\pm1,\pm2,...\}$ is the purely random sequence with parameters $E(X_t) = 0$ and $Var(X_t) = 1$.

Determine the covariance function and sketch the correlation function of the autoregressive sequence of order 1 $\{Y_t; t = 0,\pm1,\pm2,...\}$.

6.13) Let an autoregressive sequence of order 2 $\{Y_t; t = 0,\pm1,\pm2,...\}$ be given by
$$Y_t - 1.6Y_{t-1} + 0.68Y_{t-2} = 2X_t; \quad t = 0,\pm1,\pm2,...,$$
where $\{X_t; t = 0,\pm1,\pm2,...\}$ is the same purely random sequence as in the previous exercise.

(1) Is the the sequence $\{Y_t; t = 0,\pm1,\pm2,...\}$ weakly stationary?

(2) Determine its covariance and correlation function.

6.14) Let an autoregressive sequence of order 2 $\{Y_t; t = 0,\pm1,\pm2,...\}$ be given by
$$Y_t - 0.8Y_{t-1} - 0.09Y_{t-2} = X_t; \quad t = 0,\pm1,\pm2,....$$
where $\{X_t; t = 0,\pm1,\pm2,...\}$ is the same purely random sequence as in exercise (6.12).

(1) Check whether the sequence $\{Y_t; t = 0,\pm1,\pm2,...\}$ is weakly stationary. If yes, then determine its covariance function and its correlation function.

(2) Sketch its correlation function and compare its graph with the one obtained in exercise (6.12).

CHAPTER 7

Random Point Processes

7.1 BASIC CONCEPTS

A *point process* is a sequence of real numbers $\{t_1, t_2, ...\}$ with properties

$$t_1 < t_2 < \cdots \quad \text{and} \quad \lim_{i \to \infty} t_i = +\infty. \tag{7.1}$$

That means, a point process is a strictly increasing sequence of real numbers, which does not have a finite limit point. In practice, point processes occur in numerous situations: arrival time points of customers at service stations (workshops, filling stations, supermarkets, ...), failure time points of machines, time points of traffic accidents, occurrence of natural disasters, occurrence of supernovas,.... Generally, at time point t_i a certain *event* happens. Hence, the t_i are called *event times*. With regard to the arrival of customers at service stations, the t_i are also called *arrival times*. If not stated otherwise, the assumption $t_1 \geq 0$ is made.

Although the majority of applications of point processes refer to sequences of time points, there are other interpretations as well. For instance, sequences $\{t_1, t_2, ...\}$ can be generated by the location of potholes at a road. Then t_i denotes the distance of the i th pothole from the beginning of the road. Or, the location is measured, at which a beam, which is randomly directed at a forest stand, hits trees. (This is the base of the *Bitterlich method* for estimating the total number of trees in a forest stand.) All these applications deal with finite lengths (time or other). To meet assumption (7.1), they have to be considered finite samples from the respective point processes.

A point process $\{t_1, t_2, ...\}$ can equivalently be represented by the sequence of its *interevent* (*interarrival*) times

$$\{y_1, y_2, ...\} \text{ with } y_i = t_i - t_{i-1}; \ i = 1, 2, ...; \ t_0 = 0.$$

Counting Process Frequently, the event times are of less interest than the number of events, which occur in an interval $(0, t]$, $t > 0$. This number is denoted as $n(t)$:

$$n(t) = \max \{n, \ t_n \leq t\}.$$

For obvious reasons, $\{n(t), t \geq 0\}$ is said to be the *counting process* belonging to the point process $\{t_1, t_2, ...\}$. Here and in what follows, it is assumed that more than one event cannot occur at a time. Point processes with this property are called *simple*. The number of events, which occur in an interval $(s, t]$, $s < t$, is

$$n(s, t) = n(t) - n(s).$$

To be able to count the number $n(A)$ of events which occur in an arbitrary subset A of $[0, \infty)$ the indicator function of the event 't_i belongs to A' is introduced:

$$I_i(A) = \begin{cases} 1 & \text{if } t_i \in A \\ 0 & \text{otherwise} \end{cases}. \tag{7.2}$$

Then,

$$n(A) = \sum_{i=0}^{\infty} I_i(A).$$

Example 7.1 Let a finite sample from a point process be given:

$$S = \{2, 4, 10, 18, 24, 31, 35, 38, 40, 44, 45, 51, 57, 59\}.$$

These figures indicate the times (in seconds) at which within a time span of a minute cars pass a speed check point. In particular, in the interval $A = (30, 45]$

$$n(30, 45) = n(45) - n(30) = 11 - 5 = 6$$

cars passed this check point. Or, in terms of the indicator function of the event $A = (30, 45]$,

$$I_{31}(A) = I_{35}(A) = I_{38}(A) = I_{40}(A) = I_{44}(A) = I_{45}(A) = 1,$$

$$I_i(A) = 0 \text{ for } i \in S \backslash A.$$

Hence,

$$n(30, 45) = \sum_{i=0}^{\infty} I_i(A) = \sum_{i=0}^{60} I_i(A) = 6. \qquad \square$$

Recurrence Times The *forward recurrence time* of a point process $\{t_1, t_2, ...\}$ with respect to time point t is defined as

$$a(t) = t_{n+1} - t \text{ for } t_n \leq t < t_{n+1}; \ n = 0, 1, ..., t_0 = 0. \tag{7.3}$$

Hence, $a(t)$ is the time span from t (usually interpreted as the 'presence') to the occurrence of the next event. A simpler way of characterizing $a(t)$ is

$$a(t) = t_{n(t)+1} - t. \tag{7.4}$$

$t_{n(t)}$ is the largest event time before t and $t_{n(t)+1}$ is the smallest event time after t.

The *backward recurrence time* $b(t)$ with respect to time point t is

$$b(t) = t - t_{n(t)}. \tag{7.5}$$

Thus, $b(t)$ is the time which has elapsed from the last event time before t to time t.

Marked Point Processes Frequently, in addition to their arrival times, events come with another piece of information. For instance: If t_i is the time point the ith customer arrives at a supermarket, then the customer will spend there a certain amount of money m_i. If t_i is the failure time point of a machine, then the time (or cost) m_i necessary for repairing the machine may be assigned to t_i. If t_i denotes the time of the ith bank robbery in a town, then the amount m_i the robbers got away with is of interest. If t_i is the arrival time of the ith claim at an insurance company, then the size

m_i of this claim is important to the company. If t_i is the time of the ith supernova in a century, then its light intensity m_i is of interest to astronomers, and so on. This leads to the concept of a marked point process: Given a point process $\{t_1, t_2, ...\}$, a sequence of two-dimensional vectors

$$\{(t_1, m_1), (t_2, m_2), ...\} \tag{7.6}$$

with m_i being element of a *mark space* **M** is called a *marked point process*. In most applications, as in the four examples above, the mark space **M** is a subset of the real axis $(-\infty, +\infty)$ with the respective units of measurements attached.

Random Point Processes Usually the event times are random variables. A sequence of random variables $\{T_1, T_2, ...\}$ with

$$T_1 < T_2 < \cdots \quad \text{and} \quad P(\lim_{i \to \infty} T_i = +\infty) = 1 \tag{7.7}$$

is a *random point process*. By introducing the *random interevent (interarrival) times*

$$Y_i = T_i - T_{i-1}; \ i = 1, 2, ...; \ T_0 = 0,$$

a random point process can equivalently be defined as a sequence of positive random variables $\{Y_1, Y_2, ...\}$ with property

$$P(\lim_{n \to \infty} \Sigma_{i=0}^{n} Y_i = \infty) = 1.$$

With the terminology introduced in section 6.1, a random point process is a discrete-time stochastic process with state space $\mathbf{Z} = [0, +\infty)$. Thus, a point process (7.1) is a *sample path (realization)* of a random point process. A random point process is called *simple* if at any time point t not more than one event can occur.

Recurrent Point Processes A random point process $\{T_1, T_2, ...\}$ is said to be *recurrent* if its corresponding sequence of interarrival times $\{Y_1, Y_2, ...\}$ is a sequence of independent, identically distributed random variables. The most important recurrent point processes are *homogeneous Poisson processes* and *renewal processes* (sections 7.2 and 7.3).

Random Counting Processes Let

$$N(t) = \max \{n, \ T_n \leq t\}$$

be the random number of events occurring in the interval $(0, t]$. Then the continuous-time stochastic process $\{N(t), t \geq 0\}$ with state space $\mathbf{Z} = \{0, 1, ...\}$ is called the *random counting process* belonging to the random point process $\{T_1, T_2, ...\}$. Any counting process $\{N(t), t \geq 0\}$ has properties

1) $N(0) = 0$,

2) $N(s) \leq N(t)$ for $s \leq t$,

3) For any s, t with $0 \leq s < t$, the increment $N(s, t) = N(t) - N(s)$ is equal to the number of events which occur in $(s, t]$.

Conversely, every stochastic process $\{N(t), t \geq 0\}$ in continuous time having these three properties is the counting process of a certain random point process $\{T_1, T_2, ...\}$. Thus, from the statistical point of view, the stochastic processes

$$\{T_1, T_2, ...\}, \quad \{Y_1, Y_2, ...\}, \quad \text{and} \quad \{N(t), t \geq 0\}$$

are equivalent. For that reason, a random point process is frequently defined as a continuous-time stochastic process $\{N(t), t \geq 0\}$ with properties 1 to 3. Note that

$$N(t) = N(0, t).$$

The most important characteristic of a counting process $\{N(t), t \geq 0\}$ is the probability distribution of its increments $N(s, t) = N(t) - N(s)$, which determines for all intervals $[s, t)$, $s < t$, the probabilities

$$p_k(s, t) = P(N(s, t) = k); \quad k = 0, 1, ...$$

The mean numbers of events in $(s, t]$ is

$$m(s, t) = m(t) - m(s) = E(N(s, t)) = \sum_{k=0}^{\infty} k p_k(s, t). \tag{7.8}$$

With
$$p_k(t) = p_k(0, t),$$

the trend function of the counting process $\{N(t), t \geq 0\}$ is

$$m(t) = E(N(t)) = \sum_{k=0}^{\infty} k p_k(t), \quad t \geq 0. \tag{7.9}$$

A random counting process is called *simple* if the underlying point process is simple. Figure 7.1 shows a possible sample path of a simple random counting process.

Note In what follows the attribute 'random' is usually omitted if it is obvious from the notation or the context that random point processes or random counting processes are being dealt with.

Definition 7.1 (*stationarity*) A random point process $\{T_1, T_2, ...\}$ is called *stationary* if its sequence of interarrival times $\{Y_1, Y_2, ...\}$ is strongly stationary (section 6.3, page 230), that is if for any sequence of integers $i_1, i_2, ..., i_k$ with property

$$1 \leq i_1 < i_2 < \cdots < i_k, \ k = 1, 2, ...$$

and for any $\tau = 0, 1, 2, ...$, the joint distribution functions of the following two random vectors coincide:

$$\{Y_{i_1}, Y_{i_2}, ..., Y_{i_k}\} \quad \text{and} \quad \{Y_{i_1+\tau}, Y_{i_2+\tau}, ..., Y_{i_k+\tau}\}. \qquad \bullet$$

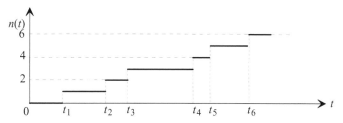

Figure 7.1 Sample path of a simple counting process

It is an easy exercise to show that if the sequence $\{Y_1, Y_2, ...\}$ is strongly stationary, the corresponding counting process $\{N(t), t \geq 0\}$ has homogeneous increments and vice versa. This implies the following corollary from definition 7.1:

Corollary A point process $\{T_1, T_2, ...\}$ is *stationary* if and only if its corresponding counting process $\{N(t), t \geq 0\}$ has homogeneous increments.

Therefore, the probability distribution of any increment $N(s, t)$ of a stationary point process depends only on the difference $\tau = t - s$:

$$p_k(\tau) = P(N(s, s + \tau) = k); \quad k = 0, 1, ...; \ s \geq 0, \ \tau > 0. \tag{7.10}$$

Thus, for a stationary point process,

$$m(\tau) = m(s, s + \tau) = m(s + \tau) - m(s) \quad \text{for all } s \geq 0, \ \tau \geq 0. \tag{7.11}$$

For having increasing sample paths, neither the point process $\{T_1, T_2, ...\}$ nor its corresponding counting process $\{N(t), t \geq 0\}$ can be strongly or weakly stationary as defined in section 6.3. In particular, since only simple point processes are considered, the sample paths of $\{N(t), t \geq 0\}$ are step functions with jump heights equal to 1.

Remark Sometimes it is more convenient or even necessary to define random point processes as doubly infinite sequences

$$\{..., T_{-2}, T_1, T_0, T_1, T_2, ...\},$$

which tend to infinity to the left and to the right with probability 1. Then their sample paths are also doubly infinite sequences: $\{..., t_{-2}, t_1, t_0, t_1, t_2, ...\}$ and only the increments of the corresponding counting process over finite intervals are finite.

Intensity of Random Point Processes For stationary point processes, the mean number of events occurring in $[0, 1]$ is called the *intensity* of the process and will be denoted as λ. By making use of notation (7.9),

$$\lambda = m(1) = \sum_{k=0}^{\infty} k p_k(1). \tag{7.12}$$

In view of the stationarity, λ is equal to the mean number of events occurring in any interval of length 1:

$$\lambda = m(s, s + 1), \ s \geq 0.$$

The mean number of events occurring in any interval $(s, t]$ of length $\tau = t - s$ is

$$m(s, t) = \lambda (t - s) = \lambda \tau.$$

Given a sample path $\{t_1, t_2, ...\}$ of a stationary random point process, λ is estimated by the number of events occurring in $[0, t]$ divided by the length of this interval:

$$\hat{\lambda} = n(t)/t.$$

In example 7.1, an estimate of the intensity of the underlying point process (assumed to be stationary) is $\hat{\lambda} = 14/60 \approx 0.233$.

In case of a nonstationary point process, the role of the constant intensity λ is taken over by an *intensity function* $\lambda(t)$. This function allows to determine the mean number of events $m(s,t)$ occurring in an interval $(s,t]$: For any s,t with $0 \le s < t$,

$$m(s,t)) = \int_s^t \lambda(x)\, dx.$$

Specifically, the mean number of events in $[0,t]$ is the trend function of the corresponding counting process:

$$m(t) = m(0,t) = \int_0^t \lambda(x)\, dx, \quad t \ge 0. \tag{7.13}$$

Hence, for $\Delta t \to 0$,

$$\Delta m(t) = \lambda(t)\,\Delta t + o(\Delta t), \tag{7.14}$$

so that for small Δt the product $\lambda(t)\,\Delta t$ is approximately the mean number of events occurring in $(t, t + \Delta t]$. Another interpretation of (7.14) is: If Δt is sufficiently small, then $\lambda(t)\,\Delta t$ is approximately equal to the probability of the occurrence of an event in the interval $[t, t + \Delta t]$. Hence, the intensity function $\lambda(t)$ is the *arrival rate* of events at time t. (For *Landau's order symbol* $o(x)$, see equation (2.100), page 89.)

Random Marked Point Processes Let $\{T_1, T_2, ...\}$ be a random point process with random marks M_i assigned to the event times T_i. Then the sequence

$$\{(T_1, M_1), (T_2, M_2), ...\} \tag{7.15}$$

is called a *random marked point process*. Its (2-dimensional) sample paths are given by (7.6). The shot noise process $\{(T_n, A_n); n = 1, 2, ...\}$ considered in example 6.5 is a special marked point process.

Random marked point processes are dealt with in full generality in *Matthes et al.* (1974); see also *Stigman* (1995).

Compound Stochastic Processes Let $\{(T_1, M_1), (T_2, M_2), ...\}$ be a marked point process and $\{N(t), t \ge 0\}$ be the counting process belonging to the point process $\{T_1, T_2, ...\}$. The stochastic process $\{C(t), t \ge 0\}$ defined by

$$C(t) = \begin{cases} 0 & \text{for } 0 \le t < T_1 \\ \sum_{i=1}^{N(t)} M_i & \text{for } \quad t \ge T_1 \end{cases}$$

is called a *compound, cumulative,* or *aggregate stochastic process*, and $C(t)$ is called a *compound random variable*. According to the underlying point process, there are e.g. *compound Poisson processes* and *compound renewal processes*. If $\{T_1, T_2, ...\}$ is a claim arrival process and M_i the size of the ith claim, then $C(t)$ is the total claim amount in $[0, t)$. If T_i is the time of the ith breakdown of a machine, and M_i is the corresponding repair cost, then $C(t)$ is the total repair cost in $[0, t)$.

7.2 POISSON PROCESSES

7.2.1 Homogeneous Poisson Processes

7.2.1.1 Definition and Properties

In the theory of stochastic processes, and maybe even more in its applications, the homogeneous Poisson process is just as popular as the exponential distribution in probability theory. Moreover, there is a close relationship between the homogeneous Poisson process and the exponential distribution (theorem 7.2).

Definition 7.2 (*homogeneous Poisson process*) A counting process $\{N(t), t \geq 0\}$ is a *homogeneous Poisson process with intensity* λ, $\lambda > 0$, if it has properties

1) $N(0) = 0$,

2) $\{N(t), t \geq 0\}$ is a stochastic process with independent increments, and

3) its increments $N(s, t) = N(t) - N(s)$, $0 \leq s < t$, have a Poisson distribution with parameter $\lambda(t - s)$:

$$P(N(s,t) = i) = \frac{(\lambda(t-s))^i}{i!} e^{-\lambda(t-s)}; \quad i = 0, 1, \dots, \tag{7.16}$$

or, equivalently, introducing the length $\tau = t - s$ of the interval $[s, t]$, for all $\tau > 0$,

$$P(N(s, s+\tau) = i) = \frac{(\lambda\tau)^i}{i!} e^{-\lambda\tau}; \quad i = 0, 1, \dots. \tag{7.17}$$

●

Formula (7.16) implies that the homogeneous Poisson process has homogeneous increments. Thus, the corresponding *Poisson point process* $\{T_1, T_2, \dots\}$ is stationary in the sense of definition 7.1

Theorem 7.1 A counting process $\{N(t), t \geq 0\}$ with $N(0) = 0$ is a homogeneous Poisson process with intensity λ if and only if it has the following properties:

a) $\{N(t), t \geq 0\}$ has homogeneous and independent increments.

b) The process is *simple,* i.e. $P(N(t, t+h) \geq 2) = o(h)$.

c) $P(N(t, t+h) = 1) = \lambda h + o(h)$.

Proof To prove that definition 7.2 implies properties a), b,) and c), it is only necessary to show that a homogeneous Poisson process satisfies properties b) and c).

b) The simplicity of the Poisson process easily results from (7.17):

$$P(N(t, t+h) \geq 2) = e^{-\lambda h} \sum_{i=2}^{\infty} \frac{(\lambda h)^i}{i!} = \lambda^2 h^2 e^{-\lambda h} \sum_{i=0}^{\infty} \frac{(\lambda h)^i}{(i+2)!} \leq \lambda^2 h^2 = o(h).$$

c) Another application of (7.17) and the simplicity of the Poisson process imply that

$$P(N(t, t+h) = 1) = 1 - P(N(t, t+h) = 0) - P(N(t, t+h) \geq 2)$$

$$= 1 - e^{-\lambda h} + o(h) = 1 - (1 - \lambda h) + o(h)$$

$$= \lambda h + o(h).$$

Conversely, it needs to be shown that a stochastic process with properties a), b), and c) is a homogeneous Poisson process. In view of the assumed homogeneity of the increments, it is sufficient to prove the validity of (7.17) for $s = 0$. Letting

$$p_i(t) = P(N(0, t) = i) = P(N(t) = i); \quad i = 0, 1, \ldots$$

it is to show that

$$p_i(t) = \frac{(\lambda t)^i}{i!} e^{-\lambda t}; \quad i = 0, 1, \ldots. \tag{7.18}$$

From a),

$$p_0(t+h) = P(N(t+h) = 0) = P(N(t) = 0, N(t, t+h) = 0)$$

$$= P(N(t) = 0) P(N(t, t+h) = 0) = p_0(t) p_0(h).$$

In view of b) and c), this result implies

$$p_0(t+h) = p_0(t)(1 - \lambda h) + o(h)$$

or, equivalently,

$$\frac{p_0(t+h) - p_0(t)}{h} = -\lambda p_0(t) + o(h).$$

Taking the limit as $h \to 0$ yields

$$p_0'(t) = -\lambda p_0(t).$$

Since $p_0(0) = 1$, the solution of this differential equation is

$$p_0(t) = e^{-\lambda t}, \quad t \geq 0,$$

so that (7.18) holds for $i = 0$.

Analogously, for $i \geq 1$,

$$p_i(t+h) = P(N(t+h) = i)$$

$$= P(N(t) = i, N(t+h) - N(t) = 0) + P(N(t) = i-1, N(t+h) - N(t) = 1)$$

$$+ \sum_{k=2}^{i} P(N(t) = k, N(t+h) - N(t) = i-k).$$

Because of c), the sum in the last row is $o(h)$. Using properties a) and b),

$$p_i(t+h) = p_i(t) p_0(h) + p_{i-1}(t) p_1(h) + o(h)$$

$$= p_i(t)(1 - \lambda h) + p_{i-1}(t) \lambda h + o(h),$$

or, equivalently,

$$\frac{p_i(t+h)-p_i(t)}{h} = -\lambda[p_i(t)-p_{i-1}(t)]+o(h).$$

Taking the limit as $h \to 0$ yields a system of linear differential equations in the $p_i(t)$:

$$p_i'(t) = -\lambda[p_i(t)-p_{i-1}(t)]; \quad i = 1, 2, \dots.$$

Starting with $p_0(t) = e^{-\lambda t}$, the solution (7.18) is obtained by induction. ∎

The practical importance of theorem 7.1 is that the properties a), b), and c) can be ver- ified without any quantitative investigations, only by qualitative reasoning based on the physical or other nature of the process. In particular, the simplicity of the homo- geneous Poisson process implies that the occurrence of more than one event at the same time point has probability 0.

Note Throughout this chapter, those events, which are generated by a Poisson process, will be called *Poisson events*.

Let $\{T_1, T_2, \dots\}$ be the point process, which belongs to the homogeneous Poisson process $\{N(t), t \geq 0\}$, i.e. T_n is the random time point at which the nth Poisson event occurs. The obvious relationship

$$T_n \leq t \text{ if and only if } N(t) \geq n \tag{7.19}$$

implies

$$P(T_n \leq t) = P(N(t) \geq n). \tag{7.20}$$

Therefore, T_n has the distribution function

$$F_{T_n}(t) = P(N(t) \geq n) = \sum_{i=n}^{\infty} \frac{(\lambda t)^i}{i!} e^{-\lambda t}; \quad n = 1, 2, \dots. \tag{7.21}$$

Differentiation of $F_{T_n}(t)$ with respect to t yields the density of T_n:

$$f_{T_n}(t) = \lambda e^{-\lambda t} \sum_{i=n}^{\infty} \frac{(\lambda t)^{i-1}}{(i-1)!} - \lambda e^{-\lambda t} \sum_{i=n}^{\infty} \frac{(\lambda t)^i}{i!}.$$

On the right-hand side of this equation, all terms but one cancel:

$$f_{T_n}(t) = \lambda \frac{(\lambda t)^{n-1}}{(n-1)!} e^{-\lambda t}; \quad t \geq 0, \ n = 1, 2, \dots. \tag{7.22}$$

Thus, T_n has an Erlang distribution with parameters n and λ (page 75). In particular, T_1 has an exponential distribution with parameter λ, and the interarrival (interevent) times $Y_i = T_i - T_{i-1}; i = 1, 2, \dots; T_0 = 0$, are independent and identically distributed as T_1. Moreover,

$$T_n = \sum_{i=1}^{n} Y_i.$$

These results yield the most simple and, at the same time, the most important charac- terization of the homogeneous Poisson process:

Theorem 7.2 Let $\{N(t), t \geq 0\}$ be a counting process and $\{Y_1, Y_2, ...\}$ be the corresponding sequence of interarrival times. Then $\{N(t), t \geq 0\}$ is a homogeneous Poisson process with intensity λ if and only if the $Y_1, Y_2, ...$ are independent, exponentially with parameter λ distributed random variables.   ■

The random counting process $\{N(t), t \geq 0\}$ is statistically equivalent to both its corresponding point process $\{T_1, T_2, ...\}$ of event times and the sequence of interarrival times $\{Y_1, Y_2, ...\}$. Hence, $\{T_1, T_2, ...\}$ and $\{Y_1, Y_2, ...\}$ are also called Poisson processes.

Example 7.2 From previous observations it is known that the number of traffic accidents $N(t)$ in an area over the time interval $[0, t)$ can be modeled by a homogeneous Poisson process $\{N(t), t \geq 0\}$. On an average, there is one accident within 4 hours, i.e. the intensity of the process is $\lambda = 0.25 \, [h^{-1}]$.

(1) What is the probability p of the event (time unit: hour)

"at most one accident in $[0, 10)$, at least two accidents in $[10, 16)$, and no accident in $[16, 24)$"?

This probability is

$$p = P(N(10) - N(0) \leq 1, \ N(16) - N(10) \geq 2, \ N(24) - N(16) = 0).$$

In view of the independence and the homogeneity of the increments of $\{N(t), t \geq 0\}$, p can be determined as follows:

$$p = P(N(10) - N(0) \leq 1) P(N(16) - N(10) \geq 2) P(N(24) - N(16) = 0)$$

$$= P(N(10) \leq 1) P(N(6) \geq 2) P(N(8) = 0).$$

Now,

$$P(N(10) \leq 1) = P(N(10) = 0) + P(N(10) = 1)$$

$$= e^{-0.25 \cdot 10} + 0.25 \cdot 10 \cdot e^{-0.25 \cdot 10} = 0.2873,$$

$$P(N(6) \geq 2) = 1 - e^{-0.25 \cdot 6} - 0.25 \cdot 6 \cdot e^{0.25 \cdot 6} = 0.4422,$$

$$P(N(8) = 0) = e^{-0.25 \cdot 8} = 0.1353.$$

Hence, the desired probability is

$$p = 0.0172.$$

(2) What is the probability that the second accident occurs not before 5 hours?

Since T_2, the random time to the occurrence of the second accident, has an Erlang distribution with parameters $n = 2$ and $\lambda = 0.25$,

$$P(T_2 > 5) = 1 - F_{T_2}(5) = e^{-0.25 \cdot 5}(1 + 0.25 \cdot 5)$$

so that $P(T_2 > 5) = 0.6446$.   □

The following examples make use of the hyperbolic sine and cosine functions:

$$\sinh x = \frac{e^x - e^{-x}}{2}, \quad \cosh x = \frac{e^x + e^{-x}}{2}, \quad x \in (-\infty, +\infty).$$

Example 7.3 (*random telegraph signal*) A random signal $X(t)$ has structure

$$X(t) = Y(-1)^{N(t)}, \quad t \geq 0, \tag{7.23}$$

where $\{N(t), t \geq 0\}$ is a homogeneous Poisson process with intensity λ and Y is a binary random variable with

$$P(Y = 1) = P(Y = -1) = 1/2,$$

which is independent of $N(t)$ for all t. Signals of this structure are called *random telegraph signals*. Random telegraph signals are basic modules for generating signals of more complicated structure. Obviously, $X(t) = 1$ or $X(t) = -1$, and Y determines the sign of $X(0)$. Figure 7.2 shows a sample path $x = x(t)$ of the process $\{X(t), t \geq 0\}$ on condition $Y = 1$ and $T_n = t_n; \; n = 1, 2, \dots$.

$\{X(t), t \geq 0\}$ is a weakly stationary process. To see this, firstly note that

$$|X(t)|^2 = 1 < \infty \quad \text{for all } t \geq 0.$$

Hence, $\{X(t), t \geq 0\}$ is a second-order process. With

$$I(t) = (-1)^{N(t)},$$

its trend function is $m(t) = E(X(t)) = E(Y) E(I(t))$. Since $E(Y) = 0$,

$$m(t) \equiv 0.$$

It remains to show that the covariance function $C(s, t)$ of this process depends only on $|t - s|$. This requires knowledge of the probability distribution of $I(t)$:

A transition from $I(t) = -1$ to $I(t) = +1$ or, conversely, a transition from $I(t) = +1$ to $I(t) = -1$ occurs at those time points, at which Poisson events occur:

$$P(I(t) = 1) = P(\text{even number of jumps in } (0, t])$$

$$= e^{-\lambda t} \sum_{i=0}^{\infty} \frac{(\lambda t)^{2i}}{(2i)!} = e^{-\lambda t} \cosh \lambda t.$$

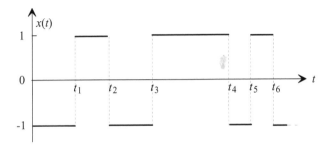

Figure 7.2 Sample path of the random telegraph signal

Analogously,

$$P(I(t) = -1) = P(\text{odd number of jumps in } [0, t])$$

$$= e^{-\lambda t} \sum_{i=0}^{\infty} \frac{(\lambda t)^{2i+1}}{(2i+1)!} = e^{-\lambda t} \sinh \lambda t.$$

Hence the mean value of $I(t)$ is

$$E[I(t)] = 1 \cdot P(I(t) = 1) + (-1) \cdot P(I(t) = -1)$$
$$= e^{-\lambda t} [\cosh \lambda t - \sinh \lambda t] = e^{-2\lambda t}.$$

Since

$$C(s, t) = Cov[X(s), X(t)]$$
$$= E[(X(s) X(t))] = E[Y I(s) Y I(t)]$$
$$= E[Y^2 I(s) I(t)] = E(Y^2) E[I(s) I(t)]$$

and $E(Y^2) = 1$, the covariance function of $\{X(t), \ t \geq 0\}$ has structure

$$C(s, t) = E[I(s) I(t)].$$

In order to evaluate $C(s, t)$, the joint distribution of $(I(s), I(t))$ has to be determined: From (1.22), page 24, and the homogeneity of the increments of $\{N(t), \ t \geq 0\}$, assuming $s < t$,

$$p_{1,1} = P(I(s) = 1, I(t) = 1) = P(I(s) = 1)P(I(t) = 1 | I(s) = 1)$$
$$= e^{-\lambda s} \cosh \lambda s \, P(\text{even number of jumps in } (s, t])$$
$$= e^{-\lambda s} \cosh \lambda s \, e^{-\lambda(t-s)} \cosh \lambda(t - s)$$
$$= e^{-\lambda t} \cosh \lambda s \, \cosh \lambda(t - s).$$

Analogously,

$$p_{1,-1} = P(I(s) = 1, I(t) = -1) \quad = e^{-\lambda t} \, \cosh \lambda s \, \sinh \lambda(t - s),$$
$$p_{-1,1} = P(I(s) = -1, I(t) = 1) \quad = e^{-\lambda t} \, \sinh \lambda s \, \sinh \lambda(t - s),$$
$$p_{-1,-1} = P(I(s) = -1, I(t) = -1) = e^{-\lambda t} \, \sinh \lambda s \, \cosh \lambda(t - s).$$

Now

$$E[I(s)I(t)] = p_{1,1} + p_{-1,-1} - p_{1,-1} - p_{-1,1},$$

so that

$$C(s, t) = e^{-2\lambda(t-s)}, \ s < t.$$

Since the roles of s and t can be changed,

$$C(s, t) = e^{-2\lambda|t-s|}.$$

Hence, the random telegraph signal $\{X(t), \ t \geq 0\}$ is a weakly stationary process. □

Theorem 7.3 Let $\{N(t), t \geq 0\}$ be a homogeneous Poisson process with intensity λ. Then the random number of Poisson events, which occur in the interval $(0, s]$ on condition that exactly n events occur in $(0, t]$, $s < t$; $i = 0, 1, ..., n$; has a binomial distribution with parameters $p = s/t$ and n.

Proof In view of the homogeneity and independence of the increments of the Poisson process $\{N(t), t \geq 0\}$,

$$P(N(s) = i \mid N(t) = n) = \frac{P(N(s) = i, N(t) = n)}{P(N(t) = n)}$$

$$= \frac{P(N(s) = i, N(s, t) = n - i)}{P(N(t) = n)}$$

$$= \frac{P(N(s) = i) P(N(s, t) = n - i)}{P(N(t) = n)} = \frac{\frac{(\lambda s)^i}{i!} e^{-\lambda s} \frac{[\lambda(t-s)]^{n-i}}{(n-i)!} e^{-\lambda(t-s)}}{\frac{(\lambda s)^n}{n!} e^{-\lambda t}}$$

$$= \binom{n}{i} \left(\frac{s}{t}\right)^i \left(1 - \frac{s}{t}\right)^{n-i}; \quad i = 0, 1, ..., n. \tag{7.24}$$

This proves the theorem. ∎

7.2.1.2 Homogeneous Poisson Process and Uniform Distribution

Theorem 7.3 implies that on condition $'N(t) = 1'$ the random time T_1 to the first and only event occurring in $[0, t]$ is uniformly distributed over this interval, since, from (7.24), for $s < t$,

$$P(T_1 \leq s \mid T_1 \leq t) = P(N(s) = 1 \mid N(t) = 1) = \frac{s}{t}.$$

This relationship between the homogeneous Poisson process and the uniform distribution is a special case of a more general result. To prove it, the joint probability density of the random vector $(T_1, T_2, ..., T_n)$ is needed.

Theorem 7.4 The joint probability density of the random vector $(T_1, T_2, ..., T_n)$ is

$$f(t_1, t_2, ..., t_n) = \begin{cases} \lambda^n e^{-\lambda t_n} & \text{for } 0 \leq t_1 < t_2 < \cdots < t_n \\ 0 & \text{elsewhere} \end{cases}. \tag{7.25}$$

Proof For $0 \leq t_1 < t_2$, the joint distribution function of (T_1, T_2) is

$$P(T_1 \leq t_1, T_2 \leq t_2) = \int_0^{t_1} P(T_2 \leq t_2 \mid T_1 = t) f_{T_1}(t) dt.$$

By theorem 7.2, the interarrival times

$$Y_i = T_i - T_{i-1}; \quad i = 1, 2, ...,$$

are independent, identically distributed random variables, which have an exponential distribution with parameter λ.

Hence, since $T_1 = Y_1$,

$$P(T_1 \leq t_1, T_2 \leq t_2) = \int_0^{t_1} P(T_2 \leq t_2 | T_1 = t) \lambda e^{-\lambda t} dt.$$

Given '$T_1 = t$', the random events

$$'T_2 \leq t_2' \text{ and } 'Y_2 \leq t_2 - t'$$

are equivalent. Thus, the desired two-dimensional distribution function is

$$F(t_1, t_2) = P(T_1 \leq t_1, T_2 \leq t_2) = \int_0^{t_1} (1 - e^{-\lambda(t_2 - t)}) \lambda e^{-\lambda t} dt$$

$$= \lambda \int_0^{t_1} (e^{-\lambda t} - e^{-\lambda t_2}) dt.$$

Therefore,

$$F(t_1, t_2) = 1 - e^{-\lambda t_1} - \lambda t_1 e^{-\lambda t_2}, \quad t_1 < t_2.$$

Partial differentiation yields the corresponding two-dimensional probability density

$$f(t_1, t_2) = \begin{cases} \lambda^2 e^{-\lambda t_2} & \text{for } 0 \leq t_1 < t_2 \\ 0 & \text{elsewhere} \end{cases}.$$

The proof of the theorem is now easily completed by induction. ∎

The formulation of the following theorem requires a result from the theory of ordered samples: Let $\{X_1, X_2, ..., X_n\}$ be a random sample taken from X, i.e. the X_i are independent, identically as X distributed random variables. The corresponding ordered sample is denoted as

$$(X_1^*, X_2^*, \cdots, X_n^*), \quad 0 \leq X_1^* \leq X_2^* \leq \cdots \leq X_n^*.$$

Given that X has a uniform distribution over $[0, x]$, the joint probability density of the random vector $\{X_1^*, X_2^*, ..., X_n^*\}$ is

$$f^*(x_1^*, x_2^*, ..., x_n^*) = \begin{cases} n!/x^n, & 0 \leq x_1^* < x_2^* < \cdots < x_n^* \leq x, \\ 0, & \text{elsewhere.} \end{cases} \quad (7.26)$$

For the sake of comparison: The joint probability density of the original (unordered) sample $\{X_1, X_2, ..., X_n\}$ is

$$f(x_1, x_2, ..., x_n) = \begin{cases} 1/x^n, & 0 \leq x_i \leq x, \\ 0, & \text{elsewhere.} \end{cases} \quad (7.27)$$

Theorem 7.5 Let $\{N(t), t \geq 0\}$ be a homogeneous Poisson process with intensity λ, and let T_i be i th event time; $i = 1, 2, ...; T_0 = 0$. Given $N(t) = n$; $n = 1, 2, ...,$ the random vector $\{T_1, T_2, ..., T_n\}$ has the same joint probability density as an ordered random sample taken from a uniform distribution over $[0, t]$.

Proof By definition, for disjoint, but otherwise arbitrary subintervals $[t_i, t_i + h_i]$ of $[0, t]$, the joint probability density of $\{T_1, T_2, ..., T_n\}$ on condition $N(t) = n$ is

$$f(t_1, t_2, ..., t_n | N(t) = n)$$

$$= \lim_{\max(h_1, h_2, ..., h_n) \to 0} \frac{P(t_i \le T_i < t_i + h_i; \ i = 1, 2, ..., n | N(t) = n)}{h_1 h_2 \cdots h_n}.$$

Since the event '$N(t) = n$' is equivalent to $T_n \le t < T_{n+1}$,

$$P(t_i \le T_i < t_i + h_i; \ i = 1, 2, ..., n | N(t) = n)$$

$$= \frac{P(t_i \le T_i < t_i + h_i, \ i = 1, 2, ..., n; \ t < T_{n+1})}{P(N(t) = n)}$$

$$= \frac{\displaystyle\int_t^\infty \int_{t_n}^{t_n + h_n} \int_{t_{n-1}}^{t_{n-1} + h_{n-1}} \cdots \int_{t_1}^{t_1 + h_1} \lambda^{n+1} e^{-\lambda x_{n+1}} \, dx_1 \cdots dx_n \, dx_{n+1}}{\dfrac{(\lambda t)^n}{n!} e^{-\lambda t}}$$

$$= \frac{h_1 h_2 \cdots h_n \, \lambda^n e^{-\lambda t}}{\dfrac{(\lambda t)^n}{n!} e^{-\lambda t}} = \frac{h_1 h_2 \cdots h_n}{t^n} \, n! \, .$$

Hence, the desired conditional joint probability density is

$$f(t_1, t_2, ..., t_n | N(t) = n) = \begin{cases} \dfrac{n!}{t^n}, & 0 \le t_1 < t_2 < \cdots < t_n \le t, \\ 0, & \text{elsewhere.} \end{cases} \tag{7.28}$$

Apart from the notation of the variables, this is the joint density (7.26). ∎

The relationship between homogeneous Poisson processes and the uniform distribution proved in this theorem motivates the common phrase that a homogeneous Poisson process is a *purely random process*, since on condition $N(t) = n$, the event times $T_1, T_2, ..., T_n$ are 'purely randomly' distributed over $[0, t]$.

Example 7.4 (*shot noise*) Shot noise processes have been formally introduced in example 6.5 (page 229). Now an application is discussed in detail:

In the circuit, depicted in Figure 7.3, a light source is switched on at time $t = 0$. A current pulse is initiated in the circuit as soon as the cathode emits a photoelectron due to the light falling on it. Such a current pulse can be quantified by a function $h(t)$ with properties

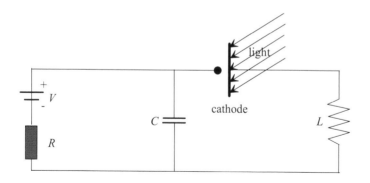

Figure 7.3 Photodetection circuit (Example 7.4)

$$h(t) \geq 0, \quad h(t) = 0 \text{ for } t < 0 \text{ and } \int_0^\infty h(t)\,dt < \infty. \tag{7.29}$$

Let $T_1, T_2, \dots$ be the sequence of random time points, at which the cathode emits photoelectrons and $\{N(t), t \geq 0\}$ be the corresponding counting process. Then the total current flowing in the circuit at time t is

$$X(t) = \sum_{i=1}^\infty h(t - T_i). \tag{7.30}$$

In view of the properties (7.29) of $h(t)$, $X(t)$ can also be written in the form

$$X(t) = \sum_{i=1}^{N(t)} h(t - T_i).$$

In what follows, $\{N(t), t \geq 0\}$ is assumed to be a homogeneous Poisson process with parameter λ. For determining the trend function of this shot noise $\{X(t), t \geq 0\}$, note that according to theorem 7.5, on condition '$N(t) = n$', the $T_1, T_2, \dots, T_n$ are uniformly distributed over $[0, t]$. Hence,

$$E(h(t - T_i)|N(t) = n) = \frac{1}{t} \int_0^t h(t - x)\,dx = \frac{1}{t} \int_0^t h(x)\,dx.$$

Therefore,

$$E(X(t)|N(t) = n) = E\left(\sum_{i=1}^n h(t - T_i)\,\Big|\,N(t) = n \right)$$

$$= \sum_{i=1}^n E(h(t - T_i)|N(t) = n) = \left(\frac{1}{t} \int_0^t h(x)\,dx \right) n.$$

The total probability rule (1.7) yields

$$E(X(t)) = \sum_{n=0}^\infty E(X(t)|N(t) = n)\,P(N(t) = n)$$

$$= \frac{1}{t} \int_0^t h(x)\,dx \sum_{n=0}^\infty n \frac{(\lambda t)^n}{n!} e^{-\lambda t}$$

$$= \left(\frac{1}{t} \int_0^t h(x)\,dx \right) E(N(t)) = \left(\frac{1}{t} \int_0^t h(x)\,dx \right)(\lambda t).$$

Therefore, the trend function of this shot noise process is

$$m(t) = \lambda \int_0^t h(x)\,dx.$$ (7.31)

In order to obtain its covariance variance function, the mean value $E(X(s)X(t))$ has to be determined:

$$E(X(s)X(t)) = \sum_{i,j=1}^n E[h(s - T_i)h(t - T_j)]$$

$$= \sum_{i=1}^n E(h(s - T_i)h(t - T_i))$$

$$+ \sum_{i,j=1, i \neq j}^{\infty} E\left[h(s - T_i)h(t - T_j)\right].$$

Since, on condition ' $N(t) = n$ ', the $T_1, T_2, ..., T_n$ are uniformly distributed over $[0, t]$,

$$E(h(s - T_i)h(t - T_i)|N(t) = n) = \frac{1}{t}\int_0^t h(s - x)h(t - x)\,dx.$$

For $s < t$,

$$E(h(s - T_i)h(t - T_i)|N(t) = n) = \frac{1}{t}\int_0^s h(x)h(t - s + x)\,dx.$$

By theorem 7.5, on condition ' $N(t) = n$ ' the $T_1, T_2, ..., T_n$ are independent. Hence,

$$E(h(s - T_i)h(t - T_j)|N(t) = n) = E(h(s - T_i)|N(t) = n)\,E(h(t - T_j)|N(t) = n)$$

$$= \left(\frac{1}{t}\int_0^s h(s - x)\,dx\right)\left(\frac{1}{t}\int_0^t h(t - x)\,dx\right)$$

$$= \left(\frac{1}{t}\int_0^s h(x)\,dx\right)\left(\frac{1}{t}\int_0^t h(x)\,dx\right).$$

Thus, for $s < t$,

$$E(X(s)X(t)|N(t) = n) = \left(\frac{1}{t}\int_0^s h(x)h(t - s + x)\,dx\right)n$$

$$+ \left(\frac{1}{t}\int_0^s h(x)\,dx\right)\left(\frac{1}{t}\int_0^t h(x)\,dx\right)(n - 1)n.$$

Applying once more the total probability rule,

$$E(X(s)X(t)) = \left(\frac{1}{t}\int_0^s h(x)h(t - s + x)\,dx\right)E(N(t))$$

$$+ \left(\frac{1}{t}\int_0^s h(x)\,dx\right)\left(\frac{1}{t}\int_0^t h(x)\,dx\right)\left[E(N^2(t)) - E(N(t))\right].$$

Making use of equations (7.31) and (6.4), page 226, as well as

$$E(N(t)) = \lambda t \text{ and } E(N^2(t)) = \lambda t(\lambda t + 1),$$

yields the covariance function:

$$C(s, t) = \lambda \int_0^s h(x)h(t - s + x)\,dx, \quad s < t.$$

More generally, for any s and t, $C(s,t)$ can be written in the form

$$C(s,t) = \lambda \int_0^{\min(s,t)} h(x)\, h(|t-s|+x)\, dx. \tag{7.32}$$

Letting $s = t$ yields the variance of $X(t)$:

$$Var(X(t)) = \lambda \int_0^t h^2(x)\, dx.$$

If s tends to infinity in such a way that $|\tau| = t - s$ stays constant, trend and covariance function become

$$m = \lambda \int_0^\infty h(x)\, dx,$$

$$C(\tau) = \lambda \int_0^\infty h(x)\, h(|\tau|+x)\, dx. \tag{7.33}$$

These two formulas are known as *Cambell's theorem*. They imply that, for large t, the shot noise process $\{X(t), t \ge 0\}$ is approximately weakly stationary. For more general formulations of this theorem see *Brandt et. al.* (1990) and *Stigman* (1995).

If the current impulses induced by photoelectrons have random intensities A_i, then the total current flowing in the circuit at time t is

$$X(t) = \sum_{i=1}^{N(t)} A_i\, h(t-T_i).$$

If the A_i are identically distributed as A with $E(A^2) < \infty$, independent of each other, and independent of all T_k, then determining trend and covariance function of this generalized shot noise $\{X(t),\, t \ge 0\}$ does not give rise to principally new problems:

$$m(t) = \lambda\, E(A) \int_0^t h(x)\, dx, \tag{7.34}$$

$$C(s,t) = \lambda\, E(A^2) \int_0^{\min(s,t)} h(x)\, h(|t-s|+x)\, dx. \tag{7.35}$$

If the process of inducing current impulses by photoelectrons has already been operating for an unboundedly long time (the circuit was switched on a sufficiently long time ago), then the underlying shot noise process $\{X(t),\, t \in (-\infty, +\infty)\}$ is given by

$$X(t) = \sum_{-\infty}^{+\infty} A_i\, h(t-T_i).$$

In this case the process is a priori stationary. $\qquad\qquad\qquad\square$

Example 7.5 Customers arrive at a service station (service system, queueing system) according to a homogeneous Poisson process $\{N(t), t \ge 0\}$ with intensity λ. Hence, the arrival of a customer is a Poisson event. The number of servers in the system is assumed to be so large that an incoming customer always will find an available server. Therefore, the service system can be modeled as having an infinite number of servers. The service times of all customers are assumed to be independent random variables, which are identically distributed as Z.

Let $G(t) = P(Z \le t)$ be the distribution function of Z, and $X(t)$ be the random number of customers in the system at time t, $X(0) = 0$. The aim is to determine the *state probabilities* $p_i(t)$ of the system:

$$p_i(t) = P(X(t) = i); \quad i = 0, 1, ...; \quad t \ge 0.$$

A customer arriving at time x is still in the system at time t, $t > x$, with probability $1 - G(t - x)$, i.e. its service has not yet been finished by t. Given $N(t) = n$, the arrival times $T_1, T_2, ..., T_n$ of the n customers in the system are, by theorem 7.4, independent and uniformly distributed over $[0, t]$. For calculating the state probabilities, the order of the T_i is not relevant. Thus, the probability that any of the n customers, who arrived in $[0, t]$, is still in the system at time t is

$$p(t) = \int_0^t (1 - G(t - x)) \frac{1}{t} dx = \frac{1}{t} \int_0^t (1 - G(x)) dx.$$

Since, by assumption, the service times are independent of each other,

$$P(X(t) = i \,|\, N(t) = n) = \binom{n}{i} [p(t)]^i [1 - p(t)]^{n-i}; \quad i = 0, 1, ..., n.$$

By the total probability rule (1.24),

$$p_i(t) = \sum_{n=i}^{\infty} P(X(t) = i \,|\, N(t) = n) \cdot P(N(t) = n)$$

$$= \sum_{n=i}^{\infty} \binom{n}{i} [p(t)]^i [1 - p(t)]^{n-i} \cdot \frac{(\lambda t)^n}{n!} e^{-\lambda t}.$$

This is a mixture of binomial distributions with regard to a Poisson structure distribution. Thus, from example 2.24, page 93, if there the parameter λ is replaced with λt, the state probabilities of the system are

$$p_i(t) = \frac{[\lambda t p(t)]^i}{i!} e^{-\lambda t p(t)}; \quad i = 0, 1, ...$$

Hence, $X(t)$ has a Poisson distribution with parameter

$$E(X(t)) = \lambda t p(t)$$

so that the trend function of $\{X(t), t \ge 0\}$ becomes

$$m(t) = \lambda \int_0^t (1 - G(x)) dx, \quad t \ge 0.$$

For $t \to \infty$ the trend function tends to

$$\lim_{t \to \infty} m(t) = \frac{E(Z)}{E(Y)}, \tag{7.36}$$

where $E(Y) = 1/\lambda$ is the mean interarrival time and $E(Z)$ the mean service time of a customer:

$$E(Z) = \int_0^{\infty} (1 - G(x)) dx.$$

By letting $\rho = E(Z)/E(Y)$, the *stationary state probabilities* of the system become

$$p_i = \lim_{t \to \infty} p_i(t) = \frac{\rho^i}{i!} e^{-\rho}; \quad i = 0, 1, \ldots . \tag{7.37}$$

If Z has an exponential distribution with parameter μ, then

$$m(t) = \lambda \int_0^t e^{-\mu x}\, dx = \frac{\lambda}{\mu}(1 - e^{-\mu t}).$$

In this case, $\rho = \lambda/\mu$. □

7.2.2 Nonhomogeneous Poisson Processses

In this section a stochastic process is investigated, which, except for the homogeneity of its increments, has all the other properties listed in theorem 7.1. Abandoning the assumption of homogeneous increments implies that a time-dependent intensity function $\lambda = \lambda(t)$ takes over the role of λ. This leads to the concept of a nonhomogeneous Poisson process. As proposed in section 7.1, the following notation will be used:

$$N(s, t) = N(t) - N(s), \quad 0 \le s < t.$$

Definition 7.3 A counting process $\{N(t), t \ge 0\}$ with $N(0) = 0$ is called a *nonhomogeneous Poisson process* with *intensity function* $\lambda(t)$ if it has properties

(1) $\{N(t), t \ge 0\}$ has independent increments,

(2) $P(N(t, t + h) \ge 2) = o(h)$,

(3) $P(N(t, t + h) = 1) = \lambda(t)\, h + o(h)$. ●

Three problems will be considered:

1) Computation of the probability distribution of its increments $N(s, t)$:

$$p_i(s, t) = P(N(s, t) = i); \quad 0 \le s < t, \ i = 0, 1, \ldots .$$

2) Computation of the probability density of the random event time T_i (time point at which the *i*-th Poisson event occurs).

3) Computation of the joint probability density of $(T_1, T_2, \ldots, T_n); \ n = 1, 2, \ldots .$

1) In view of the assumed independence of the increments, for $h > 0$,

$$p_0(s, t + h) = P(N(s, t + h) = 0)$$

$$= P(N(s, t) = 0, \ N(t, t + h) = 0)$$

$$= P(N(s, t) = 0) \cdot P(N(t, t + h) = 0)$$

$$= p_0(s, t)\,[1 - \lambda(t)\, h + o(h)].$$

Thus,

$$\frac{p_0(s, t+h) - p_0(s, t)}{h} = -\lambda(t) p_0(s, t) + \frac{o(h)}{h}.$$

Letting $h \to 0$ yields a partial differential equation of the first order:

$$\frac{\partial}{\partial t} p_0(s, t) = -\lambda(t) p_0(s, t).$$

Since $N(0) = 0$ or, equivalently, $p_0(0, 0) = 1$, the solution is

$$p_0(s, t) = e^{-[\Lambda(t) - \Lambda(s)]}, \tag{7.38}$$

where

$$\Lambda(x) = \int_0^x \lambda(u) \, du; \quad x \geq 0. \tag{7.39}$$

Starting with $p_0(s, t)$, the probabilities $p_i(s, t)$ for $i \geq 1$ can be determined by induction:

$$p_i(s, t) = \frac{[\Lambda(t) - \Lambda(s)]^i}{i!} e^{-[\Lambda(t) - \Lambda(s)]}; \quad i = 0, 1, 2, \dots. \tag{7.40}$$

In particular, the absolute state probabilities

$$p_i(t) = p_i(0, t) = P(N(t) = i)$$

of the nonhomogeneous Poisson process at time t are

$$p_i(t) = \frac{[\Lambda(t)]^i}{i!} e^{-\Lambda(t)}; \quad i = 0, 1, 2, \dots. \tag{7.41}$$

Hence, the mean number of Poisson events $m(s, t) = E(N(s, t))$ occurring in the interval $(s, t]$, $s < t$, is

$$m(s, t) = \Lambda(t) - \Lambda(s) = \int_s^t \lambda(x) \, dx. \tag{7.42}$$

In particular, the trend function $m(t) = m(0, t)$ of $\{N(t), t \geq 0\}$ is

$$m(t) = \Lambda(t) = \int_0^t \lambda(x) \, dx, \quad t \geq 0.$$

2) Let $F_{T_1}(t) = P(T_1 \leq t)$ be the distribution function and $f_{T_1}(t)$ the probability density of the random time T_1 to the occurrence of the first Poisson event. Then

$$p_0(t) = p_0(0, t) = P(T_1 > t) = 1 - F_{T_1}(t).$$

From (7.38),

$$p_0(t) = e^{-\Lambda(t)}.$$

Hence,

$$F_{T_1}(t) = 1 - e^{-\Lambda(t)}, \quad f_{T_1}(t) = \lambda(t) e^{-\Lambda(t)}, \quad t \geq 0. \tag{7.43}$$

A comparison of (7.43) with formula (2.98) (page 88) shows that the intensity function $\lambda(t)$ of the nonhomogeneous Poisson process $\{N(t), t \geq 0\}$ is identical to the failure rate belonging to T_1. Since

$$F_{T_n}(t) = P(T_n \leq t) = P(N(t) \geq n),$$

the distribution function of the n th event time T_n is

$$F_{T_n}(t) = \sum_{i=n}^{\infty} \frac{[\Lambda(t)]^i}{i!} e^{-\Lambda(t)}, \quad n = 1, 2, \dots . \tag{7.44}$$

Differentiation with respect to t yields the probability density of T_n:

$$f_{T_n}(t) = \frac{[\Lambda(t)]^{n-1}}{(n-1)!} \lambda(t) e^{-\Lambda(t)}; \quad t \geq 0, \; n = 1, 2, \dots . \tag{7.45}$$

Equivalently,

$$f_{T_n}(t) = \frac{[\Lambda(t)]^{n-1}}{(n-1)!} f_{T_1}(t); \quad t \geq 0, \; n = 1, 2, \dots .$$

By formula (2.52), page 64, and formula (7.44), the mean value of T_n is

$$E(T_n) = \int_0^{\infty} e^{-\Lambda(t)} \left(\sum_{i=0}^{n-1} \frac{[\Lambda(t)]^i}{i!} \right) dt . \tag{7.46}$$

Hence, the mean time

$$E(Y_n) = E(T_n) - E(T_{n-1})$$

between the $(n-1)$ th and the n th event is

$$E(Y_n) = \frac{1}{(n-1)!} \int_0^{\infty} [\Lambda(t)]^{n-1} e^{-\Lambda(t)} dt; \quad n = 1, 2, \dots . \tag{7.47}$$

Letting $\lambda(t) \equiv \lambda$ and $\Lambda(t) \equiv \lambda t$ yields the corresponding characteristics for the homogeneous Poisson process, in particular $E(Y_n) = 1/\lambda$.

3) The conditional probability $P(T_2 \leq t_2 | T_1 = t_1)$ is equal to the probability that at least one Poisson event occurs in $(t_1, t_2]$, $t_1 < t_2$. Thus, from (7.40),

$$F_{T_2}(t_2 | t_1) = 1 - p_0(t_1, t_2) = 1 - e^{-[\Lambda(t_2) - \Lambda(t_1)]}. \tag{7.48}$$

Differentiation with respect to t_2 yields the corresponding probability density:

$$f_{T_2}(t_2 | t_1) = \lambda(t_2) e^{-[\Lambda(t_2) - \Lambda(t_1)]}, \quad 0 \leq t_1 < t_2.$$

By (3.19), page 128, the joint probability density of (T_1, T_2) is

$$f(t_1, t_2) = \begin{cases} \lambda(t_1) f_{T_1}(t_2) & \text{for } t_1 < t_2 \\ 0, & \text{elsewhere} \end{cases}.$$

Starting with $f(t_1, t_2)$, one inductively obtains the joint density of $(T_1, T_2, ..., T_n)$:

$$f(t_1, t_2, ..., t_n) = \begin{cases} \lambda(t_1)\lambda(t_2) \cdots \lambda(t_{n-1}) f_{T_1}(t_n) & \text{for } 0 \le t_1 < t_2 < \cdots < t_n, \\ 0, & \text{elsewhere.} \end{cases} \quad (7.49)$$

This result includes as a special case formula (7.25).

As with the homogeneous Poisson process, the nonhomogeneous Poisson counting process $\{N(t), t \ge 0\}$, the corresponding point process $\{T_1, T_2, ...\}$ of Poisson event times, and the sequence of interevent times $\{Y_1, Y_2, ...\}$ are statistically equivalent stochastic processes.

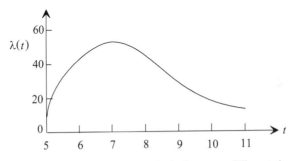

Figure 7.4 Intensity of the arrival of cars at a filling station

Example 7.6 From historical observations it is known that the number of cars arriving for petrol at a particular filling station weekdays between 5:00 and 11:00 a.m. can be modeled by a nonhomogeneous Poisson process $\{N(t), t \ge 0\}$ with intensity function (Figure 7.4)

$$\lambda(t) = 10 + 35.4\,(t-5)\,e^{-(t-5)^2/8}, \quad 5 \le t \le 11.$$

1) What is the mean number of cars arriving for petrol weekdays between 5:00 and 11:00? According to (7.42), this mean number is

$$E(N(5, 11)) = \int_5^{11} \lambda(t)\,dt = \int_0^6 \left(10 + 35.4\,t\,e^{-t^2/8} \right) dt$$

$$= \left[10\,t - 141.6\,e^{-t^2/8} \right]_0^6 = 200.$$

2) What is the probability that at least 90 cars arrive for petrol weekdays between 6:00 and 8:00? The mean number of cars arriving between 6:00 and 8:00 is

$$\int_6^8 \lambda(t)\,dt = \int_1^3 (10 + 35.4\,t\,e^{-t^2/8})\,dt$$

$$= \left[10\,t - 141.6\,e^{-t^2/8} \right]_1^3 = 99.$$

Hence, the random number of cars $N(6, 8) = N(8) - N(6)$ arriving between 6:00 and 8:00 has a Poisson distribution with parameter 99 so that the desired probability is

$$P(N(6, 8) \geq 90) = \sum_{n=90}^{\infty} \frac{99^n}{n!} e^{-0.99}.$$

By using the normal approximation to the Poisson distribution (page 213):

$$\sum_{n=90}^{\infty} \frac{99^n}{n!} e^{-0.99} \approx 1 - \Phi\left(\frac{90 - 99}{\sqrt{99}}\right) \approx 1 - 0.1827.$$

Therefore,

$$P(N(6, 8) \geq 90) = 0.8173. \qquad \square$$

7.2.3 Mixed Poisson Processes

Mixed Poisson processes had been introduced by *J. Dubourdieu* (1938) for modeling claim number processes in accident and sickness insurance. In view of their flexibility, they are now a favorite point process model for many other applications. A recent monograph on mixed Poisson processes is *Grandell* (1997).

Let $\{N(t), t \geq 0\}$ be a homogeneous Poisson process with intensity λ. To explicitly express the dependence of this process on λ, in this section the notation $\{N_\lambda(t), t \geq 0\}$ for the process $\{N(t), t \geq 0\}$ is adopted. The basic idea of Dubourdieu was to consider λ a realization of a positive random variable L, which is called the *(random) structure* or *mixing parameter*. Correspondingly, the probability distribution of L is called the *structure* or *mixing distribution* (section 2.4, pages 92 and 94).

Definition 7.4 Let L be a positive random variable with range $\mathbf{R}_L$. Then the counting process $\{N_L(t), t \geq 0\}$ is said to be a *mixed Poisson process* with structure parameter L if it has the following properties:

(1) $\{N_{L|L=\lambda}(t), t \geq 0\}$ has independent, homogeneous increments for all $\lambda \in \mathbf{R}_L$.

(2) $P(N_{L|L=\lambda}(t) = i) = \frac{(\lambda t)^i}{i!} e^{-\lambda t}$ for all $\lambda \in \mathbf{R}_L$, $i = 0, 1, \dots$.   ●

Thus, on condition $L = \lambda$, the mixed Poisson process is a homogeneous Poisson process with parameter λ:

$$\{N_{L|L=\lambda}(t), t \geq 0\} = \{N_\lambda(t), t \geq 0\}.$$

The absolute state probabilities $p_i(t) = P(N_L(t) = i)$ of the mixed Poisson process at time t are

$$P(N_L(t) = i) = E\left(\frac{(Lt)^i}{i!} e^{-Lt}\right); \quad i = 0, 1, \dots. \qquad (7.50)$$

If L is a discrete random variable with $P(L = \lambda_k) = \pi_k$; $k = 0, 1, ...$; then

$$P(N_L(t) = i) = \sum_{k=0}^{\infty} \frac{(\lambda_k t)^i}{i!} e^{-\lambda_k t} \pi_k. \tag{7.51}$$

In applications, a binary structure parameter L is particularly important. In this case,

$$P(N_L(t) = i) = \frac{(\lambda_1 t)^i}{i!} e^{-\lambda_1 t} \pi + \frac{(\lambda_2 t)^i}{i!} e^{-\lambda_2 t} (1 - \pi) \tag{7.52}$$

for $0 \le \pi \le 1$, $\lambda_1 \ne \lambda_2$.

The basic results, obtained in what follows, do not depend on the probability distribution of L. Hence, for convenience, throughout this section the assumption is made that L is a continuous random variable with density $f_L(\lambda)$. Then,

$$p_i(t) = \int_0^{\infty} \frac{(\lambda t)^i}{i!} e^{-\lambda t} f_L(\lambda) d\lambda; \quad i = 0, 1, ...$$

Obviously, the probability $p_0(t) = P(N_L(t) = 0)$ is the Laplace transform of $f_L(\lambda)$ with parameter $s = t$ (page 99):

$$p_0(t) = \hat{f}_L(t) = E(e^{-Lt}) = \int_0^{\infty} e^{-\lambda t} f_L(\lambda) d\lambda.$$

The i th derivative of $p_0(t)$ is

$$\frac{d^i p_0(t)}{d^i t} = p_0^{(i)}(t) = \int_0^{\infty} (-\lambda)^i e^{-\lambda t} f_L(\lambda) d\lambda.$$

Therefore, all state probabilities of a mixed Poisson process can be written in terms of $p_0(t)$:

$$p_i(t) = P(N_L(t) = i) = (-1)^i \frac{t^i}{i!} p_0^{(i)}(t); \quad i = 1, 2, ... \tag{7.53}$$

Mean value and variance of $N_L(t)$ are (compare with the parameters of the mixed Poisson distribution given by formulas (2.108), page 94):

$$E(N_L(t)) = t E(L), \quad Var(N_L(t)) = t E(L) + t^2 Var(L). \tag{7.54}$$

The following theorem lists two important properties of mixed Poisson processes.

Theorem 7.6 (1) A mixed Poisson process $\{N_L(t), t \ge 0\}$ has homogeneous increments.

(2) If L is not a constant (i.e. the structure distribution is not *degenerate*), then the increments of the mixed Poisson process $\{N_L(t), t \ge 0\}$ are not independent.

Proof (1) Let $0 = t_0 < t_1 < \cdots < t_n$; $n = 1, 2, ...$ Then, for any nonnegative integers $i_1, i_2, ..., i_n$,

$$P(N_L(t_{k-1} + \tau, \ t_k + \tau) = i_k; \ k = 1, 2, ..., n)$$

$$= \int_0^\infty P(N_\lambda(t_{k-1} + \tau, \ t_k + \tau) = i_k; \ k = 1, 2, ..., n) f_L(\lambda) d\lambda$$

$$= \int_0^\infty P(N_\lambda(t_{k-1}, t_k) = i_k; \ k = 1, 2, ..., n) f_L(\lambda) d\lambda$$

$$= P(N_L(t_{k-1}, t_k) = i_k; \ k = 1, 2, ..., n).$$

(2) Let $0 \le t_1 < t_2 < t_3$. Then,

$$P(N_L(t_1, t_2) = i_1, \ N_L(t_2, t_3) = i_2)$$

$$= \int_0^\infty P(N_\lambda(t_1, t_2) = i_1, \ N_\lambda(t_2, t_3) = i_2) f_L(\lambda) d\lambda$$

$$= \int_0^\infty P(N_\lambda(t_1, t_2) = i_1) \ P(N_\lambda(t_2, t_3) = i_2) f_L(\lambda) d\lambda$$

$$\neq \int_0^\infty P(N_\lambda(t_1, t_2) = i_1) f_L(\lambda) d\lambda \int_0^\infty P(N_\lambda(t_2, t_3) = i_2) f_L(\lambda) d\lambda$$

$$= P(N_L(t_1, t_2) = i_1) P(N_L(t_2, t_3) = i_2).$$

This proves the theorem if the mixing parameter L is a continuous random variable. If L is discrete, the same pattern applies.   ■

Multinomial Criterion Let $0 = t_0 < t_1 < \cdots < t_n$; $n = 1, 2, ...$. Then, for any nonnegative integers $i_1, i_2, ..., i_n$ with $i = i_1 + i_2 + \cdots + i_n$,

$$P(N_L(t_{k-1}, t_k) = i_k; \ k = 1, 2, ..., n | N_L(t_n) = i)$$

$$= \frac{i!}{i_1! i_2! \cdots i_n!} \left(\frac{t_1}{t_n}\right)^{i_1} \left(\frac{t_2 - t_1}{t_n}\right)^{i_2} \cdots \left(\frac{t_n - t_{n-1}}{t_n}\right)^{i_n}. \tag{7.55}$$

Interestingly, this conditional probability does not depend on the structure distribution (compare to theorem 7.5). Although the derivation of the multinomial criterion is elementary, it is not done here (Exercise 7.17).

As an application of the multinomial criterion (7.55), the joint distribution of the increments $N_L(0, t) = N_L(t)$ and $N_L(t, t + \tau)$ will be derived:

$$P(N_L(t) = i, \ N_L(t, t + \tau) = k)$$

$$= P(N_L(t) = i | N_L(t + \tau) = i + k) P(N_L(t + \tau) = i + k)$$

$$= \frac{(i + k)!}{i! k!} \left(\frac{t}{t + \tau}\right)^i \left(\frac{\tau}{t + \tau}\right)^k \int_0^\infty \frac{[\lambda(t + \tau)]^{i+k}}{(i + k)!} e^{-\lambda(t+\tau)} f_L(\lambda) d\lambda.$$

Hence, the joint distribution is for $i, k = 0, 1, ...$,

$$P(N_L(0, t) = i, \ N_L(t, t + \tau) = k) = \frac{t^i \tau^k}{i! k!} \int_0^\infty \lambda^{i+k} e^{-\lambda(t+\tau)} f_L(\lambda) d\lambda. \tag{7.56}$$

Since a mixed Poisson process has dependent increments, it is important to get information on the nature and strength of the statistical dependence between two neighboring increments. As a first step into this direction, the mean value of the product of the increments $N_L(t) = N_L(0,t)$ and $N_L(t, t+\tau)$ has to be determined. From formula (7.56),

$$E([N_L(t)] [N_L(t, t+\tau)]) = \sum_{i=1}^{\infty} \sum_{k=1}^{\infty} ik \frac{t^i \tau^k}{i! k!} \int_0^{\infty} \lambda^{i+k} e^{-\lambda(t+\tau)} f_L(\lambda) d\lambda$$

$$= t\tau \int_0^{\infty} \lambda^2 \sum_{i=0}^{\infty} \frac{(\lambda t)^i}{i!} \sum_{k=0}^{\infty} \frac{(\lambda \tau)^k}{k!} e^{-\lambda(t+\tau)} f_L(\lambda) d\lambda$$

$$= t\tau \int_0^{\infty} \sum_{i=0}^{\infty} \lambda^2 e^{\lambda t} e^{\lambda \tau} e^{-\lambda(t+\tau)} f_L(\lambda) d\lambda$$

$$= t\tau \int_0^{\infty} \lambda^2 f_L(\lambda) d\lambda$$

so that

$$E([N_L(t)] [N_L(t, t+\tau)]) = t\tau E(L^2). \tag{7.57}$$

Hence, in view of formula (6.4), page 226,

$$Cov(N_L(\tau), N_L(\tau, \tau+t)) = t\tau Var(L).$$

Thus, two neighboring increments of a mixed Poisson process are positively correlated. Consequently, a large number of events in an interval will on average induce a large number of events in the following interval ('large' relative to the respective lengths of these intervals). This property of a stochastic process is also called *positive contagion*.

Pólya Process A mixed Poisson process with a gamma distributed structure parameter L is called a *Pólya process* (or *Pólya-Lundberg process*).

Let the gamma density of L be

$$f_L(\lambda) = \frac{\beta^\alpha}{\Gamma(\alpha)} \lambda^{\alpha-1} e^{-\beta\lambda}, \quad \lambda > 0, \ \alpha > 0, \ \beta > 0.$$

Then, proceeding as in example 2.24 (page 95) yields

$$P(N_L(t) = i) = \int_0^{\infty} \frac{(\lambda t)^i}{i!} e^{-\lambda t} \frac{\beta^\alpha}{\Gamma(\alpha)} \lambda^{\alpha-1} e^{-\beta\lambda} d\lambda$$

$$= \frac{\Gamma(i+\alpha)}{i! \Gamma(\alpha)} \frac{t^i \beta^\alpha}{(\beta+t)^{i+\alpha}}.$$

Hence,

$$P(N_L(t) = i) = \binom{i-1+\alpha}{i} \left(\frac{t}{\beta+t}\right)^i \left(\frac{\beta}{\beta+t}\right)^\alpha; \quad i = 0, 1, \dots \tag{7.58}$$

Thus, the one-dimensional distribution of the Pólya process $\{ N_L(t), t \geq 0 \}$ is a negative binomial distribution with parameters $r = \alpha$ and $p = t/(\beta + t)$. In particular, for an exponential structure distribution ($\alpha = 1$), $N_L(t)$ has a geometric distribution with parameter $p = t/(t + \beta)$.

To determine the n-dimensional distribution of the Pólya process the multinomial criterion (7.55) and the absolute state distribution (7.58) are used:

For $0 = t_0 < t_1 < \cdots < t_n$; $n = 1, 2, \ldots$ and $i_0 = 0$,

$$P(N_L(t_k) = i_k; k = 1, 2, \ldots, n)$$

$$= P(N_L(t_k) = i_k; k = 1, 2, \ldots, n | N_L(t_n) = i_n) P(N_L(t_n) = i_n)$$

$$= P(N_L(t_{k-1}, t_k) = i_k - i_{k-1}; k = 1, 2, \ldots, n | N_L(t_n) = i_n) P(N_L(t_n) = i_n)$$

$$= \frac{i_n!}{\prod_{k=1}^{n}(i_k - i_{k-1})!} \prod_{k=1}^{n} \left(\frac{t_k - t_{k-1}}{t_n} \right)^{i_k - i_{k-1}} \binom{i_n - 1 + \alpha}{i_n} \left(\frac{t_n}{\beta + t_n} \right)^{i_n} \left(\frac{\beta}{\beta + t_n} \right)^{\alpha}.$$

After some algebra, the n-dimensional distribution of the Pólya process becomes

$$P(N_L(t_k) = i_k; k = 1, 2, \ldots, n)$$

$$= \frac{i_n!}{\prod_{k=1}^{n}(i_k - i_{k-1})!} \binom{i_n - 1 + \alpha}{i_n} \left(\frac{\beta}{\beta + t_n} \right)^{\alpha} \prod_{k=1}^{n} \left(\frac{t_k - t_{k-1}}{\beta + t_n} \right)^{i_k - i_{k-1}}. \qquad (7.59)$$

For the following three reasons its is not surprising that the Pólya process is increasingly used for modeling real-life point processes, in particular customer flows:

1) The finite dimensional distributions of this process are explicitly available.

2) Dependent increments occur more frequently than independent ones.

3) The two free parameters α and β of this process allow its adaptation to a wide variety of data sets.

Example 7.7 An insurance company analyzed the incoming flow of claims and found that the arrival intensity λ is subject to random fluctuations, which can be modeled by the probability density $f_L(\lambda)$ of a gamma distributed random variable L with mean value $E(L) = 0.24$ and variance $Var(L) = 0.16$ (unit: working hour). The parameters α and β of this gamma distribution are obtained from

$$E(L) = 0.24 = \alpha/\beta, \qquad Var(L) = 0.16 = \alpha/\beta^2.$$

Hence, $\alpha = 0.36$ and $\beta = 1.5$. Thus, L has density

$$f_L(\lambda) = \frac{(1.5)^{0.36}}{\Gamma(0.36)} \lambda^{-0.64} e^{-(1.5)\lambda}, \quad \lambda > 0.$$

In time intervals, in which the arrival rate was nearly constant, the flow of claims behaved like a homogeneous Poisson process. Hence, the insurance company modeled the incoming flow of claims by a Pólya process $\{N_L(t), t \geq 0\}$ with the one-dimensional probability distribution

$$P(N_L(t) = i) = \binom{i - 0.64}{i} \left(\frac{t}{1.5 + t}\right)^i \left(\frac{1.5}{1.5 + t}\right)^{0.36}; \quad i = 0, 1, \dots .$$

By (7.54), mean value and variance of $N_L(t)$ are

$$E(N_L(t)) = 0.24 \, t, \quad Var(N_L(t)) = 0.24 \, t + 0.16 \, t^2.$$

As illustrated by this example, the Pólya process (as any other mixed Poisson process) is a more appropriate model than a homogeneous Poisson process with intensity $\lambda = E(L)$ for fitting claim number developments, which exhibit an increasing variability with increasing t. □

Doubly Stochastic Poisson Process The mixed Poisson process generalizes the homogeneous Poisson process by replacing its parameter λ with a random variable L. The corresponding generalization of the nonhomogeneous Poisson process leads to the concept of a doubly stochastic Poisson process. A *doubly stochastic Poisson process* $\{N_{L(\cdot)}(t), t \geq 0\}$ can be thought of as a nonhomogeneous Poisson process the intensity function $\lambda(t)$ of which has been replaced with a stochastic process $\{L(t), t \geq 0\}$ called *intensity process*. Thus, a sample path of a doubly stochastic Poisprocess $\{N_{L(\cdot)}(t), t \geq 0\}$ can be generated as follows:

1) A sample path $\{\lambda(t), t \geq 0\}$ of a given intensity process $\{L(t), t \geq 0\}$ is simulated according to the probability distribution of $\{L(t), t \geq 0\}$.

2) Given $\{\lambda(t), t \geq 0\}$, the process $\{N_{L(\cdot)}(t), t \geq 0\}$ evolves like a nonhomogeneous Poisson process with intensity function $\lambda(t)$.

Thus, a doubly stochastic Poisson process $\{N_{L(\cdot)}(t), t \geq 0\}$ is generated by two independent 'stochastic mechanisms'.

The absolute state probabilities of the doubly stochastic Poisson process at time t are

$$P(N_{L(\cdot)}(t) = i) = \frac{1}{i!} E\left(\left[\int_0^t L(x) \, dx\right]^i e^{-\int_0^t L(x) dx}\right); \quad i = 0, 1, \dots . \quad (7.60)$$

In this formula, the mean value operation 'E' eliminates the randomness generated by the intensity process in $[0, t]$.

The trend function of $\{N_{L(\cdot)}(t), t \geq 0\}$ is

$$m(t) = E\left(\int_0^t L(x) \, dx\right) = \int_0^t E(L(x)) \, dx, \quad t \geq 0.$$

A nonhomogeneous Poisson process with intensity function $\lambda(t) = E(L(t))$ can be used as an approximation to the doubly stochastic Poisson process $\{N_{L(\cdot)}(t), t \geq 0\}$.

The doubly stochastic Poisson process becomes

1. the homogeneous Poisson process if $L(t)$ is equal to a constant λ for all $t > 0$,

2. the nonhomogeneous process if $L(t)$ is a deterministic function $\lambda(t)$, $t \geq 0$,

3. the mixed Poisson process if $L(t)$ is a random variable L, which does not depend on t.

The two 'degrees of freedom', a doubly stochastic Poisson process has, make this process a universal point process model. The term 'doubly stochastic Poisson process' was introduced by *R. Cox*, who was the first to investigate this class of point processes. Hence, these processes are also called *Cox processes*. For detailed treatments and applications in engineering, insurance, and in other fields see *Snyder* (1975) and *Grandell* (1997).

7.2.4 Superposition and Thinning of Poisson Processes

7.2.4.1 Superposition

Assume that a service station recruits its customers from n independent sources. For instance, a branch of a bank serves customers from n different towns, or a car workshop repairs and maintains n different makes of cars, or the service station is a watering place in a game reserve, which is visited by n different species of animals. Each town, each make of cars, and each species generates its own arrival process. Let

$$\{N_i(t), t \geq 0\}; \quad i = 1, 2, ..., n,$$

be the corresponding counting processes. Then, the total number of customers arriving at the service station in $[0, t]$ is

$$N(t) = N_1(t) + N_2(t) + \cdots + N_n(t).$$

$\{N(t), t \geq 0\}$ can be thought of as the counting process of a marked point process, where the marks indicate from which source the customers come.

On condition that $\{N_i(t), t \geq 0\}$ is a homogeneous Poisson process with parameter λ_i; $i = 1, 2, ..., n$, what type of counting process is $\{N(t), t \geq 0\}$?

From example 4.18 (page 180) it is known that the z-transform of $N(t)$ is

$$M_{N(t)}(z) = e^{-(\lambda_1 + \lambda_2 + \cdots + \lambda_n) t (z-1)}.$$

Therefore, $N(t)$ has a Poisson distribution with parameter

$$(\lambda_1 + \lambda_2 + \cdots + \lambda_n) t.$$

Since the counting processes $\{N_i(t), t \geq 0\}$ have homogeneous and independent increments, their additive *superposition* $\{N(t), t \geq 0\}$ also has homogeneous and independent increments. This proves the following theorem.

Theorem 7.7 The additive superposition $\{N(t), t \geq 0\}$ of n independent, homogeneous Poisson processes $\{N_i(t), t \geq 0\}$ with intensities λ_i; $i = 1, 2, ..., n$; is a homogeneous Poisson process with intensity

$$\lambda = \lambda_1 + \lambda_2 + \cdots + \lambda_n.$$ ∎

Quite analogously, if $\{N_i(t), t \geq 0\}$ are independent nonhomogeneous Poisson processes with intensity functions $\lambda_i(t)$; $i = 1, 2, ..., n$; then their additive superposition $\{N(t), t \geq 0\}$ is a nonhomogeneous Poisson process with intensity function

$$\lambda(t) = \lambda_1(t) + \lambda_2(t) + \cdots + \lambda_n(t).$$

7.2.4.2 Thinning

There are many situations, in which not superposition, but the opposite operation, namely *thinning* or *splitting*, of a Poisson process occurs. For instance, a cosmic particle counter registers only α-particles and ignores other types of particles, a reinsurance company is only interested in claims, the size of which exceeds, say, one million dollars, or a game ranger counts only the number of rhinos, which arrive at a watering place per day. Formally, a marked point process $\{(T_1, M_1), (T_2, M_2), ...\}$ arrives and only events with special marks will be taken into account. It is assumed that the marks M_i are independent of each other and independent of $\{T_1, T_2, ...\}$, and that they are identically distributed as

$$M = \begin{cases} m_1 & \text{with probability } 1 - p \\ m_2 & \text{with probability } \quad p \end{cases},$$

i.e., the mark space only consists of two elements: $\mathbf{M} = \{m_1, m_2\}$. In this case, there are two different types of Poisson events: type 1-events (attached with mark m_1) and type 2-events (attached with mark m_2).

Of what kind is the arising point process if only type 1-events are counted?

Let Y be the first event time with mark m_2. If $t < T_1$, then there is surely no type 2-event in $[0, t]$, and if $T_n \leq t < T_{n+1}$, then there are exactly n events in $[0, t]$ and $(1 - p)^n$ is the probability that none of them is a type 2-event. Hence,

$$P(Y > t) = P(0 < t < T_1) + \sum_{n=1}^{\infty} P(T_n \leq t < T_{n+1})(1 - p)^n.$$

Since $P(T_n \leq t < T_{n+1}) = P(N(t) = n)$,

$$P(Y > t) = e^{-\lambda t} + \sum_{n=1}^{\infty} \left(\frac{(\lambda t)^n}{n!} e^{-\lambda t} \right)(1 - p)^n$$

$$= e^{-\lambda t} + e^{-\lambda t} \sum_{n=1}^{\infty} \frac{[\lambda(1-p)t]^n}{n!} = e^{-\lambda t} + e^{-\lambda t}\left[e^{\lambda(1-p)t} - 1 \right].$$

Hence,

$$P(Y > t) = e^{-\lambda p t}, \quad t \geq 0. \tag{7.61}$$

Hence, the interevent times between type 2-events have an exponential distribution with parameter $p\lambda$. Moreover, in view of our assumptions, these interevent times are independent. By changing the roles of type 1- and type 2-events, theorem 7.2 implies theorem 7.8:

Theorem 7.8 Consider a homogeneous Poisson process $\{N(t), t \geq 0\}$ with intensity λ and two types of Poisson events 1 and 2, which occur independently with respective probabilities $1 - p$ and p. Then $N(t)$ can be represented in the form

$$N(t) = N_1(t) + N_2(t),$$

where $\{N_1(t), t \geq 0\}$ and $\{N_2(t), t \geq 0\}$ are two independent homogeneous Poisson processes with $(1 - p)\lambda$ and $p\lambda$, which count only type 1- and type 2-events, respectively.   ■

From this theorem one obtains by induction the following corollary, which is the analogue to theorem 7.7:

Corollary Let $\{(T_1, M_1), (T_2, M_2), ...\}$ be a marked point process with the marks M_i being independent of each other and identically distributed as M:

$$P(M = m_i) = p_i; \quad i = 1, 2, ..., n, \quad \sum_{n=1}^{\infty} p_i = 1.$$

The underlying point process $\{T_1, T_2, ...\}$ is assumed to be Poisson with intensity λ. If only events with mark m_i are counted, then the arising point process is a Poisson process with intensity λp_i, $i = 1, 2, ..., n$.

Nonhomogeneous Poisson Process Now the situation is partially generalized by assuming that the underlying counting process $\{N(t), t \geq 0\}$ is a nonhomogeneous Poisson process with intensity function $\lambda(t)$. The ith Poisson event occurring at time T_i comes with a random mark M_i, where the $\{M_1, M_2, ...\}$ are independent and have the following probability distribution:

$$M_i = \begin{cases} m_1 & \text{with probability } 1 - p(t) \\ m_2 & \text{with probability } \quad p(t) \end{cases} \text{ given that } T_i = t; \; i = 1, 2, ... $$

Note that the M_i are no longer identically distributed. Again, an event coming with mark m_i is called a *type i-event*, $i = 1, 2$.

Let Y be the time to the first occurrence of a type 2- event, $G(t) = P(Y \leq t)$ its distribution function, and $\overline{G}(t) = 1 - G(t)$. Then the relationship

$$P(t < Y \leq t + \Delta t | Y > t) = p(t)\lambda(t)\Delta t + o(\Delta t)$$

implies

$$\frac{1}{\overline{G}(t)} \cdot \frac{G(t + \Delta t) - G(t)}{\Delta t} = p(t)\lambda(t) + \frac{o(\Delta t)}{\Delta t}.$$

Letting Δt tend to 0 yields

$$\frac{G'(t)}{\overline{G}(t)} = p(t)\lambda(t).$$

By integration,

$$\overline{G}(t) = e^{-\int_0^t p(x)\lambda(x)\,dx}, \quad t \ge 0. \tag{7.62}$$

If $p(x) \equiv p$, then (7.62) becomes (7.61).

Theorem 7.9 Given a nonhomogeneous Poisson process $\{N(t), t \ge 0\}$ with intensity function $\lambda(t)$ and two types of events 1 and 2, which occur independently with respective probabilities $1 - p(t)$ and $p(t)$ if t is an event time. Then $N(t)$ can be represented in the form

$$N(t) = N_1(t) + N_2(t),$$

where $\{N_1(t), t \ge 0\}$ and $\{N_2(t), t \ge 0\}$ are independent nonhomogeneous Poisson processes with intensity functions $(1 - p(t))\lambda(t)$ and $p(t)\lambda(t)$, which count only type 1- or type 2-events, respectively. ∎

7.2.5 Compound Poisson Processes

Let $\{(T_i, M_i); \ i = 1, 2, ...\}$ be a marked point process, where $\{T_i; \ i = 1, 2, ...\}$ is a Poisson point process with corresponding counting process $\{N(t), t \ge 0\}$. Then the stochastic process $\{C(t), t \ge 0\}$ defined by

$$C(t) = \sum_{i=0}^{N(t)} M_i$$

with $M_0 = 0$ is called a *compound (cumulative, aggregate) Poisson process*.

Compound Poisson processes occur in many situations:

1) If T_i is the time point at which the i th customer arrives at an insurance company and M_i is its claim size, then $C(t)$ is the total claim amount the company is confronted with in the time interval $[0, t]$.

2) If T_i is the time of the i th breakdown of a machine and M_i the corresponding repair cost, then $C(t)$ is the total repair cost in $[0, t]$.

3) If T_i is the time point the i th shock occurs and M_i the amount of (mechanical) wear, which this shock contributes to the degree of wear of an item, then $C(t)$ is the total wear the item has experienced up to time t. (For the brake discs of a car, every application of the brakes is a shock, which increases their degree of mechanical wear. For the tires of the undercarriage of an aircraft, every takeoff and every touchdown is a shock, which diminishes their tread depth.)

In what follows, $\{N(t), t \geq 0\}$ is assumed to be a homogeneous Poisson process with intensity λ. If the M_i are independent and identically distributed as M and independent of $\{T_1, T_2, ...\}$, then $\{C(t), t \geq 0\}$ has the following properties:

1) $\{C(t), t \geq 0\}$ has independent and homogeneous increments.

2) The Laplace transform of $C(t)$ is

$$\hat{C}_t(s) = e^{\lambda t[\hat{M}(s)-1]}, \tag{7.63}$$

where

$$\hat{M}(s) = E(e^{-sM})$$

is the Laplace transform of M. The proof of (7.63) is straightforward: By (2.118) at page 99,

$$\hat{C}_t(s) = E\left(e^{-s\,C(t)}\right) = E\left(e^{-s\,(M_0+M_1+M_2+\cdots+M_{N(t)})}\right)$$

$$= \sum_{n=0}^{\infty} E\left(e^{-s\,(M_0+M_1+M_2+\cdots+M_n)}\right) P(N(t)=n)$$

$$= \sum_{n=0}^{\infty} E\left(e^{-sM}\right)^n \frac{(\lambda t)^n}{n!}\, e^{-\lambda t} = e^{-\lambda t} \sum_{n=0}^{\infty} \frac{[\lambda t\,\hat{M}(s)]^n}{n!}$$

$$= e^{\lambda t[\hat{M}(s)-1]}.$$

From $\hat{C}_t(s)$, all the moments of $C(t)$ can be obtained by making use of (2.119). In particular, mean value and variance of $C(t)$ are

$$E(C(t)) = \lambda\, t\, E(M), \quad Var(C(t)) = \lambda\, t\, E(M^2). \tag{7.64}$$

Hint These formulas can also be derived by formulas (4.74) and (4.75), page 194.

Now the compound Poisson process is considered on condition that M has a Bernoulli distribution:

$$M = \begin{cases} 1 & \text{with probability} \quad p \\ 0 & \text{with probability} \quad 1-p \end{cases}.$$

Then $M_1 + M_2 + \cdots + M_n$ as a sum of independent and Bernoulli distributed random variables is binomially distributed with parameters n and p (page 49). Hence,

$$P(C(t)=k) = \sum_{0=1}^{n} P(M_0 + M_1 + \cdots + M_n = k | N(t)=n) P(N(t)=n)$$

$$= \sum_{n=0}^{\infty} \binom{n}{k} p^k (1-p)^{n-k} \frac{(\lambda t)^n}{n!}\, e^{-\lambda t}.$$

This is a mixture of binomial distributions with regard to a Poisson structure distribution. Hence, by example 2.24 (page 93), $C(t)$ has a Poisson distribution with parameter $\lambda p t$:

$$P(C(t)=k) = \frac{(\lambda p t)^n}{n!}\, e^{-\lambda p t}; \quad k = 0, 1, ... $$

Corollary If the marks of a compound Poisson process $\{C(t), t \geq 0\}$ have a Bernoulli distribution with parameter p, then $\{C(t), t \geq 0\}$ arises by thinning a homogeneous Poisson process with parameter λ.

If the underlying counting process $\{N(t), t \geq 0\}$ is a nonhomogeneous Poisson process with intensity function $\lambda(t)$ and integrated intensity function $\Lambda(t) = \int_0^t \lambda(x)\,dx$, then (7.63) and (7.64) become in this order

$$\hat{C}_t(s) = e^{\Lambda(t)[\hat{M}(s)-1]},$$
$$E(C(t)) = \Lambda(t)\,E(M),$$
$$Var(C(t)) = \Lambda(t)\,E(M^2).$$
(7.65)

Again, these formulas are an immediate consequence of (4.74) and (4.75).

7.2.6 Applications to Maintenance

The nonhomogeneous Poisson process is an important mathematical tool for modeling and optimizing the maintenance of technical systems with respect to cost and reliability criteria by applying proper *maintenance policies* (*strategies*). *Maintenance policies* prescribe when to carry out (preventive) repairs, replacements, inspections, or other maintenance measures. *Repairs* after system failures usually only tackle the causes which triggered off the failures. A *minimal repair* performed after a failure enables the system to continue its work but it does not affect the failure rate (2.56) (page 88) of the system. In other words, after a minimal repair the failure rate of the system has the same value as immediately before a failure. For example, if a failure of a complicated electronic system is caused by a defective plug and socket connection, then removing this cause of failure can be considered a minimal repair. *Preventive replacements* (*renewals*) and *preventive repairs* are not initiated by system failures, but they are carried out to prevent or at least to postpone future failures. Preventive minimal repairs make no sense with regard to the survival probability of systems.

Minimal Repair Policy Every system failure is (and can be) removed by a minimal repair.

Henceforth it is assumed that all renewals and repairs take only negligibly small times and that, after completing a renewal or a repair, the system immediately resumes its work. The random lifetime $T = T_1$ of the system has probability density $f(t)$, distribution function $F(t)$, survival probability $\overline{F}(t) = 1 - F(t)$, and failure rate $\lambda(t)$.

Theorem 7.10 A system is subject to a minimal repair policy. Let T_i be the time at which its ith failure (minimal repair) takes place. Then the sequence $\{T_1, T_2, ...\}$ is a nonhomogeneous Poisson process, the intensity function of which is given by the failure rate $\lambda(t)$ of the system.

Proof The first failure of the system, which starts working at time $t = 0$, occurs at the random time $T = T_1$ with density

$$f_{T_1}(t) = \lambda(t)\,e^{-\Lambda(t)}; \; t \geq 0.$$

The same density one gets from (7.45) or (7.49) for $n = 1$. Now let us assume that a failure (minimal repair) occurs at time point $T_1 = t_1$. Then the failure probability of the system in $[t_1, t_2)$ with $t_1 < t_2$ is nothing else than the conditional failure probability of a system, which has survived the interval $[0, t_1]$ (in either case the system has failure rate $\lambda(t_1)$ at time t_1). Hence, by formula (2.98):

$$P(T_2 < t_2 | T_1 = t_1) = 1 - e^{-[\Lambda(t_2 + t_1) - \Lambda(t_1)]}.$$

But this is formula (7.48) and just as there it can be concluded that the joint density of the random vector (T_1, T_2) is given by (7.49) with $n = 2$. Finally, induction yields that the joint density of the random vector $(T_1, T_2, ..., T_n)$ is for all $n = 1, 2, ...$ given by (7.49), where $\lambda(t)$ is the failure rate of the system.   ■

The minimal repair policy provides the theoretical fundament for analyzing a number of more sophisticated maintenance policies including preventive replacements. To justify preventive replacements, the assumption has to be made that the underlying system is aging, i.e. its failure rate is increasing (pages 87–89).

The criterion for evaluating the efficiency of maintenance policies will be the average maintenance cost per unit time over an infinite time span. To establish this criterion, the time axis is partitioned into *replacement cycles*, i.e. into the times between two neighboring replacements. Let L_i be the random length of the i th replacement cycle and C_i the total random maintenance cost (replacement + repair cost) in the i th replacement cycle. It is assumed that the L_i are independent and identically distributed as L. This assumption implies that a replaced system is as good as the previous one ('as good as new') from the point of view of its lifetime. The C_i are assumed to be independent, identically distributed as C, and independent of the L_i. Then the *maintenance cost per unit time over an infinite time span* is

$$K = \lim_{n \to \infty} \frac{\sum_{i=1}^{n} C_i}{\sum_{i=1}^{n} L_i} = \lim_{n \to \infty} \frac{\frac{1}{n}\sum_{i=1}^{n} C_i}{\frac{1}{n}\sum_{i=1}^{n} L_i}.$$

The strong law of the large numbers implies

$$K = \frac{E(C)}{E(L)}. \tag{7.66}$$

For the sake of brevity, K is referred to as the *(long-run) maintenance cost rate*. Thus, the maintenance cost rate is equal to the mean maintenance cost per cycle divided by the mean cycle length. In what follows, c_p denotes the cost of a preventive replacement, and c_m is the cost of a minimal repair; c_p, c_m constants.

Policy 1 A system is preventively replaced at fixed times $\tau, 2\tau, ...$ Failures between replacements are removed by minimal repairs.

This policy reflects the common approach of preventively overhauling complicated systems after fixed time periods whilst in between only the absolutely necessary repairs are done. With this policy, all cycle lengths are equal to τ so that in view of (7.65) the mean cost per cycle is equal to $c_p + c_m \Lambda(\tau)$. Hence, the corresponding maintenance cost rate is

$$K_1(\tau) = \frac{c_p + c_m \Lambda(\tau)}{\tau}.$$

A replacement interval $\tau = \tau*$, which minimizes $K_1(\tau)$, satisfies the condition

$$\tau \lambda(\tau) - \Lambda(\tau) = c_p/c_m.$$

If $\lambda(t)$ tends to infinity as $t \to \infty$, then there exists a unique solution $\tau = \tau*$ of this equation. The corresponding minimal maintenance cost rate is

$$K_1(\tau*) = c_m \lambda(\tau*).$$

Policy 2 A system is replaced at the first failure which occurs <u>after</u> a fixed time τ. Failures which occur between replacements are removed by minimal repairs.

This policy makes use fully of the system lifetime so that, from this point of view, it is preferable to policy 1. The partial uncertainty, however, about the times of replacements leads to larger replacement costs than with policy 1. The replacement is no longer purely preventative so that its cost are denoted as c_r. Thus, in practice the maintenance cost rate of policy 2 may actually exceed the one of policy 1 if c_r is sufficiently larger than the c_p used in policy 1. The residual lifetime T_τ of the system after time point τ, when having survived interval $[0, \tau]$, has according to (2.93) mean value

$$\mu(\tau) = \frac{1}{F(\tau)} \int_\tau^\infty \bar{F}_\tau(x) \, dx. \tag{7.67}$$

The mean maintenance cost per cycle is $c_r + c_m \Lambda(\tau)$, and the mean replacement cycle length is $\tau + \mu(\tau)$ so that the corresponding maintenance cost rate is

$$K_2(\tau) = \frac{c_r + c_m \Lambda(\tau)}{\tau + \mu(\tau)}.$$

An optimal $\tau = \tau*$ satisfies the necessary condition $dK_2(\tau)/d\tau = 0$, i.e.,

$$\left[\Lambda(\tau) + \frac{c_r}{c_m} - 1 \right] \mu(\tau) = \tau.$$

Example 7.8 Let the system lifetime T have a Rayleigh distribution with failure rate $\lambda(t) = 2t/\theta^2$. The corresponding mean residual lifetime of the system after having survived $[0, \tau]$ is

$$\mu(\tau) = \theta \sqrt{\pi} \, e^{(\tau/\theta)^2} \left[1 - \Phi\left(\frac{\sqrt{2}}{\theta} \tau \right) \right].$$

If $\theta = 100 \, [h^{-1}]$, $c_m = 1$, and $c_r = 5$, the optimal parameters are

$$\tau* = 180 \, [h], \quad K_2(\tau*) = 0.0402. \qquad \square$$

Policy 3 The first $n-1$ failures are removed by minimal repairs. At the time point T_n of the nth failure, a replacement is carried out.

The random cycle length is $L = T_n$. Hence, the maintenance cost rate is

$$K_3(n) = \frac{c_r + (n-1)c_m}{E(T_n)},$$

where the mean cycle length $E(T_n)$ is given by (7.46). By analyzing the behavior of the difference $K_3(n) - K_3(n-1)$, an optimal $n = n*$ is seen to be the smallest integer n satisfying

$$E(T_n) - [n - 1 + c_r/c_m]E(Y_{n+1}) \geq 0; \quad n = 1, 2, ..., \tag{7.68}$$

where the mean time $E(Y_n)$ between the $(n-1)$th and the nth minimal repair is given by formula (7.47).

Example 7.9 Let the system lifetime T have a Weibull distribution:

$$\lambda(t) = \frac{\beta}{\theta}\left(\frac{t}{\theta}\right)^{\beta-1}, \quad \Lambda(t) = \left(\frac{t}{\theta}\right)^{\beta}, \quad \beta > 1. \tag{7.69}$$

Under this assumption condition (7.68) becomes

$$\beta n - [n - 1 + c_r/c_m] \geq 0.$$

Hence, if $c_r > c_m$,

$$n* = \left\|\frac{1}{\beta-1}\left(\frac{c_r}{c_m} - 1\right)\right\| + 1,$$

where $\|x\|$ is the largest integer being less or equal to x. (If $x < 0$, then $\|x\| = 0$.) If the aging process of the system proceeds fast (β large), then $n*$ is small.   □

7.2.7 Applications To Risk Analysis

Random point processes are key tools for quantifying the financial risk in virtually all branches of industry. This section uses the terminology for analyzing the financial risk in the insurance industry. A risky situation for an insurance company arises if it has to pay out a total claim amount, which exceeds its total premium income plus initial capital. To be able to establish the corresponding mathematical risk model, next the concept of a risk process has to be introduced: An insurance company starts its business at time $t = 0$. Claims arrive at random time points $T_1, T_2, ...$ and come with the respective random claim sizes $M_1, M_2, ...$. Thus, the insurance company is subjected to a random marked point process

$$\{(T_1, M_1), (T_2, M_2), ...\},$$

called *risk process*. The two components of the risk process are the *claim arrival process* $\{T_1, T_2, ...\}$ and the *claim size process* $\{M_1, M_2, ...\}$. Let $\{N(t), t \geq 0\}$ be the random counting process, which belongs to the claim arrival process. Then the total claim size $C(t)$, the company is faced with in the interval $[0,t]$, is a compound random variable of structure

$$C(t) = \begin{cases} \sum_{i=1}^{N(t)} M_i & \text{if } N(t) \geq 1, \\ 0 & \text{if } N(t) = 0. \end{cases} \tag{7.70}$$

The compound Poisson process

$$\{C(t), t \geq 0\}$$

is the main ingredient of the risk model to be analyzed in this section.

To equalize the loss caused by claims and to eventually make a profit, an insurance company imposes a premium on its clients. Let $\kappa(t)$ be the total premium income of the insurance company in $[0,t]$. In case $C(t) < \kappa(t)$, the company has made a profit of

$$\kappa(t) - C(t)$$

in the interval $[0,t]$ (not taking into account staff and other running costs of the company).

With an *initial capital* or an *initial reserve x*, which the company has at its disposal at the start, the *risk reserve* at time t is defined as

$$R(t) = x + \kappa(t) - C(t) \tag{7.71}$$

The corresponding (stochastic) *risk reserve process* is $\{R(t), t \geq 0\}$. If the sample path of $\{R(t), t \geq 0\}$ becomes negative at a time point t_r, the financial expenses of the company in $[0,t_r]$ exceed its available capital of $x + \kappa(t_r)$ at the time point t_r. This leads to the definition of the *ruin probability $p(x)$* of the company:

$$p(x) = P(\text{there is a positive, finite } t \text{ so that } R(t) < 0). \tag{7.72}$$

Correspondingly, the *non-ruin probability* or *survival probability* of the company is

$$q(x) = 1 - p(x).$$

These probabilities refer to an infinite time horizon. The *ruin probability* of the company with regard to a finite time horizon τ is

$$p(x,\tau) = P(\text{there is a finite } t \text{ with } 0 < t \leq \tau \text{ so that } R(t) < 0).$$

The ruin probabilities $p(x)$ and $p(x,\tau)$ decrease with increasing initial capital x.

Since ruin can only occur at the arrival time points of claims (Figure 7.5), $p(x)$ and $p(x,\tau)$ can also be defined in the following way:

$$p(x) = P(\text{there is a positive, finite integer } n \text{ so that } R(T_n) < 0). \tag{7.73}$$

$$p(x,\tau) = P(\text{there is a positive, finite integer } n \text{ with } T_n \leq \tau \text{ so that } R(T_n) < 0),$$

where $R(T_n)$ is understood to be $R(T_n + 0)$, i.e. the value of the risk reserve process at time point T_n includes the effect of the nth claim.

Note In the actuarial literature, claim sizes are frequently denoted as U_i, the initial capital as u, and the ruin probability as $\psi(u)$.

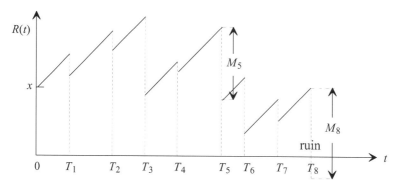

Figure 7.5 Sample path of a risk process leading to ruin

In this section, the problem of determining the ruin probability is dealt with under the so-called 'classical assumptions:'

1) $\{N(t), t \geq 0\}$ is a homogeneous Poisson process with parameter λ.

2) The claim sizes $M_1, M_2, \ldots$ are independent, identically as M distributed random variables. They are independent of the $T_1, T_2, \ldots$.

3) The premium income is a linear function in t: $\kappa(t) = \kappa t$. The constant parameter κ is called the *premium rate*.

4) The time horizon is infinite ($\tau = \infty$).

Under asumptions 1 and 2, risk analysis is subjected to a homogeneous portfolio, i.e. claim sizes are independent, differences in the claim sizes are purely random, and the arrival rate of claims is constant. For instance, consider a portfolio which only includes policies covering burgleries in houses. If the houses are in a demarcated area, have about the same security standards and comparable valuables inside, then this portfolio may be considered a homogeneous one. Generally, an insurance company tries to establish its portfolios in such a way that they are approximately homogeneous. Regardless of the terminology adopted, the subsequent risk analysis will not apply to an insurance company as a whole, but to its basic operating blocks, the homogeneous portfolios.

By assumption 1 and theorem 7.2, the interarrival times between neighboring claims are independent and identical as Y distributed random variables, where Y has an exponential distribution with parameter $\lambda = 1/\mu$. The mean claim size is denoted as v :

$$\mu = E(Y) \quad \text{and} \quad v = E(M). \tag{7.74}$$

By (7.64), under the assumptions 1 and 2, the trend function of the total claim size process $\{C(t), t \geq 0\}$ is a linear function in time:

$$E(C(t)) = \frac{v}{\mu} t, \quad t \geq 0. \tag{7.75}$$

This justifies assumption 3, namely a linear premium income in time.

In the longrun, an insurance company, however large its initial capital may be, cannot be successful if the average total claim cost in any interval $[0, t]$ exceeds the premium income in $[0, t]$. Hence, in what follows the assumption

$$\kappa\mu - \nu > 0 \qquad (7.76)$$

is made. This inequality requires that the average premium income between the arrival of two neighboring claims is larger than the mean claim size. The difference $\kappa\mu - \nu$ is called *safety loading* and will be denoted as σ:

$$\sigma = \kappa\mu - \nu.$$

Let distribution function and density of the claim size be

$$B(y) = P(M \le y) \quad \text{and} \quad b(y) = dB(y)/dy.$$

Derivation of an Integro-Differential Equation for $q(x)$ To derive an integro-differential equation for the survival probability, consider what may happen in the time interval $[0, \Delta t]$:

1) No claim arrives in $[0, \Delta t]$. Under this condition, the survival probability is

$$q(x + \kappa\,\Delta t).$$

This is because at the end of the interval $[0, \Delta t]$ the capital of the company has increased by $\kappa\Delta t$ units. So the 'new' initial capital at time point Δt is $x + \kappa\,\Delta t$.

2) One claim arrives in $[0, \Delta t]$ and the risk reserve remains positive. Under this condition, the survival probability is

$$\int_0^{x+\kappa\,\Delta t} q(x + \kappa\,\Delta t - y)\, b(y)\, dy.$$

To understand this integral, remember that '$b(y)\,dy$' can be interpreted as the 'probability' that the claim size is equal to y (see comment after formula (2.50) at page 61).

3) One claim arrives in $[0, \Delta t]$ and the risk reserve becomes negative (ruin occurs). Under this condition, the survival probability is 0.

4) At least two claims arrive in $[0, \Delta t]$. Since the Poisson process is simple, the probability of this event is $o(\Delta t)$.

To get the unconditional survival probability, the conditional survival probabilities $1 - 4$ have to be multiplied by the probabilities of their respective conditions and added. By theorem 7.1, the probability that there is one claim in $[0, \Delta t]$, is

$$P(N(0, \Delta t) = 1) = \lambda\Delta t + o(\Delta t),$$

and, correspondingly, the probability that there is no claim in $[0, \Delta t]$ is

$$P(N(0, \Delta t) = 0) = 1 - \lambda\Delta t + o(\Delta t).$$

Therefore, given the initial capital x,

$$q(x) = [1 - \lambda\,\Delta t + o(\Delta t)]\, q(x + \kappa\,\Delta t)$$

$$+ [\lambda\,\Delta t + o(\Delta t)] \int_0^{x+\kappa\,\Delta t} q(x + \kappa\,\Delta t - y)\, b(y)\, dy + o(\Delta t).$$

From this, letting $h = \kappa \Delta t$, by some simple algebra,

$$\frac{q(x+h) - q(x)}{h} = \frac{\lambda}{\kappa} q(x+h) - \frac{\lambda}{\kappa} \int_0^{x+h} q(x+h-y) b(y) dy + \frac{o(h)}{h}.$$

Assuming that $q(x)$ is differentiable, letting $h \to 0$ yields

$$q'(x) = \frac{\lambda}{\kappa} \left[q(x) - \int_0^x q(x-y) b(y) dy \right]. \tag{7.77}$$

A solution can be obtained in terms of Laplace transforms, since the integral in (7.77) is the convolution of $q(x)$ and $b(y)$: Let $\hat{q}(s)$ and $\hat{b}(s)$ be the Laplace transforms of $q(x)$ and $b(y)$, respectively. Then, applying the Laplace transformation to (7.77), using its properties (2.123) and (2.127) (page 100) and replacing λ with $1/\mu$ yields a simple algebraic equation for $\hat{q}(s)$

$$s \hat{q}(s) - q(0) = \frac{1}{\mu \kappa} \left[\hat{q}(s) - \hat{q}(s) \hat{b}(s) \right].$$

Solving for $\hat{q}(s)$ gives

$$\hat{q}(s) = \frac{1}{s - \frac{1}{\kappa \mu}[1 - \hat{b}(s)]} q(0). \tag{7.78}$$

This representation of $\hat{q}(s)$ involves the survival probability of the company $q(0)$ on condition that it has no initial capital.

Example 7.10 Let the claim size M have an exponential distribution with mean value $E(M) = v$. Then M has density

$$b(y) = \frac{1}{v} e^{-y/v}, \quad y \geq 0,$$

so that

$$\hat{b}(s) = \int_0^\infty e^{-sy} \frac{1}{v} e^{-(1/v)y} dy = \frac{1}{vs+1}.$$

Inserting $\hat{b}(s)$ in (7.78) gives the Laplace transform of the survival probability:

$$\hat{q}(s) = \frac{vs+1}{\mu \kappa s (vs+1) - vs} q(0) \mu \kappa.$$

By introducing the coefficient

$$\alpha = \frac{\mu \kappa - v}{\mu \kappa} = \frac{\sigma}{\mu \kappa}, \quad 0 < \alpha < 1, \tag{7.79}$$

$\hat{q}(s)$ simplifies to

$$\hat{q}(s) = \left[\frac{1}{s + \alpha/v} + \frac{1}{vs} \cdot \frac{1}{s + \alpha/v} \right] q(0).$$

Retransformation yields (Table 2.5, page 105)

$$q(x) = \left[e^{-\frac{\alpha}{v} x} + \frac{1}{\alpha} - \frac{1}{\alpha} e^{-\frac{\alpha}{v} x} \right] q(0). \tag{7.80}$$

If the company has infinite initial capital, then it can never experience ruin. Therefore, $q(\infty) = 1$ so that, from (7.80), survival and ruin probability without initial capital are

$$q(0) = \alpha \text{ and } p(0) = 1 - \alpha. \tag{7.81}$$

This gives the final formulas for the survival- and ruin probability:

$$q(x) = 1 - (1 - \alpha)e^{-\frac{\alpha}{v}x}, \quad p(x) = (1 - \alpha)e^{-\frac{\alpha}{v}x}. \tag{7.82}$$

Figure 7.6 shows the graph of the ruin probability in dependence on the initial capital $x[\$10^4]$ for $\alpha = 0.1$ and $\alpha = 0.2$. In both cases, $v = 0.4[\$10^4]$. From (7.79) one gets that for $\alpha = 0.1$ the safety loading is $\sigma = 0.0\underline{4}$, and for $\alpha = 0.2$ it is $\sigma = 0.1$.   ◻

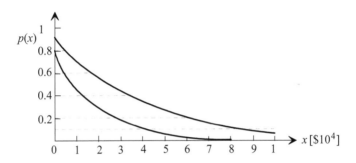

Figure 7.6 Comparison of ruin probabilities for example 7.10

Cramér-Lundberg Approximation If the explicit retransformation of $\hat{q}(s)$ as given by (7.78) is not possible for a given claim size distribution, then the *Cramér-Lundberg approximation* for the ruin probability $p(x)$ is an option to get reliable information on the ruin probability if the initial capital x is large compared to the mean claim size:

$$p(x) \approx \frac{\alpha}{r\gamma}e^{-rx}, \tag{7.83}$$

where the *Lundberg-coefficient r* is defined as solution of the equation

$$\frac{1}{\mu\kappa}\int_0^\infty e^{ry}\overline{B}(y)\,dy = 1, \tag{7.84}$$

and the parameter γ is given by

$$\gamma = \frac{1}{\mu\kappa}\int_0^\infty y e^{ry}\overline{B}(y)\,dy.$$

Note that in view of (7.84) $\frac{1}{\mu\kappa}e^{ry}\overline{B}(y)$ can be interpreted as the probability density of a nonnegative random variable, and the parameter γ is the mean value of this random variable (for a proof of (7.83) see, e.g., *Grandell* (1991)).

A solution r of equation (7.84) exists if the probability density of the claim size $b(y)$ has a 'short tail' to the right, which implies that large values of the claim size occur fairly seldom.

It is interesting to compare the exact value of the ruin probability under an exponential claim size distribution (7.82) with the corresponding approximation (7.83): For

$$B(y) = 1 - e^{-(1/\nu)y}, \ y \geq 0,$$

equation (7.84) becomes

$$\int_0^\infty e^{-(1/\nu - r)y} \, dy = \frac{1}{\frac{1}{\nu} - r} = \mu\kappa$$

so that $r = \alpha/\nu$. The corresponding parameter γ is

$$\gamma = \frac{1}{\mu\kappa} \int_0^\infty y e^{-(1/\nu - r)y} \, dy = \frac{1}{\mu\kappa(1/\nu - r)} \int_0^\infty y(1/\nu - r)e^{-(1/\nu - r)y} \, dy$$

$$= \frac{1}{\mu\kappa(1/\nu - r)^2}.$$

After some simple algebra:

$$\frac{\alpha}{r\gamma} = 1 - \alpha.$$

By comparing (7.82) and (7.83):

> *The Cramér-Lundberg approximation gives the exact value of the ruin probability if the claim sizes are exponentially distributed.*

Lundberg Inequality Assuming the existence of the Lundberg exponent r as defined by equation (7.84), the ruin probability is bounded by e^{-rx} :

$$p(x) \leq e^{-rx}. \tag{7.85}$$

This is the famous *Lundberg inequality*. A proof will be given in chapter 10, page 490, by applying martingale techniques.

Both *F. Lundberg* and *H. Cramér* did their pioneering research in *collective risk analysis* in the first third of the twentieth century; see *Lundberg* (1964).

Example 7.11 As in example 7.10, let $\nu = 0.4 \ [\$10^4]$, but M is assumed to have a Rayleigh distribution:

$$\overline{B}(y) = P(M > y) = e^{-(y/\theta)^2}, \ y \geq 0.$$

Since $\nu = E(M) = \theta\sqrt{\pi/4} = 0.4$, the parameter θ must be equal to $0.8/\sqrt{\pi}$. Again the case $\alpha = 0.1$ is considered, i.e. $\mu\kappa = 4/9 = 0.\underline{4}$ and $\sigma = 2/45 = 0.0\underline{4}$. The corresponding Lundberg exponent is solution of $\frac{9}{4}\int_0^\infty e^{ry} e^{-\pi(y/0.8)^2} \, dy = 1$, which gives

$$r = 0.398 \ \text{ and } \ \gamma = \frac{9}{4}\int_0^\infty y e^{0.398y} e^{-\pi(y/0.8)^2} \, dy = 0.2697.$$

Figure 7.7 shows the graphs of the Cramér-Lundberg approximation (7.83) and the upper bound (7.85) for the ruin probability $p(x)$ in dependency of the initial capital x:

$$p(x) \approx 0.9316 \cdot e^{-0.398 \cdot x}, \quad p(x) \le e^{-0.398 \cdot x}, \quad x \ge 0.$$

Although (7.83) yields best results only for large x, the graph of the approximation is everywhere lower than the upper bound (7.85). The dotted line shows once more the exact ruin probability for exponentially distributed claim sizes with the same mean and α–values as in Figure 7.6. Obviously, the distribution type of the claim size has a significant influence on $p(x)$ under otherwise the same assumptions.   □

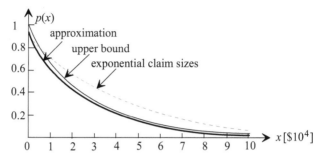

Figure 7.7 Approximation and upper bound for the ruin probability for Example 7.11

7.3 RENEWAL PROCESSES

7.3.1 Definitions and Examples

The motivation for this chapter is a simple maintenance policy: A system is replaced on every failure by a statistically equivalent new one in negligible time and, after that, the new system (or the 'renewed system') immediately starts operating. In this context, the replacements of failed systems are also called *renewals*. The sequence of the system lifetimes after renewals generates renewal process:

Definition 7.5 An *ordinary renewal process* is a sequence of nonnegative, independent, and identically distributed random variables $\{Y_1, Y_2, ...\}$.   ●

Thus, Y_i is the time between the $(i-1)$th and the ith renewal; Renewal processes do not only play an important role in engineering, but also in the natural, economical, and social sciences. They are a basic stochastic tool for modeling particle counting, population development, and arrivals of customers at a service station. In the latter context, Y_i is the random time between the arrival of the $(i-1)$th and the ith customer. Renewal processes are particularly important in actuarial risk analysis, namely for modeling the arrival of claims at an insurance company, since they are a straightforward generalization of homogeneous Poisson processes. In this section a terminology is adopted, which refers to the 'simple maintenance policy'.

If the observation of a renewal process starts at time $t = 0$ and the process had been operating for a while before that time point, then the lifetime of the system operating at time $t = 0$ is a 'residual lifetime' as introduced in section 2.3.4 (page 86) and will, therefore, usually not have the same probability distribution as the lifetime of a system after a renewal. Hence it makes sense to define a generalized renewal process by assuming that only the $Y_2, Y_3, ...$ are identically distributed. This leads to

Definition 7.6 Let $\{Y_1, Y_2, ...\}$ be a sequence of nonnegative, independent random variables with property that Y_1 has distribution function

$$F_1(t) = P(Y_1 \le t),$$

whereas the random variables $Y_2, Y_3, ...$ are identically distributed as Y with distribution function

$$F(t) = P(Y \le t).$$

Then $\{Y_1, Y_2, ...\}$ is called a *delayed renewal process*. ●

The random time point at which the nth renewal takes place is

$$T_n = \sum_{i=1}^{n} Y_i; \quad n = 1, 2, ...$$

The random point process $\{T_1, T_2, ...\}$ is called the *process of the time points of renewals*. The time intervals between two neighboring renewals are *renewal cycles*. The corresponding counting process $\{N(t), t \ge 0\}$, defined by

$$N(t) = \begin{cases} \max(n; \ T_n \le t) \\ 0 \quad \text{for} \quad t < T_1 \end{cases},$$

is called *renewal counting process*. Note that $N(t)$ is the random number of renewals in $(0, t]$, i.e., a possible renewal at time point $t = 0$ is not counted. The relationship

$$N(t) \ge n \text{ if and only if } T_n \le t \tag{7.86}$$

implies

$$F_{T_n}(t) = P(T_n \le t) = P(N(t) \ge n). \tag{7.87}$$

Because of the independence of the Y_i, the distribution function $F_{T_n}(t)$ is the convolution of $F_1(t)$ with the $(n-1)$th convolution power of F (page 190):

$$F_{T_n}(t) = F_1 * F^{*(n-1)}(t), \ F^{*(0)}(t) \equiv 1, \ t \ge 0; \ n = 1, 2, ... \tag{7.88}$$

If the densities

$$f_1(t) = F_1'(t) \text{ and } f(t) = F'(t)$$

exist, then the density of T_n is

$$f_{T_n}(t) = f_1 * f^{*(n-1)}(t), \ f^{*(0)}(t) \equiv 1, \ t \ge 0; \ n = 1, 2, ... \tag{7.89}$$

Using (7.87) and

$$P(N(t) \geq n) = P(N(t) = n) + P(N(t) \geq n + 1),$$

the probability distribution of $N(t)$ is seen to be

$$P(N(t) = n) = F_{T_n}(t) - F_{T_{n+1}}(t), \quad F_{T_0}(t) \equiv 1; \quad n = 0, 1, \dots \quad (7.90)$$

Example 7.12 Let $\{Y_1, Y_2, \dots\}$ be an ordinary renewal process with property that the renewal cycle lengths Y_i have an exponential distribution with parameter λ :

$$F(t) = P(Y \leq t) = 1 - e^{-\lambda t}, \quad t \geq 0.$$

Then, by theorem 7.2, the corresponding counting process $\{N(t), t \geq 0\}$ is the homogeneous Poisson process with intensity λ. In particular, by (7.21), T_n has an Erlang distribution with parameters n and λ:

$$F_{T_n}(t) = P(T_n \leq t) = e^{-\lambda t} \sum_{i=n}^{\infty} \frac{(\lambda t)^i}{i!} . \qquad \square$$

Apart from the homogeneous Poisson process, there are two other important ordinary renewal processes for which the convolution powers of the renewal cycle length distributions explicitly exist so that the distribution functions of the renewal time points T_n can be given:

1) Erlang Distribution The renewal cycle length Y have an Erlang distribution with parameters m and λ. Then T_n is the sum of mn independent, identically distributed exponential random variables with parameter λ. Therefore, T_n has an Erlang distribution with parameters mn and λ :

$$F^{*(n)}(t) = P(T_n \leq t) = e^{-\lambda t} \sum_{i=mn}^{\infty} \frac{(\lambda t)^i}{i!}, \quad t \geq 0. \quad (7.91)$$

This result is of general importance, since the probability distribution of any nonnegative random variable can be arbitrarily accurately approximated by an Erlang distribution by proper choice of the parameters of this distribution.

2) Normal Distribution Let the renewal cycle length Y have a normal distribution with parameters μ and σ, $\mu > 3\sigma$. The assumption $\mu > 3\sigma$ is necesssary for making sure that the cycle lengths are practically nonnegative. (Renewal theory, however, has been extended to negative 'cycle lengths' as well.) Since the sum of independent, normally distributed random variables is again normally distributed, where the parameters of the sum are obtained by summing up the parameters of the summands (formula (4.72), page 191), T_n has distribution function

$$F^{*(n)}(t) = P(T_n \leq t) = \Phi\left(\frac{t - n\mu}{\sigma \sqrt{n}}\right), \quad t \geq 0. \quad (7.92)$$

This result has a more general potential for applications: Since T_n is the sum of n independent, identically distributed random variables, then, by the central limit theorem (theorem 5.6), T_n has approximately the distribution function (7.92) if n is sufficiently large:

$$T_n \approx N(n\mu, \sigma^2 n) \text{ if } n \geq 20.$$

Example 7.13 The distribution function of T_n can be used to solve the *spare part problem*: How many spare parts (spare systems) are necessary for making sure that the renewal process can be maintained over the interval $[0, t]$ with probability $1 - \alpha$?

This requires to determine the smallest integer n satisfying

$$1 - F_{T_n}(t) = P(N(t) \leq n) \geq 1 - \alpha.$$

For instance, let $\mu = E(Y) = 8$ and $\sigma^2 = Var(Y) = 25$. If $t = 200$ and $1 - \alpha = 0.99$, then

$$1 - F_{T_n}(200) = 1 - \Phi\left(\frac{200 - 8n}{5\sqrt{n}}\right) \geq 1 - \alpha = 0.99$$

is equivalent to

$$z_{0.01} = 2.32 \leq \frac{8n - 200}{5\sqrt{n}}.$$

Thus, at least $n_{min} = 34$ spare parts have to be in stock to ensure that with probability 0.99 every failed part can be replaced by a new one over the interval $(0, 200]$. In view of $n_{min} \geq 20$, the application of the normal approximation to the distribution of T_n is justified. □

7.3.2   Renewal Function

7.3.2.1 Renewal Equations

The mean number of renewals which occur in a given time interval is of great practical and theoretical importance.

Definition 7.7 The mean value of the random number $N(t)$ of renewals occurring in $(0, t]$ as a function of t is called *renewal function*.   ●

Thus, with the terminology and the notation introduced in section 6.2, the renewal function is the trend function of the renewal counting process $\{N(t), t \geq 0\}$:

$$m(t) = E(N(t)), \quad t \geq 0.$$

To be, however, in line with the majority of publications on renewal theory, in what follows, the renewal functions belonging to an ordinary and a delayed renewal process are denoted as $H(t)$ and $H_1(t)$, respectively. If not stated otherwise, it is assumed throughout section 7.3 that the densities of Y and Y_1 exist:

$$dF(t) = f(t)dt \text{ and } dF_1(t) = f_1(t)dt. \tag{7.93}$$

In this case, the first derivatives of $H_1(t)$ and $H(t)$ also exist:

$$h_1(t) = \frac{dH_1(t)}{dt}, \quad h(t) = \frac{dH(t)}{dt}.$$

The functions $h_1(t)$ and $h(t)$ are the *renewal densities* of a delayed and of an ordinary renewal process, respectively. From (2.9) (page 46), a sum representation of the renewal function is

$$H_1(t) = E(N(t)) = \sum_{n=1}^{\infty} P(N(t) \geq n). \tag{7.94}$$

In view of (7.87) and (7.94),

$$H_1(t) = \sum_{n=1}^{\infty} F_1 * F^{*(n-1)}(t). \tag{7.95}$$

In particular, the renewal function of an ordinary renewal process is

$$H(t) = \sum_{n=1}^{\infty} F^{*(n)}(t). \tag{7.96}$$

By differentiation of (7.95) and (7.96) with respect to t, one obtains sum representations of the respective renewal densities:

$$h_1(t) = \sum_{n=1}^{\infty} f_1 * f^{*(n-1)}(t), \quad h(t) = \sum_{n=1}^{\infty} f^{*(n)}(t).$$

Remark These sum representations allow a useful probabilistic interpretation of the renewal density: For Δt sufficiently small,

$$h_1(t) \Delta t \text{ or } h(t) \Delta t,$$

respectively, are approximately the probabilities of the occurrence of a renewal in the interval $[t, t + \Delta t]$. (Compare to the remark after formula (2.50), page 61.)

By (7.95) and the definition of the convolution power of distribution functions,

$$H_1(t) = \sum_{n=0}^{\infty} F_1 * F^{*(n)}(t)$$

$$= F_1(t) + \sum_{n=1}^{\infty} \int_0^t F_1 * F^{*(n-1)}(t-x) \, dF(x)$$

$$= F_1(t) + \int_0^t \sum_{n=1}^{\infty} \left(F_1 * F^{*(n-1)}(t-x) \right) dF(x).$$

Again by (7.95), the integrand is equal to $H_1(t-x)$. Hence, $H_1(t)$ satisfies

$$H_1(t) = F_1(t) + \int_0^t H_1(t-x) \, dF(x). \tag{7.97}$$

By assumption (7.93), the integral in (7.97) is the convolution $H_1 * f$ of the renewal function H_1 with f. In particular, the renewal function $H(t)$ of an ordinary renewal process satisfies the integral equation

$$H(t) = F(t) + \int_0^t H(t-x) \, dF(x). \tag{7.98}$$

A heuristic derivation of formula (7.98) can be done by conditioning with regard to the time point of the first renewal: Given the first renewal occurs at time x, the mean number of renewals in $[0, t]$ is

$$[1 + H(t - x)], \quad 0 < x \leq t.$$

Since the first renewal occurs at time x with 'probability' $dF(x) = f(x)\,dx$, taking into account all possible values of x in $[0, t]$ yields (7.98). The same argument yields an integral equation for the renewal function of a delayed renewal process:

$$H_1(t) = F_1(t) + \int_0^t H(t - x)\,dF_1(x). \tag{7.99}$$

This is because after the first renewal at time x the process develops in $(x, t]$ as an ordinary renewal process. Since the convolution is a commutative operation, the renewal equations can be rewritten. For instance, integral equation (7.97) is equivalent to

$$H_1(t) = F_1(t) + \int_0^t F(t - x)\,dH_1(x). \tag{7.100}$$

The equations (7.97)–(7.100) are called *renewal equations.*

By differentiating the renewal equations (7.97) to (7.99) with respect to t, one obtains analogous integral equations for $h_1(t)$ and $h(t)$:

$$h_1(t) = f_1(t) + \int_0^t h_1(t - x) f(x)\,dx, \tag{7.101}$$

$$h(t) = f(t) + \int_0^t h(t - x) f(x)\,dx, \tag{7.102}$$

$$h_1(t) = f_1(t) + \int_0^t h(t - x) f_1(x)\,dx. \tag{7.103}$$

Generally, solutions of the renewal equations including equations (7.101) to (7.103) can only be obtained by numerical methods. Since, however, all these integral equations involve convolutions, it is easily possible to find their solutions in the image space of the Laplace transformation. To see this, let $\hat{h}_1(s)$, $\hat{h}(s)$, $\hat{f}_1(s)$, and $\hat{f}(s)$ in this order be the Laplace transforms of $h_1(t), h(t), f_1(t),$ and $f(t)$. Then, by (2.127), applying the Laplace transformation to (7.101) and (7.102) yields algebraic equations for $\hat{h}_1(s)$ and $\hat{h}(s)$:

$$\hat{h}_1(s) = \hat{f}_1(s) + \hat{h}_1(s) \cdot \hat{f}(s), \quad \hat{h}(s) = \hat{f}(s) + \hat{h}(s) \cdot \hat{f}(s).$$

The solutions are

$$\hat{h}_1(s) = \frac{\hat{f}_1(s)}{1 - \hat{f}(s)}, \quad \hat{h}(s) = \frac{\hat{f}(s)}{1 - \hat{f}(s)}. \tag{7.104}$$

Thus, for ordinary renewal processes there is a one-to-one correspondence between the renewal function and the probability distribution of the cycle length. By (2.120), the Laplace transforms of the corresponding renewal functions are

$$\hat{H}_1(s) = \frac{\hat{f}_1(s)}{s(1 - \hat{f}(s))}, \quad \hat{H}(s) = \frac{\hat{f}(s)}{s(1 - \hat{f}(s))}. \tag{7.105}$$

Integral Equations of Renewal Type The renewal equations (7.97) to (7.100) and other, equivalent ones derived from these belong to the broader class of integral equations of renewal type. A function $Z(t)$ is said to satisfy an *integral equation of renewal type* if for any function $a(t)$, which is integrable on $[0, \infty)$, and for any probability density $f(x)$ of a nonnegative random variable,

$$Z(t) = a(t) + \int_0^t Z(t-x)f(x)\,dx. \tag{7.106}$$

A function $Z(t)$ satisfying (7.106) need not be the trend function of a renewal counting process; see example 7.17. As proved in *Feller* (1971), the general solution of the integral equation (7.106) has the unique structure

$$Z(t) = g(t) + \int_0^t g(t-x)h(x)\,dx,$$

where $h(t)$ is the renewal density of the ordinary renewal process belonging to $f(x)$.

Example 7.14 Let $f_1(t) = f(t) = \lambda e^{-\lambda t},\ t \geq 0$. The Laplace transform of $f(t)$ is

$$\hat{f}(s) = \frac{\lambda}{s+\lambda}.$$

By the right equation in (7.105),

$$\hat{H}(s) = \frac{\lambda}{s+\lambda} \Big/ \left(s - \frac{\lambda s}{s+\lambda}\right) = \frac{\lambda}{s^2}.$$

The corresponding preimage (Table 2.5, page 105) is $H(t) = \lambda t$. Thus, an ordinary renewal process has exponentially with parameter λ distributed cycle lengths if and only if its renewal function is given by $H(t) = \lambda t$. $\qquad\qquad\square$

Example 7.15 Let the cycle length of an ordinary renewal process be a mixture of two exponential distributions:

$$f(t) = p\lambda_1 e^{-\lambda_1 t} + (1-p)\lambda_2 e^{-\lambda_2 t}$$

with $0 \leq p \leq 1,\ \lambda_1 > 0,\ \lambda_2 > 0,\ t \geq 0$. With its three free parameters, this distribution can be expected to provide a good fit to many lifetime data sets. The Laplace transform of $f(t)$ is

$$\hat{f}(s) = \frac{p\lambda_1}{s+\lambda_1} + \frac{(1-p)\lambda_2}{s+\lambda_2}.$$

Hence, the right formula of (7.104) yields the Laplace transform of the corresponding renewal density

$$\hat{h}(s) = \frac{\dfrac{p\lambda_1}{s+\lambda_1} + \dfrac{(1-p)\lambda_2}{s+\lambda_2}}{1 - \dfrac{p\lambda_1}{s+\lambda_1} - \dfrac{(1-p\lambda_2)}{s+\lambda_2}}.$$

From this, by identical transformations,

$$\hat{h}(s) = \frac{[p\lambda_1 + (1-p)\lambda_2]s + \lambda_1\lambda_2}{(s+\lambda_1)(s+\lambda_2) - [p\lambda_1 + (1-p)\lambda_2]s - \lambda_1\lambda_2}$$

$$= \frac{[p\lambda_1 + (1-p)\lambda_2]s + \lambda_1\lambda_2}{s^2 + (1-p)\lambda_1 s + p\lambda_2 s}$$

$$= \frac{p\lambda_1 + (1-p)\lambda_2}{s + (1-p)\lambda_1 + p\lambda_2} + \frac{\lambda_1\lambda_2}{s[s + (1-p)\lambda_1 + p\lambda_2]}.$$

Retransformation is easily done by making use of Table 2.5 (page 105)

$$h(t) = \frac{\lambda_1\lambda_2}{(1-p)\lambda_1 + p\lambda_2}$$

$$+ \left[p\lambda_1 + (1-p)\lambda_2 - \frac{\lambda_1\lambda_2}{(1-p)\lambda_1 + p\lambda_2} \right] e^{-[(1-p)\lambda_1 + p\lambda_2]t}, \quad t \geq 0.$$

After some algebra,

$$h(t) = \frac{\lambda_1\lambda_2}{(1-p)\lambda_1 + p\lambda_2} + p(1-p)\frac{(\lambda_1 - \lambda_2)^2}{(1-p)\lambda_1 + \lambda_2 p} e^{-[(1-p)\lambda_1 + p\lambda_2]t}, \quad t \geq 0.$$

Integration yields the renewal function:

$$H(t) = \frac{\lambda_1\lambda_2}{(1-p)\lambda_1 + p\lambda_2} t$$

$$+ p(1-p)\left(\frac{\lambda_1 - \lambda_2}{(1-p)\lambda_1 + \lambda_2 p} \right)^2 \left(1 - e^{-[(1-p)\lambda_1 + p\lambda_2]t} \right).$$

Mean value $\mu = E(Y)$ and variance $\sigma^2 = Var(Y)$ of the renewal cycle length Y are

$$\mu = \frac{p}{\lambda_1} + \frac{1-p}{\lambda_2} = \frac{(1-p)\lambda_1 + p\lambda_2}{\lambda_1\lambda_2},$$

$$\sigma^2 = \frac{p}{\lambda_1^2} + \frac{1-p}{\lambda_2^2} = \frac{(1-p)\lambda_1^2 + p\lambda_2^2}{\lambda_1^2\lambda_2^2}.$$

With these parameters, the representation of the renewal function can be simplified:

$$H(t) = \frac{t}{\mu} + \left(\frac{\sigma^2}{\mu^2} - 1 \right)\left(1 - e^{-[(1-p)\lambda_1 + p\lambda_2]t} \right), \quad t \geq 0.$$

For $\lambda_1 = \lambda_2 = \lambda$ and $p = 1$ this representation of $H(t)$ reduces to $H(t) = \lambda t$. □

More explicit formulas for the renewal function of ordinary renewal processes exist for the following two classes of cycle length distributions:

1) Erlang Distribution Let the cycle lengths be Erlang distributed with parameters m and λ. Then, by (7.87) and (7.91),

$$H(t) = e^{-\lambda t} \sum_{n=1}^{\infty} \sum_{i=mn}^{\infty} \frac{(\lambda t)^i}{i!}. \tag{7.107}$$

In particular,

$m = 1:$ $H(t) = \lambda t$ (*homogeneous Poisson process*)

$m = 2:$ $H(t) = \dfrac{1}{2}\left[\lambda t - \dfrac{1}{2} + \dfrac{1}{2} e^{-2\lambda t} \right]$

$m = 3:$ $H(t) = \dfrac{1}{3}\left[\lambda t - 1 + \dfrac{2}{\sqrt{3}} e^{-1.5\lambda t} \sin\left(\dfrac{\sqrt{3}}{2} \lambda t + \dfrac{\pi}{3} \right) \right]$

$m = 4:$ $H(t) = \dfrac{1}{4}\left[\lambda t - \dfrac{3}{2} + \dfrac{1}{2} e^{-2\lambda t} + \sqrt{2}\, e^{-\lambda t} \sin\left(\lambda t + \dfrac{\pi}{4} \right) \right]$.

2) Normal Distribution Let the cycle lengths be normally distributed with mean value μ and variance σ^2, $\mu > 3\sigma^2$. From (7.87) and (7.92),

$$H(t) = \sum_{n=1}^{\infty} \Phi\left(\frac{t - n\mu}{\sigma\sqrt{n}} \right). \tag{7.108}$$

This sum representation is very convenient for numerical computations, since already the sum of the first few terms approximates the renewal function with sufficient accuracy.

As shown in example 7.14, an ordinary renewal process has renewal function

$$H(t) = \lambda t = t/\mu \ \text{ if and only if } \ f(t) = \lambda e^{-\lambda t}, t \geq 0,$$

where $\mu = E(Y)$. An interesting question is, whether for given $F(t)$ a delayed renewal process exists which also has renewal function $H_1(t) = t/\mu$.

Theorem 7.11 Let $\{Y_1, Y_2, ...\}$ be a delayed renewal process with cycle lengths $Y_2, Y_3, ...$ being identically distributed as Y. If Y has finite mean value μ and distribution function $F(t) = P(Y \leq t)$, then $\{Y_1, Y_2, ...\}$ has renewal function

$$H_1(t) = t/\mu \tag{7.109}$$

if and only if the length of the first renewal cycle Y_1 has density $f_1(t) \equiv f_S(t)$, where

$$f_S(t) = \frac{1}{\mu}(1 - F(t)), \quad t \geq 0. \tag{7.110}$$

Equivalently, $\{Y_1, Y_2, ...\}$ has renewal function (7.109) if and only if Y_1 has distribution function $F_1(t) \equiv F_S(t)$ with

$$F_S(t) = \frac{1}{\mu} \int_0^t (1 - F(x))\, dx, \quad t \geq 0. \tag{7.111}$$

Proof Let $\hat{f}(s)$ and $\hat{f}_S(s)$ be the respective Laplace transforms of $f(t)$ and $f_S(t)$. By applying the Laplace transformation to both sides of (7.110),

$$\hat{f}_S(s) = \frac{1}{\mu s}(1 - \hat{f}(s)).$$

Replacing in the left equation of (7.105) $\hat{f}_1(s)$ with $\hat{f}_S(s)$ yields the Laplace transform of the corresponding renewal function $H_1(t) = H_S(t)$:

$$\hat{H}_S(s) = 1/(\mu s^2).$$

Retransformation of $\hat{H}_S(s)$ gives the desired result: $H_S(t) = t/\mu$.   ∎

The first two moments of S are

$$E(S) = \frac{\mu^2 + \sigma^2}{2\mu} \quad \text{and} \quad E(S^2) = \frac{\mu_3}{3\mu}, \tag{7.112}$$

where $\sigma^2 = Var(Y)$ and $\mu_3 = E(Y^3)$.

The random variable S with density (7.110) plays an important role in characterizing stationary renewal processes (section 7.3.5).

7.3.2.2 Bounds on the Renewal Function

Generally, integral equations of renewal type have to be solved by numerical methods. Hence, bounds on $H(t)$, which only require information on one or more numerical parameters of the cycle length distribution, are of special interest. This section presents bounds on the renewal function of ordinary renewal processes.

1) Elementary Bounds By definition of T_n,

$$\max_{1 \le i \le n} Y_i \le \sum_{i=1}^{n} Y_i = T_n.$$

Hence, for any t with $F(t) < 1$,

$$F^{*(n)}(t) = P(T_n \le t) \le P(\max_{1 \le i \le n} Y_i \le t) = [F(t)]^n.$$

Summing from $n = 1$ to ∞ on both sides of this inequality, the sum representation of the renewal function (7.96) and the geometric series (2.16) at page 48 yield

$$F(t) \le H(t) \le \frac{F(t)}{1 - F(t)}.$$

The left-hand side of this inequality is the first term of the sum (7.96). These bounds are only useful for small t.

2) Marshall-Bounds Let $\mathbf{F} = \{t; t \ge 0, F(t) < 1\}$, $\mu = E(Y)$, $\overline{F}(t) = 1 - F(t)$, and

$$a_0 = \inf_{t \in \mathbf{F}} \frac{F(t) - F_S(t)}{\overline{F}(t)}, \quad a_1 = \sup_{t \in \mathbf{F}} \frac{F(t) - F_S(t)}{\overline{F}(t)},$$

where $F_S(t)$ is given by (7.111). Then,

$$\frac{t}{\mu} + a_0 \le H(t) \le \frac{t}{\mu} + a_1. \tag{7.113}$$

The derivation of these bounds is straightforward and very instructive: According to the definition of a_0 and a_1,

$$a_0 \overline{F}(t) \le F(t) - F_S(t) \le a_1 \overline{F}(t).$$

Convolution of both sides with $F^{*(n)}(t)$ leads to

$$a_0 \left[F^{*(n)}(t) - F^{*(n+1)}(t) \right] \le F^{*(n+1)}(t) - F_S * F^{*(n)}(t) \le a_1 \left[F^{*(n)}(t) - F^{*(n+1)}(t) \right].$$

In view of (7.96) and theorem 7.11, summing up from $n = 0$ to ∞ on both sides of this inequality proves (7.113). Since

$$\frac{F(t) - F_S(t)}{\overline{F}(t)} \ge -F_S(t) \ge -1 \text{ for all } t \ge 0,$$

formula (7.113) implies a simpler lower bound on $H(t)$:

$$H(t) \ge \frac{t}{\mu} - F_S(t) \ge \frac{t}{\mu} - 1.$$

Let $\lambda_S(t) = f_S(t)/\overline{F}_S(t)$ be the failure rate belonging to $F_S(t)$:

$$\lambda_S(t) = \frac{\overline{F}(t)}{\int_t^\infty \overline{F}(x)\, dx}.$$

Then a_0 and a_1 can be rewritten as follows:

$$a_0 = \frac{1}{\mu} \inf_{t \in F} \frac{1}{\lambda_S(t)} - 1 \quad \text{and} \quad a_1 = \frac{1}{\mu} \sup_{t \in F} \frac{1}{\lambda_S(t)} - 1.$$

Thus, (7.113) becomes

$$\frac{t}{\mu} + \frac{1}{\mu} \inf_{t \in F} \frac{1}{\lambda_S(t)} - 1 \le H(t) \le \frac{t}{\mu} + \frac{1}{\mu} \sup_{t \in F} \frac{1}{\lambda_S(t)} - 1. \tag{7.114}$$

Since

$$\inf_{t \in F} \lambda(t) \le \inf_{t \in F} \lambda_S(t) \text{ and } \sup_{t \in F} \lambda(t) \ge \sup_{t \in F} \lambda_S(t),$$

the bounds (7.114) can be simplified:

$$\frac{t}{\mu} + \frac{1}{\mu} \inf_{t \in F} \frac{1}{\lambda(t)} - 1 \le H(t) \le \frac{t}{\mu} + \frac{1}{\mu} \sup_{t \in F} \frac{1}{\lambda(t)} - 1. \tag{7.115}$$

3) Lorden's Upper Bound If $\mu = E(Y)$ and $\mu_2 = E(Y^2)$, then

$$H(t) \le \frac{t}{\mu} + \frac{\mu_2}{\mu^2} - 1. \tag{7.116}$$

4) Brown's Upper Bound If $F(t)$ is *IFR*, then (7.116) can be improved:

$$H(t) \le \frac{t}{\mu} + \frac{\mu_2}{2\mu^2} - 1.$$

5) Barlow and Proschan Bounds If $F(t)$ is *IFR*, then

$$\frac{t}{\int_0^t \overline{F}(x)\,dx} - 1 \le H(t) \le \frac{t F(t)}{\int_0^t \overline{F}(x)\,dx}. \qquad (7.117)$$

Example 7.16 Let

$$F(t) = (1 - e^{-t})^2, \quad t \ge 0,$$

be the distribution function of the cycle length Y of an ordinary renewal process. In this case, $\mu = E(Y) = 3/2$ and

$$\overline{F}_S(t) = \frac{1}{\mu} \int_t^\infty \overline{F}(x)\,dx = \frac{2}{3}\left(2 - \frac{1}{2}e^{-t}\right)e^{-t}, \quad t \ge 0.$$

Therefore, the failure rates belonging to $F(t)$ and $F_S(t)$ are (Figure 7.8)

$$\lambda(t) = \frac{2(1 - e^{-t})}{2 - e^{-t}}, \quad \lambda_S(t) = 2\frac{2 - e^{-t}}{4 - e^{-t}}, \quad t \ge 0.$$

Both failure rates are strictly increasing in t and have properties

$$\lambda(0) = 0, \; \lambda(\infty) = 1 \text{ and } \lambda_S(0) = 2/3, \; \lambda_S(\infty) = 1.$$

Hence, the respective bounds (7.114) and (7.115) are (Figure 7.9)

$$\tfrac{2}{3}t - \tfrac{1}{3} \le H(t) \le \tfrac{2}{3}t \text{ and } \tfrac{2}{3}t - \tfrac{1}{3} \le H(t) \le \infty.$$

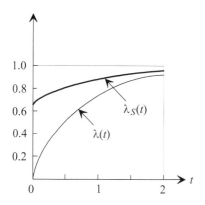

Figure 7.8 Failure rates

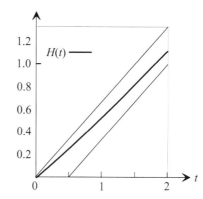

Figure 7.9 Bounds for the renewal function

In this case, the upper bound in (7.115) contains no information on the renewal function. Figure 7.9 compares the bounds (7.114) with the exact graph of the renewal function given in example 7.15 The deviation of the lower bound from $H(t)$ is negligibly small for $t \geq 3$. $\qquad\Box$

7.3.3 Asymptotic Behavior

This section investigates the behavior of the renewal counting process $\{N(t), t \geq 0\}$ and its trend function $H(t)$ as $t \to \infty$. The results allow the construction of estimates of the renewal function and of the probability distribution of $N(t)$ if t is sufficiently large. Throughout this section, it is assumed that both $\mu_1 = E(Y_1)$ and $\mu = E(Y)$ are finite. Some of the key results require that the cycle length Y or, equivalently, its distribution function, is *nonarithmetic* (see definition 5.3, page 216), i.e., that there is no positive constant b with property that the possible values of Y are multiples of b. A continuous random variable is always nonarithmetic.

A simple consequence of the strong law of the large numbers is

$$P\left(\lim_{t \to \infty} \frac{N(t)}{t} = \frac{1}{\mu} \right) = 1. \tag{7.118}$$

To avoid technicalities, the verification of (7.118) is done for an ordinary renewal process: The inequality $T_{N(t)} \leq t < T_{N(t)+1}$ implies that

$$\frac{T_{N(t)}}{N(t)} \leq \frac{t}{N(t)} < \frac{T_{N(t)+1}}{N(t)} = \frac{T_{N(t)+1}}{N(t)+1} \frac{N(t)+1}{N(t)}$$

or, equivalently, that

$$\frac{1}{N(t)} \sum_{i=1}^{N(t)} Y_i \leq \frac{t}{N(t)} < \left[\frac{1}{N(t)+1} \sum_{i=1}^{N(t)+1} Y_i \right] \frac{N(t)+1}{N(t)}.$$

Since by assumption $\mu = E(Y) < \infty$, $N(t)$ tends to infinity as $t \to \infty$. Hence, theorem 5.4 yields the desired result (7.118). For μ being the mean distance between two renewals, this result is quite intuitive.

The following theorem considers the corresponding limit behavior of the mean value of $N(t)$. As with the subsequent theorems 7.13 and 7.14, no proof is given.

Theorem 7.12 (*elementary renewal theorem*) The renewal function satisfies

$$\lim_{t \to \infty} \frac{H_1(t)}{t} = \frac{1}{\mu}. \qquad\blacksquare$$

Corollary For large t,

$$H_1(t) \approx t/\mu.$$

The theorem shows that for $t \to \infty$ the influence of the first renewal interval with possibly $\mu_1 \neq \mu$ fades away. (For this property to be valid, the assumption $\mu_1 < \infty$ had to be made.) In terms of the renewal density, the analogue to theorem 7.12 is

$$\lim_{t \to \infty} h_1(t) = \frac{1}{\mu}.$$

Note that (7.118) does not imply theorem 7.12. The following theorem was called *fundamental* or *key renewal theorem* by its discoverer *W. L. Smith*.

Theorem 7.13 (*fundamental renewal theorem*) If $F(t)$ is nonarithmetic and $g(t)$ an integrable function on $[0, \infty)$, then

$$\lim_{t \to \infty} \int_0^t g(t - x)\, dH_1(x) = \frac{1}{\mu} \int_0^\infty g(x)\, dx. \qquad \blacksquare$$

The fundamental renewal theorem (or *key renewal theorem*, *theorem of Smith*) has proved a useful tool for solving many problems in stochastic modeling. With

$$g(x) = \begin{cases} 1 & \text{for } 0 \le x \le h, \\ 0 & \text{elsewhere,} \end{cases}$$

the fundamental renewal theorem implies

Blackwell's renewal theorem: If $F(t)$ is nonarithmetic, then, for any $h > 0$,

$$\lim_{t \to \infty} [H_1(t + h) - H_1(t)] = \frac{h}{\mu}. \qquad (7.119)$$

Whereas the elementary renewal theorem refers to 'a global transition' into the stationary regime, Blackwell's renewal theorem refers to the corresponding 'local behavior' in a time interval of length h.

Theorem 7.14 gives another variant of the fundamental renewal theorem. It refers to the integral equation of renewal type (7.106).

Theorem 7.14 Let $a(x)$ be an integrable function on $[0, \infty)$ and $f(x)$ a probability density. If a function $Z(t)$ satisfies the renewal type equation

$$Z(t) = a(t) + \int_0^t Z(t - x) f(x)\, dx, \qquad (7.120)$$

then

$$\lim_{t \to \infty} Z(t) = \frac{1}{\mu} \int_0^\infty a(x)\, dx. \qquad \blacksquare$$

As mentioned previously, the function $Z(t)$ in (7.130) need not be a renewal function. Proofs of the now 'classic' theorems 7.12 to 7.14 can be found in *Tijms* (1994).

In the following example, theorem 7.14 is used to sketch the proof the Cramer-Lundberg approximation for the ruin probability (7.83); for details see *Grandell* (1991).

Example 7.17 The integro-differential equation (7.77)

$$q'(x) = \frac{\lambda}{\kappa}\left[q(x) - \int_0^x q(x-y)\, b(y)\, dy \right]$$

for the survival probability $q(x)$ of an insurance company can be transformed by integration on both sides and some routine manipulations to an integral equation for the ruin probability $p(x) = 1 - q(x)$

$$p(x) = a_0(x) + \int_0^x p(x-y)\, g_0(y)\, dy \qquad (7.121)$$

with

$$a_0(x) = 1 - \alpha - \frac{1}{\mu\kappa}\int_0^x \overline{B}(y)\, dy \quad \text{and} \quad g_0(y) = \frac{1}{\mu\kappa}\overline{B}(y),$$

where α is given by (7.79). Equation (7.121) is not of type (7.120), since $g_0(y)$ is only an 'incomplete' probability density:

$$\frac{1}{\mu\kappa}\int_0^\infty \overline{B}(y)dy = \frac{\nu}{\mu\kappa} = 1 - \alpha < 1.$$

For this reason, equation (7.121) is multiplied by the factor $e^{rx} = e^{r(x-y)} \cdot e^{ry}$, which transforms equation (7.121) into an integral equation for $p_r(x) = e^{rx}p(x)$:

$$p_r(x) = a(x) + \int_0^x p_r(x-y)\, g(y)\, dy, \qquad (7.122)$$

where $a(x) = e^{rx}a_0(x)$, $g(y) = e^{ry}g_0(y)$, and r is such that $g(y)$ is a probability density, i.e.,

$$\int_0^\infty g(y)dy = \frac{1}{\mu\kappa}\int_0^\infty e^{ry}\overline{B}(y)\, dy = 1.$$

This is the definition of the Lundberg-exponent r according to (7.84). Now (7.132) is a renewal type equation and theorem 7.14 can be applied: With

$$\gamma = \int_0^\infty y\, g(y)dy = \frac{1}{\mu\kappa}\int_0^\infty y\, e^{ry}\overline{B}(y)\, dy \quad \text{and} \quad \int_0^\infty a(x)\, dx = \frac{\alpha}{r},$$

theorem 7.14 yields

$$\lim_{x\to\infty} p_r(x) = \lim_{x\to\infty} e^{rx}p(x) = \frac{\alpha}{\gamma r}$$

so that for large x

$$p(x) \approx \frac{\alpha}{\gamma r}e^{-rx}. \qquad \square$$

Theorem 7.15 If $F(t)$ is nonarithmetic and $\sigma^2 = Var(Y) < \infty$, then

$$\lim_{t\to\infty}\left(H_1(t) - \frac{t}{\mu} \right) = \frac{\sigma^2}{2\mu^2} - \frac{\mu_1}{\mu} + \frac{1}{2}. \qquad (7.123)$$

Proof The renewal equation (7.99) is equivalent to

$$H_1(t) = F_1(t) + \int_0^t F_1(t-x)\, dH(x). \qquad (7.124)$$

If $F_1(t) \equiv F_S(t)$, then, by theorem 7.11 this integral equation becomes

$$\frac{t}{\mu} = F_S(t) + \int_0^t F_S(t-x)\, dH(x). \tag{7.125}$$

By subtracting integral equation (7.125) from integral equation (7.124),

$$H_1(t) - \frac{t}{\mu} = \overline{F}_S(t) - \overline{F}_1(t) + \int_0^t \overline{F}_S(t-x)\, dH(x) - \int_0^t \overline{F}_1(t-x)\, dH(x).$$

Applying the fundamental renewal theorem yields

$$\lim_{t\to\infty} \left(H_1(t) - \frac{t}{\mu} \right) = \frac{1}{\mu}\int_0^\infty \overline{F}_S(x)\, d(x) - \frac{1}{\mu}\int_0^\infty \overline{F}_1(x)\, d(x).$$

Now the desired results follows from (2.52) and (7.112). ∎

For ordinary renewal processes, (7.123) simplifies to

$$\lim_{t\to\infty} \left(H_1(t) - \frac{t}{\mu} \right) = \frac{1}{2}\left(\frac{\sigma^2}{\mu^2} - 1 \right). \tag{7.126}$$

Corollary Under the assumptions of theorem 7.15, the fundamental renewal theorem implies the elementary renewal theorem.

Theorem 7.16 For an ordinary renewal process, the integrated renewal function has property

$$\lim_{t\to\infty} \left\{ \int_0^t H(x)\, dx - \left[\frac{t^2}{2\mu} + \left(\frac{\mu_2}{2\mu^2} - 1 \right) t \right] \right\} = \frac{\mu_2^2}{4\mu^3} - \frac{\mu_3}{6\mu^2}$$

with $\mu_2 = E(Y^2)$ and $\mu_3 = E(Y^3)$. ∎

For a proof see, for instance, *Tijms* (1994). The following theorem is basically a consequence of the central limit theorem; for details see *Karlin, Taylor* (1981).

Theorem 7.17 The random number $N(t)$ of renewals in $[0, t]$ satisfies

$$\lim_{t\to\infty} P\left(\frac{N(t) - t/\mu}{\sigma\sqrt{t\mu^{-3}}} \le x \right) = \Phi(x). ∎$$

Corollary For t sufficiently large, $N(t)$ is approximately normally distributed with mean value t/μ and variance $\sigma^2 t/\mu^3$:

$$N(t) \approx N(t/\mu,\ \sigma^2 t/\mu^3). \tag{7.127}$$

Hence, theorem 7.17 can be used to construct approximate intervals, which contain $N(t)$ with a given probability: If t is sufficiently large, then

$$P\left(\frac{t}{\mu} - z_{\alpha/2}\,\sigma\sqrt{t\mu^{-3}} \le N(t) \le \frac{t}{\mu} + z_{\alpha/2}\,\sigma\sqrt{t\mu^{-3}} \right) = 1 - \alpha. \tag{7.128}$$

As usual, $z_{\alpha/2}$ is the $(1-\alpha/2)$–percentile of the standard normal distribution.

Example 7.18 Let $t = 1000$, $\mu = 10$, $\sigma = 2$, and $\alpha = 0.05$. Since $z_{0.025} \approx 2$,

$$P(96 \le N(t) \le 104) = 0.95 \,. \qquad \qquad \square$$

Knowledge of the asymptotic distribution of $N(t)$ makes it possible, without knowing the exact distribution of Y, to approximately answer a question which already arose in section 7.3.1: How many spare systems (spare parts) are necessary for guaranteeing that the (ordinary) renewal process can be maintained over an interval $[0, t]$ with a given probability of $1 - \alpha$? Since with probability $1 - \alpha$ approximately

$$\frac{N(t) - t/\mu}{\sigma\sqrt{t\mu^{-3}}} \le z_\alpha \,,$$

for large t the required number n_{min} is approximately equal to

$$n_{min} \approx \frac{t}{\mu} + z_\alpha \, \sigma \sqrt{t\mu^{-3}} \,. \tag{7.129}$$

The same numerical parameters as in example 7.13 are considered:

$$t = 200, \ \mu = 8, \ \sigma^2 = 25, \ \text{and} \ \alpha = 0.01.$$

Since $z_{0.01} = 2.32$,

$$n_{min} \ge \frac{200}{8} + 2.32 \cdot 5\sqrt{200 \cdot 8^{-3}} = 32.25.$$

Thus, 33 spare parts are at least needed to make sure that with probability 0.99 the renewal process can be maintained over a period of 200 time units. Remember, formula (7.92) applied in example 7.13 yielded $n_{min} = 34$. $\qquad \square$

7.3.4 Recurrence Times

For any point processes, recurrence times have been defined by (7.3) and (7.5). In particular, if $\{Y_1, Y_2, ...\}$ is a renewal process and $\{T_1, T_2, ...\}$ is the corresponding process of renewal time points, then its (random) *forward recurrence time* $A(t)$ is

$$A(t) = T_{N(t)+1} - t$$

and its (random) *backward recurrence time* $B(t)$ is

$$B(t) = t - T_{N(t)}.$$

With the interpretation of renewal processes adopted in this chapter, $A(t)$ is the *residual lifetime* and $B(t)$ the *age* of the system operating at time t in the sense of terminology introduced in section 2.3.4 (Figure 7.10). The stochastic processes

$$\{Y_1, Y_2, ...\} \,, \{T_1, T_2, ...\} \,, \{N(t), t \ge 0\}, \ \{A(t), t \ge 0\}, \ \text{and} \ \{B(t), t \ge 0\}$$

are statistically equivalent, since there is a one to one correspondence between their sample paths, i.e., each of these five processes can be used to define a renewal process (Figure 7.11).

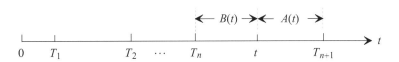

Figure 7.10 Illustration of the recurrence times

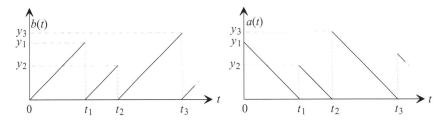

Figure 7.11 Sample paths of the backward and forward recurrence time processes

Let

$$F_{A(t)}(x) = P(A(t) \le x) \quad \text{and} \quad F_{B(t)}(x) = P(B(t) \le x)$$

be the distribution functions of the forward and the backward recurrence times. Then, for $0 < x < t$, making use of (7.95),

$$F_{A(t)}(x) = P(T_{N(t)+1} - t \le x)$$

$$= \sum_{n=0}^{\infty} P(T_{N(t)+1} \le t + x, \, N(t) = n)$$

$$= F_1(t+x) - F_1(t) + \sum_{n=1}^{\infty} P(T_n \le t < T_{n+1} \le t+x)$$

$$= F_1(t+x) - F_1(t) + \sum_{n=1}^{\infty} \int_0^t [F(x+t-y) - F(t-y)] \, dF_{T_n}(y)$$

$$= F_1(t+x) - F_1(t) + \int_0^t [F(x+t-y) - F(t-y)] \sum_{n=1}^{\infty} dF_{T_n}(y)$$

$$= F_1(t+x) - F_1(t) + \int_0^t [F(x+t-y) - F(t-y)] \sum_{n=1}^{\infty} d(F_1 * F^{*(n-1)}(y))$$

$$= F_1(t+x) - F_1(t) + \int_0^t [F(x+t-y) - F(t-y)] \, d\left(\sum_{n=1}^{\infty} F_1 * F^{*(n-1)}(y) \right)$$

$$= F_1(t+x) - F_1(t) + \int_0^t [F(x+t-y) - F(t-y)] \, dH_1(y).$$

This representation of $F_{A(t)}$ can be simplified by combining it with (7.100). The result is

$$F_{A(t)}(x) = F_1(t+x) - \int_0^t \overline{F}(x+t-y) \, dH_1(y); \quad x, t \ge 0. \qquad (7.130)$$

Differentiation yields the probability density of $A(t)$:

$$f_{A(t)}(x) = f_1(t+x) + \int_0^t f(x+t-y) h_1(y) \, dy; \quad x, t \ge 0. \qquad (7.131)$$

The probability that the system, which is working at time t, does not fail in $(t, t+x]$ is

$$\overline{F}_{A(t)}(x) = 1 - F_{A(t)}(x).$$

$\overline{F}_{A(t)}(x)$ is sometimes called *interval reliability*.

For determining the mean value of the forward recurrence time of an ordinary renewal process, $A(t)$ is written in the form

$$A(t) = \sum_{i=1}^{N(t)+1} Y_i - t,$$

where the $Y_1, Y_2, \ldots$ are independent and identically distributed as Y with $\mu = E(Y)$. Wald's identity (4.74) at page 194 cannot be applied to obtain $E(A(t))$, since $N(t) + 1$ is surely not independent of the sequence $Y_1, Y_2, \ldots$. However, $N(t) + 1$ is a stopping time for the sequence $Y_1, Y_2, \ldots$:

$$'N(t) + 1 = n' = 'N(t) = n - 1' = 'Y_1 + Y_2 + \cdots + Y_{n-1} \le t < Y_1 + Y_2 + \cdots + Y_n.'$$

Thus, the event $'N(t) + 1 = n'$ is independent of all $Y_{n+1}, Y_{n+2}, \ldots$ so that, by definition 4.2, $N(t) + 1$ is a stopping time for the sequence $Y_1, Y_2, \ldots$ Hence, the mean value of $A(t)$ can be obtained from (4.76) at page 195 with $N = N(t) + 1$:

$$E(A(t)) = \mu [E(N(t) + 1)] - t.$$

Thus, the mean forward recurrence time of an ordinary renewal process is

$$E(A(t)) = \mu [H(t) + 1] - t.$$

The probability distribution of the backward recurrence time is obtained as follows:

$$F_{B(t)}(x) = P(t - x \le T_{N(t)})$$
$$= \sum_{n=1}^{\infty} P(t - x \le T_n, N(t) = n)$$
$$= \sum_{n=1}^{\infty} P(t - x \le T_n \le t < T_{n+1})$$
$$= \sum_{n=1}^{\infty} \int_{t-x}^{t} \overline{F}(t - u) \, dF_{T_n}(u)$$
$$= \int_{t-x}^{t} \overline{F}(t - u) \, d\left(\sum_{n=1}^{\infty} F_1 * F^{*(n)}\right)$$
$$= \int_{t-x}^{t} \overline{F}(t - u) \, dH_1(u).$$

Hence, the distribution function of $B(t)$ is

$$F_{B(t)}(x) = \begin{cases} \int_{t-x}^{t} \overline{F}(t - u) \, dH_1(u) & \text{for} \quad 0 \le x \le t \\ 1 & \text{for} \quad t > x \end{cases} . \qquad (7.132)$$

Differentiation yields the probability density of $B(t)$:

$$f_{B(t)}(x) = \begin{cases} \overline{F}(x) h_1(t - x) & \text{for} \quad 0 \le x \le t, \\ 0 & \text{for} \quad t < x. \end{cases} \qquad (7.133)$$

One easily verifies that the forward and backward recurrence times of an ordinary renewal process, whose cycle lengths are exponentially distributed with parameter λ, are also exponentially distributed with parameter λ :

$$f_{A(t)}(x) = f_{B(t)}(x) = \lambda e^{-\lambda x} \quad \text{for all } t \geq 0.$$

In view of the *memoryless property* of the exponential distribution (example 2.21, page 87), this result is not surprising.

A direct consequence of the fundamental renewal theorem is that $F_S(t)$, as defined by (7.111), is the limiting distribution function of both backward and forward recurrence time as t tends to infinity:

$$\lim_{t \to \infty} F_{A(t)}(x) = \lim_{t \to \infty} F_{B(t)}(x) = F_S(x), \quad x \geq 0. \tag{7.144}$$

Paradox of Renewal Theory In view of the definition of the forward recurrence time, one may suppose that the following equation is true:

$$\lim_{t \to \infty} E(A(t)) = \mu/2.$$

However, according to (7.134) and (7.112),

$$\lim_{t \to \infty} E(A(t)) = \int_0^\infty \overline{F}_S(t)\, dt = E(S) = \frac{\mu^2 + \sigma^2}{2\mu} > \frac{\mu}{2}.$$

This 'contradiction' is known as the *paradox of renewal theory*. The intuitive explanation of this phenomenon is that on average the 'reference time point' t is to be found more frequently in longer renewal cycles than in shorter ones.

7.3.5 Stationary Renewal Processes

By definition 7.1, a renewal process $\{Y_1, Y_2, ...\}$ is stationary if for all $k = 1, 2, ...$ and any sequence of integers $i_1, i_2, ..., i_k$ with $1 \leq i_1 < i_2 < \cdots < i_k$ and any $\tau = 0, 1, ...$ the joint distribution functions of the vectors

$$(Y_{i_1}, Y_{i_2}, ..., Y_{i_k}) \text{ and } (Y_{i_1+\tau}, Y_{i_2+\tau}, ..., Y_{i_k+\tau})$$

coincide, $k = 1, 2, ...$. According to the corollary after definition 7.1, $\{Y_1, Y_2, ...\}$ is stationary if and only if the corresponding renewal counting process $\{N(t), t \geq 0\}$ has homogeneous increments. A third way of defining the stationarity of a renewal process $\{Y_1, Y_2, ...\}$ makes use of the statistical equivalence between $\{Y_1, Y_2, ...\}$ and the corresponding processes $\{A(t), t \geq 0\}$ or $\{B(t), t \geq 0\}$, respectively.

> A renewal process is stationary if and only if the process of its forward (backward) recurrence times $\{A(t), t \geq 0\}$ ($\{B(t), t \geq 0\}$) is strongly stationary.

The stochastic process in continuous time $\{B(t), t \geq 0\}$ is a Markov process. This is quite intuitive, but a strict proof will not be given here. By theorem 7.1, a Markov process $\{X(t), t \in \mathbf{T}\}$ is strongly stationary if and only if its one-dimensional distribution functions $F_t(x) = P(X(t) \le x)$ do not depend on t. Hence, a renewal process is stationary if and only if there is a distribution function $F(x)$ so that

$$F_{A(t)}(x) = P(A(t) \le x) = F(x) \text{ for all } x \ge 0 \text{ and } t \ge 0.$$

Theorem 7.18 yields a simple criterion for the stationarity of renewal processes:

Theorem 7.18 Let $F(x) = P(Y \le x)$ be nonarithmetic and $\mu = E(Y) < \infty$. Then a delayed renewal process given by $F_1(x)$ and $F(x)$ is stationary if and only if

$$H_1(t) = t/\mu. \tag{7.135}$$

Equivalently, as a consequence of theorem 7.11, a delayed renewal process is stationary if and only if

$$F_1(x) = F_S(x) = \frac{1}{\mu} \int_0^x \overline{F}(y)\,dy \quad \text{for all } x \ge 0. \tag{7.136}$$

Proof If (7.136) holds, then (7.135) as well, so that, from (7.130),

$$\begin{aligned} F_{A(t)}(x) &= \frac{1}{\mu} \int_0^{t+x} \overline{F}(y)\,dy - \frac{1}{\mu} \int_0^t \overline{F}(x+t-y)\,dy \\ &= \frac{1}{\mu} \int_0^{t+x} \overline{F}(y)\,dy - \frac{1}{\mu} \int_x^{t+x} \overline{F}(y)\,dy \\ &= \frac{1}{\mu} \int_0^x \overline{F}(y)\,dy. \end{aligned}$$

Hence, $F_{A(t)}(x)$ does not depend on t.

Conversely, if $F_{A(t)}(x)$ does not depend on t, then (7.134) implies

$$F_{A(t)}(x) \equiv F_S(x) \text{ for all } t.$$

This completes the proof of the theorem.   ∎

As a consequence from theorem 7.87 and the elementary renewal theorem: After a sufficiently large time span (*transient response time*) every renewal process with nonarithmetic distribution function $F(t)$ and finite mean cycle length $\mu = E(Y)$ behaves as a stationary renewal process.

7.3.6 Alternating Renewal Processes

So far it has been assumed that renewals take only negligibly small amounts of time. In order to be able to model practical situations, in which this assumption is not fulfilled, the concept of a renewal process has to be generalized in the following way: The renewal time of the system after its ith failure is assumed to be a positive random variable Z_i; $i = 1, 2, \ldots$. Immediately after a renewal the system starts operating. In this way, a marked point process $\{(Y_i, Z_i); i = 1, 2, \ldots\}$ is generated, where Y_i as before denotes the lifetime of the system after the ith renewal.

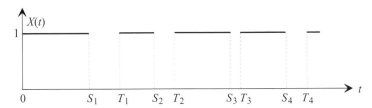

Figure 7.12 Sample path of an alternating renewal process

Definition 7.8 (*alternating renewal process*) If $\{Y_1, Y_2, ...\}$ and $\{Z_1, Z_2, ...\}$ are two independent sequences of independent, nonnegative random variables, then the marked point process $\{(Y_1, Z_1), (Y_2, Z_2), ...\}$ is said to be an *alternating renewal process* if the Y_i and the Z_i have the meanings given above. ●

The random variables

$$S_1 = Y_1; \quad S_n = \sum_{i=1}^{n-1}(Y_i + Z_i) + Y_n; \quad n = 2, 3, ...,$$

are the time points, at which failures occur and the random variables

$$T_n = \sum_{i=1}^{n-1}(Y_i + Z_i); \quad n = 1, 2, ...$$

are the time points at which a new system starts operating. If an operating system is assigned a '1' and a failed system a '0', a binary indicator variable of the system state is

$$X(t) = \begin{cases} 0 & \text{if } t \in [S_n, T_n), \quad n = 1, 2, ..., \\ 1 & \text{elsewhere.} \end{cases} \tag{7.137}$$

Obviously, an alternating renewal process can equivalently be defined by the stochastic process in continuous time $\{X(t), t \geq 0\}$ with $X(t)$ given by (7.137) (Figure 7.12).

In what follows, all Y_i and Z_i are assumed to be distributed as Y and Z with distribution functions $F_Y(y) = P(Y \leq y)$ and $F_Z(z) = P(Z \leq z)$, respectively. By agreement,

$$P(X(+0) = 1) = 1.$$

Analogously to the concept of a delayed renewal process, the alternating renewal process can be generalized by assigning to the random lifetime Y_1 a probability distribution different from that of Y. This way of generalization and some other possibilities will not be discussed here, although no principal difficulties would arise.

Let $N_f(t)$ and $N_r(t)$ be the respective numbers of failures and renewals in $(0, t]$. Since S_n and T_n are sums of independent random variables,

$$F_{S_n}(t) = P(S_n \leq t) = P(N_f(t) \geq n) = F_Y * (F_Y * F_Z)^{*(n-1)}(t), \tag{7.138}$$

$$F_{T_n}(t) = P(T_n \leq t) = P(N_r(t) \geq n) = (F_Y * F_Z)^{*(n)}(t). \tag{7.139}$$

Analogously to (7.95) and (7.96), sum representations of the mean values

$$H_f(t) = E(N_f(t)) \text{ and } H_r(t) = E(N_r(t))$$

are

$$H_f(t) = \sum_{n=1}^{\infty} F_Y * (F_Y * F_Z)^{*(n-1)}(t),$$

and

$$H_r(t) = \sum_{n=1}^{\infty} (F_Y * F_Z)^{*(n)}(t).$$

$H_f(t)$ and $H_r(t)$ are referred to as the *renewal functions* of the alternating renewal process. Since $H_f(t)$ can be interpreted as the renewal function of a delayed renewal process, whose first system lifetime is distributed as Y, whereas the following 'system lifetimes' are identically distributed as $Y + Z$ it satisfies renewal equation (7.97) with

$$F_1(t) \equiv F_Y(t) \text{ and } F(t) = F_Y * F_Z(t).$$

Analogously, $H_r(t)$ can be interpreted as the renewal function of an ordinary renewal process whose cycle lengths are identically distributed as $Y + Z$. Therefore, $H_r(t)$ satisfies renewal equation (7.98) with $F(t)$ replaced by $F_Y * F_Z(t)$.

Let R_t be the residual lifetime of the system if it is operating at time t. Then

$$P(X(t) = 1, R_t > x)$$

is the probability that the system is working at time t and does not fail in the interval $(t, t+x]$. This probability is called *interval availability* or *interval reliability*, and it is denoted as $A_x(t)$. It can be obtained as follows:

$$A_x(t) = P(X(t) = 1, R_t > x)$$

$$= \sum_{n=0}^{\infty} P(T_n \le t, \ T_n + Y_{n+1} > t + x)$$

$$= \overline{F}_Y(t+x) + \int_0^t P(t+x-u < Y) \, d \sum_{n=1}^{\infty} (F_Y * F_Z)^{*(n)}(u).$$

Hence,

$$A_x(t) = \overline{F}_Y(t+x) + \int_0^t \overline{F}_Y(t+x-u) \, dH_r(u). \tag{7.140}$$

Note In this section 'A' does no longer refer to forward recurrence time.

Let $A(t)$ be the probability that the system is operating (available) at time t:

$$A(t) = P(X(t) = 1). \tag{7.141}$$

This important characteristic of an alternating renewal process is obtained from (7.140) by letting there $x = 0$:

$$A(t) = \overline{F}_Y(t) + \int_0^t \overline{F}_Y(t-u) \, dH_r(u). \tag{7.142}$$

$A(t)$ is called *availability* of the system, *system availability*, or, more exactly, *point availability* of the system, since it refers to a specific time point t. It is equal to the mean value of the indicator variable of the system state:

$$E(X(t)) = 1 \cdot P(X(t) = 1) + 0 \cdot P(X(t) = 0) = P(X(t) = 1) = A(t).$$

The *average availability* of the system in the interval $[0, t]$ is

$$\overline{A}(t) = \frac{1}{t} \int_0^t A(x) \, dx.$$

The random *total operating time* $U(t)$ of the system in the interval $[0, t]$ is

$$U(t) = \int_0^t X(x) \, dx. \tag{7.143}$$

By changing the order of integration

$$E(U(t)) = E\left(\int_0^t X(x) \, dx\right) = \int_0^t E(X(x)) \, dx.$$

Thus,

$$E(U(t)) = \int_0^t A(x) \, dx = t \, \overline{A}(t).$$

The following theorem provides information on the limiting behavior of the interval reliability and the point availability as $t \to \infty$. A proof of the assertions need not be given, since they are immediate consequences of theorem 7.13.

Theorem 7.19 If $E(Y) + E(Z) < \infty$ and the distribution function $(F_Y * F_Z)(t)$ of the sum is nonarithmetic, then

$$A_x = \lim_{t \to \infty} A_x(t) = \frac{1}{E(Y) + E(Z)} \int_x^\infty \overline{F}_Y(u) \, du,$$

$$A = \lim_{t \to \infty} A(t) = \lim_{t \to \infty} \overline{A}(t) = \frac{E(Y)}{E(Y) + E(Z)}. \tag{7.144}$$

∎

A_x is said to be the *long-run* or *stationary interval availability* (*reliability*) with regard to an interval of length x, and A is called the *long-run* or *stationary availability*. Clearly, $A = A_0$. If, analogously to renewal processes, the time between two neighboring time points at which a new system starts operating is called a *renewal cycle*, then the long-run availability is equal to the mean share of the operating time of a system in the mean renewal cycle length. Equation (7.144) is also valid if within renewal cycles Y_i and Z_i depend on each other.

Example 7.19 Life- and renewal times have exponential distributions with densities

$$f_Y(y) = \lambda e^{-\lambda y}, \ y \ge 0, \quad \text{and} \quad f_Z(z) = \mu e^{-\mu z}, \ z \ge 0.$$

The Laplace transforms of these densities and of

$$\overline{F}(y) = e^{-\lambda y}, \ y \ge 0,$$

are

$$\hat{f}_Y(s) = \frac{\lambda}{s + \lambda}, \quad \hat{f}_Z(s) = \frac{\mu}{s + \mu}, \quad \text{and} \quad L\{\overline{F}_Y, s\} = \frac{1}{s + \lambda}.$$

Application of the Laplace transform to the integral equation (7.142) yields

$$\hat{A}(s) = L\{\overline{F}_Y, s\} + L\{\overline{F}_Y, s\} \cdot \hat{h}_r(s) = \frac{1}{s+\lambda}\left[1 + \hat{h}_r(s)\right]. \qquad (7.145)$$

By (2.127), the Laplace transform of the convolution $(f_Y * f_Z)(t)$ is

$$L\{f_Y * f_Z, s\} = \hat{f}_Y(s) \cdot \hat{f}_Z(s) = \frac{\lambda\mu}{(s+\lambda)(s+\mu)}.$$

From the second equation of (7.104)

$$\hat{h}_r(s) = \frac{\lambda\mu}{s(s+\lambda+\mu)}.$$

By inserting $\hat{h}_r(s)$ into (7.145) and expanding $\hat{A}(s)$ into partial fractions,

$$\hat{A}(s) = \frac{1}{s+\lambda} + \frac{\lambda}{s(s+\lambda)} - \frac{\lambda}{s(s+\lambda+\mu)}.$$

Retransformation (use Table 2.5, page 105) yields the point availability

$$A(t) = \frac{\mu}{\lambda+\mu} + \frac{\lambda}{\lambda+\mu} e^{-(\lambda+\mu)t}, \quad t \geq 0.$$

Since

$$E(Y) = 1/\lambda \text{ and } E(Z) = 1/\mu,$$

taking in $A(t)$ the limit as $t \to \infty$ verifies relationship (7.144). On the other hand, if $\lambda \neq \mu$, as derived in example 4.14 (page 174),

$$E\left(\frac{Y}{Y+Z}\right) = \frac{\mu}{\mu-\lambda}\left(1 + \frac{\lambda}{\mu-\lambda} \ln\frac{\lambda}{\mu}\right).$$

For instance, if $E(Z) = 0.25\, E(Y)$, then

$$A = \frac{E(Y)}{E(Y)+E(Z)} = 0.800$$

and

$$E\left(\frac{Y}{Y+Z}\right) = 0.717.$$

Hence, in general,

$$E\left(\frac{Y}{Y+Z}\right) \neq \frac{E(Y)}{E(Y)+E(Z)}. \qquad \square$$

Usually, numerical methods have to be applied to determine interval and point availability when applying formulas (7.140) and (7.142). This is again due to the fact that there are either no explicit or rather complicated representations of the renewal function for most of the common lifetime distributions. These formulas can, however, be applied for obtaining approximate values for interval and point availability if they are used in conjunction with the bounds and approximations for the renewal function given in sections 7.3.2.2 and 7.3.3.

7.3.7   Compound Renewal Processes

7.3.7.1   Definition and Properties

Compound stochastic processes arise by additive superposition of random variables at random time points. (For motivation, see section 7.2.5.)

Definition 7.9 Let $\{(T_1, M_1),\ (T_2, M_2),\ ...\}$ be a random marked point process with property that $\{T_1, T_2, ...\}$ is the sequence of renewal time points of a renewal process $\{Y_1, Y_2, ...\}$, and let $\{N(t),\ t \geq 0\}$ be the corresponding renewal counting process. Then the stochastic process $\{C(t),\ t \geq 0\}$ defined by

$$C(t) = \begin{cases} \sum_{i=1}^{N(t)} M_i & \text{if } N(t) \geq 1 \\ 0 & \text{if } N(t) = 0 \end{cases} \qquad (7.146)$$

is called a *compound (aggregate, cumulative) renewal process*, and $C(t)$ is called a *compound random variable.*   ●

The compound Poisson process defined in section 7.2.5 is a compound renewal process with property that the renewal cycle lengths $Y_i = T_i - T_{i-1}$, $i = 1, 2, ...$, are independent and identically exponentially distributed (theorem 7.2).

A compound renewal process is also called a *renewal reward process*, in particular, if M_i is a 'profit' of any kind made at the renewal time points. In most applications, however, M_i is a 'loss', for instance, replacement cost, repair time, or claim size. But it also can represent a 'loss' or 'gain', which accumulates over the ith renewal cycle (maintenance cost, profit by operating the system). In any case, $C(t)$ is the total loss (gain), which has accumulated over the interval $(0, t]$. The sample paths of a compound renewal process are step functions. Jumps occur at times T_i and the respective jump heights are M_i (Figure 7.13).

In this section, compound renewal processes are considered under the following assumptions:

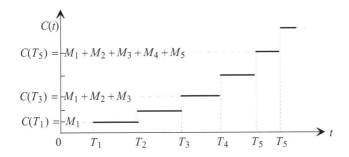

Figure 7.13 Sample path of a compound process with positive increments

1) $\{N(t), t \geq 0\}$ is a renewal counting process, which belongs to an <u>ordinary</u> renewal process $\{Y_1, Y_2, ...\}$.

2) The sequences $\{M_1, M_2, ...\}$ and $\{Y_1, Y_2, ...\}$ are independent of each other and consist each of independent, nonnegative random variables, which are identically distributed as M and Y, respectively. M_i and Y_j are allowed to depend on each other if $i = j$, i.e., if they refer to the same renewal cycle.

3) The mean values of Y and M are finite and positive.

Under these assumptions, Wald's equation (4.74) yields the trend function of the compound renewal process $\{C(t), t \geq 0\}$:

$$m(t) = E(C(t)) = E(M) H(t), \tag{7.147}$$

where $H(t) = E(N(t))$ is the renewal function, which belongs to the underlying renewal process $\{Y_1, Y_2, ...\}$. Formula (7.147) and theorem 7.12, the elementary renewal theorem, imply an important asymptotic property of the trend function of compound renewal processes:

$$\lim_{t \to \infty} \frac{E(C(t))}{t} = \frac{E(M)}{E(Y)}. \tag{7.148}$$

Equation (7.148) means that the average long-run (stationary) loss or profit per unit time is equal to the average loss or profit per unit time within a renewal cycle. The 'stochastic analog' to (7.148) is: With probability 1,

$$\lim_{t \to \infty} \frac{C(t)}{t} = \frac{E(M)}{E(Y)}. \tag{7.149}$$

To verify (7.149), consider the obvious relationship

$$\sum_{i=1}^{N(t)} M_i \leq C(t) \leq \sum_{i=1}^{N(t)+1} M_i.$$

From this,

$$\left(\frac{1}{N(t)} \sum_{i=1}^{N(t)} M_i \right) \frac{N(t)}{t} \leq \frac{C(t)}{t} \leq \left(\frac{1}{N(t)+1} \sum_{i=1}^{N(t)} M_i \right) \frac{N(t)+1}{t}.$$

Now the strong law of the large numbers (theorem 5.4) and (7.118) imply (7.149). The relationships (7.148) and (7.149) are called *renewal reward theorems*.

Distribution of $C(t)$ If M has distribution function $G(t)$, then, given $N(t) = n$, the compound random variable $C(t)$ has distribution function

$$P(C(t) \leq x | N(t) = n) = G^{*(n)}(x),$$

where $G^{*(n)}(x)$ is the nth convolution power of $G(t)$. Hence, by the total probability rule,

$$F_{C(t)}(x) = P(C(t) \leq x) = \sum_{n=1}^{\infty} G^{*(n)}(x) P(N(t) = n), \tag{7.150}$$

where the probabilities $P(N(t) = n)$ are given by (7.90). (With the terminology of section 2.4, $F_{C(t)}$ is a mixture of the probability distribution functions $G^{*(1)}, G^{*(2)}, ...$ If Y has an exponential distribution with parameter λ, then $C(t)$ has distribution function

$$F_{C(t)}(x) = e^{-\lambda t} \sum_{n=0}^{\infty} G^{*(n)}(x) \frac{(\lambda t)^n}{n!}; \quad G^{*(0)}(x) \equiv 1, \ x > 0, \ t > 0. \qquad (7.151)$$

If, in addition, M has a normal distribution with $E(M) \geq 3\sqrt{Var(M)}$, then

$$F_{C(t)}(x) = e^{-\lambda t} \left[1 + \sum_{n=1}^{\infty} \Phi\left(\frac{x - n E(M)}{\sqrt{n\,Var(M)}} \right) \frac{(\lambda t)^n}{n!} \right]; \quad x > 0, \ t > 0. \qquad (7.152)$$

The distribution function $F_{C(t)}$, for being composed of convolution powers of G and F, is usually not tractable and useful for numerical applications. Hence, much effort has been put into constructing bounds on $F_{C(t)}$ and into establishing asymptotic expansions. For surveys, see, e.g. *Rolski et al.* (1999) and *Willmot, Lin* (2001). The following result of *Gut* (1990) is particularly useful.

Theorem 7.20 If

$$\gamma^2 = Var\{E(Y) M - E(M) Y\} > 0, \qquad (7.153)$$

then

$$\lim_{t \to \infty} P\left(\frac{C(t) - \dfrac{E(M)}{E(Y)} t}{[E(Y)]^{-3/2} \gamma \sqrt{t}} \leq x \right) = \Phi(x),$$

where $\Phi(x)$ is the distribution function of the standardized normal distribution.　■

This theorem implies that for large t the compound variable $C(t)$ has approximately a normal distribution with mean value and variance

$$\frac{E(M)}{E(Y)} t \ \text{ and } [E(Y)]^{-3} \gamma^2 t,$$

respectively:

$$C(t) \approx N\left(\frac{E(M)}{E(Y)} t, \ [E(Y)]^{-3} \gamma^2 t \right). \qquad (7.154)$$

If M and Y are independent, then the parameter γ^2 can be written in the following form:

$$\gamma^2 = [E(Y)]^2 \, Var(M) + [E(M)]^2 \, Var(Y). \qquad (7.155)$$

In this case, in view of assumption 3, condition (7.153) is always fulfilled. Condition (7.153) actually only excludes the case $\gamma^2 = 0$, i.e. linear dependence between Y and M. The following examples present applications of theorem 7.20.

Example 7.20 For an alternating renewal process $\{(Y_i, Z_i); \; i = 1, 2, ...\}$, the total renewal time in $(0, t]$ is given by (a possible renewal time running at time t is neglected)

$$C(t) = \sum_{i=1}^{N(t)} Z_i,$$

where

$$N(t) = \max_n \; \{n, \; T_n < t\}.$$

(Notation and assumptions as in section 7.3.6.) Hence, the development of the total renewal time is governed by a compound stochastic process. In order to investigate the asymptotic behaviour of $C(t)$ as $t \to \infty$ by means of theorem 7.20, M has to be replaced with Z and Y with $Y + Z$. Consequently, if t is sufficiently large, then $C(t)$ has approximately a normal distribution with parameters

$$E(X(t)) = \frac{E(Z)}{E(Y) + E(Z)} t \quad \text{and} \quad Var(X(t)) = \frac{\gamma^2}{[E(Y) + E(Z)]^3} t.$$

Because of the independence of Y and Z,

$$\gamma^2 = Var[Z E(Y + Z) - (Y + Z) E(Z)]$$
$$= Var[Z E(Y) - Y E(Z)]$$
$$= [E(Y)]^2 Var(Z) + [E(Z)]^2 Var(Y) > 0$$

so that assumption (7.153) is satisfied. In particular, let (all parameters in *hours*)

$$E(Y) = 120, \; \sqrt{Var(Y)} = 40, \; \text{and} \; E(Z) = 4, \; \sqrt{Var(Z)} = 2.$$

Then,

$$\gamma^2 = 120^2 \cdot 4 + 16 \cdot 1600 = 83\,200 \quad \text{and} \quad \gamma = 288.4.$$

Consider for example the total renewal time in the interval $[0, 10^4 \, hours]$. The probability that $C(10^4)$ does not exceed a nominal value of 350 *hours* is

$$P(C(10^4) \le 350) = \Phi\left(\frac{350 - \frac{4}{124} 10^4}{124^{-3/2} \cdot 288.4 \cdot \sqrt{10^4}}\right) = \Phi(1.313).$$

Hence,

$$P(C(10^4) \le 350) = 0.905. \qquad \Box$$

Example 7.21 (normal approximation to risk processes) Let the sequence of the claim interarrival times $Y_1, Y_2, ...$ be an ordinary renewal process. This includes the homogeneous Poisson arrival process, to which section 7.2.7 is restricted. Otherwise, assumptions 2 to 4 (page 294) and the notation introduced there will be retained. Then, by theorem 7.20, if t is sufficiently large compared to $\mu = E(Y)$, the total claim size arising in $[0, t]$ has approximately a normal distribution with mean value $\frac{\nu}{\mu} t$ and variance $\mu^{-3} \gamma^2 t$:

$$C(t) \approx N\left(\tfrac{v}{\mu} t, \, \mu^{-3}\gamma^2 t\right), \tag{7.156}$$

where

$$\gamma^2 = \mu^2 Var(M) + v^2 Var(Y).$$

The random profit $G(t)$ the insurance company has made in $[0, t]$ is given by

$$G(t) = \kappa t - C(t).$$

By (7.156), $G(t)$ has approximately a normal distribution with parameters

$$E(G(t)) = (\kappa - \tfrac{v}{\mu})t \quad \text{and} \quad Var(G(t)) = \mu^{-3}\gamma^2 t.$$

Note that the situation considered here refers to the situation that, when being 'in red numbers' (ruin has happened), the company continues operating until it reaches a profitable time period and so on. In case of a positive safety loading the company will leave 'loss periods' with probability 1.

As a numerical special case, let us consider a risk process $\{(Y_1, M_1), (Y_2, M_2), ...\}$ with

$$\mu = E(Y) = 2 \ [h], \quad Var(Y) = 3 \ [h^2],$$

$$v = E(M) = 900 \ [\$], \quad Var(M) = 360\,000 \ [\$^2].$$

(1) What minimal premium per hour κ_α has the insurance company to take in so that it will achieve a profit of at least $\$10^6$ within 10^3 hours with probability $\alpha = 0.95$?
Since $\gamma = 1967.2$,

$$P(G(10^4) \geq 10^6) = P(C(t) < 10^4 (\kappa_{0.95} - 100))$$

$$= \Phi\left(\frac{(\kappa_{0.95} - 100) - 450}{2^{-1.5} \cdot 19.672}\right).$$

Since the 0.95-percentile of the standardized normal distribution is $z_{0.95} = 1.64$, the desired premium per hour $\kappa_{0.95}$ satisfies equation

$$\frac{\kappa_{0.95} - 550}{6.955} = 1.64.$$

Hence, $\kappa_{0.95} = 561 \ [\$/h]$.

This result does not take into account the fact that in reality the premium size has an influence on the claim flow.

(2) Let the premium income of the company be $\kappa = 460 \ [\$/h]$. Thus, the company has a positive safety loading of $\sigma = 10 \ [\$]$. Given an initial capital of $x = 10^4 \ [\$]$, what is the probability of the company to be in the state of ruin at time $t = 1000 \ [h]$?
This probability is given by

$$P(G(10^3) < -10^{-4}) = \Phi\left(\frac{-10^4 - (460 - 450)\,10^3}{2^{-1.5} \cdot 1967.2 \cdot \sqrt{1000}}\right)$$

$$= \Phi(-0.910) = 0.181. \qquad \square$$

7.3.7.2 First Passage Time

Example 7.21 motivates the investigation of the random time $L(x)$, at which the compound renewal process $\{C(t),\, t \geq 0\}$ ($C(t)$ not necessarily a cost criterion) exceeds a given nominal value x for the first time:

$$L(x) = \inf_{t} \{t,\, C(t) > x\}. \tag{7.157}$$

If, for instance, x is the critical wear limit of an item, then crossing level x is commonly referred to as the occurrence of a *drift failure*. Hence, in this case it is justified to denote L as the lifetime of the system (Figure 7.14).

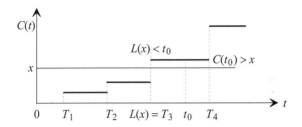

Figure 7.14 Level crossing of a compound stochastic process

Since the M_i are nonnegative random variables, the compound renewal process $\{C(t),\, t \geq 0\}$ has nondecreasing sample paths. In such a case, the following relationship between the distribution function of the *first passage time $L(x)$* and the distribution function of the compound random variable $C(t)$) is obvious (Figure 7.14):

$$P(L(x) \leq t) = P(C(t) > x). \tag{7.158}$$

Specifically, if $\{N(t), t \geq 0\}$ is the homogeneous Poisson process, then, from formulas (7.151) and (7.158),

$$P(L(x) > t) = e^{-\lambda t} \sum_{n=0}^{\infty} G^{*(n)}(x)\frac{(\lambda t)^n}{n!}; \quad t \geq 0,$$

with $x,\, x > 0,$ fixed. The probability distribution of $L(x)$ is generally not explicitly available. Hence the following theorem (*Gut* (1990)) is important for applications, since it provides information on the asymptotic behavior of the distribution of $L(x)$ as $x \to \infty$. The analogy of this theorem to theorem 7.20 is obvious.

Theorem 7.21 If $\gamma^2 = \mu^2 Var(M) + v^2 Var(Y) > 0$, then

$$\lim_{x \to \infty} P\left(\frac{L(x) - \dfrac{E(Y)}{E(M)}\, x}{[E(M)]^{-3/2}\, \gamma\, \sqrt{x}} \leq t \right) = \Phi(t),$$

where $\Phi(t)$ is the distribution function of the standardized normal distribution. ∎

Actually, in view of our assumption that the compound process $\{C(t), t \geq 0\}$ has non-decreasing sample paths, condition (7.158) implies that theorems 7.20 and 7.21 are equivalent.

A consequence of theorem 7.21 is that, for large x, the first passage time $L = L(x)$ has approximately a normal distribution with parameters

$$E(L(x)) = \frac{E(Y)}{E(M)} x \quad \text{and} \quad Var(L(x)) = [E(M)]^{-3} \gamma^2 x,$$

i.e.

$$L(x) \approx N\left(\frac{E(Y)}{E(M)} x, [E(M)]^{-3} \gamma^2 x \right), \quad x > 0. \qquad (7.159)$$

The probability distribution given by (7.159) is called *Birnbaum-Saunders distribution*.

Example 7.22 Mechanical wear of an item is caused by shocks. (For instance, for the brake discs of a car, every application of the brakes is a shock.) After the ith shock the degree of wear of the item increases by M_i units. The $M_1, M_2, \ldots$ are supposed to be independent random variables, which are identically normally distributed as M with parameters

$$E(M) = 9.2 \quad \text{and} \quad \sqrt{Var(M)} = 2.8 \text{ [in } 10^{-4} mm].$$

The initial degree of wear of the item is zero. The item is replaced by an equivalent new one if the total degree of wear exceeds a critical level of $0.1\, mm$.

(1) What is the probability p_{100} that the item has to be replaced before or at the occurrence of the 100th shock? The degree of wear after 100 shocks is

$$C_{100} = \sum_{i=1}^{100} M_i$$

and has approximately the distribution function (unit of x: $10^{-4} mm$)

$$P(C_{100} \leq x) = \Phi\left(\frac{x - 9.2 \cdot 100}{\sqrt{2.8^2 \cdot 100}} \right) = \Phi\left(\frac{x - 920}{28} \right).$$

Thus, the item survives the first 100 shocks with probability

$$p_{100} = P(C_{100} \leq 1000) = \Phi(2.86).$$

Hence, $p_{100} = 0.979$.

(2) In addition to the parameters of M, the random cycle Y is assumed to have mean value and variance

$$E(Y) = 6 \quad \text{and} \quad \sqrt{Var(Y)} = 2 \text{ [hours]}.$$

What is the probability that the nominal value of $0.1\, mm$ is not exceeded within the time interval $[0, 600]$ (*hours*)?

To answer this question, theorem 7.21 can be applied since 0.1 *mm* is sufficiently large in comparison to the shock parameter $E(M)$. Provided M and Y are independent, the parameter γ is $\gamma = 0.0024916$. Hence,

$$P(L(0.1) > 600) = 1 - \Phi\left(\frac{600 - \frac{6}{9.2}\, 10^3}{(9.2)^{-3/2} \cdot 2491.6 \cdot \sqrt{0.1}}\right)$$

$$= 1 - \Phi(-1.848).$$

Thus, the desired probability is $P(L(0.1) > 600) = 0.967$.   □

Example 7.23 Let the risk process $\{(Y_1, M_1), (Y_2, M_2), ...\}$ have the parameters

$$\mu = E(Y) = 5 \ [h], \quad Var(Y) = 25 \ [h^2],$$

$$\nu = E(M) = 1000 \ [\$], \quad Var(M) = 640\,000 \ [\$^2].$$

What is the probability that the total claim reaches level $a = 10^6$ [$\$$] before or at time point $t = 5500$ [h]?

a) Since $\gamma = 6403$, because of (7.159),

$$P(L(10^6) \le 5500) \le \Phi\left(\frac{5500 - \frac{5 \cdot 10^6}{1000}}{1000^{-1.5} \cdot 6403 \cdot 10^6}\right) = \Phi(2.4694)$$

so that

$$P(L(10^6) \le 5500) = 0.993.$$

b) Now the same question is answered by making use of (7.156) and (7.158):

$$P(L(10^6) \le 5500) = P(C(5500) > 10^6)$$

$$= 1 - P(C(5500) \le 10^6)$$

$$= 1 - \Phi\left(\frac{10^6 - \frac{1000 \cdot 5500}{5}}{5^{-1.5} \cdot 6403 \cdot \sqrt{5500}}\right) = 1 - \Phi(-2.354)$$

so that

$$P(L(10^6) \le 5500) = P(C(5500) > 10^6) \approx 0.991.$$

Taking into account the piecewise constant sample paths of the compound process $\{C(t), t \ge 0\}$, there is an excellent correspondence between the results obtained under a) and b).   □

7.4 EXERCISES

Sections 7.1 and 7.2

7.1) The occurrence of catastrophic accidents at Sosal & Sons follows a homogeneous Poisson process with intensity $\lambda = 3$ a year.

(1) What is the probability $p_{\geq 2}$ that at least two catastrophic accidents will occur in the second half of the current year?

(2) Determine the same probability given that two catastrophic accidents have occurred in the first half of the current year.

7.2) By making use of the independence and homogeneity of the increments of a homogeneous Poisson process with intensity λ, show that its covariance function is given by
$$C(s, t) = \lambda \min(s, t).$$

7.3) The number of cars which pass a certain intersection daily between 12:00 and 14:00 follows a homogeneous Poisson process with intensity $\lambda = 40$ per hour. Among these there are 2.2% which disregard the stop sign. The car drivers behave independently with regard to ignoring stop signs.

(1) What is the probability that at least two cars disregard the stop sign between 12:30 and 13:30?

(2) A car driver, who ignores the stop sign at this interection, causes an accident there with probability 0.05. What is the probability of one or more accidents at this intersection between 12:30 and 13:30, caused by a driver, who ignores the stop sign?

7.4) A Geiger counter is struck by radioactive particles according to a homogeneous Poisson process with intensity $\lambda = 1$ per 12 seconds. On average, the Geiger counter only records 4 out of 5 particles.

(1) What is the probability $p_{\geq 2}$ that the Geiger counter records at least 2 particles a minute?

(2) What are mean value and variance of the random time Y between the occurrence of two successively recorded particles?

7.5) The location of trees in an even, rectangular forest stand of size $200m \times 500m$ follows a homogeneous Poisson distribution with intensity $\lambda = 1$ per $25m^2$. The diameters of the stems of all trees at a distance of $130cm$ to the ground is assumed to be $24cm$. From outside, a shot is vertically fired at a $500m$ side of the forest stand (parallel to the ground at level $130cm$). What is the probability that a bullet with diameter $1cm$ hits no tree?

Hint With regard to the question, the location of a tree is fully determined by the coordinates of the center of the cross-section of its stem at level $130cm$.

7.6) An electronic system is subject to two types of shocks, which occur independently of each other according to homogeneous Poisson processes with intensities

$$\lambda_1 = 0.002 \text{ and } \lambda_2 = 0.01 \text{ per hour,}$$

respectively. A shock of type 1 always causes a system failure, whereas a shock of type 2 causes a system failure with probability 0.4.

What is the probability that the system fails within 24 hours due to a shock?

7.7) A system is subjected to shocks of types 1, 2, and 3, which are generated by independent homogeneous Poisson processes with respective intensities per hour $\lambda_1 = 0.2$, $\lambda_2 = 0.3$, and $\lambda_3 = 0.4$. A type 1-shock causes a system failure with probability 1, a type 2-shock causes a system failure with probability 0.4, ana d shock of type 3 causes a system failure with probability 0.2. The shocks occur permanently, whether the system is operating or not.

(1) On condition that three shocks arrive in the interval $[0, 10\,h]$, determine the probability that the system does not experience a failure in this interval.

(2) What is the (unconditional) probability that the system fails in $[0, 10\,h]$ due to a shock?

7.8) Claims arrive at a branch of an insurance company according a homogeneous Poisson process with an intensity of $\lambda = 0.4$ per working hour. The claim size Z has an exponential distribution so that 80% of the claim sizes are below \$100 000, where as 20% are equal or larger than \$100 000.

(1) What is the probability that the fourth claim does not arrive in the first two working hours of a day?

(2) What is the mean size of a claim?

(3) Determine approximately the probability that the sum of the sizes of 10 consecutive claims exceeds \$800 000.

7.9) Consider two independent homogeneous Poisson processes 1 and 2 with respective intensities λ_1 and λ_2. Determine the mean value of the random number of events of process 2, which occur between any two successive events of process 1.

7.10) Let $\{N(t), t \geq 0\}$ be a homogeneous Poisson process with intensity λ.

Prove that for an arbitrary, but fixed, positive h the stochastic process $(X(t), t \geq 0)$ defined by $X(t) = N(t+h) - N(t)$ is weakly stationary.

7.11) Let a homogeneous Poisson process have intensity λ, and let T_i be the time point at which the i th Poisson event occurs. For $t \to \infty$, determine and sketch the covariance function $C(\tau)$ of the shot noise process $\{X(t), t \geq 0\}$ given by

$$X(t) = \sum_{i=1}^{N(t)} h(t - T_i) \quad \text{with} \quad h(t) = \begin{cases} \sin t & \text{for} \quad 0 \leq t \leq \pi \\ 0, & \text{elsewhere} \end{cases}.$$

7.12) Statistical evaluation of a large sample justifies to model the number of cars which arrive daily for petrol between 0:00 and 4:00 a.m. at a particular filling station by an inhomogeneous Poisson process $\{N(t), t \geq 0\}$ with intensity function

$$\lambda(t) = 8 - 4t + 3t^2 \quad [h^{-1}], \quad 0 \leq t \leq 4.$$

(1) How many cars arrive on average between 0:00 and 4:00 a.m.?

(2) What is the probability that at least 40 cars arrive between 2:00 and 4:00?

7.13) Let $\{N(t), t \geq 0\}$ be an inhomogeneous Poisson process with intensity function
$$\lambda(t) = 0.8 + 2t, \quad t \geq 0.$$
Determine the probability that at least 500 Poisson events occur in $[20, 30]$.

7.14)* Let $\{N(t), t \geq 0\}$ be a nonhomogeneous Poisson process with trend function $\Lambda(t)$ and arrival time point T_i of the ith Poisson event.

Given $N(t) = n$, show that the random vector $(T_1, T_2, ..., T_n)$ has the same probability distribution as n ordered, independent and identically distributed random variables with distribution function

$$F(x) = \begin{cases} \dfrac{\Lambda(x)}{\Lambda(t)} & \text{for } 0 \leq x < t, \\ 1, & t \leq x. \end{cases}$$

Hint Compare to theorem 7.5 (page 268).

7.15) Clients arrive at an insurance company according to a mixed Poisson process the structure parameter L of which has a uniform distribution over the interval $[0, 1]$.

(1) Determine the state probabilities of this process at time t.

(2) Determine trend and variance function of this process.

(3) For what values of α and β are trend and variance function of a Pólya arrival process identical to the ones obtained under (2)?

7.16) A system is subjected to shocks of type 1 and type 2, which are generated by independent Pólya processes $\{N_{L_1}(t), t \geq 0\}$ and $\{N_{L_2}(t), t \geq 0\}$ with respective trend and variance functions

$$E(N_{L_1}(t)) = t, \quad Var(N_{L_1}(t)) = t + 0.5t^2,$$
$$E(N_{L_2}(t)) = 0.5t, \quad Var(N_{L_2}(t)) = 0.5t + 0.125t^2$$

(time unit: *hour*). A shock of any type causes a system failure with probability 1. What is the probability that the system fails within 2 hours due to a shock?

7.17)* Prove the multinomial criterion (formula 7.55, page 280).

7.18) An insurance company has a premium income of \$106 080 per day. The claim sizes are iid random variables and have an exponential distribution with variance $4 \cdot 10^6 [\$^2]$. On average, 2 claims arrive per hour according to a homogeneous Poisson process. The time horizon is assumed to be infinite.

(1) What probability distribution have the interarrival times between two neighboring claims?

(2) Calculate the company's ruin probability if its initial capital is $x = \$20\,000$.

(3) What minimal initial capital should the company have to make sure that its ruin probability does not exceed 0.01?

7.19) Pramod is setting up an insurance policy for low-class cars (homogeneous portfolio) over an infinite time horizon. Based on previous statistical work, he expects that claims will arrive according to a homogeneous Poisson process with intensity $\lambda = 0.8\,[h^{-1}]$, and that the claim size will be iid distributed as an exponentially distributed random variable M with mean value $v = E(M) = \$3000$. He reckons with a total premium income of $\$2800\,[h^{-1}]$.

(1) Given that these assumptions are correct, has Pramod a chance to be financially successful with this portfolio over an infinite period of time?

(2) What is the minimal initial capital x_0 Pramod has to invest to make sure that the lower bound for the survival probability of this portfolio derived from the Lundberg inequality is 0.96?

(3) For the sake of comparison, determine the exact value of the survival probability of this company for an initial capital of $x_0/3$.

7.20) The lifetime L of a system has a Weibull-distribution with distribution function
$$F(t) = P(L \le t) = 1 - e^{-0.1t^3}, \quad t \ge 0.$$

(1) Determine its failure rate $\lambda(t)$ and its integrated failure rate $\Lambda(t)$.

(2) The system is maintained according to Policy 1 (page 290, bottom) over an infinite time span. The cost of a minimal repair is $c_m = 40\,[\$]$, and the cost of a preventive replacement is $c_p = 2000\,[\$]$.

Determine the cost-optimum replacement interval $\tau*$ and the corresponding minimal maintenance cost rate $K_1(\tau*)$.

7.21) A system is maintained according to Policy 3 (page 292, top) over an infinite time span. It has the same lifetime distribution and minimal repair cost parameter as in exercise 7.20. As with exercise 7.20, let $c_r = 2000$.

(1) Determine the optimum integer $n = n*$, and the corresponding maintenance cost rate $K_3(n*)$.

(2) Compare $K_3(n*)$ to $K_1(\tau*)$ (exercise 7.20) and try to intuitively explain the result.

Sections 7.3 and 7.4

Note Exercises 7.22 to 7.31 refer to ordinary renewal processes. The functions $f(t)$ and $F(t)$ denote density and distribution function; the parameters μ and μ_2 are mean value and second moment of the cycle length Y. $N(t)$ is the (random) renewal counting function, and $H(t)$ denotes the corresponding renewal function.

7.22) A system starts working at time $t = 0$. Its lifetime has approximately a normal distribution with mean value $\mu = 125$ hours and standard deviation $\sigma = 40$ hours. After a failure, the system is replaced with an equivalent new one in negligible time, and it immediately takes up its work. All system lifetimes are independent.

(1) What is the minimal number of systems, which must be available, in order to be able to maintain the replacement process over an interval of length 500 hours with probability 0.99?

(2) Solve the same problem on condition that the system lifetime has an exponential distribution with mean value $\mu = 125$.

7.23) (1) Use the Laplace transformation to find the renewal function $H(t)$ of an ordinary renewal process whose cycle lengths have an Erlang distribution with parameters $n = 2$ and λ.

(2) For $\lambda = 1$, sketch the exact graph of the renewal function and the bounds (7.117) in the interval $0 \le t \le 6$. Make sure the bounds (7.117) are applicable.

7.24) An ordinary renewal function has the renewal function $H(t) = t/10$. Determine the probability $P(N(10) \ge 2)$.

7.25) A system is preventively replaced by an identical new one at time points $\tau, 2\tau, ...$ If failures happen in between, then the failed system is replaced by an identical new one as well. The latter replacement actions are called *emergency replacements*. This replacement policy is called *block replacement*. The costs for preventive and emergency replacements are c_p and c_e, $0 < c_p < c_e$, respectively. The lifetime L of a system is assumed to have distribution function

$$F(t) = P(L \le t) = (1 - e^{-\lambda t})^2, \quad t \ge 0.$$

(1) Determine the renewal function $H(t)$ of the ordinary renewal process with cycle length distribution function $F(t)$.

(2) Based on the renewal reward theorem (7.148), give a formula for the long-run maintenance cost rate $K(\tau)$ under the block replacement policy.

(3) Determine an optimal $\tau = \tau*$ with regard to $K(\tau)$ for

$$\lambda = 0.1, \quad c_e = 900, \quad c_p = 100.$$

(4) Under otherwise the same assumptions, determine the cost rate if the system is only replaced after failures and compare it with the one obtained under (3).

7.26) Given the existence of the first three moments of the cycle length Y of an ordinary renewal process, verify the formulas (7.112).

7.27) (1) Verify that the probability $p(t) = P(N(t)$ is odd) satisfies

$$p(t) = F(t) - \int_0^t p(t-x)f(x)\, dx, \quad f(x) = F'(x).$$

(2) Determine this probability if the cycle lengths are exponential with parameter λ.

7.28)* Verify that the second moment of $N(t)$, denoted as $H_2(t) = E(N^2(t))$, satisfies the integral equation

$$H_2(t) = 2H(t) - F(t) + \int_0^t H_2(t-x)f(x)\, dx.$$

Hint Verify the equation directly or by applying the Laplace transformation.

7.29) The times between the arrivals of successive particles at a counter generate an ordinary renewal process. Its random cycle length Y has distribution function $F(t)$ and mean value $\mu = E(Y)$. After having recorded 10 particles, the counter is blocked for τ time units. Particles arriving during a blocked period are not registered.

What is the distribution function of the time from the end of a blocked period to the arrival of the first particle after this period if $\tau \to \infty$?

7.30) The cycle length distribution of an ordinary renewal process is given by the distribution function $F(t) = 1 - e^{-t^2}, \quad t \ge 0$ (Rayleigh distribution).

(1) What is the statement of theorem 7.13 if $g(x) = (x+1)^{-2}, \quad x \ge 0$?

(2) What is the statement of theorem 7.15?

7.31) Let be $A(t)$ the forward and $B(t)$ the backward recurrence times of an ordinary renewal process at time t. For $x > y/2$, determine functional relationships between $F(t)$ and the conditional probabilities

(1) $P(A(t) > y - t | B(t) = t - x), \quad 0 \le x < t < y,$

(2) $P(A(t) \le y | B(t) = x), \quad 0 \le x < t, \quad y > 0.$

7.32) Let (Y, Z) be the typical cycle of an alternating renewal process, where Y and Z have an Erlang distribution with joint parameter λ and parameters $n = 2$ and $n = 1$, respectively. For $t \to \infty$, determine the probability that the system is in state 1 at time t and that it stays in this state over the entire interval $[t, t+x], x > 0$ (process states as introduced in section 7.3.6).

7.33) The time intervals between successive repairs of a system generate an ordinary renewal process $\{Y_1, Y_2, ...\}$ with typical cycle length Y. The costs of repairs are mutually independent and independent of $\{Y_1, Y_2, ...\}$.

Let M be the typical repair cost and

$$\mu = E(Y) = 180 \,[\text{days}] \quad \text{and} \quad \sigma = \sqrt{Var(Y)} = 30,$$

$$v = E(M) = 200 \,[\$] \quad \text{and} \quad \sqrt{Var(M)} = 40.$$

Determine approximately the probabilities that

(1) the total repair costs arising in [0, 3600 days] do not exceed $\$4500$, and

(2) a total repair cost of $\$3000$ is not exceeded before 2200 days.

7.34) (1) Determine the ruin probability $p(x)$ of an insurance company with an initial capital of $x = \$20\,000$ and operating parameters

$$1/\mu = 2 \,[h^{-1}], \; v = \$800 \; \text{and} \; \kappa = 1700 \,[\$/h].$$

(2) Under otherwise the same conditions, draw the the graphs of the ruin probability for $x = 20\,000$ and $x = 0$ in dependence on κ over the interval $1600 \le \kappa \le 1800$.

(3) With the numerical parameters given under (1), determine the upper bound e^{-rx} for $p(x)$ given by the Lundberg inequality (7.85).

(4) Under otherwise the same conditions, draw the graph of e^{-rx} with $x = 20\,000$ in dependence on κ over the interval $1600 \le \kappa \le 1800$ and compare to the corresponding graph obtained under (2).

Note For problems (1) to (4), the model assumptions made in example 7.10 apply.

7.35) Under otherwise the same assumptions as made in example 7.10, determine the ruin probability if the random claim size M has density

$$b(y) = \lambda^2 y \, e^{-\lambda y}, \; \lambda > 0, \; y \ge 0.$$

This is an Erlang-distribution with parameters λ and $n = 2$.

7.36) Claims arrive at an insurance company according to an ordinary renewal process $\{Y_1, Y_2, ...\}$. The corresponding claim sizes $M_1, M_2, ...$ are independent and identically distributed as M and independent of $\{Y_1, Y_2, ...\}$. Let the Y_i be distributed as Y; i.e., Y is the typical interarrival interval. Then (Y, M) is the typical interarrival cycle. From historical observations it is known that

$$\mu = E(Y) = 1 \,[h], \; Var(Y) = 0.25, \; v = E(M) = \$800, \; Var(M) = 250.000.$$

Find approximate answers to the following problems:

(1) What minimum premium per unit time $\kappa_{min,\alpha}$ has the insurance company to take in so that it will make a profit of at least $\$10^6$ within 20 000 hours with probability $\alpha = 0.99$?

(2) What is the probability that the total claim reaches level $\$10^5$ within $135 \, h$?

Note Before possibly reaching its goals, the insurance company may have experienced one or more ruins with subsequent 'red number periods'.

CHAPTER 8

Discrete-Time Markov Chains

8.1 FOUNDATIONS AND EXAMPLES

This chapter is subjected to discrete-time stochastic processes $\{X_0, X_1, ...\}$ with discrete state space $\mathbf{Z}$ which have the Markov property. That is, on condition $X_n = x_n$ the random variable X_{n+1} is independent of all $X_0, X_1, ..., X_{n-1}$. However, without this condition, X_{n+1} may very well depend on all the other X_i, $i \le n$.

Definition 8.1 Let $\{X_0, X_1, ...\}$ be a stochastic process in discrete time with discrete state space $\mathbf{Z}$. Then $\{X_0, X_1, ...\}$ is a *discrete-time Markov chain* if for all vectors $x_0, x_1, ..., x_{n+1}$ with $x_k \in \mathbf{Z}$ and for all $n = 1, 2, ...,$

$$P(X_{n+1} = x_{n+1} \mid X_n = x_n, ..., X_1 = x_1, X_0 = x_0) = P(X_{n+1} = x_{n+1} \mid X_n = x_n). \quad (8.1)$$

●

Condition (8.1) is called the *Markov property*. It can be interpreted as follows: If time time point $t = n$ is the present, then $t = n + 1$ is a time point in the future, and the time points $t = n - 1$, ..., 1, 0 are in the past. Thus,

> *The future development of a discrete-time Markov chain depends only on its present state, but not on its evolution in the past.*

For the special class of stochastic processes considered in this chapter, definition 8.1 is equivalent to the definition of the Markov property via (6.23) at page 233. It usually requires much effort to check by statistical methods, whether a particular stochastic process has the Markov property (8.1). Hence one should first try to confirm or to reject this hypothesis by considering properties of the underlying technical, physical, economical, or other practical background. For instance, the final profit of a gambler usually depends on his present profit, but not on the way the gambler has obtained it. If it is known that at the end of the n th month a manufacturer has sold a total of $X_n = x_n$ personal computers, then for predicting the total number of computers X_{n+1} sold a month later knowledge about the number of computers sold within the first n months will make no difference. A car driver checks the tread depth of his tires after every $5000\, km$. For predicting the tread depth after a further $5000\ km$, the driver will only need the present tread depth, not how the tread depth has evolved to its present level. For predicting, however, the future concentration of noxious substances in the air, it has been proved necessary to take into account not only the present value of the concentration, but also the past development leading to this value. In this chapter it will be assumed that the state space of the Markov chain is $\mathbf{Z} = \{0, \pm 1, \pm 2, ...\}$ or a subset of it. Generally, states will be denoted as $i, j, k, ...$.

Transition Probabilities The conditional probabilities

$$p_{ij}(n) = P(X_{n+1} = j | X_n = i); \quad n = 0, 1, \dots$$

are the *one-step transition probabilities* of the Markov chain. A Markov chain is called *homogeneous* if it has homogeneous increments. Thus, a Markov chain is homogeneous if and only if its one-step transition probabilities do not depend on n:

$$p_{ij}(n) = p_{ij} \quad \text{for all} \quad n = 0, 1, \dots.$$

Note This chapter only deals with homogeneous Markov chains. For the sake of brevity, the attribute *homogeneous* is generally omitted.

The one-step transition probabilities are combined in the *matrix of the one-step transition probabilities* (shortly: *transition matrix*) **P**:

$$\mathbf{P} = \begin{pmatrix} p_{00} & p_{01} & p_{02} & \cdots \\ p_{10} & p_{11} & p_{12} & \cdots \\ \vdots & \vdots & \vdots & \cdots \\ p_{i0} & p_{i1} & p_{i2} & \cdots \\ \vdots & \vdots & \vdots & \cdots \end{pmatrix}.$$

p_{ij} is the probability of a transition from state i to state j in one step (or, equivalently, *in one time unit, in one jump*). With probability p_{ii} the Markov chain remains in state i for another time unit. The one-step transition probabilities have some obvious properties:

$$p_{ij} \geq 0, \quad \sum_{j \in \mathbf{Z}} p_{ij} = 1; \quad i, j \in \mathbf{Z}. \tag{8.2}$$

The *m-step transition probabilities* of a Markov chain are defined as

$$p_{ij}^{(m)} = P(X_{n+m} = j | X_n = i); \quad m = 1, 2, \dots. \tag{8.3}$$

Thus, $p_{ij}^{(m)}$ is the probability that the Markov chain, starting from state i, will be after m steps in state j. However, in between the Markov chain may already have arrived at state j. Note that $p_{ij} = p_{ij}^{(1)}$.

It is convenient to introduce the notation

$$p_{ij}^{(0)} = \delta_{ij} = \begin{cases} 1 & \text{if} \quad i = j, \\ 0 & \text{if} \quad i \neq j. \end{cases} \tag{8.4}$$

δ_{ij} defined in this way is called the *Kronecker symbol.*

The following relationship between the multi-step transition probabilities of a discrete-time Markov chain is called the

Chapman-Kolmogorov equations:

$$p_{ij}^{(m)} = \sum_{k \in \mathbf{Z}} p_{ik}^{(r)} p_{kj}^{(m-r)}; \quad r = 0, 1, ..., m. \tag{8.5}$$

The proof is easy: Conditioning with regard to the state, which the Markov chain assumes after r steps, $0 \le r \le m$, and making use of the Markov property yields

$$p_{ij}^{(m)} = P(X_m = j | X_0 = i) = \sum_{k \in \mathbf{Z}} P(X_m = j, X_r = k | X_0 = i)$$

$$= \sum_{k \in \mathbf{Z}} P(X_m = j | X_r = k, X_0 = i) P(X_r = k | X_0 = i)$$

$$= \sum_{k \in \mathbf{Z}} P(X_m = j | X_r = k) P(X_r = k | X_0 = i)$$

$$= \sum_{k \in \mathbf{Z}} p_{ik}^{(r)} p_{kj}^{(m-r)}.$$

This proves formula (8.5).

It simplifies notation, when introducing the *matrix of the m-step transition probabilities* of the Markov chain:

$$\mathbf{P}^{(m)} = \left(\left(p_{ij}^{(m)} \right) \right); \quad m = 0, 1, ...$$

Then the Chapman-Kolmogorov equations can be written in the elegant form

$$\mathbf{P}^{(m)} = \mathbf{P}^{(r)} \mathbf{P}^{(m-r)}; \quad r = 1, 2, ..., m.$$

This relationship implies that

$$\mathbf{P}^{(m)} = \mathbf{P}^m.$$

Thus, the matrix of the m-step transition probabilities is equal to the m-fold product of the matrix of the one-step transition probabilities.

A probability distribution $\mathbf{p}^{(0)}$ of X_0 is said to be an *initial distribution* of the Markov chain:

$$\mathbf{p}^{(0)} = \left\{ p_i^{(0)} = P(X_0 = i), \; i \in \mathbf{Z}, \; \sum_{i \in \mathbf{Z}} p_i^{(0)} = 1 \right\}. \tag{8.6}$$

A Markov chain is completely characterized by its transition matrix $\mathbf{P}$ and an initial distribution $\mathbf{p}^{(0)}$. In order to prove this one has to show that, given $\mathbf{P}$ and $\mathbf{p}^{(0)}$, all its finite-dimensional probabilities can be determined: By the Markov property, for any finite set of states $i_0, i_1, ..., i_n,$

$$P(X_0 = i_0, X_1 = i_1, ..., X_n = i_n)$$

$$= P(X_n = i_n | X_0 = i_0, X_1 = i_1, ..., X_{n-1} = i_{n-1}) \cdot P(X_0 = i_0, X_1 = i_1, ..., X_{n-1} = i_{n-1})$$

$$= P(X_n = i_n | X_{n-1} = i_{n-1}) \cdot P(X_0 = i_0, X_1 = i_1, ..., X_{n-1} = i_{n-1})$$

$$= p_{i_{n-1} i_n} \cdot P(X_0 = i_0, X_1 = i_1, ..., X_{n-1} = i_{n-1}).$$

The second factor in the last line is now treated in the same way. Continuing in this way yields

$$P(X_0 = i_0, X_1 = i_1, ..., X_n = i_n) = p_{i_0}^{(0)} \cdot p_{i_0 i_1} \cdot p_{i_1 i_2} \cdot \ \cdots \ p_{i_{n-1} i_n}. \qquad (8.7)$$

This proves the assertion. The *absolute* or *one-dimensional state probabilities* of the Markov chain after m steps are denoted as

$$p_j^{(m)} = P(X_m = j), \quad j \in \mathbf{Z}.$$

The set $\left\{ p_j^{(m)}, \ j \in \mathbf{Z} \right\}$ is the *absolute probability distribution of the Markov chain after m steps*, $m = 0, 1, ...$. Given an initial distribution $\mathbf{p}^{(0)} = \{ p_i^{(0)}, \ i \in \mathbf{Z} \}$, by the total probability rule,

$$p_j^{(m)} = \sum_{i \in \mathbf{Z}} p_i^{(0)} \, p_{ij}^{(m)}, \quad m = 1, 2, ... \qquad (8.8)$$

Definition 8.2 An initial distribution $\{ \pi_i = P(X_0 = i); \ i \in \mathbf{Z} \}$ is called *stationary* if it satisfies the system of linear equations

$$\pi_j = \sum_{i \in \mathbf{Z}} \pi_i \, p_{ij}; \quad j \in \mathbf{Z}, \qquad (8.9)$$

$$1 = \sum_{i \in \mathbf{Z}} \pi_i. \qquad (8.10)$$

●

It can be shown by induction that, starting with a stationary initial distribution, the absolute state distributions of the Markov chain for any number m of steps coincide with the stationary initial distribution, i.e., for all $j \in \mathbf{Z}$,

$$p_j^{(m)} = \sum_{i \in \mathbf{Z}} \pi_i \, p_{ij}^{(m)} = \pi_j, \quad m = 1, 2, ... \qquad (8.11)$$

In this case, the Markov chain is said to be in a *(global) state of equilibrium,* and the probabilities π_i are also called *equilibrium state probabilities* of the Markov chain.

If a stationary initial distribution exists, then the structure (8.7) of the n-dimensional state probabilities of the Markov chain verifies theorem 6.1:

| *A homogeneous Markov chain is strictly stationary if and only if its one-dimensional) absolute state probabilities do not depend on time.*

Markov chains in discrete time virtually occur in all fields of science, engineering, operations research, economics, risk analysis, and finance. In what follows, this will be illustrated by some examples.

Example 8.1 (*unbounded symmetric random walk*) A particle moves along the real axis in one step from an integer-valued coordinate i either to $i+1$ or to $i-1$ with equal probabilities. The steps occur independently of each other. If X_0 is the start position of the particle and X_n its position after n steps, then $\{ X_0, X_1, ... \}$ is a discrete-time Markov chain with state space $\mathbf{Z} = \{0, \pm 1, \pm 2, \cdots\}$ and one-step transition probabilities

$$p_{ij} = \begin{cases} 1/2 & \text{for } j = i+1 \text{ or } j = i-1 \\ 0 & \text{otherwise} \end{cases}.$$

It is quite intuitive that the unbounded symmetric random walk cannot have a stationary initial distribution. An exact argument will be given later. □

Example 8.2 (*random walk with reflecting barriers–Ehrenfest's diffusion model*)
For a given positive integer z, the state space of a Markov chain is $\mathbf{Z} = \{0, 1, \cdots, 2z\}$.
A particle moves from position i to position j in one step with probability

$$p_{ij} = \begin{cases} \frac{2z-i}{2z} & \text{for } j = i+1, \\ \frac{i}{2z} & \text{for } j = i-1, \\ 0 & \text{otherwise.} \end{cases} \tag{8.12}$$

Thus, the greater the distance of the particle from the central point z of $\mathbf{Z}$, the greater the probability that the particle moves in the next step into the direction of the central point. Once the particle has arrived at one of the end points $x = 0$ or $x = 2z$, it will return in the next step with probability 1 to position $x = 1$ or $x = 2z - 1$, respectively. (Hence the terminology *reflecting barriers*.) If the particle is at $x = z$, then the probabilities of moving to the left or to the right in the next step are equal, namely 1/2. In this sense, the particle is at $x = z$ in an *equilibrium state*. This situation may be thought of as caused by a force, which is situated at the central point. Its attraction to a particle linearly increases with the particle's distance from this point.

A stationary state distribution exists and satisfies the corresponding system of linear equations (8.9):

$$\pi_0 = \pi_1 p_{10},$$
$$\pi_j = \pi_{j-1} p_{j-1,j} + \pi_{j+1} p_{j+1,j}; \quad j = 1, 2, ..., 2z - 1,$$
$$\pi_{2z} = \pi_{2z-1} p_{2z-1,2z}.$$

The solution, taking into account the normalizing condition (8.10), is

$$\pi_j = \binom{2z}{j} 2^{-2z}; \quad j = 0, 1, ..., 2z.$$

As expected, state z has the greatest stationary probability.

P. and *T. Ehrenfest* (1907) came across this random walk with reflecting barriers when investigating the following diffusion model: In a closed container there are exactly $2z$ molecules of a particular type. The container is separated into two equal parts by a membrane, which is permeable to these molecules. Let X_n be the random number of molecules in one part of the container after n transitions of any molecule from one part of the container to the other one. If X_0 denotes the initial number of molecules in the specified part of the container, then they observed that the random sequence $\{X_0, X_1, ...\}$ behaves approximately as a Markov chain with transition pro-babilities (8.12). Hence, the more molecules are in one part of the container, the more they want to move into the other part. In other words, the system tends to the equilib-rium state, i.e. to equal numbers of particles in each part of the container.   □

Example 8.3 (*random walk with two absorbing barriers*) The movement of a particle within the state space $\mathbf{Z} = \{0, 1, \cdots, z\}$, $z > 1$, is controlled by a discrete-time Markov chain $\{X_0, X_1, ...\}$ with transition probabilities

$$p_{ij} = \begin{cases} p & \text{for } j = i+1, \ 1 \le i \le z-1, \\ q & \text{for } j = i-1, \ 1 \le i \le z-1, \\ 0 & \text{otherwise.} \end{cases}$$

Hence, $x = 0$ and $x = z$ are *absorbing states* ('barriers'), i.e., if the particle arrives at state 0 or at state z, it cannot leave these states anymore: $p_{00} = 1$, $p_{zz} = 1$. The matrix of the one-step transition probabilities is

$$\mathbf{P} = \begin{pmatrix} 1 & 0 & 0 & 0 & 0 & \cdots & 0 \\ q & 0 & p & 0 & 0 & \cdots & 0 \\ 0 & q & 0 & p & 0 & \cdots & 0 \\ 0 & 0 & q & 0 & p & 0 & \cdots \\ 0 & 0 & 0 & q & 0 & p & 0 \\ \vdots & \vdots & \vdots & \vdots & \vdots & \vdots & \vdots \\ 0 & 0 & 0 & 0 & 0 & 0 & 1 \end{pmatrix}.$$

This random walk cannot have a stationary initial distribution, since given any initial distribution the Markov chain will arrive at an absorbing barrier with probability 1 in finite time.

Absorption It is an interesting and important exercise to determine the probabilities of absorption of the particle at $x = 0$ and $x = z$, respectively. Let $a(n)$ be the probabil-ity of absorption at $x = 0$ if the particle starts moving from $x = n$, $0 < n < z$. On condi-tion that the particle moves from n to the right, its absorption probability at $x = 0$ is $a(n+1)$ if $n+1 < z$. On condition that the particle moves from n to the left, the ab-sorption probability at $x = 0$ is $a(n-1)$ if $n-1 \ge 0$. Hence, in view of the formula of total probability (1.24), $a(n)$ satisfies the system of linear equations

$$a(n) = p \cdot a(n+1) + q \cdot a(n-1); \quad n = 1, 2, \cdots, z-1. \tag{8.13}$$

The boundary conditions are

$$a(0) = 1, \quad a(z) = 0. \tag{8.14}$$

Replacing $a(n)$ in (8.13) with $p\, a(n) + q\, a(n)$ yields the following algebraic system of equations for the $a(n)$:

$$[a(n) - a(n+1)] = \frac{q}{p}[a(n-1) - a(n)], \quad n = 1, 2, ..., z-1. \tag{8.15}$$

Starting with $n = 1$, repeated application of (8.15) yields

$$
\begin{aligned}
a(0) - a(1) &= \quad\quad [1 - a(1)] \\
a(1) - a(2) &= (q/p)[1 - a(1)] \\
a(2) - a(3) &= (q/p)^2 [1 - a(1)] \\
&\quad\vdots \\
a(z-1) - a(z) &= (q/p)^{z-1} [1 - a(1)].
\end{aligned}
\tag{8.16}
$$

By taking into account the boundary conditions (8.14),

$$\sum_{n=1}^{z}[a(n-1) - a(n)]$$

$$= [1 - a(1)] + [a(1) - a(2)] + \cdots + [a(z-2) - a(z-1)] + [a(z-1) - 0] = 1. \quad 8.17)$$

Using the finite geometrical series (2.18) at page 48, equations (8.16) yield for $p \neq q$

$$\sum_{n=1}^{z} [a(n-1) - a(n)] = [1 - a(1)] \sum_{n=1}^{z} (q/p)^{n-1} = [1 - a(1)] \frac{1 - (q/p)^z}{1 - q/p} = 1.$$

Solving this equation for $a(1)$ gives

$$a(1) = \frac{(q/p)^z - q/p}{(q/p)^z - 1}.$$

Starting with $a(0) = 1$ and $a(1)$, the systems of equations (8.16) or (8.13), respectively, provide the complete set of absorption probabilities at state 0:

$$a(n) = \frac{(q/p)^z - (q/p)^n}{(q/p)^z - 1}, \quad n = 1, 2, ..., z, \ p \neq q. \tag{8.18}$$

If $p = q = 1/2$, equations (8.16) show that all the differences $a(n-1) - a(n)$ are equal to $1 - a(1)$. Hence, equation (8.17) implies

$$a(n) = 1 - \frac{n}{z} = \frac{z-n}{z}, \quad n = 0, 1, ..., z, \ p = 1/2.$$

The absorption probabilities $b(n)$ of the particle at state z, when starting from state n, are given by

$$b(n) = 1 - a(n), \quad n = 0, 1, 2, ..., z.$$

Time till absorption Let $m(n)$ be the mean time till the particle reaches one of the absorbing states 0 or z, when starting from state n, $1 \leq n \leq z - 1$. If the first jump goes from the starting point n to the right, then the mean time till absorption is $1 + m(n+1)$. When the first jump goes to the left, then the meantime till absorption is $1 + m(n-1)$. Hence, the $m(n)$ satisfy the system of equations

$$m(n) = p[1 + m(n+1)] + q[1 + m(n-1)]; \quad n = 1, 2, ..., z-1, \tag{8.19}$$

with the boundary conditions

$$m(0) = m(z) = 0.$$

(8.19) is equivalent to

$$[m(n) - m(n+1)] = \frac{q}{p}[m(n-1) - m(n)] + \frac{1}{p}, \quad n = 1, 2, ..., z-1,$$

This system of equations can be solved analogously to (8.15), see exercise 8.7. (The random walk considered here is actually a special case of a more general one analysed in detail in section 8.4.2, page 365.) Taking into account the boundary conditions $m(0) = m(z) = 0$, its solution is for $n = 1, 2, ..., z-1$

$$m(n) = \frac{1}{p-q}\left[z\left(\frac{1-(q/p)^n}{1-(q/p)^z}\right) - n\right] \quad \text{if} \ \ p \neq q, \tag{8.20}$$

$$m(n) = n(z-n) \quad \text{if} \ \ p = q = 1/2$$

Table 8.1 shows some numerical results. In particular for large z, even small changes in p have a significant impact on the absorption probabilities.

z	10			80		
p	$a(5)$	$b(5)$	$m(5)$	$a(40)$	$b(40)$	$m(40)$
0.50	0.500	0.500	25	0.500	0.500	1600
0.51	0.450	0.550	24.9	0.168	0.832	1328
0.52	0.401	0.599	24.7	0.039	0.961	922

Table 8.1 Probabilities and mean times to absorption for example 8.3

Gambler's ruin: The random walk with two absorbing barriers has a famous interpretation: A gambler has an initial capital of $\$n$. After each game his capital has increased by $\$1$ with probability p (win) or decreased by $\$1$ (loss) with probability q. The gambler has decided to stop gambling when having lost his initial capital or when having reached a total capital of $\$z$, $0 < n < z$. When following this strategy, the gambler will lose all of his initial capital with probability $a(n)$ given by (8.18) or will walk away with a total capital of z with probability $b(n) = 1 - a(n)$. ☐

Example 8.4 (*electron orbits*) Depending on its energy, an electron circles around the atomic nucleus in one of the countably infinite sets of trajectories $\{1, 2, ...\}$. The one-step transition from trajectory i to trajectory j occurs with probability

$$p_{ij} = a_i e^{-b|i-j|}, \ \ b > 0.$$

Hence, the two-step transition probabilities are

$$p_{ij}^{(2)} = a_i \sum_{k=1}^{\infty} a_k e^{-b(|i-k|+|k-j|)}.$$

The a_i cannot be chosen arbitrarily. In view of (8.2), they must satisfy the condition

$$a_i \left(e^{-b(i-1)} + e^{-b(i-2)} + \cdots + e^{-b} \right) + a_i \sum_{k=0}^{\infty} e^{-bk} = 1,$$

or, equivalently,
$$a_i \left(e^{-b} \frac{1 - e^{-b(i-1)}}{1 - e^{-b}} + \frac{1}{1 - e^{-b}} \right) = 1.$$

Therefore,
$$a_i = \frac{e^b - 1}{1 + e^b - e^{-b(i-1)}}; \quad i = 1, 2, \dots.$$

The structure of the p_{ij} implies that $a_i = p_{ii}$ for all $i = 1, 2, \dots$. $\quad\square$

Example 8.5 (dynamics of traffic accidents) Let X_n denote the number of traffic accidents over a period of n weeks in a particular area, and let Y_i be the corresponding number in the ith week. Then, $X_n = \sum_{i=1}^{n} Y_i$.

The Y_i are assumed to be independent and identically distributed as a random variable Y with probability distribution $\{q_k = P(Y = k); \ k = 0, 1, \dots\}$. Then $\{X_1, X_2, \dots\}$ is a Markov chain with state space $\mathbf{Z} = \{0, 1, \dots\}$ and transition probabilities

$$p_{ij} = \begin{cases} q_k & \text{if } j = i + k; \ k = 0, 1, \dots, \\ 0 & \text{otherwise.} \end{cases} \quad\square$$

Example 8.6 (reproduction of diploid cells) Chromosomes determine the hereditary features of higher organisms. Essentially they consist of strings of genes. The position of a gene within a chromosome is called its *locus*. The different types of genes, which can be found at a locus, are called *alleles*. The chromosomes of mammals occur in pairs (two strings of chromosomes 'in parallel', i.e. *diploid* chromosomes). If, in the diploid case, the possible alleles are g and G, then at a locus the combinations (g,g), (g,G), or (G,G) are possible. Such a combination is called a *genotype*. Note that

$$(g, G) = (G, g).$$

Consider a one-sex population with an infinite (very large) number of individuals. All of them have genotype (g,g), (g, G), or (G,G). Each individual is equally likely to pair with any other member of the population, and, when pairing, each individual randomly gives one of its alleles to its offspring. Genotypes (g,g) and (G,G) can only contribute g or G, respectively, whereas (g, G) contributes g or G with probability $1/2$ each to the offspring.

Let $\alpha_0, \beta_0,$ and γ_0 with $\alpha_0 + \beta_0 + \gamma_0 = 1$ be the probabilities that an individual, randomly selected from the first generation, belongs to genotype (g,g), (g, G), or (G,G), respectively. By the formula of the total probability, a randomly chosen allele from the first generation is of type g with probability

$$P_1(g) = P_1(g|gg)\alpha_0 + P_1(g|gG)\beta_0 + P_1(g|GG)\gamma_0 = \alpha_0 + \beta_0/2,$$

since $P_1(g|gg) = 1$, $P_1(g|gG) = 1/2$, and $P_1(g|GG) = 0$.

By changing the roles of g and G,

$$P_1(G) = P_1(G|GG)\gamma_0 + P_1(G|gG)\beta_0 + P_1(G|gg)\alpha_0 = \gamma_0 + \beta_0/2.$$

Hence, a randomly selected individual of the second generation has genotype (g,g), (g,G), or (G,G) with respective probabilities α, β, and γ, $\alpha + \beta + \gamma = 1$, given by

$$\alpha = (\alpha_0 + \beta_0/2)^2,$$
$$\beta = 2(\alpha_0 + \beta_0/2)(\gamma_0 + \beta_0/2), \qquad (8.21)$$
$$\gamma = (\gamma_0 + \beta_0/2)^2,$$

Thus, the respective probabilities that a randomly from the second generation chosen allele is of type g or G are

$$P_2(g) = \alpha + \beta/2 = (\alpha_0 + \beta_0/2)^2 + (\alpha_0 + \beta_0/2)(\gamma_0 + \beta_0/2) = \alpha_0 + \beta_0/2 = P_1(g),$$
$$P_2(G) = \gamma + \beta/2 = (\gamma_0 + \beta_0/2)^2 + (\alpha_0 + \beta_0/2)(\gamma_0 + \beta_0/2) = \gamma_0 + \beta_0/2 = P_1(G).$$

Corollary Under the assumption of random mating, the respective percentages of the population belonging to genotype (g,g), (g, G), or (G,G) stay at levels $\alpha\,[100\%]$, $\beta\,[100\%]$, and $\gamma\,[100\%]$ in all successive generations.

In the literature on population genetics, this result is known as the ***Hardy-Weinberg law***; see *Hardy* (1908). A relationship between this law and discrete-time Markov chains is readily established: Let X_2 be the genotype of a randomly from the second generation chosen individual, and $X_3, X_4, \ldots$ be the genotypes of its offspring in the following generations. Then the state space of the Markov chain $\{X_2, X_3, \ldots\}$ is

$$Z = \{z_1 = gg, \ z_2 = gG, \ z_3 = GG\}$$

with the absolute state probabilities

$$\alpha = P(X_i = z_1), \quad \beta = P(X_i = z_2), \quad \gamma = P(X_i = z_3), \quad i = 2, 3, \ldots.$$

The one-step transition probabilities p_{ij}, $i, j = 1, 2, 3$, are determined by conditioning with regard to the genotype M of the randomly selected mate, e.g.:

$$p_{11} = (p_{11}|M = z_1) \cdot P(M = z_1) + (p_{11}|M = z_2) \cdot P(M = z_2) + (p_{11}|M = z_3) \cdot P(M = z_3)$$
$$= 1 \cdot \alpha + \beta/2 + 0 \cdot \gamma = \alpha + \beta/2.$$

$$p_{12} = (p_{12}|M = z_1) \cdot P(M = z_1) + (p_{12}|M = z_2) \cdot P(M = z_2) + (p_{12}|M = z_3) \cdot P(M = z_3)$$
$$= 0 + \beta/2 + \gamma = \gamma + \beta/2.$$

$$p_{13} = 1 - p_{11} - p_{12} = 1 - \alpha - \beta/2 - \gamma - \beta/2 = 0.$$

$$p_{21} = (p_{21}|M = z_1) \cdot P(M = z_1) + (p_{21}|M = z_2) \cdot P(M = z_2) + (p_{21}|M = z_3) \cdot P(M = z_3)$$
$$= \alpha/2 + \beta/4 + 0 \cdot \gamma = \alpha/2 + \beta/4.$$

$$p_{22} = \alpha/2 + \beta/2 + \gamma/2 = 1/2 \ (\text{since } \alpha + \beta + \gamma = 1).$$

$$p_{23} = 1 - p_{21} - p_{22} = 1 - \alpha/2 - \beta/4 - 1/2 = \beta/4 + \gamma/2.$$

The complete one-step transition matrix of the Markov chain $\{X_2, X_3, ...\}$ is

$$\begin{pmatrix} \alpha + \beta/2 & \gamma + \beta/2 & 0 \\ \alpha/2 + \beta/4 & 1/2 & \beta/4 + \gamma/2 \\ 0 & \alpha + \beta/2 & \gamma + \beta/2 \end{pmatrix}. \tag{8.22}$$

In view of its property to generate the same absolute state probabilities in all generations following the first one,

$$\pi = \{\pi_1 = \alpha, \ \pi_2 = \beta, \ \pi_3 = \gamma\}$$

is a stationary initial distribution of the homogeneous Markov chain $\{X_2, X_3, ...\}$. This can be verified by showing that π satisfies the system of linear equations (8.9) with the transition matrix (8.22), making use of (8.21). $\square$

Example 8.7 (*sequence of moving averages*) Let $\{Y_i; \ i = 0, 1, ...\}$ be a sequence of independent, identically distributed binary random variables with

$$P(Y_i = 1) = P(Y_i = -1) = 1/2.$$

Moving averages X_n are defined as follows (see also page 240):

$$X_n = \tfrac{1}{2}(Y_n + Y_{n-1}); \ \ n = 1, 2, ...$$

X_n has range $\{-1, 0, +1\}$ and probability distribution

$$\left\{ P(X_n = -1) = \tfrac{1}{4}, \ P(X_n = 0) = \tfrac{1}{2}, \ P(X_n = +1) = \tfrac{1}{4} \right\}.$$

Since X_n and X_{n+m} are independent for $m > 1$, the corresponding matrix of the m-step transition probabilities

$$p_{ij}^{(m)} = P(X_{n+m} = j | X_n = i)$$

is

$$\mathbf{P}^{(m)} = \begin{pmatrix} 1/4 & 1/2 & 1/4 \\ 1/4 & 1/2 & 1/4 \\ 1/4 & 1/2 & 1/4 \end{pmatrix}.$$

The matrix of the one-step transition probabilities $p_{ij} = P(X_{n+1} = j | X_n = i)$ is

$$\mathbf{P}^{(1)} = \mathbf{P} = \begin{pmatrix} 1/2 & 1/2 & 0 \\ 1/4 & 1/2 & 1/4 \\ 0 & 1/2 & 1/2 \end{pmatrix}.$$

Since

$$\mathbf{P}^{(1)} \cdot \mathbf{P}^{(1)} \neq \mathbf{P}^{(2)},$$

the Chapman-Kolmogorov equations do not hold. Therefore, the sequence of moving averages $\{X_1, X_2, ...\}$ cannot be a Markov chain. $\square$

8.2   CLASSIFICATION OF STATES

8.2.1 Closed Sets of States

A subset C of the state space Z of a Markov chain is said to be *closed* if

$$\sum_{j \in C} p_{ij} = 1 \quad \text{for all } i \in C. \tag{8.23}$$

If a Markov chain is in a closed set of states, then it cannot leave this set since (8.23) is equivalent to $p_{ij} = 0$ for all $i \in C$, $j \notin C$. Furthermore, (8.23) implies that

$$p_{ij}^{(m)} = 0 \quad \text{for all } i \in C, \ j \notin C \text{ and } m \geq 1. \tag{8.24}$$

For $m = 2$ formula (8.24) can be proved as follows: From (8.5),

$$p_{ij}^{(2)} = \sum_{k \in C} p_{ik} p_{kj} + \sum_{k \notin C} p_{ik} p_{kj} = 0,$$

since $j \notin C$ implies $p_{kj} = 0$ in the first sum and $p_{ik} = 0$ in the second sum. Now formula (8.24) follows inductively from the Chapman-Kolmogorov equations.

A closed set of states is called *minimal* if it does not contain a proper closed subset. In particular, a Markov chain is said to be *irreducible* if its state space Z is minimal. Otherwise the Markov chain is *reducible*.

A state i is said to be absorbing if $p_{ii} = 1$. Thus, if a Markov chain has arrived in at absorbing state, it cannot leave this state anymore. Hence, an absorbing state is a minimal closed set of states. Absorbing barriers of a random walk (example 8.3) are absorbing states.

Example 8.8 Let $Z = \{1,2,3,4,5\}$ be the state space of a Markov chain with transition matrix

$$P = \begin{pmatrix} 0.2 & 0 & 0.5 & 0.3 & 0 \\ 0.1 & 0 & 0.9 & 0 & 0 \\ 0 & 1 & 0 & 0 & 0 \\ 0.4 & 0.1 & 0.2 & 0 & 0.3 \\ 0 & 0 & 0 & 0 & 1 \end{pmatrix}.$$

It is helpful to illustrate the possible transitions between the states of a Markov chain by *transition graphs*. The nodes of these graphs represent the states of the Markov chain. A directed edge from node i to node j exists if and only if $p_{ij} > 0$, that is if a one-step transition from state i to state j is possible. The corresponding one-step transition probabilities are attached to the edges. Figure 8.1 shows that $\{1,2,3,4\}$ is not a closed set of states since condition (8.24) is not fulfilled for $i = 4$. State 5 is absorbing so that $\{5\}$ is a minimal closed set of states. This Markov chain is, therefore, reducible. □

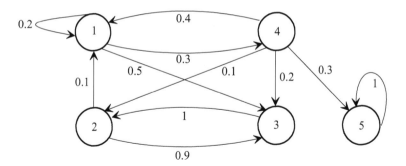

Figure 8.1 Transition graph in example 8.8

8.2.2 Equivalence Classes

State j is said to be *accessible* from state i (symbolically: $i \Rightarrow j$) if there exists an $m \geq 1$ such that $p_{ij}^{(m)} > 0$. The relation '$\Rightarrow$' is transitive:

If $i \Rightarrow k$ and $k \Rightarrow j$, there exist $m > 0$ and $n > 0$ with $p_{ik}^{(m)} > 0$ and $p_{kj}^{(n)} > 0$. Hence,

$$p_{ij}^{(m+n)} = \sum_{r \in \mathbf{Z}} p_{ir}^{(m)} p_{rj}^{(n)} \geq p_{ik}^{(m)} p_{kj}^{(n)} > 0.$$

Consequently, $i \Rightarrow k$ and $k \Rightarrow j$ imply $i \Rightarrow j$, i.e., the transitivity of '$\Rightarrow$.'

The set $\mathbf{M}(i) = \{k, \ i \Rightarrow k\}$ consisting of all those states which are accessible from i is closed. To prove this assertion it is to show that $k \in \mathbf{M}(i)$ and $j \notin \mathbf{M}(i)$ imply $k \not\Rightarrow j$. The proof is carried out indirectly: If under the assumptions stated $k \Rightarrow j$, then $i \Rightarrow k$ and the transitivity would imply $i \Rightarrow j$. But this contradicts the definition of $\mathbf{M}(i)$.

If both $i \Rightarrow j$ and $j \Rightarrow i$ hold, then i and j are said to *communicate* (symbolically: $i \Leftrightarrow j$). Communication '$\Leftrightarrow$' is an *equivalence relation* since it satisfies the three characteristic properties:

(1) $i \Leftrightarrow i$. *reflexivity*

(2) If $i \Leftrightarrow j$, then $j \Leftrightarrow i$. *commutativity*

(3) If $i \Leftrightarrow j$ and $j \Leftrightarrow k$, then $i \Leftrightarrow k$. *associativity*

Properties (1) and (2) are an immediate consequence of the definition of " $\Leftrightarrow$ ". To verify property (3), note that $i \Leftrightarrow j$ and $j \Leftrightarrow k$ imply the existence of m and n so that $p_{ij}^{(m)} > 0$ and $p_{jk}^{(n)} > 0$, respectively. Hence, by (8.5),

$$p_{ik}^{(m+n)} = \sum_{r \in \mathbf{Z}} p_{ir}^{(m)} p_{rk}^{(n)} \geq p_{ij}^{(m)} p_{jk}^{(n)} > 0.$$

Likewise, there exist M and N with

$$p_{ki}^{(M+N)} \geq p_{kj}^{(M)} p_{ji}^{(N)} > 0$$

so that the associativity is proved.

The equivalence relation '$\Leftrightarrow$' partitions state space $\mathbf{Z}$ into disjoint, but not necessarily closed classes in the following way: *Two states i and j belong to the same class if and only if they communicate.* In what follows, the class containing state i is denoted as $\mathbf{C}(i)$. Clearly, any state in a class can be used to characterize this class. All properties of states introduced in what follows will be *class properties*, i.e. if state i has one of these properties, all states in $\mathbf{C}(i)$ have this property as well.

A state i is called *essential* if any state j which is accessible from i has the property that i is also accessible from j. In this case, $\mathbf{C}(i)$ is called an *essential class*.

A state i is called *inessential* if it is not essential. In this case, $\mathbf{C}(i)$ is called an *inessential class*. If i is inessential, then there exists a state j for which $i \Rightarrow j$ and $j \not\Rightarrow i$.

It is easily verified that *essential* and *inessential* are indeed class properties. In example 8.8, the states 1, 2, 3 and 4 are inessential since state 5 is accessible from each of these states but none of the states 1, 2, 3 or 4 is accessible from state 5.

Theorem 8.1 (1) Essential classes are minimal closed classes. (2) Inessential classes are not closed.

Proof (1) The assertion is a direct consequence of the definition of essential classes. (2) If i is inessential, then there is a state j with $i \Rightarrow j$ and $j \not\Rightarrow i$. Hence, $j \notin \mathbf{C}(i)$.

Assuming $\mathbf{C}(i)$ is closed implies that $p_{kj}^{(m)} = 0$ for all $m \geq 1$, $k \in \mathbf{C}(i)$ and $j \notin \mathbf{C}(i)$.

Therefore, $\mathbf{C}(i)$ cannot be closed. (According to the definition of the relation $i \Rightarrow j$, there exists a positive integer m with $p_{ij}^{(m)} > 0$.) ∎

Let $p_i^{(m)}(\mathbf{C})$ be the probability that the Markov chain, starting from state i, is in state set $\mathbf{C}$ after m time units:

$$p_i^{(m)}(\mathbf{C}) = \sum_{j \in \mathbf{C}} p_{ij}^{(m)}.$$

Furthermore, let $\mathbf{C}_w$ and $\mathbf{C}_u$ be the sets of all essential and inessential states of a Markov chain. The following theorem asserts that a Markov chain with finite state space, which starts from an inessential state, will leave the set of inessential states with probability 1 and never return (for a proof see e.g. *Chung* (1960)). This theorem justifies the notation *essential* and *inessential states*. However, depending on the transition probabilites, the Markov chain may in the initial phase return more or less frequently to the set of inessential states if it has started there.

Theorem 8.2 Let the state space set $\mathbf{Z}$ be finite. Then,

$$\lim_{m \to \infty} p_i^{(m)}(\mathbf{C}_u) = 0.$$ ∎

8.2.3 Periodicity

Let d_i be the greatest common divisor of those indices $m \geq 1$ for which $p_{ii}^{(m)} > 0$. Then d_i is said to be the *period* of state i. If

$$p_{ii}^{(m)} = 0 \text{ for all } m > 0,$$

then the period of i is defined to be infinite. A state i is said to be *aperiodic* if $d_i = 1$.

If i has period d_i, then $p_{ii}^{(m)} > 0$ holds if and only if m can be represented in the form $m = n \cdot d_i$; $n = 1, 2, \ldots$. Hence, returning to state i is only possible after such a number of steps which is a multiple of d_i. The following theorem shows that the period is a class property.

Theorem 8.3 All states of a class have the same period.

Proof Let $i \Leftrightarrow j$. Then there exist integers m and n with $p_{ij}^{(m)} > 0$ and $p_{ji}^{(n)} > 0$. If the inequality $p_{ii}^{(r)} > 0$ holds for a positive integer r, then, from (8.5),

$$p_{jj}^{(n+r+m)} \geq p_{ji}^{(n)} p_{ii}^{(r)} p_{ij}^{(m)} > 0.$$

Since $p_{ii}^{(2r)} \geq p_{ii}^{(r)} \cdot p_{ii}^{(r)} > 0$, this inequality also holds if r is replaced with $2r$:

$$p_{jj}^{(n+2r+m)} > 0.$$

Thus, d_j divides the difference $(n + 2r + m) - (n + r + m) = r$. Since this holds for all r for which $p_{ii}^{(r)} > 0$, d_j must divide d_i. Changing the roles of i and j shows that d_i also divides d_j. Thus, $d_i = d_j$, which completes the proof. ■

Example 8.9 A Markov chain has state space $Z = \{0, 1, \ldots, 6\}$ and transition matrix

$$P = \begin{pmatrix} 1/3 & 2/3 & 0 & 0 & 0 & 0 & 0 \\ 1/3 & 1/3 & 1/3 & 0 & 0 & 0 & 0 \\ 1 & 0 & 0 & 0 & 0 & 0 & 0 \\ 0 & 1/3 & 0 & 1/3 & 1/3 & 0 & 0 \\ 0 & 0 & 0 & 0 & 1 & 0 & 0 \\ 0 & 0 & 0 & 0 & 0 & 1/2 & 1/2 \\ 0 & 0 & 0 & 0 & 1/2 & 0 & 1/2 \end{pmatrix}.$$

Clearly, $\{0, 1, 2\}$ is a closed set of essential states. State 4 is absorbing, so $\{4\}$ is another closed set. Having once arrived in a closed set of states the Markov chain cannot leave it anymore. $\{3, 5, 6\}$ is a set of inessential states. When starting in one of its sets of inessential states, the Markov chain will at some stage leave this set and never return. All states in $\{0, 1, 2\}$ have period 1. □

Theorem 8.4 (*Chung* (1960)) The state space **Z** of an irreducible Markov chain with period $d > 1$ can be partitioned into disjoint subsets $\mathbf{Z}_1, \mathbf{Z}_1, ..., \mathbf{Z}_d$ in such a way that from any state $i \in \mathbf{Z}_k$ a transition can only be made to a state $j \in \mathbf{Z}_{k+1}$. (By agreement, $j \in \mathbf{Z}_1$ if $i \in \mathbf{Z}_d$.) ∎

Example 8.10 Theorem 8.4 is illustrated by a discrete-time Markov chain with state space $\mathbf{Z} = \{0, 1, ..., 5\}$ and transition matrix

$$\mathbf{P} = \begin{pmatrix} 0 & 0 & 2/5 & 3/5 & 0 & 0 \\ 0 & 0 & 1 & 0 & 0 & 0 \\ 0 & 0 & 0 & 0 & 1/2 & 1/2 \\ 0 & 0 & 0 & 0 & 2/3 & 1/3 \\ 1/2 & 1/2 & 0 & 0 & 0 & 0 \\ 1/4 & 3/4 & 0 & 0 & 0 & 0 \end{pmatrix}.$$

This Markov chain has period $d = 3$. One-step transitions between the states are possible in the order $\mathbf{Z}_1 = \{0, 1\} \rightarrow \mathbf{Z}_2 = \{2, 3\} \rightarrow \mathbf{Z}_1 = \{4, 5\} \rightarrow \mathbf{Z}_1$. The three-step transition matrix $\mathbf{P}^{(3)} = \mathbf{P}^3$ is

$$\mathbf{P}^{(3)} = \begin{pmatrix} 2/5 & 3/5 & 0 & 0 & 0 & 0 \\ 3/8 & 5/8 & 0 & 0 & 0 & 0 \\ 0 & 0 & 31/40 & 9/40 & 0 & 0 \\ 0 & 0 & 3/4 & 1/4 & 0 & 0 \\ 0 & 0 & 0 & 0 & 11/20 & 9/20 \\ 0 & 0 & 0 & 0 & 21/40 & 19/40 \end{pmatrix}.$$
□

8.2.4 Recurrence and Transience

This section deals with the return of a Markov chain to an initial state. Such returns are controlled by the *first-passage time probabilities*

$$f_{ij}^{(m)} = P(X_m = j; X_k \neq j; k = 1, 2, ..., m - 1 | X_0 = i); \quad i, j \in \mathbf{Z}.$$

Thus, $f_{ij}^{(m)}$ is the probability that the Markov chain, starting from state i, makes its first transition into state j after m steps. Recall that $p_{ij}^{(m)}$ is the probability that the Markov chain, starting from state i, is in state j after m steps, but it may have been in state j in between. For $m = 1$,

$$f_{ij}^{(1)} = p_{ij}^{(1)}.$$

The total probability rule yields a relationship between the m-step transition probabilities and the first-passage time probabilities

$$p_{ij}^{(m)} = \sum_{k=1}^{m} f_{ij}^{(k)} p_{jj}^{(m-k)} ,$$

where, by convention

$$p_{ij}^{(0)} = 1 \text{ for all } j \in \mathbf{Z}.$$

Thus, the first-passage time probability can be determined recursively from the following formula

$$f_{ij}^{(m)} = p_{ij}^{(m)} - \sum_{k=1}^{m-1} f_{ij}^{(k)} p_{jj}^{(m-k)} ; \quad m = 2, 3, \dots . \tag{8.25}$$

The random variable Y_{ij} with probability distribution $\left\{ f_{ij}^{(m)} ; m = 1, 2, \dots \right\}$ is a *first-passage time*. Its mean value is

$$\mu_{ij} = E(Y_{ij}) = \sum_{m=1}^{\infty} m f_{ij}^{(m)}.$$

The probability of ever making a transition into state j if the process starts in state i is

$$f_{ij} = \sum_{m=1}^{\infty} f_{ij}^{(m)} . \tag{8.26}$$

In particular, f_{ii} is the probability of ever returning to state i. This motivates the introduction of the following concepts:

| A state i is said to be *recurrent* if $f_{ii} = 1$ and *transient* if $f_{ii} < 1$.

Clearly, if state i is transient, then $\mu_{ii} = \infty$. But, if i is recurrent, then $\mu_{ii} = \infty$ is also possible. Therefore, recurrent states are subdivided as follows:

| A recurrent state i is said to be *positive recurrent* if $\mu_{ii} < \infty$ and *null recurrent* if $\mu_{ii} = \infty$. An aperiodic and positive recurrent state is called *ergodic*.

The random time points

$$T_{i,n} ; \quad n = 1, 2, \dots ,$$

at which the nth return into starting state i occurs, are *renewal points* within a Markov chain. By convention, $T_{i,0} = 0$. The time spans between neighboring renewal points

$$T_{i,n} - T_{i,n-1} ; \quad n = 1, 2, \dots$$

are called *recurrence times*. They are independent and identically distributed as Y_{ii}. Therefore, the sequence of recurrence times constitutes an ordinary renewal process. Let

$$N_i(t) = \max(n; T_{i,n} \le t) \text{ and } N_i(\infty) = \lim_{t \to \infty} N_i(t)$$

with corresponding mean values

$$H_i(t) = E(N_i(t)) \text{ and } H_i(\infty) = \lim_{t \to \infty} H_i(t).$$

Theorem 8.5 State i is recurrent if and only if

(1) $H_i(\infty) = \infty$, or

(2) $\sum_{m=1}^{\infty} p_{ii}^{(m)} = \infty$.

Proof (1) If i is recurrent, then $P(T_{i,n} = \infty) = 0$ for $n = 1, 2, ...$. The limit $N_i(\infty)$ is finite if and only if there is a finite n with $T_{i,n} = \infty$. Therefore,

$$P(N_i(\infty) < \infty) \leq \sum_{i=1}^{\infty} P(T_{i,n} = \infty) = 0.$$

Thus, assumption $f_{ii} = 1$ implies $N_i(\infty) = \infty$ with probability 1 so that $H_i(\infty) = \infty$.

On the other hand, if $f_{ii} < 1$, then the Markov chain will not return to state i with positive probability $1 - f_{ii}$. In this case $N_i(\infty)$ has a geometric distribution with mean value

$$E(N_i(\infty)) = H_i(\infty) = \frac{f_{ii}}{1 - f_{ii}} < \infty.$$

Both results together prove part (1) of the theorem.

(2) Let the indicator variable for the random event that the Markov chain is in state i at time $t = m$ be

$$I_{m,i} = \begin{cases} 1 & \text{for } X_m = i, \\ 0 & \text{for } X_m \neq i, \end{cases} \qquad m = 1, 2, ...$$

Then,

$$N_i(\infty) = \sum_{m=1}^{\infty} I_{m,i} .$$

Hence,

$$H_i(\infty) = E\left(\sum_{m=1}^{\infty} I_{m,i} \right)$$
$$= \sum_{m=1}^{\infty} E(I_{m,i})$$
$$= \sum_{m=1}^{\infty} P(I_{m,i} = 1)$$
$$= \sum_{m=1}^{\infty} p_{ii}^{(m)} .$$

Now assertion (2) follows from (1). ∎

By adding up both sides of (8.25) from $m = 1$ to ∞ and changing the order of summation according to formula (2.115) at page 99, theorem 8.5 implies the

Corollary If state j is transient, then, for any $i \in \mathbf{Z}$,

$$\sum_{m=1}^{\infty} p_{ij}^{(m)} < \infty,$$

and, therefore,

$$\lim_{m \to \infty} p_{ij}^{(m)} = 0. \tag{8.27}$$

Theorem 8.6 Let i be a recurrent state and $i \Leftrightarrow j$. Then state j is recurrent, too.

Proof By definition of the equivalence relation "$i \Leftrightarrow j$", there are integers m and n with

$$p_{ij}^{(m)} > 0 \text{ and } p_{ji}^{(n)} > 0.$$

By (8.5),

$$p_{jj}^{n+r+m} \geq p_{ji}^{(n)} p_{ii}^{(r)} p_{ij}^{(m)}$$

so that

$$\sum_{r=1}^{\infty} p_{jj}^{n+r+m} \geq p_{ij}^{(m)} p_{ji}^{(n)} \sum_{r=1}^{\infty} p_{ii}^{(r)} = \infty.$$

The assertion is now a consequence of theorem 8.5. ∎

Corollary Recurrence and transience are class properties. Hence, an irreducible Markov chain is either *recurrent* or *transient*. In particular, an irreducible Markov chain with finite state space is recurrent.

It is easy to see that an inessential state is transient. Therefore, each recurrent state is essential. But not each essential state is recurrent. This assertion is proved by the following example.

Example 8.11 (*unbounded random walk*) Starting from $x = 0$, a particle jumps a unit distance along the x-axis to the right with probability p or to the left with probability $1 - p$. The transitions occur independently of each other. Let X_n denote the location of the particle after the nth jump under the initial condition $X_0 = 0$. Then the Markov chain $\{X_0, X_1, ...\}$ has period $d = 2$. Thus,

$$p_{00}^{(2m+1)} = 0; \quad m = 0, 1, ...$$

To return to state $x = 0$ after $2m$ steps, the particle must jump m times to the left and m times to the right. There are $\binom{2m}{m}$ sample paths which satisfy this condition. Hence,

$$p_{00}^{(2m)} = \binom{2m}{m} p^m (1-p)^m; \quad m = 1, 2, ...$$

Letting $y = p(1-p)$ and making use of the well-known series

$$\sum_{m=0}^{\infty} \binom{2m}{m} y^m = \frac{1}{\sqrt{1-4y}}, \quad -1/4 < y < 1/4,$$

yields

$$\sum_{m=0}^{\infty} p_{00}^{(m)} = \frac{1}{\sqrt{(1-2p)^2}} = \frac{1}{|1-2p|}, \quad p \neq 1/2.$$

Thus,

$$\sum_{m=0}^{\infty} p_{00}^{(m)} < \infty \text{ for all } p \neq 1/2.$$

Hence, by theorem 8.5, state 0 is transient. But for any p with $0 < p < 1$ all states are essential, since there is always a positive probability of making a transition to any state irrespective of the starting position. By the corollary from theorem 8.6, the Markov chain $\{X_0, X_1, ...\}$ is transient, since it is irreducible.

If $p = 1/2$ (*symmetric random walk*), then

$$\sum_{m=0}^{\infty} p_{00}^{(m)} = \lim_{p \to 1/2} \frac{1}{|1 - 2p|} = \infty. \tag{8.28}$$

Therefore, in this case all states are recurrent.   □

The symmetric random walk along a straight line can easily be generalized to n-dimensional Euclidian spaces: In the plane, the particle jumps one unit to the West, South, East, or North, respectively, each with probability 1/4. In the 3-dimensional Euclidian space, the particle jumps one unit to the West, South, East, North, up- or downwards, respectively, each with probability 1/6. When analyzing these random walks analogously to the one-dimensional case, an interesting phenomenon becomes visible: the symmetric two-dimensional random walk (more exactly, the underlying Markov chain) is recurrent like the one-dimensional symmetric random walk, but all n-dimensional symmetric random walks with $n > 2$ are transient. Thus, there is a positive probability that Jim, who randomly chooses one of the six possibilities in a 3-dimensional labyrinth, each with probability 1/6, will never return to his starting position.

Example 8.12 A particle jumps from $x = i$ to $x = 0$ with probability p_i or to $i + 1$ with probability

$$1 - p_i, \quad 0 < p_i < 1, \ i = 0, 1, ...$$

The jumps are independent of each other. In terms of population dynamics, a population increases by one individual at each jump with positive probability $1 - p_i$ if before the jump it comprised i individuals (state i). But at any state i a disaster can wipe out the whole population with probability p_i. (State 0 is, however, not absorbing.)

Let X_n be the position of the particle after the nth jump. Then the transition matrix of the Markov chain $\{X_0, X_1, ...\}$ is

$$\mathbf{P} = \begin{pmatrix} p_0 & 1-p_0 & 0 & 0 & 0 & \cdots & 0 & 0 & \cdots \\ p_1 & 0 & 1-p_1 & 0 & 0 & \cdots & 0 & 0 & \cdots \\ p_2 & 0 & 0 & 1-p_2 & 0 & \cdots & 0 & 0 & \cdots \\ \vdots & \vdots & \vdots & \vdots & \vdots & \cdots & 0 & 0 & \cdots \\ p_i & 0 & \cdots & \cdots & 0 & \cdots & 1-p_i & 0 & \cdots \\ \vdots & \vdots & \vdots & \vdots & \vdots & \cdots & \vdots & \vdots & \vdots \end{pmatrix}.$$

The Markov chain $\{X_0, X_1, ...\}$ is irreducible and aperiodic. Hence, for finding the conditions under which this Markov chain is recurrent or transient it is sufficient to consider state 0, say. It is not difficult to determine $f_{00}^{(m)}$:

Starting with

$$f_{00}^{(1)} = p_0,$$

the m-step first return probabilities are

$$f_{00}^{(m)} = \left(\prod_{i=0}^{m-2} (1-p_i) \right) p_{m-1}; \quad m = 2, 3, \ldots$$

If p_{m-1} is replaced with $(1 - (1 - p_{m-1}))$, then $f_{00}^{(m)}$ becomes

$$f_{00}^{(m)} = \left(\prod_{i=0}^{m-2} (1-p_i) \right) - \left(\prod_{i=0}^{m-1} (1-p_i) \right); \quad m = 2, 3, \ldots$$

so that

$$\sum_{n=1}^{m+1} f_{00}^{(n)} = 1 - \left(\prod_{i=0}^{m} (1-p_i) \right), \quad m = 1, 2, \ldots.$$

Thus, state 0 is recurrent if and only if

$$\lim_{m \to \infty} \prod_{i=0}^{m} (1-p_i) = 0. \tag{8.29}$$

Proposition Condition (8.29) is true if and only if

$$\sum_{i=0}^{\infty} p_i = \infty. \tag{8.30}$$

To prove this proposition, note that

$$1 - p_i \le e^{-p_i}; \quad i = 0, 1, \ldots.$$

Hence,

$$\prod_{i=0}^{m} (1-p_i) \le \exp\left(-\sum_{i=0}^{m} p_i \right).$$

Letting $m \to \infty$ proves that (8.29) follows from (8.30).

The converse direction is proved indirectly: The assumption that (8.29) is true and (8.30) is wrong implies the existence of a positive integer k satisfying

$$0 < \sum_{i=k}^{m} p_i < 1.$$

By induction

$$\prod_{i=k}^{m} (1-p_i) > 1 - p_k - p_{k+1} - \cdots - p_m = 1 - \sum_{i=k}^{m} p_i.$$

Therefore,

$$\lim_{m \to \infty} \prod_{i=k}^{m} (1-p_i) > \lim_{m \to \infty} \left(1 - \sum_{i=k}^{m} p_i \right) > 0.$$

This contradicts the assumption that condition (8.29) is true, and, hence, completes the proof of the proposition.

Thus, state 0 and with it the Markov chain are recurrent if and only if condition (8.30) is true. This is the case, for instance, if $p_i = p > 0$; $i = 0, 1, \ldots$. $\square$

8.3 LIMIT THEOREMS AND STATIONARY DISTRIBUTION

Theorem 8.7 Let state i and j communicate, i.e. $i \Leftrightarrow j$. Then,

$$\lim_{n\to\infty} \frac{1}{n} \sum_{m=1}^{n} p_{ij}^{(m)} = \frac{1}{\mu_{jj}}. \tag{8.31}$$

Proof Analogously to the proof of theorem 8.5 it can be shown that, given the Markov chain is at state i at time $t = 0$, the sum

$$\sum_{m=1}^{n} p_{ij}^{(m)}$$

is equal to the mean number of transitions into state j in the time interval $(0, n]$. The theorem is, therefore, a direct consequence of the elementary renewal theorem (theorem 7.12, page 311). (If $i \neq j$, the corresponding renewal process is delayed.) ∎

If the limit

$$\lim_{m\to\infty} p_{ij}^{(m)}$$

exists, then it coincides with the limit at the right-hand side of equation (8.31). Since it can be shown that for an irreducible Markov chain these limits exist for all $i, j \in \mathbf{Z}$, theorem 8.7 implies the

Corollary Let $p_{ij}^{(m)}$ be the m-step transition probabilities of an irreducible, aperiodic Markov chain. Then,

$$\lim_{m\to\infty} p_{ij}^{(m)} = \frac{1}{\mu_{jj}}.$$

If state j is transient or null-recurrent, then

$$\lim_{m\to\infty} p_{ij}^{(m)} = 0.$$

If the irreducible Markov chain has period $d > 1$, then

$$\lim_{m\to\infty} p_{ij}^{(m)} = \frac{d}{\mu_{jj}}.$$

To see this, switch from the one-step transition matrix $\mathbf{P}$ to the d-step transition matrix $\mathbf{P}^d$. A proof of the following theorem is e.g. given in *Feller* (1968).

Theorem 8.8 For any irreducible, aperiodic Markov chain, there are two possibilities:

(1) If the Markov chain is transient or null recurrent, then a stationary distribution does not exist.

(2) If the Markov chain is positive recurrent, then there exists a unique stationary distribution $\{\pi_j, j \in \mathbf{Z}\}$, which for any $i \in \mathbf{Z}$ is given by

$$\pi_j = \lim_{m\to\infty} p_{ij}^{(m)} = \frac{1}{\mu_{jj}}.$$ ∎

Example 8.13 A particle moves along the real axis. Starting from position (state) i it jumps to state $i+1$ with probability p and to state $i-1$ with probability $q = 1-p$, $i = 1, 2, \ldots$. When the particle arrives at state 0, it remains there for a further time unit with probability q or jumps to state 1 with probability p. Let X_n denote the position of the particle after the nth jump (time unit). Under which condition has the Markov chain $\{X_0, X_1, \ldots\}$ a stationary distribution?

Since $p_{00} = q$, $p_{i\,i+1} = p$, and $p_{i\,i-1} = q = 1-p$; $i = 1, 2, \ldots$, the system (8.9) is

$$\pi_0 = \pi_0 q + \pi_1 q$$
$$\pi_i = \pi_{i-1} p + \pi_{i+1} q; \quad i = 1, 2, \ldots.$$

By recursively solving this system of equations,

$$\pi_i = \left(\frac{p}{q}\right)^i \pi_0; \quad i = 0, 1, \ldots.$$

To ensure that $\sum_{i=0}^{\infty} \pi_i = 1$, condition $p < q$ or, equivalently, $p < 1/2$, must hold. In this case,

$$\pi_i = \frac{q-p}{q}\left(\frac{p}{q}\right)^i; \quad i = 0, 1, \ldots. \tag{8.32}$$

The necessary condition $p < 1/2$ for the existence of a stationary distribution is intuitive, since otherwise the particle would tend to drift to infinity. But then no time-invariant behavior of the Markov chain can be expected. □

Theorem 8.9 Let $\{X_0, X_1, \ldots\}$ be an irreducible, recurrent Markov chain with state space **Z** and stationary state probabilities π_i, $i \in \mathbf{Z}$. If g is any bounded function on **Z**, then

$$\lim_{n \to \infty} \frac{1}{n} \sum_{j=0}^{n} g(X_j) = \sum_{i \in \mathbf{Z}} \pi_i g(i). \quad \blacksquare$$

For example, if $c_i = g(i)$ is the 'profit' which arises when the Markov chain makes a transition to state i, then

$$\sum_{i \in \mathbf{Z}} \pi_i c_i$$

is the mean profit in the long-run resulting from a state change of the Markov chain. Thus, theorem 8.9 is the analog to the renewal reward theorem (formula (7.148) at page 325) for compound renewal processes. In particular, let

$$g(i) = \begin{cases} 1 & \text{for } i = k \\ 0 & \text{for } i \neq k \end{cases}.$$

If changes of state of the Markov chain occur after unit time intervals, then the limit

$$\lim_{n \to \infty} \frac{1}{n} \sum_{j=0}^{n} g(X_j)$$

is equal to the mean percentage of time the system is in state k. By theorem 8.9, this percentage coincides with π_k. This property of the stationary state distribution illustrates once more that it refers to an equilibrium state of the Markov chain. A proof of theorem 8.9 under weaker assumptions can be found in *Tijms* (1994).

Example 8.14 A system can be in one of the three states 1, 2, and 3: In state 1 it operates most efficiently. In state 2 it is still working but its efficiency is lower than in state 1. State 3 is the *down state*, the system is no longer operating and has to be maintained. State changes can only occur after a fixed time unit of length 1. Transitions into the same state are allowed. If X_n denotes the state of the system at time n, then $\{X_0, X_1, ...\}$ is assumed to be a Markov chain with transition matrix

$$\mathbf{P} = \begin{pmatrix} 0.8 & 0.1 & 0.1 \\ 0 & 0.6 & 0.4 \\ 0.8 & 0 & 0.2 \end{pmatrix}.$$

Note that from state 3 the system most likely makes a transition to state 1, but it may also stay in state 3 for one or more time units (for example, if a maintenance action has not been successful). The corresponding stationary state probabilities satisfy the system of linear equations

$$\pi_1 = 0.8\,\pi_1 \qquad\qquad + 0.8\,\pi_3$$
$$\pi_2 = 0.1\,\pi_1 + 0.6\,\pi_2$$
$$\pi_3 = 0.1\,\pi_1 + 0.4\,\pi_2 + 0.2\,\pi_3.$$

Only two of these equations are linearly independent. Together with the normalizing constraint $\pi_1 + \pi_2 + \pi_3 = 1$, the unique solution is

$$\pi_1 = \tfrac{4}{6}, \quad \pi_2 = \pi_3 = \tfrac{1}{6}. \tag{8.33}$$

The profits the system makes per unit time in states 1 and 2 are

$$g(1) = \$\,1000, \quad g(2) = \$\,600,$$

wheras, when in state 3, the system causes a loss of

$$g(3) = \$100$$

per unit time. According to theorem 8.9, after an infinite (sufficiently long) running time, the mean profit per unit time is

$$\sum_{i=1}^{3} \pi_i g(i) = 1000 \cdot \frac{4}{6} + 600 \cdot \frac{1}{6} - 100 \cdot \frac{1}{6} = 250 \ [\$ \text{ per unit time}].$$

Now, let Y be the random time, in which the system is in the profitable states 1 and 2. According to the structure of the transition matrix, such a time period must begin with state 1. Further, let Z be the random time in which the system is in the unprofitable state 3. The mean values $E(Y)$ and $E(Z)$ are to be determined. The random vector (Y, Z) characterizes the typical cycle of an alternating renewal process. Therefore, by formula (7.144), page 322, the ratio

$$E(Y)/[E(Y)+E(Z)]$$

is equal to the mean percentage of time the system is in states 1 or 2. As pointed out after theorem 8.9, this percentage must be equal to $\pi_1 + \pi_2$:

$$\frac{E(Y)}{E(Y)+E(Z)} = \pi_1 + \pi_2. \qquad (8.34)$$

Since the mean time between transitions into state 3 is equal to $E(Y)+E(Z)$, the ratio $1/[E(Y)+E(Z)]$ is equal to the rate of transitions to state 3. On the other hand, this rate is $\pi_1 p_{13} + \pi_2 p_{23}$. Hence,

$$\frac{1}{E(Y)+E(Z)} = \pi_1 p_{13} + \pi_2 p_{23}. \qquad (8.35)$$

From (8.34) and (8.35)

$$E(Y) = \frac{\pi_1 + \pi_2}{\pi_1 p_{13} + \pi_2 p_{23}}, \quad E(Z) = \frac{\pi_3}{\pi_1 p_{13} + \pi_2 p_{23}}.$$

Substituting the numerical values (8.33) gives $E(Y) = 6.25$ and $E(Z) = 1.25$. Hence, the percentage of time, the system is in the profit-generating states 1 and 2 is

$$6.25/7.50 \,[100\%] = 83{,}\underline{3}\,[\%]. \qquad \square$$

Example 8.15 An insurer knows that the total annual claim size X of a client in a certain portfolio is exponentially distributed with mean value $E(X) = \$1000$, i.e.

$$F(x) = P(X \le x) = 1 - e^{-x/1000}, \; x \ge 0.$$

The insurer partitions his clients into classes 1, 2, and 3 depending on the annual amounts they claim, and the class they belong to: A client, who is in class 1 in the current year, will make a transition to class 1, 2 or 3 next year, when his respective total claims are between 0 and 600, 600 and 1200, or greater than 1200 in the current year. A client, who is in class 2 in the current year, will make a transition to class 1, 2, or 3 next year if his respective total claim sizes are between 0 and 500, 500 and 1100, or more than 1100. A client, who is in class 3 and claims between 0 and 1100 or at least 1100 in the current year, will be in class 2 or in class 3 next year, respectively. In this case, a direct transition from class 3 to class 1 is not possible. When in class 1, 2, or 3, the clients will pay the respective premiums 600, 1200, or 1400 a year. The one-step transition probabilities p_{ij} are

$$p_{11} = F(600) = 0.4512, \quad p_{12} = F(1200) - F(600) = 0.2476,$$
$$p_{21} = F(500) = 0.3935, \quad p_{22} = F(1100) - F(500) = 0.2736,$$
$$p_{31} = 0, \quad p_{32} = F(1100) = 0.6671.$$

Taking into account $p_{i1} + p_{i2} + p_{i3} = 1$, $i = 1, 2, 3$, the complete matrix of the one-step transition probabilities is

$$\mathbf{P} = \begin{pmatrix} 0.4512 & 0.2476 & 0.3012 \\ 0.3935 & 0.2736 & 0.2736 \\ 0.0000 & 0.6671 & 0.3329 \end{pmatrix}.$$

By (8.9), the stationary state probabilities satisfy the system of linear equations (note that one of the equations (8.9) is redundant, i.e., linearly dependent on the other two equations, and must be replaced by the normalizing equation (8.10)):

$$\pi_1 = 0.4512\,\pi_1 + 0.3935\,\pi_2$$
$$\pi_2 = 0.2476\,\pi_1 + 0.2736\,\pi_2 + 0.6671\,\pi_3,$$
$$1 = \pi_1 + \pi_2 + \pi_3.$$

The solution is

$$\pi_1 = 0.2823, \quad \pi_2 = 0.3938, \quad \pi_3 = 0.3239.$$

Hence, the average annual long-run premium a client has to pay is

$$\sum_{i=1}^{3} \pi_i\, g(i) = 0.2823 \cdot 600 + 0.3938 \cdot 1200 + 0.3239 \cdot 1400 = 1095.4$$

so that the long-run average profit of the insurer per client and year is $\$\,95.4$. $\qquad\square$

8.4 BIRTH AND DEATH PROCESSES

8.4.1 Introduction

In some of the examples considered so far only direct transitions to 'neighboring' states were possible. More exactly, if starting at state i and not staying there for one or more time units, only transitions to states $i-1$ or $i+1$ could be made in one step. In these cases, the positive one-step transition probabilities have structure (Figure 8.2)

$$p_{i\,i+1} = p_i, \quad p_{i\,i-1} = q_i, \quad p_{ii} = r_i \quad \text{with} \quad p_i + q_i + r_i = 1. \qquad (8.36)$$

A discrete Markov chain with state space $\mathbf{Z} = \{0, 1, ..., z\}$, $z \le \infty$, and transition probabilities (8.36) is called a *birth and death process*. The state space implies $q_0 = 0$. $r_i = 1 - p_i - q_i$ is the probability that the process stays for another time unit at state i. The term *birth and death process* results from the application of these processes to describing the development in time of biological populations. In this context, X_n is the number of individuals of a population at time n assuming that the population does not increase or decrease by more than one individual per unit time. Correspondingly, the p_i and the q_i are called *birth* and *death probabilities*, respectively.

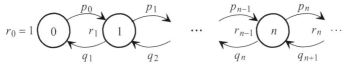

Figure 8.2 Transition graph of a birth and death process with infinite state space

A birth and death process is called a *pure birth process* if all the q_i are 0 (no deaths are possible), and a *pure death process* if all the p_i are 0 (no births are possible).

To make sure that a birth and death process is irreducible, the assumptions (8.36) have to be supplemented by

$$p_i > 0 \quad \text{for } i = 0, 1, \dots \text{ and } q_i > 0 \text{ for } i = 1, 2, \dots . \tag{8.37}$$

For instance, the random walk of example 8.13 is a birth- and death process with

$$p_i = p, \quad q_i = q, \quad r_i = 0 \text{ for } i = 1, 2, \dots; \quad p_0 = p, \quad q_0 = 0, \quad r_0 = q = 1 - p.$$

The unbounded random walk in example 8.11 also makes direct transitions only to neighboring states. But its state space is $\mathbf{Z} = \{0, \pm 1, \pm 2, \dots\}$ so that this random walk is not a birth and death process.

8.4.2 General Random Walk with two Absorbing Barriers

In generalizing example 8.3, a random walk with state space $Z = \{0, 1, \dots, z\}$ and transition probabilities (8.36) is considered, which satisfy the additional conditions

$$r_0 = r_z = 1, \quad p_i > 0 \text{ and } q_i > 0 \text{ for } i = 1, 2, \dots, z - 1. \tag{8.38}$$

Thus, states 0 and z are absorbing (Figure 8.3).

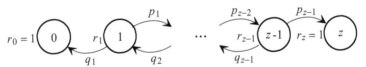

Figure 8.3 Transition graph of a birth and death process with absorbing barriers

Let $a(n)$ be the probability that the random walk is absorbed by state 0 when starting from n; $n = 1, 2, \dots, z - 1$. (Since z is absorbing as well, the process cannot have been in state z before arriving at state 0.) It is obvious that

$$1 = a(0) > a(1) > \cdots > a(z - 1) > a(z) = 0. \tag{8.39}$$

From the total probability rule (1.24),

$$a(n) = p_n \cdot a(n + 1) + q_n \cdot a(n - 1) + r_n \, a(n), \tag{8.40}$$

or, equivalently, when replacing r_n with $r_n = 1 - p_n - q_n$,

$$a(n) - a(n + 1) = \frac{q_n}{p_n} [a(n - 1) - a(n)]; \quad n = 1, 2, \dots, z - 1.$$

Repeated application of these difference equations gives

$$a(n) - a(n + 1) = A_n [a(0) - a(1)]; \quad n = 0, 1, \dots, z - 1, \tag{8.41}$$

with

$$A_n = \frac{q_1 \, q_2 \cdots q_n}{p_1 \, p_2 \cdots p_n}; \quad n = 1, 2, \dots, z - 1; \quad A_0 = 1, \tag{8.42}$$

and $a(0) = 1$ and $a(z) = 0$.

Summing equations (8.41) from $n = k$ to $n = z - 1$ yields

$$a(k) = \sum_{n=k}^{z-1} [a(n) - a(n+1)] = [a(0) - a(1)] \sum_{n=k}^{z-1} A_n .$$

In particular, for $k = 0$,

$$1 = [a(0) - a(1)] \sum_{n=0}^{z-1} A_n .$$

By combining the last two equations,

$$a(k) = \frac{\sum_{n=k}^{z-1} A_n}{\sum_{n=0}^{z-1} A_n} ; \quad k = 0, 1, ..., z - 1; \quad a(z) = 0, \; A_0 = 1 . \tag{8.43}$$

The probability of absorption at state z if the particle starts at k is $b(k) = 1 - a(k)$.

Gambler's ruin problem: The probabilities $a(k)$ can be interpreted as follows (compare to example 8.3): Two gamblers begin a game with stakes of sizes k and $z - k$, respectively; k, z integers with $0 < k < z$. After each move a gambler either wins or loses \$1 or the gambler's stake remains constant. These possibilities are controlled by transition probabilities satisfying (8.36) and (8.38). The game is finished if a gambler has won the entire stake of the other one or, equivalently, if one gambler has lost her/his entire stake.

Mean time to absorption Let $m(n)$ be the mean number of time units (steps) till the particle arrives at <u>any</u> of the absorbing states 0 or z, when it has started at location n, $0 < n < z$. If the particle moves from the starting point n to the right, then the mean time till absorption is $1 + m(n+1)$; if the particle jumps to the left, then the mean time till absorption is $1 + m(n-1)$, and if the particle stays at position n a further time unit, then the mean time to absorption is $1 + m(n)$. Hence, analogously to (8.19), the $m(n)$ satisfy the system of equations

$$m(n) = p_n \cdot [1 + m(n+1)] + q_n \cdot [1 + m(n-1)] + r_n \cdot [1 + m(n)], \tag{8.44}$$

or, when replacing r_n with $r_n = 1 - p_n - q_n$, the system of the equations (8.44) for the $m(n)$ becomes a system of equations for the differences $d(n) = m(n) - m(n-1)$:

$$d(n+1) = \frac{q_n}{p_n} d(n) - \frac{1}{p_n}; \quad n = 1, 2, ..., z - 1. \tag{8.45}$$

The boundary conditions are $m(0) = m(z) = 0$ so that $d(1) = m(1)$.

k-fold application of the recursive equations (8.45) starting with $n = 1$ yields

$$d(2) = \frac{q_1}{p_1} m(1) - \frac{1}{p_1},$$

$$d(3) = \frac{q_2}{p_2} \left(\frac{q_1}{p_1} m(1) - \frac{1}{p_1} \right) - \frac{1}{p_2} = \frac{q_1 q_2}{p_1 p_2} m(1) - \frac{q_2}{p_1 p_2} - \frac{1}{p_2},$$

$$d(4) = \frac{q_3}{p_3} d_3 - \frac{1}{p_3} = \frac{q_1 q_2 q_3}{p_1 p_2 p_3} m(1) - \frac{q_2 q_3}{p_1 p_2 p_3} - \frac{q_3}{p_2 p_3} - \frac{1}{p_3},$$

and, finally,

$$d(k) = A_{k-1} m(1) - \sum_{i=2}^{k-1} \frac{q_i q_{i+1} \cdots q_{k-1}}{p_{i-1} p_i \cdots p_{k-1}} - \frac{1}{p_{k-1}}; \quad k = 3, 4, ..., z, \qquad (8.46)$$

where the A_{k-1} are given by (8.42) with $n = k - 1$. The desired mean values $m(n)$ are simply obtained by summation of the $d(k)$:

$$m(n) = \sum_{k=1}^{n} d(k) = \sum_{k=1}^{n} [m(k) - m(k-1)], \quad n = 1, 2, ..., z. \qquad (8.47)$$

The still unknown $m(1)$, which occurs as a factor in each of the $d(k)$, can be determined from (8.47) by making use of the boundary condition $m(z) = 0$, i.e. from

$$m(z) = 0 = \sum_{n=1}^{z} d(n).$$

The result is

$$m(1) = \frac{\sum_{k=2}^{z-1} \left(\sum_{i=2}^{k} \frac{q_i q_{i+1} \cdots q_k}{p_{i-1} p_i \cdots p_k} \right) + \sum_{k=1}^{z-1} \frac{1}{p_k}}{1 + \sum_{k=1}^{z-1} A_k}. \qquad (8.48)$$

k	0	1	2	3	4	5	6
p_k	0	0.8		0.4	0.3	0.1	0
q_k	0	0.1	0.3	0.4	0.5	0.8	0
r_k	1	0.1	0.2	0.2	0.2	0.1	1
A_k	1	0.1250	0.075	0.075	0.125	1.0	
$a(k)$	1	0.5833	0.5313	0.5000	0.4687	0.4167	0
$b(k)$	0	0.4167	0.4687	0.5000	0.5313	0.5833	1
$m(k)$	0	53.54	58.95	60.50	58.95	53.54	0

Table 8.2 Numerical results for example 8.16

Example 8.16 A random walk with state space $\mathbf{Z} = \{0, 1, 2, ..., 6\}$ and the absorbing barriers 0 and 6 is considered. Table 8.2 shows the birth and death probabilities p_n and q_n, the corresponding r_n, the ratios A_k defined by (8.42), the absorption probabilities $a(k)$ and $b(k)$ with regard to locations 0 and 6, respectively, and the mean times to absorption $m(k)$ at any of the locations 0 or 6. From (8.48),

$$m(1) = 53.54.$$

Now the mean times to absorption $m(2)$, $m(3)$, $\cdots$, $m(6)$ can be obtained from (8.47). For manual calculations, it is most efficient to determine the $d(k)$ recursively by (8.45). In view of the symmetric structure of the birth and death probabilities, the absorption probabilities $a(k)$ and $b(6-k)$, $k = 0, 1, 2, 3$, coincide. $\square$

8.4.3 General Random Walk with One Absorbing Barrier

The same situation as in section 8.4.2 is considered except that state z is no longer assumed to be absorbing (Figure 8.4), i.e. the corresponding transition probabilities have properties

$$r_0 = 1, \; p_z = 0, \; q_z > 0, \; r_z = 1 - q_z; \; p_i > 0 \text{ and } q_i > 0 \text{ for } i = 1, 2, ..., z - 1.$$

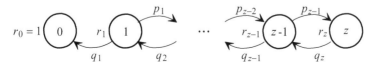

Figure 8.4 Transition graph for a random walk with absorption at 0

These transition probabilities imply that state 0 is absorbing, whereas from state z transitions to state $z - 1$ are possible. The states $1, 2, ..., z - 1$ are transient so that after a random number of time units the particle will arrive at location 0 with probability 1. Again, jumps of the particle (possibly to the same location) always occur after one time unit. Since the boundary condition $m(0) = 0$ is the same as in in the previous section, formulas (8.46) and (8.47) stay valid for $k = 1, 2, ..., z - 1$. Since $p_z = 0$, equation (8.44) yields for $n = z$ the boundary condition

$$m(z) = q_z \cdot [1 + m(z - 1)] + (1 - q_z) \cdot [1 + m(z)],$$

or, equivalently, $$m(z) - m(z - 1) = \frac{1}{q_z} . \qquad (8.49)$$

Letting $n = z - 1$ in (8.47) and combining the resulting equation with (8.49) leads to an equation for $m(1)$, the solution of which is

$$m(1) = \frac{\displaystyle\sum_{i=2}^{z-1} \frac{q_i q_{i+1} \cdots q_{z-1}}{p_{i-1} p_i \cdots p_{z-1}} + \frac{1}{p_{z-1}} + \frac{1}{q_z}}{Q_{z-1}}$$

or, equivalently,

$$m(1) = \frac{1}{q_1} + \sum_{i=2}^{z} \frac{p_1 p_2 \cdots p_{i-1}}{q_1 q_2 \cdots q_i} . \qquad (8.50)$$

Now the $m(2), m(3), ..., m(z)$ can be recursively determined by (8.45) or (8.46), respectively, or directly by (8.47). After some algebra, a more elegant representation of $m(k)$ is obtained by inserting (8.50) into (8.47) (*Nisbet, Gurney* (1982)):

$$m(k) = m(1) + \sum_{n=1}^{k-1} \left(\frac{q_1 q_2 \cdots q_n}{p_1 p_2 \cdots p_n} \sum_{i=n+1}^{z} \frac{p_1 p_2 \cdots p_{i-1}}{q_1 q_2 \cdots q_i} \right); \; k = 2, 3, ..., z.$$

Mean Time to Extinction $m(k)$ can be interpreted as the mean time to the extinction of a finite population under the following assumptions: The maximal possible number of individuals the environment can sustain is z. If the population has k members, it will grow per time unit by one individual with probability p_k, $1 \le k \le z - 1$, it will decrease per time unit by one individual with probability q_k, $1 \le k \le z$, or the number of members does not change per time unit with probability $r_k = 1 - p_k - q_k$. In addition, $q_0 = p_0 = p_z = 0$. No immigration occurs. One jump per time unit (possibly to the same state) is realistic if the time unit is chosen small enough. If this birth and death process arrives at the absorbing state 0, the population is extinct.

Example 8.17 Consider a population with a maximal size of $z = 6$ individuals and transition probabilities with regard to a unit time given by Table 8.3. Then, by (8.50),

$$m(1) = \frac{1}{q_1} + \frac{p_1}{q_1 q_2} + \frac{p_1 p_2}{q_1 q_2 q_3} + \cdots + \frac{p_1 p_2 \cdots p_5}{q_1 q_2 \cdots q_6} = 155 .$$

Table 8.3 shows the mean times to extinction $m(1)$, $m(2)$, ..., $m(6)$. Condition (8.49) is satisfied. □

k	0	1	2	3	4	5	6
p_k	0	0.8	0.5	0.4	0.2	0.1	0
q_k	0	0.1	0.2	0.4	0.5	0.6	0.8
r_k	1	0.1	0.3	0.2	0.3	0.3	2
$d(k)$		155	18.125	5.250	2.750	1.875	1.250
$m(k)$	0	155	173.125	178.375	181.125	183.000	184.250

Table 8.3 Numerical results for example 8.17

Theorem 8.10 Under the additional assumptions (8.37) on its transition probabilities (8.36), a birth- and death process is recurrent if and only if

$$\sum_{n=1}^{\infty} \frac{q_1 q_2 \cdots q_n}{p_1 p_2 \cdots p_n} = \infty . \tag{8.51}$$

Proof It is sufficient to show that state 0 is recurrent. This can be established by using the result (8.43) referring to a general random walk with two absorbing barriers, since

$$\lim_{z \to \infty} p(k) = f_{k0} ; \quad k = 1, 2, \dots ,$$

where the first passage time probabilities f_{k0} are given by (8.26). If state 0 is recurrent, then, from the irreducibility of the Markov chain, $f_{00} = 1$ and $f_{k0} = 1$. However, $f_{k0} = 1$ if and only if (8.51) is valid. Conversely, let (8.51) be true. Then, by the total probability rule,

$$f_{00} = p_{00} + p_{01} f_{10} = r_0 + p_0 \cdot 1 = 1 . \qquad \blacksquare$$

Discrete-time birth and death processes have significance on their own, but may also serve as approximations to the more important continuous-time birth and death processes, which are the subject of section 9.6.

8.5 DISCRETE-TIME BRANCHING PROCESSES

8.5.1 Introduction

Closely related to pure birth processes are branching processes. In this section, the simplest branching process, the *Galton-Watson process*, is considered. The terminology applied refers to population dynamics. The Galton-Watson process $\{X_0, X_1, ...\}$ is characterized by the following properties (For illustration, see a tree-representation of a sample path of this process on condition $X_0 = 1$ in Figure 8.5):

1) The population starts with X_0 individuals. They constitute the *zeroth generation.*

2) Each individual i of the zeroth generation has $Y_{i,0}$ offspring; $i = 0, 1, 2, ...$ The $Y_{i,0}$ are independent and identically distributed as a random variable Y with

$$p_k = P(Y = k); \; k = 0, 1, ..., \; \sum_{k=0}^{\infty} p_k = 1; \; \mu = E(Y) \text{ and } \sigma^2 = Var(Y). \quad (8.52)$$

The set of all offspring of individuals of the zeroth generation constitutes the *first generation*. The total number of all individuals in the first generation is denoted as X_1:

$$X_1 = \sum_{i=1}^{X_0} Y_{i,0}.$$

3) Generally, each member i of the $(n-1)th$ generation produces a random number $Y_{i,n-1}$ of offspring, and all $Y_{i,n-1}$ are independent and identically distributed as Y. In addition, the $Y_{i,n-1}$ are independent of all previous offspring figures

$$Y_{i,n-2}, ..., Y_{i,0}; \; n = 2, 3, ...$$

The set of offspring generated by the $(n-1)th$ generation constitutes the nth generation with a total of X_n individuals, $n = 0, 1, ...$

4) All individuals of a generation are of the same type.

According to its construction, the random sequence $\{X_0, X_1, ...\}$ is a discrete-time Markov chain. Given $X_0 = i$, its m-step transition probabilities (8.3) are equal to the absolute state probabilities $p_j^{(m)} = P(X_m = j)$ of X_m:

$$p_{ij}^{(m)} = P(X_m = j | X_0 = i).$$

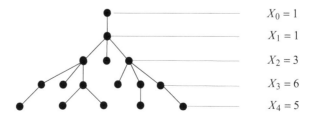

$X_0 = 1$
$X_1 = 1$
$X_2 = 3$
$X_3 = 6$
$X_4 = 5$

Figure 8.5 Piece of a sample path of a Galton-Watson process

The first motive for dealing with branching processes was to determine the duration of (noble) families. The French statistician *L. F. Benoiston de Châteauneuf* (1776–1856) estimated their average duration to be 300 years (according to *Moser* (1839)). As pointed out by *Heyde, Seneta* (1972), *I. J. Bienaymé* (1796–1878) was very likely able to determine the probability of the extinction of family names based on the extinction of male offspring, but, unfortunately, did not leave behind any written account. *Sir F. Galton* (1822–1911) and *H.W. Watson* (1822–1900) formulated the mathematical problem, but could not fully solve it; see *Galton, Watson* (1875). This was done by the Danish actuary *J.F. Steffenson* only in 1930 (*Steffenson* (1930)). Other applications of branching processes are among else in mutant genes dynamics, nuclear chain reactions, electron multipliers to boost a current of electrons, and cell kinetics. There are numerous generalizations of the Galton–Watson process, e.g., multi-type branching processes, continuous-time branching processes, and age dependent branching processes. Recent monographs on theory and applications of branching processes are *Haccou et al.* (2011), *Kimmel, Axelrod* (2015), and *Durret* (2015). Pioneering classics are *Harris* (1963) and *Sevastyanov* (1971).

8.5.2 Generating Function and Distribution Parameters

In what follows, the assumption is made that the development of the population starts with one individual, i.e, $X_0 = 1$. The respective z-transforms (moment generating functions) of Y and X_n are denoted as (section 2.5, page 96)

$$M(z) = E(z^Y) = \sum_{k=0}^{\infty} p_k z^k,$$

$$M_n(z) = E(z^{X_n}) = \sum_{k=0}^{\infty} P(X_n = k) z^k; \quad n = 0, 1, \dots.$$

In particular, $\qquad M_0(z) = z$ and $M_1(z) = M(z).$ $\qquad\qquad$ (8.53)

According to the notation introduced,

$$X_n = \sum_{i=1}^{X_{n-1}} Y_{i,n-1},$$

where the random variables $Y_{1,n-1}, Y_{2,n-1}, \dots, Y_{X_{n-1},n-1}$ are independent and identically distributed as Y. Hence, by formula (2.116), page 99, on condition $X_{n-1} = m$ the z-transform of X_n is

$$M_n(z|X_{n-1} = m) = [M(z)]^m; \quad m = 0, 1, 2, \dots.$$

Now, by using this result and the formula of the total probability

$$M_n(z) = \sum_{k=0}^{\infty} P(X_n = k) z^k$$

$$= \sum_{k=0}^{\infty} \sum_{m=0}^{\infty} P(X_n = k | X_{n-1} = m) P(X_{n-1} = m) z^k$$

$$= \sum_{m=0}^{\infty} P(X_{n-1} = m) \sum_{k=0}^{\infty} z^k P(X_n = k | X_{n-1} = m)$$

$$= \sum_{m=0}^{\infty} P(X_{n-1} = m) [M(z)]^m.$$

The last row is the z-transform of X_{n-1} with the variable z replaced by the variable $M(z)$, i.e. the following recursive equation for the $M_n(z)$ is valid:

$$M_n(z) = M_{n-1}(M(z)), \quad n = 1, 2, \dots. \tag{8.54}$$

A similar recursive equation for $M_n(z)$ is

$$M_n(z) = M(M_{n-1}(z)), \quad n = 1, 2, \dots, \tag{8.55}$$

which easily follows from (8.54) by induction:

For $n = 2$ formula (8.55) is true since by (8.53) and (8.54),

$$M_2(z) = M_1(M(z)) = M(M_1(z)).$$

Now assume $M_{n-1}(z) = M(M_{n-2}(z))$ is true. Then, by (8.54),

$$M_n(z) = M_{n-1}(M(z)) = M(M_{n-2}(M(z))) = M(M_{n-1}(z)),$$

which proves (8.55).

The first and second derivative of $M_n(z)$ given by (8.55) with regard to z are

$$M_n'(z) = M'(M_{n-1}(z)) \cdot M_{n-1}'(z), \tag{8.56}$$

$$M_n''(z) = M''(M_{n-1}(z)) \cdot [M_{n-1}'(z)]^2 + M'(M_{n-1}(z)) \cdot M_{n-1}''(z). \tag{8.57}$$

Now let $z = 1$. Then, since $M_n(1) = 1$ for all $n = 0, 1, \dots$ and $\mu = E(Y) = M'(1)$, formula (8.56) yields $M_n'(1) = \mu \cdot M_{n-1}'(1)$, or, equivalently, $M_n'(1) = E(X_n)$. Therefore,

$$E(X_n) = \mu E(X_{n-1}), \quad n = 1, 2, \dots.$$

By repeated application of this relation,

$$E(X_n) = \mu^n, \quad n = 1, 2, \dots. \tag{8.58}$$

Thus, if $\mu < 1$, i.e. there is on average less than one offspring per individual, the population will eventually sooner or later become extinct, since in this case

$$\lim_{n \to \infty} E(X_n) = 0.$$

From (8.57),

$$M_n''(1) = M''(1) \cdot [M_{n-1}'(1)]^2 + M'(1) \cdot M_{n-1}''(1), \quad n = 1, 2, \dots$$

or, taking into account (8.53)

$$M_n''(1) = M''(1) \cdot \mu^{2(n-1)} + \mu \cdot M_{n-1}''(1), \quad n = 1, 2, \dots.$$

Repeated application of this recursive equation for the $M_k''(1)$ gives

$$M_n''(1) = M''(1) [\mu^{2n-2} + \mu^{2n-3} + \dots + \mu^{n-1}].$$

By (2.112), page 96,

$$M''(1) = \sigma^2 - \mu + \mu^2 \quad \text{and} \quad M_n''(1) = Var(X_n) - \mu^n + \mu^{2n}.$$

After some algebra, $Var(X_n)$ becomes

$$Var(X_n) = M_n''(1) + \mu^n - \mu^{2n} = \sigma^2 \left[\mu^{2n-2} + \mu^{2n-3} + \cdots + \mu^{n-1} \right]$$

$$= \sigma^2 \mu^{n-1} \left[\mu^{n-1} + \mu^{n-2} + \cdots + \mu^0 \right], \quad n = 1, 2, \dots.$$

By making use of the finite exponential series (2.18) (page 48), the final result is

$$Var(X_n) = \begin{cases} \sigma^2 \mu^{n-1} \dfrac{\mu^n - 1}{\mu - 1} & \text{for } \mu \neq 1, \\ n\sigma^2 & \text{for } \mu = 1. \end{cases}$$

The variance of X_n increases linearly with increasing n if $\mu = 1$. For $\mu > 1$, this variance increases, and for $\mu < 1$ it decreases with increasing n. Clearly, this increase/decrease occurs the faster the larger σ^2, which is the variance of the number of offspring a member of the population has.

8.5.3 Probability of Extinction and Examples

A population can only become extinct if the probability p_0 (an individual has no offspring) is positive. Hence, let us assume in this section that

$$0 < p_0 < 1.$$

As in the previous section, let $X_0 = 1$. Then the probability of extinction π_0 is formally given as the limit of the m-step transition probabilities

$$\pi_0 = \lim_{m \to \infty} p_{10}^{(m)} = \lim_{m \to \infty} P(X_m = 0 | X_0 = 1).$$

By equations (2.9) (page 46) and (8.58),

$$E(X_n) = \mu^n = \sum_{i=1}^{\infty} i \, P(X_n = i) = \sum_{i=1}^{\infty} P(X_n \geq i) \geq P(X_n \geq 1).$$

Thus, if $\mu < 1$, then $\lim_{n \to \infty} \mu^n = 0$ so that $\lim_{n \to \infty} P(X_n \geq 1) = 1 - \pi_0 = 0$. Hence, if $\mu < 1$, then $\pi_0 = 1$. Moreover, it can be shown that $\pi_0 = 1$ even if $\mu = 1$. Since

$$p_{10}^{(n)} = P(X_n = 0 | X_0 = 0) = M_n(0), \quad n = 1, 2, \dots$$

equation (8.55) implies that

$$\pi_0 = \lim_{n \to \infty} p_{10}^{(n)} = \lim_{n \to \infty} M_n(0) = M(\lim_{n \to \infty} M_{n-1}(0)) = M(\pi_0), \quad n = 1, 2, \dots.$$

Thus, the probability of extinction π_0 satisfies the equation

$$z = M(z). \tag{8.59}$$

In view of $M(1) = 1$, the integer $z_1 = 1$ is always a solution. Without proof:

> *The desired probability of extinction π_0 is the smallest positive solution*
> *of the equation $M(z) = z$. Such a solution exists if $\mu = E(Y) \geq 1$.*

Let T be the random time to extinction. Then T is the smallest integer n with property $X_n = 0$, i.e., $T = \min\limits_{n}\{n, X_n = 0\}$.

The values of the distribution function $F_T(n)$ of T at the 'jump points' n are

$$F_T(n) = P(T \leq n) = P(X_n = 0) = M_n(0), \quad n = 1, 2, \dots.$$

Furthermore, $P(T \leq n) = P(T \leq n-1) + P(T = n)$ so that

$$P(T = n) = M_n(0) - M_{n-1}(0), \quad n = 1, 2, \dots. \tag{8.60}$$

Given $\lim\limits_{n\to\infty} P(X_n = 0) = 1$, by formula (2.9), page 46, the mean time to expiration is

$$E(T) = \sum_{n=1}^{\infty}[1 - M_{n-1}(0)].$$

A sufficient condition for $\lim\limits_{n\to\infty} P(X_n = 0) = 1$ is $\mu \leq 1$.

Example 8.18 A standard example for an application of the Galton-Watson process is due to *Lotka* (1931): Alfred Lotka investigated the random number Y of male offspring per male of the white population in the USA in 1920. (Some male offspring may arise out of wedlock so that Y need not refer to a married couple.) He found that Y has approximately a modified geometric distribution with z-transform

$$M(z) = \frac{0.482 - 0.041\,z}{1 - 0.559\,z}.$$

From this it follows that with probability $p_0 = P(Y = 0) = M(0) = 0.482$ a male has no male offspring. The first and second derivatives of $M(z)$ are

$$M'(z) = \frac{0.2284}{(1 - 0.559z)^2}, \quad M''(z) = \frac{0.2554 - 0.0714z}{(1 - 0.559z)^2}$$

so that $M'(1) = 1.1744$ and $M''(1) = 0.9461$. Hence, by formulas (2.112),

$$E(Y) = M'(1) = 1.1744, \quad Var(Y) = 0.7413, \quad \text{and} \quad \sqrt{Var(Y)} = 0.8610.$$

Thus, a male produces on average 1.1744 male offspring with a fairly high standard deviation of 0.8610. In this case, formula (8.59) is a quadratic equation:

$$0.559 z^2 - 1.041 z + 0.482 = 0.$$

$z_1 = 1$ is surely a solution. The second solution is $z_2 = 0.86$, which is the desired probability of extinction: $\pi_0 = 0.86$.

Lotka found that the geometric distribution as given by formula (2.27), page 50, did not fit well to his data set. Hence he estimated $p_0 = 0.482$ from his data and calculated the $p_1, p_2, \dots$ in such a way that their sum is $1 - p_0 = 0.518$:

$$p_i = 0.518 \cdot (1 - 0.559) \cdot 0.559^{i-1}; \quad i = 1, 2, \dots.$$

Generally, for any fixed $p_0 = P(Y = 0)$ with $0 < p_0 < 1$, the p_0-*modified geometric distribution* is given by the probability mass function

$$p_i = P(Y = i) = (1 - p_0)p(1 - p)^{i-1}; \quad i = 1, 2, \dots \tag{8.61}$$

By formula (2.16), page 48, $\sum_{i=1}^{\infty}(1 - p)^{i-1} = \sum_{i=0}^{\infty}(1 - p)^i = 1/p$ so that indeed

$$\sum_{i=0}^{\infty} p_i = 1. \qquad \square$$

Some individuals have the potential to produce a huge number of offspring (locusts, turtles, fish), even if only a few of them may reach adulthood (defined by the time when being capable of reproduction). In these cases a distribution allowing for infinite offspring is a suitable model. For human populations, a truncated distribution (page 71) can be expected to provide best results. For instance, consider the truncated p_0-modified geometrical distribution with upper limit m, i.e., m is the maximal number of offspring an individual can produce. The probability p_0, for being directly estimated from the sample, is not subject to truncation. Given the probabilities (8.61), making use of the series (2.118), the truncated p_0-modified geometric distribution $\{p_0, p_1, \dots, p_m\}$ is for any p_0 with $0 < p_0 < 1$ defined by

$$p_0, \quad p_i = \frac{1 - p_0}{1 - (1 - p)^m} p(1 - p)^{i-1}, \quad i = 1, 2, \dots, m. \tag{8.62}$$

Example 8.19 A female thrush produces up to 4 eggs a year from which adult birds arise. The random number Y of such eggs has the distribution $p_i = P(Y = i)$ with

$$p_0 = 0.32, \ p_1 = 0.24, \ p_2 = 0.28, \ p_3 = 0,10, \ p_4 = 0.06.$$

The corresponding mean value is $E(Y) = 1.34$, and the z-transform is

$$M(z) = 0.32 + 0.24z + 0.28z^2 + 0,10z^3 + 0.06z^4.$$

The probability of extinction of the whole offspring of the zeroth generation thrush in one of the subsequent generations is the smallest solution of the equation $M(z) = z$. This solution $\pi_0 = 0.579$. $\qquad \square$

Example 8.20 Let the random number of offspring Y have a mixed Poisson distribution with continuous structure parameter L with density $f_L(\lambda)$. Then Y has the z-transform (see page 98)

$$M(z) = \int_0^{\infty} e^{\lambda(z-1)} f_L(\lambda) \, d\lambda.$$

The structure parameter L is supposed to have a Gamma distribution with density given by (2.74) (page 75):

$$f_L(\lambda) = \frac{\beta^\alpha}{\Gamma(\alpha)} \lambda^{\alpha-1} e^{-\beta\lambda}; \quad \lambda > 0, \ \alpha > 0, \ \beta > 0.$$

Then $M(z)$ becomes

$$M(z) = \int_0^{\infty} e^{\lambda(z-1)} f_L(\lambda) \, d\lambda = \frac{\beta^\alpha}{\Gamma(\alpha)} \int_0^{\infty} e^{-\lambda(\beta+1-z)} \lambda^{\alpha-1} \, d\lambda.$$

Substituting $x = (\beta + 1 - z)\lambda$ gives the final form of $M(z)$:

$$M(z) = \left(\frac{\beta}{\beta + 1 - z}\right)^{\alpha}.$$

From formula (7.58) at page 281 we know that this is the z-transfom of a negative binomial distribution with parameters α and β. Its first derivative is

$$M'(z) = \frac{\alpha\,\beta^{\alpha}}{(\beta + 1 - z)^{\alpha+1}}.$$

Hence, the mean number of offspring is $E(Y) = M'(1) = \alpha/\beta$. A general solution, different to 1, of equation $M(z) = z$ has a complicated structure. Hence, only two special cases are considered.

1) $\alpha = 1$: In this case the structure parameter L has an exponential distribution with parameter β. The equation $M(z) = z$ becomes

$$z^2 - (\beta + 1)z + \beta = 0,$$

and the solutions are $z_1 = 1$ and $z_2 = \beta$. Hence, the probability of extinction will be

$$\pi_0 = 1 \text{ for } \beta \geq 1 \text{ and } \pi_0 = \beta \text{ for } \beta < 1.$$

This result is in line with $E(Y) = 1/\beta \leq 1$ for $\beta \geq 1$.

2) $\alpha = 2$, $\beta = 1.2$: In this case equation $M(z) = z$ becomes

$$z^3 - 4.4z^2 + 4.84z - 1.44 = 0.$$

The solutions are $z_1 = 1$ and $z_2 = 0.496$. Hence, the probability of extinction is

$$\pi_0 = 0.496. \qquad\qquad \square$$

8.6 EXERCISES

8.1) A Markov chain $\{X_0, X_1, ...\}$ has state space $\mathbf{Z} = \{0, 1, 2\}$ and transition matrix

$$\mathbf{P} = \begin{pmatrix} 0.5 & 0 & 0.5 \\ 0.4 & 0.2 & 0.4 \\ 0 & 0.4 & 0.6 \end{pmatrix}.$$

(1) Determine $P(X_2 = 2 | X_1 = 0, X_0 = 1)$ and $P(X_2 = 2, X_1 = 0 | X_0 = 1)$.

(2) Determine $P(X_2 = 2, X_1 = 0 | X_0 = 0)$ and, for $n > 1$,

$$P(X_{n+1} = 2, X_n = 0 | X_{n-1} = 0).$$

(3) Assuming the initial distribution

$$P(X_0 = 0) = 0.4; \quad P(X_0 = 1) = P(X_0 = 2) = 0.3,$$

determine $P(X_1 = 2)$ and $P(X_1 = 1, X_2 = 2)$.

8.2) A Markov chain $\{X_0, X_1, ...\}$ has state space $\mathbf{Z} = \{0, 1, 2\}$ and transition matrix

$$\mathbf{P} = \begin{pmatrix} 0.2 & 0.3 & 0.5 \\ 0.8 & 0.2 & 0 \\ 0.6 & 0 & 0.4 \end{pmatrix}.$$

(1) Determine the matrix of the 2-step transition probabilities $\mathbf{P}^{(2)}$.

(2) Given the initial distribution $P(X_0 = i) = 1/3$; $i = 0, 1, 2$; determine the probabilities $P(X_2 = 0)$ and $P(X_0 = 0, X_1 = 1, X_2 = 2)$.

8.3) A Markov chain $\{X_0, X_1, ...\}$ has state space $\mathbf{Z} = \{0, 1, 2\}$ and transition matrix

$$\mathbf{P} = \begin{pmatrix} 0 & 0.4 & 0.6 \\ 0.8 & 0 & 0.2 \\ 0.5 & 0.5 & 0 \end{pmatrix}.$$

(1) Given the initial distribution $P(X_0 = 0) = P(X_0 = 1) = 0.4$ and $P(X_0 = 2) = 0.2$, determine $P(X_3 = 2)$.

(2) Draw the corresponding transition graph.

(3) Determine the stationary distribution.

8.4) Let $\{Y_0, Y_1, ...\}$ be a sequence of independent, identically distributed binary random variables with $P(Y_i = 0) = P(Y_i = 1) = 1/2$; $i = 0, 1, ...$ Define a sequence of random variables $\{X_1, X_2, ...\}$ by $X_n = \frac{1}{2}(Y_n - Y_{n-1})$; $n = 1, 2, ...$

Check whether the random sequence $\{X_1, X_2, ...\}$ has the Markov property.

8.5) A Markov chain $\{X_0, X_1, ...\}$ has state space $\mathbf{Z} = \{0, 1, 2, 3\}$ and transition matrix

$$\mathbf{P} = \begin{pmatrix} 0.1 & 0.2 & 0.4 & 0.3 \\ 0.2 & 0.3 & 0.1 & 0.4 \\ 0.4 & 0.1 & 0.3 & 0.2 \\ 0.3 & 0.4 & 0.2 & 0.1 \end{pmatrix}.$$

(1) Draw the corresponding transition graph.

(2) Determine the stationary distribution of this Markov chain.

8.6) Let $\{X_0, X_1, ...\}$ be an irreducible Markov chain with state space $\mathbf{Z} = \{1, 2, ..., n\}$, $n < \infty$, and with the doubly stochastic transition matrix $\mathbf{P} = ((p_{ij}))$, i.e.,

$$\sum_{j \in \mathbf{Z}} p_{ij} = 1 \text{ for all } i \in \mathbf{Z} \text{ and } \sum_{i \in \mathbf{Z}} p_{ij} = 1 \text{ for all } j \in \mathbf{Z}.$$

(1) Prove that the stationary distribution of $\{X_0, X_1, ...\}$ is $\{\pi_j = 1/n, j \in \mathbf{Z}\}$.

(2) Can $\{X_0, X_1, ...\}$ be a transient Markov chain?

8.7) Prove formulas (8.20), page 346, for the mean times to absorption in a random walk with two absorbing barriers (example 8.3).

8.8) Show that the vector $\pi = (\pi_1 = \alpha, \pi_2 = \beta, \pi_3 = \gamma)$, determined in example 8.6, is a stationary initial distribution with regard to a Markov chain which has the one-step transition matrix (8.22) (page 349).

8.9) A source emits symbols 0 and 1 for transmission to a receiver. Random noises $S_1, S_2, \ldots$ successively and independently affect the transmission process of a symbol in the following way: if a '0' ('1') is to be transmitted, then S_i distorts it to a '1' ('0') with probability p (q); $i = 1, 2, \ldots$ Let $X_0 = 0$ or $X_0 = 1$ denote whether the source has emitted a '0' or a '1' for transmission. Further, let $X_i = 0$ or $X_i = 1$ denote whether the attack of noise S_i implies the transmission of a '0' or a '1'; $i = 1, 2, \ldots$ The random sequence $\{X_0, X_1, \ldots\}$ is an irreducible Markov chain with state space $\mathbf{Z} = \{0, 1\}$ and transition matrix

$$\mathbf{P} = \begin{pmatrix} 1-p & p \\ q & 1-q \end{pmatrix}.$$

(1) Verify: On condition $0 < p + q \le 1$, the m-step transition matrix is given by

$$\mathbf{P}^{(m)} = \frac{1}{p+q} \begin{pmatrix} q & p \\ q & p \end{pmatrix} + \frac{(1-p-q)^m}{p+q} \begin{pmatrix} p & -p \\ -q & q \end{pmatrix}.$$

(2) Let $p = q = 0.1$. The transmission of the symbols 0 and 1 is affected by the random noises $S_1, S_2, \ldots, S_5$. Determine the probability that a '0' emitted by the source is actually received.

8.10) Weather is classified as (predominantly) sunny (S) and (predominantly) cloudy (C), where C includes rain. For the town of Musi, a fairly reliable prediction of tomorrow's weather can only be made on the basis of today's and yesterday's weather. Let (C,S) indicate that the weather yesterday was cloudy and today's weather is sunny and so on. Based on past observations it is known that, given the constellation (S,S) today, the weather tomorrow will be sunny with probability 0.8 and cloudy with probability 0.2; given (S,C) today, the weather tomorrow will be sunny with probability 0.4 and cloudy with probability 0.6; given (C,S) today, the weather tomorrow will be sunny with probability 0.6 and cloudy with probability 0.4; given (C,C) today, the weather tomorrow will be cloudy with probability 0.8 and sunny with probability 0.2.

(1) Illustrate graphically the transition between the states

$$1 = (\text{S,S}), \ 2 = (\text{S,C}), \ 3 = (\text{C,S}), \text{ and } 4 = (\text{C,C}).$$

(2) Determine the matrix of the transition probabilities of the corresponding discrete-time Markov chain and its stationary state distribution.

8.11) A supplier of toner cartridges of a certain brand checks her stock every Monday. If the stock is less than or equal to s cartridges, she orders an amount of $S - s$ cartridges, which will be available the following Monday, $0 \leq s < S$. The weekly demands of cartridges D are independent and identically distributed according to

$$p_i = P(D = i); \quad i = 0, 1, \dots.$$

Let X_n be the number of cartridges on stock on the n th Sunday (no business over weekends) given that the supplier starts her business on a Monday.

(1) Is $\{X_1, X_2, \dots\}$ a Markov chain?

(2) If yes, determine its transition probabilities.

8.12) A Markov chain has state space $\mathbf{Z} = \{0, 1, 2, 3, 4\}$ and transition matrix

$$\mathbf{P} = \begin{pmatrix} 0.5 & 0.1 & 0.4 & 0 & 0 \\ 0.8 & 0.2 & 0 & 0 & 0 \\ 0 & 1 & 0 & 0 & 0 \\ 0 & 0 & 0 & 0.9 & 0.1 \\ 0 & 0 & 0 & 1 & 0 \end{pmatrix}.$$

(1) Determine the minimal closed sets.

(2) Identify essential and inessential states.

(3) What are the recurrent and transient states?

8.13) A Markov chain has state space $\mathbf{Z} = \{0, 1, 2, 3\}$ and transition matrix

$$\mathbf{P} = \begin{pmatrix} 0 & 0 & 1 & 0 \\ 1 & 0 & 0 & 0 \\ 0.4 & 0.6 & 0 & 0 \\ 0.1 & 0.4 & 0.2 & 0.3 \end{pmatrix}.$$

Determine the classes of essential and inessential states.

8.14) A Markov chain has state space $\mathbf{Z} = \{0, 1, 2, 3, 4\}$ and transition matrix

$$\mathbf{P} = \begin{pmatrix} 0 & 0.2 & 0.8 & 0 & 0 \\ 0 & 0 & 0 & 0.9 & 0.1 \\ 0 & 0 & 0 & 0.1 & 0.9 \\ 1 & 0 & 0 & 0 & 0 \\ 1 & 0 & 0 & 0 & 0 \end{pmatrix}.$$

(1) Draw the transition graph.

(2) Verify that this Markov chain is irreducible with period 3.

(3) Determine the stationary distribution.

8.15) A Markov chain has state space $\mathbf{Z} = \{0, 1, 2, 3, 4\}$ and transition matrix

$$
\mathbf{P} = \begin{pmatrix}
0 & 1 & 0 & 0 & 0 \\
1 & 0 & 0 & 0 & 0 \\
0.2 & 0.2 & 0.2 & 0.4 & 0 \\
0.2 & 0.8 & 0 & 0 & 0 \\
0.4 & 0.1 & 0.1 & 0 & 0.4
\end{pmatrix}.
$$

(1) Find the essential and inessential states.

(2) Find the recurrent and transient states.

8.16) Determine the stationary distribution of the random walk considered in example 8.12 on condition $p_i = p$, $0 < p < 1$.

8.17) The weekly power consumption of a town depends on the weekly average temperature in that town. The weekly average temperature, observed over a long time span in the month of August, has been partitioned in 4 classes (in C^0):

$$1 = [10 - 15), \quad 2 = [15 - 20), \quad 3 = [20 - 25), \quad 4 = [25 - 30].$$

The weekly average temperature fluctuations between the classes in August follow a homogeneous Markov chain with transition matrix

$$
\begin{pmatrix}
0.1 & 0.5 & 0.3 & 0.1 \\
0.2 & 0.3 & 0.3 & 0.2 \\
0.1 & 0.4 & 0.4 & 0.1 \\
0 & 0.2 & 0.5 & 0.3
\end{pmatrix}.
$$

When the weekly average temperatures are in class 1, 2, 3 or 4, the respective average power consumption per week is 1.5, 1.3, 1.2, and 1.3 [in MW]. (The increase from class 3 to class 4 is due to air conditioning.)

What is the average power consumption in the longrun in August?

8.18) A household insurer knows that the total annual claim size X of clients in a certain portfolio hasy a normal distribution with mean value $800 and standard deviation $260. The insurer partitions his clients into classes 1, 2, and 3 depending on the annual amounts they claim, and the class they belong to (all costs in $):

A client, who is in class 1 in the current year, will make a transition to class 1, 2, or 3 next year, when his respective total claims are between 0 and 600, 600 and 1000, or greater than 1000 in the current year.

A client, who is in class 2 in the current year, will make a transition to class 1, 2, or 3 next year if his respective total claim sizes are between 0 and 500, 500 and 900, or more than 900.

A client, who is in class 3 and claims between 0 and 400, between 400 and 800, or at least 800 in the current year, will be in class 1, 2, or in class 3 next year, respectively.

When in class 1, 2, or 3, the clients will pay the premiums 500, 800, or 1000 a year, respectively.

(1) What is the average annual contribution of a client in the longrun?

(2) Does the insurer make any profit under this policy in the longrun?

8.19) Two gamblers 1 and 2 begin a game with stakes of sizes $3 and $4, respectively. After each move a gambler either wins or loses $1 or the gambler's stake remains constant. These possibilities are controlled by the transition probabilities

$$p_0 = 0, \; p_1 = 0.5, \; p_2 = 0.4, \; p_3 = 0.3, \; p_4 = 0.3, \; p_5 = 0.4, \; p_6 = 0.5, \; p_7 = 0,$$
$$q_0 = 0, \; q_1 = 0.5, \; q_2 = 0.4, \; q_3 = 0.3, \; q_4 = 0.3, \; q_5 = 0.4, \; q_6 = 0.5, \; q_7 = 0.$$

(According to Figure 8.3 there is $p_i = p_{ii+1}$ and $q_i = p_{ii-1}$.) The game is finished as soon as a gambler has won the entire stake of the other one or, equivalently, if one gambler has lost her/his entire stake.

(1) Determine the respective probabilities that gambler 1 or gambler 2 wins.

(2) Determine the mean time of the game.

8.20) Analogously to example 8.17 (page 369), consider a population with a maximal size of $z = 5$ individuals, which comprises at the beginning of its observation 3 individuals. Its birth and death probabilities with regard to a time unit are

$$p_0 = 0, \; p_1 = 0.6, \; p_2 = 0.4, \; p_3 = 0.2, \; p_4 = 0.4, \; p_5 = 0,$$
$$q_0 = 0, \; q_1 = 0.4, \; q_2 = 0.4, \; q_3 = 0.6, \; q_4 = 0.5, \; q_5 = 0.8.$$

(1) What is the probability of extinction of this population?

(2) Determine its mean time to extinction.

8.21) Let the transition probabilities of a birth and death process be given by

$$p_i = \frac{1}{1 + [i/(i+1)]^2} \quad \text{and} \quad q_i = 1 - p_i; \; i = 1, 2, \ldots; \; p_0 = 1.$$

Show that the process is transient.

8.22) Let i and j be two different states with $f_{ij} = f_{ji} = 1$. Show that both i and j are recurrent.

8.23) The respective transition probabilities of two irreducible Markov chains 1 and 2 with common state space $\mathbf{Z} = \{0, 1, \ldots\}$ are for all $i = 0, 1, \ldots$,

(1) $p_{ii+1} = \dfrac{1}{i+2}, \; p_{i0} = \dfrac{i+1}{i+2}$ and (2) $p_{ii+1} = \dfrac{i+1}{i+2}, \; p_{i0} = \dfrac{1}{i+2}.$

Check whether these Markov chains are transient, null recurrent, or positive recurrent.

8.24) Let N_i be the random number of time periods a discrete-time Markov chain stays in state i (sojourn time of the Markov chain in state i).
Determine $E(N_i)$ and $Var(N_i)$.

8.25) A Galton-Watson process starts with one individual. The random number of offspring Y of this individual has the z-transform
$$M(z) = (0.6z + 0.4)^3.$$
(1) What type of probability distribution has Y (see section 2.5.1)?
(2) Determine the probabilities $P(Y = k)$.
(3) What is the corresponding probability of extinction?
(4) Let T be the random time to extinction. Determine the probability $P(T = 2)$ by applying formula (8.60). Verify this result by applying the total probability rule to $P(T = 2)$.

8.26) A Galton-Watson process starts with one individual. The random number of offspring Y of this individual has the z-transform
$$M(z) = e^{1.5(z-1)}.$$
(1) What is the underlying probability distribution of Y?
(2) Determine the corresponding probability of extinction.
(3) Let T be the random time to extinction. Determine the probability $P(T = 3)$ by applying formula (8.60).

8.27) (1) Determine the z-transform of the truncated, p_0 - modified geometric distribution given by formula (8.62).
(2) Determine the corresponding probability of extinction π_0 if
$$m = 6, \ p_0 = 0.482, \text{ and } p = 0.441.$$
(3) Compare this π_0 with the probability of extinction obtained in example (8.18) without truncation, but under otherwise the same assumptions.

8.28) Assume a Galton-Watson process starts with $X_0 = n > 1$ individuals.
Determine the corresponding probability of extinction given that the same Galton-Watson process, when starting with one individual, has probability of extinction π_0.

8.29) Given $X_0 = 1$, show that the probability of extinction π_0 satisfies equation
$$M(\pi_0) = \pi_0$$
by applying the total probability rule. (Make use of the answer to exercise 8.28.)

CHAPTER 9

Continuous-Time Markov Chains

9.1 BASIC CONCEPTS AND EXAMPLES

This chapter deals with Markov processes which have parameter set $\mathbf{T} = [0, \infty)$ and state space $\mathbf{Z} = \{0, \pm1, \pm2, \ldots\}$ or subsets of it. According to the terminology introduced in section 6.3, for having a discrete parameter space, this class of Markov processes is called *Markov chains*.

Definition 9.1 A stochastic process $\{X(t),\ t \geq 0\}$ with parameter set $\mathbf{T}$ and discrete state space $\mathbf{Z}$ is called a *continuous-time Markov chain* or a *Markov chain in continuous time* if, for any $n \geq 1$ and arbitrary sequences

$\{t_0, t_1, \ldots, t_{n+1}\}$ with $t_0 < t_1 < \cdots < t_{n+1}$ and $\{i_0, i_1, \ldots, i_{n+1}\}$, $i_k \in \mathbf{Z}$,

the following relationship holds:

$$P(X(t_{n+1}) = i_{n+1} | X(t_n) = i_n, \ldots, X(t_1) = i_1, X(t_0) = i_0) \tag{9.1}$$

$$= P(X(t_{n+1}) = i_{n+1} | X(t_n) = i_n). \qquad \bullet$$

The intuitive interpretation of the *Markov property* (9.1) is the same as for discrete-time Markov chains:

> *The future development of a continuous-time Markov chain depends only on its present state and not on its evolution in the past.*

The conditional probabilities

$$p_{ij}(s, t) = P(X(t) = j | X(s) = i); \quad s < t;\ i, j \in \mathbf{Z};$$

are the *transition probabilities of the Markov chain*. A Markov chain is said to be *homogeneous* if for all $s, t \in \mathbf{T}$ and $i, j \in \mathbf{Z}$ the transition probabilities $p_{ij}(s, t)$ depend only on the difference $t - s$:

$$p_{ij}(s, t) = p_{ij}(0, t - s).$$

In this case the transition probabilities depend only on one variable:

$$p_{ij}(t) = p_{ij}(0, t).$$

Note This chapter only considers homogeneous Markov chains. Hence no confusion can arise if only *Markov chains* are referred to.

The transition probabilities are comprised in the *matrix of transition probabilities* $\mathbf{P}$ (simply: *transition matrix*):

$$\mathbf{P}(t) = ((p_{ij}(t))); \quad i,j \in \mathbf{Z}.$$

Besides the trivial property $p_{ij}(t) \geq 0$, transition probabilities are generally assumed to satisfy the conditions

$$\sum_{j \in \mathbf{Z}} p_{ij}(t) = 1; \quad t \geq 0, \ i \in \mathbf{Z}. \tag{9.2}$$

Comment It is theoretically possible that, for some $i \in \mathbf{Z}$,

$$\sum_{j \in \mathbf{Z}} p_{ij}(t) < 1; \quad t > 0, \ i \in \mathbf{Z}. \tag{9.3}$$

In this case, unboundedly many transitions between the states may occur in any finite time interval $[0, t)$ with positive probability

$$1 - \sum_{j \in \mathbf{Z}} p_{ij}(t).$$

This situation approximately applies to nuclear chain reactions and population explosions of certain species of insects (e.g., locusts). Henceforth it is assumed that

$$\lim_{t \to +0} p_{ii}(t) = 1. \tag{9.4}$$

By (9.2), this assumption is equivalent to

$$p_{ij}(0) = \lim_{t \to +0} p_{ij}(t) = \delta_{ij}; \quad i,j \in \mathbf{Z}. \tag{9.5}$$

The *Kronecker symbol* δ_{ij} is defined by formula (8.4), page 340.

Analogously to (8.5), the ***Chapman-Kolmogorov equations*** are

$$p_{ij}(t+\tau) = \sum_{k \in \mathbf{Z}} p_{ik}(t) p_{kj}(\tau) \tag{9.6}$$

for any $t \geq 0$, $\tau \geq 0$, and $i,j \in \mathbf{Z}$. By making use of the total probability rule, the homogeneity, and the Markov property, formula (9.6) is proved as follows:

$$p_{ij}(t+\tau) = P(X(t+\tau) = j | X(0) = i) = \frac{P(X(t+\tau) = j, X(0) = i)}{P(X(0) = i)}$$

$$= \sum_{k \in \mathbf{Z}} \frac{P(X(t+\tau) = j, X(t) = k, X(0) = i)}{P(X(0) = i)}$$

$$= \sum_{k \in \mathbf{Z}} \frac{P(X(t+\tau) = j | X(t) = k, X(0) = i) P(X(t) = k, X(0) = i)}{P(X(0) = i)}$$

$$= \sum_{k \in \mathbf{Z}} \frac{P(X(\tau+t) = j | X(t) = k) P(X(t) = k | X(0) = i) P(X(0) = i)}{P(X(0) = i)}$$

$$= \sum_{k \in \mathbf{Z}} P(X(\tau) = j | X(0) = k) P(X(t) = k | X(0) = i)$$

$$= \sum_{k \in \mathbf{Z}} p_{ik}(t) p_{kj}(\tau).$$

Absolute and Stationary Distributions Let $p_i(t) = P(X(t) = i)$ be the probability that the Markov chain is in state i at time t. $p_i(t)$ is called the *absolute state probability* (of the Markov chain) at time t. Hence, $\{p_i(t), i \in \mathbf{Z}\}$ is said to be the *absolute (one-dimensional) probability distribution* of the Markov chain at time t. In particular, $\{p_i(0); i \in \mathbf{Z}\}$ is called an *initial (probability) distribution* of the Markov chain. By the total probability rule, given an initial distribution, the absolute probability distribution of the Markov chain at time t is

$$p_j(t) = \sum_{i \in \mathbf{Z}} p_i(0) \, p_{ij}(t), \quad j \in \mathbf{Z}. \tag{9.7}$$

For determining the *multidimensional distribution* of the Markov chain at time points $t_0, t_1, ..., t_n$ with $0 \le t_0 < t_1 < \cdots < t_n < \infty$, only its absolute probability distribution at time t_0 and its transition probabilities need to be known. This can be proved by repeated application of the formula of the conditional probability (1.22) and by making use of homogeneity of the Markov chain:

$$P(X(t_0) = i_0, X(t_1) = i_1, ..., X(t_n) = i_n)$$
$$= p_{i_0}(t_0) p_{i_0 i_1}(t_1 - t_0) p_{i_1 i_2}(t_2 - t_1) \cdots p_{i_{n-1} i_n}(t_n - t_{n-1}). \tag{9.8}$$

Definition 9.2 An initial distribution $\{\pi_i = p_i(0), i \in \mathbf{Z}\}$ is said to be *stationary* if

$$\pi_i = p_i(t) \quad \text{for all } t \ge 0 \text{ and } i \in \mathbf{Z}. \tag{9.9}$$

$\bullet$

Thus, if at time $t = 0$ the initial state is determined by a stationary initial distribution, then the absolute state probabilities $p_j(t)$ do not depend on t and are equal to π_j. Consequently, the stationary initial probabilities π_j are the absolute state probabilities $p_j(t)$ for all $j \in \mathbf{Z}$ and $t \ge 0$. Moreover, it follows from (9.8) that in this case all n-dimensional distributions of the Markov chain, namely

$$\{P(X(t_1 + h) = i_1, X(t_2 + h) = i_2, ..., X(t_n + h) = i_n\}, \quad i_j \in \mathbf{Z} \tag{9.10}$$

do not depend on h, i.e. if the process starts with a stationary initial distribution, then the Markov chain is strictly stationary. (This result once more verifies the more general statement of theorem 6.1, page 234.) Moreover, it is justified to call $\{\pi_i, i \in \mathbf{Z}\}$ a *stationary (probability) distribution* of the Markov chain.

Example 9.1 The homogeneous Poisson process $\{N(t), t \ge 0\}$ with intensity λ is a homogeneous Markov chain with state space $\mathbf{Z} = \{0, 1, ...\}$ and transition probabilities

$$p_{ij}(t) = \frac{(\lambda t)^{j-i}}{(j-i)!} e^{-\lambda t}; \quad i \le j.$$

The sample paths of the process $\{N(t), t \ge 0\}$ are nondecreasing step-functions. Its trend function is linearly increasing: $m(t) = E(N(t)) = \lambda t$. Thus, a stationary initial distribution cannot exist. (But, by the corollary following definition 7.1 (page 259), the homogeneous Poisson process is a stationary point process.) $\square$

Example 9.2 At time $t = 0$ exactly n systems start operating. Their lifetimes are independent, identically distributed exponential random variables with parameter λ. Let $X(t)$ be the number of systems still operating at time t. Then $\{X(t), t \geq 0\}$ is a Markov chain with state space $\mathbf{Z} = \{0, 1, ..., n\}$, transition probabilities

$$p_{ij}(t) = \binom{i}{i-j} (1 - e^{-\lambda t})^{i-j} e^{-\lambda tj}, \quad n \geq i \geq j \geq 0,$$

and initial distribution $P(X(0) = n) = 1$. The structure of these transition probabilities is due to the memoryless property of the exponential distribution (see example 2.21, page 87). Of course, this Markov chain cannot be stationary. □

Example 9.3 Let $\mathbf{Z} = \{0, 1)$ be the state space and

$$\mathbf{P}(t) = \begin{pmatrix} \dfrac{1}{t+1} & \dfrac{t}{t+1} \\ \dfrac{t}{t+1} & \dfrac{1}{t+1} \end{pmatrix}$$

the transition matrix of a stochastic process $\{X(t), t \geq 0\}$. It is to check whether this process is a Markov chain. Assuming the initial distribution

$$p_0(0) = P(X(0) = 0) = 1$$

and applying formula (9.7) yields the absolute probability of state 0 at time $t = 3$:

$$p_0(3) = p_0(0) p_{00}(3) = 1/4.$$

On the other hand, applying (9.6) with $t = 2$ and $\tau = 1$ yields the (wrong) result

$$p_0(3) = p_{00}(2) p_{00}(1) + p_{01}(2) p_{10}(1) = 1/2.$$

Therefore, Chapman-Kolmogorov's equations (9.6) are not valid so that $\{X(t), t \geq 0\}$ cannot be a Markov chain. □

Classification of States The classification concepts already introduced for discrete-time Markov chains can analogously be defined for continuous-time Markov chains. In what follows, some concepts are defined, but not discussed in detail.

A state set $\mathbf{C} \subseteq \mathbf{Z}$ is called *closed* if

$$p_{ij}(t) = 0 \text{ for all } t > 0, \, i \in \mathbf{C} \text{ and } j \notin \mathbf{C}.$$

If, in particular, $\{i\}$ is a closed set, then i is called an *absorbing state*. The state j is *accessible* from i if there exists a t with $p_{ij}(t) > 0$.

If i and j are accessible from each other, then they are said to *communicate*. Thus, equivalence classes, essential, and inessential states, as well as irreducible and reducible Markov chains can be defined as in section 8.2 for discrete Markov chains.

State i is *recurrent* (*transient*) if

$$\int_0^\infty p_{ii}(t)\,dt = \infty \quad \left(\int_0^\infty p_{ii}(t)\,dt < \infty\right).$$

A recurrent state i is *positive recurrent* if the mean value of its recurrence time (time between two successive occurences of state i) is finite. Since it can easily be shown that $p_{ij}(t_0) > 0$ implies $p_{ij}(t) > 0$ for all $t > t_0$, introducing the concept of a period analogously to section 8.2.3 makes no sense.

9.2 TRANSITION PROBABILITIES AND RATES

This section discusses some structural properties of continuous-time Markov chains, which are fundamental to mathematically modeling real systems.

Theorem 9.1 On condition (9.4), the transition probabilities $p_{ij}(t)$ are differentiable in $[0, \infty)$ for all $i,j \in \mathbf{Z}$.

Proof For any $h > 0$, the Chapman-Kolmogorov equations (9.6) yield

$$p_{ij}(t+h) - p_{ij}(t) = \sum_{k \in \mathbf{Z}} p_{ik}(h)p_{kj}(t) - p_{ij}(t)$$

$$= -(1 - p_{ii}(h))p_{ij}(t) + \sum_{k \in \mathbf{Z},\, k \neq i} p_{ik}(h)p_{kj}(t).$$

Thus,

$$-(1 - p_{ii}(h)) \le -(1 - p_{ii}(h))p_{ij}(t) \le p_{ij}(t+h) - p_{ij}(t)$$

$$\le \sum_{\substack{k \in \mathbf{Z} \\ k \neq i}} p_{ik}(h)p_{kj}(t) \le \sum_{\substack{k \in \mathbf{Z} \\ k \neq i}} p_{ik}(h)$$

$$= 1 - p_{ii}(h).$$

Hence,

$$\left| p_{ij}(t+h) - p_{ij}(t) \right| \le 1 - p_{ii}(h).$$

The uniform continuity of the transition probabilities and, therefore, their differentiability for all $t \ge 0$ is now a consequence of assumption (9.4). ∎

Transition Rates The following limits play an important role in future derivations. For any $i,j \in \mathbf{Z}$, let

$$q_i = \lim_{h \to 0} \frac{1 - p_{ii}(h)}{h}, \tag{9.11}$$

$$q_{ij} = \lim_{h \to 0} \frac{p_{ij}(h)}{h}, \quad i \neq j. \tag{9.12}$$

These limits exist, since by (9.5),

$$p_{ii}(0) = 1 \text{ and } p_{ij}(0) = 0 \text{ for } i \neq j$$

so that, by theorem 9.1,

$$p'_{ii}(0) = \left.\frac{dp_{ii}(t)}{dt}\right|_{t=0} = -q_i, \tag{9.13}$$

$$p'_{ij}(0) = \left.\frac{dp_{ij}(t)}{dt}\right|_{t=0} = q_{ij}, \quad i \neq j. \tag{9.14}$$

For $h \to 0$, relationships (9.13) and (9.14) are equivalent to

$$p_{ii}(h) = 1 - q_i h + o(h) \tag{9.15}$$

$$p_{ij}(h) = q_{ij} h + o(h), \quad i \neq j, \tag{9.16}$$

respectively. The parameters q_i and q_{ij} are the *transition rates* of the Markov chain. More exactly, q_i is the *unconditional transition rate* of leaving state i for any other state, and q_{ij} is the *conditional transition rate* of making a transition from state i to state j. According to (9.2),

$$\sum_{\{j, j \neq i\}} q_{ij} = q_i, \quad i \in \mathbf{Z}. \tag{9.17}$$

Kolmogorov's Differential Equations In what follows, systems of differential equations for the transition probabilities and the absolute state probabilities of a Markov chain are derived. For this purpose, the system of the Chapman-Kolmogorov equations is written in the form

$$p_{ij}(t+h) = \sum_{k \in \mathbf{Z}} p_{ik}(h) p_{kj}(t).$$

It follows that

$$\frac{p_{ij}(t+h) - p_{ij}(t)}{h} = \sum_{k \neq i} \frac{p_{ik}(h)}{h} p_{kj}(t) - \frac{1 - p_{ii}(h)}{h} p_{ij}(t).$$

By (9.13) and (9.14), letting $h \to 0$ yields *Kolmogorov's backward equations* for the transition probabilities:

$$p'_{ij}(t) = \sum_{k \neq i} q_{ik} p_{kj}(t) - q_i p_{ij}(t), \quad t \geq 0. \tag{9.18}$$

Analogously, starting with

$$p_{ij}(t+h) = \sum_{k \in \mathbf{Z}} p_{ik}(t) p_{kj}(h)$$

yields *Kolmogorov's forward equations* for the transition probabilities:

$$p'_{ij}(t) = \sum_{k \neq j} p_{ik}(t) q_{kj} - q_j p_{ij}(t), \quad t \geq 0. \tag{9.19}$$

Let $\{p_i(0), i \in \mathbf{Z}\}$ be any initial distribution. Multiplying Kolmogorov's forward equations (9.19) by $p_i(0)$ and summing with respect to i yields

$$\sum_{i \in \mathbf{Z}} p_i(0) p'_{ij}(t) = \sum_{i \in \mathbf{Z}} p_i(0) \sum_{k \neq j} p_{ik}(t) q_{kj} - \sum_{i \in \mathbf{Z}} p_i(0) q_j p_{ij}(t)$$

$$= \sum_{k \neq j} q_{kj} \sum_{i \in \mathbf{Z}} p_i(0) p_{ik}(t) - q_j \sum_{i \in \mathbf{Z}} p_i(0) p_{ij}(t).$$

Thus, in view of (9.7), the absolute state probabilities satisfy the system of linear differential equations

$$p'_j(t) = \sum_{k \neq j} q_{kj} p_k(t) - q_j p_j(t), \quad t \geq 0, j \in \mathbf{Z}. \tag{9.20}$$

In future, the absolute state probabilities are assumed to satisfy

$$\sum_{i \in \mathbf{Z}} p_i(t) = 1. \tag{9.21}$$

This *normalizing condition* is always fulfilled if $\mathbf{Z}$ is finite.

Note If the initial distribution has structure

$$p_i(0) = 1, \; p_j(0) = 0 \text{ for } j \neq i,$$

then the absolute state probabilities are equal to the transition probabilities

$$p_j(t) = p_{ij}(t), \; j \in \mathbf{Z}.$$

Transition Times and Transition Rates It is only possible to exactly model real systems by continuous-time Markov chains if the lengths of the time periods between changes of states are exponentially distributed, since in this case the 'memoryless property' of the exponential distribution (example 2.21, page 87) implies the Markov property. If the times between transitions have known exponential distributions, then it is no problem to determine the transition rates. For instance, if the sojourn time of a Markov chain in state 0 has an exponential distribution with parameter λ_0, then, according to (9.11), the unconditional rate of leaving this state is given by

$$q_0 = \lim_{h \to 0} \frac{1 - p_{00}(h)}{h} = \lim_{h \to 0} \frac{1 - e^{-\lambda_0 h}}{h}$$

$$= \lim_{h \to 0} \frac{\lambda_0 h + o(h)}{h} = \lambda_0 + \lim_{h \to 0} \frac{o(h)}{h}.$$

Hence,

$$q_0 = \lambda_0. \tag{9.22}$$

Now, let the sojourn time of a Markov chain in state 0 have structure

$$Y_0 = \min (Y_{01}, Y_{02}),$$

where Y_{01} and Y_{02} are independent exponential random variables with respective parameters λ_1 and λ_2. If $Y_{01} < Y_{02}$, the Markov chain makes a transition to state 1 and if $Y_{01} > Y_{02}$ to state 2. Thus, by (9.12), the conditional transition rate from state 0 to state 1 is

$$q_{01} = \lim_{h \to 0} \frac{p_{01}(h)}{h} = \lim_{h \to 0} \frac{(1 - e^{-\lambda_1 h}) e^{-\lambda_2 h} + o(h)}{h}$$

$$= \lim_{h \to 0} \frac{\lambda_1 h (1 - \lambda_2 h)}{h} + \lim_{h \to 0} \frac{o(h)}{h}$$

$$= \lim_{h \to 0} (\lambda_1 - \lambda_1 \lambda_2 h) = \lambda_1.$$

Hence, since the roles of Y_{01} and Y_{02} can be interchanged,

$$q_{01} = \lambda_1, \quad q_{02} = \lambda_2, \quad q_0 = \lambda_1 + \lambda_2. \tag{9.23}$$

The results (9.22) and (9.23) will be generalized in section 9.4.

Transition Graphs Establishing the Kolmogorov equations can be facilitated by *transition graphs*. These graphs are constructed analogously to the transition graphs for discrete-time Markov chains: The nodes of a transition graph represent the states of the Markov chain. A (directed) edge from node i to node j exists if and only if $q_{ij} > 0$. The edges are weighted by their corresponding transition rates. Thus, two sets of states (possibly empty ones) can be assigned to each node i: first edges with initial node i and second edges with end node i, that is, edges which leave node i and edges which end in node i. The unconditional transition rate q_i equals the sum of the weights of all those edges leaving node i. If there is an edge ending in state i and no edge leaving state i, then i is an absorbing state.

Example 9.4 (*system with renewal*) The lifetime L of a system has an exponential distribution with parameter λ. After a failure the system is replaced by an equivalent new one. A replacement takes a random time Z, which is exponentially distributed with parameter μ. All life- and replacement times are assumed to be independent. Thus, the operation of the system can be described by an alternating renewal process (section 7.3.6) with 'typical renewal cycle' (L, Z). Consider the Markov chain $\{X(t), t \geq 0\}$ defined by

$$X(t) = \begin{cases} 1 & \text{if the system is operating} \\ 0 & \text{if the system is being replaced} \end{cases}.$$

Its state space is $\mathbf{Z} = \{0, 1\}$. The absolute state probability $p_1(t) = P(X(t) = 1)$ of this Markov chain is the *point availability* of the system at time t.

In this simple example, only state changes from 0 to 1 and from 1 to 0 are possible. Hence, by (9.22),

$$q_0 = q_{01} = \mu \quad \text{and} \quad q_1 = q_{10} = \lambda.$$

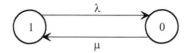

Figure 9.1 Transition graph of an alternating renewal process (example 9.4)

The corresponding Kolmogorov differential equations (9.20) are

$$p_0'(t) = -\mu p_0(t) + \lambda p_1(t),$$
$$p_1'(t) = +\mu p_0(t) - \lambda p_1(t).$$

These two equations are linearly dependent. (The sums at the left hand-sides and the right-hand sides are equal to 0.) Replacing $p_0(t)$ in the second equation by $1 - p_1(t)$ yields a first-order nonhomogeneous differential equation with constant coefficients for $p_1(t)$:

$$p_1'(t) + (\lambda + \mu)p_1(t) = \mu.$$

Given the initial condition $p_1(0) = 1$, the solution is

$$p_1(t) = \frac{\mu}{\lambda + \mu} + \frac{\lambda}{\lambda + \mu} e^{-(\lambda + \mu)t}, \quad t \ge 0.$$

The corresponding stationary availability is

$$\pi_1 = \lim_{t \to \infty} p_1(t) = \frac{\mu}{\lambda + \mu}.$$

In example 7.19, page 322) the same results have been obtained by applying the Laplace transform. (There the notation $L = Y$ is used.)   □

Example 9.5 (*two-unit redundant system, standby redundancy*) A system consists of two identical units. The system is available if and only if at least one of its units is available. If both units are available, then one of them is in standby redundancy (cold redundancy), that is, in this state it does not age and cannot fail. After the failure of a unit, the other one (if available) is immediately switched from the redundancy state to the operating state and the replacement of the failed unit begins. The replaced unit becomes the standby unit if the other unit is still operating. Otherwise it immediately resumes its work. The lifetimes and replacement times of the units are independent random variables, identically distributed as L and Z, respectively. L and Z are assumed to be exponentially distributed with respective parameters λ and μ. Let L_s denote the system lifetime, i.e. the random time to a system failure. A system failure occurs when a unit fails whilst the other unit is being replaced. A Markov chain $\{X(t), t \ge 0\}$ with state space $\mathbf{Z} = \{0, 1, 2\}$ is introduced in the following way: $X(t) = i$ if i units are unavailable at time t. Let Y_i be the unconditional sojourn time of the system in state i and Y_{ij} be the conditional sojourn time of the system in state i given that the system makes a transition from state i into state j. From state 0, the system can only

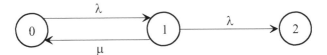

Figure 9.2 Transition graph for example 9.5 a)

make a transition to state 1. Hence, $Y_0 = Y_{01} = L$. According to (9.22), the corresponding transition rate is given by

$$q_0 = q_{01} = \lambda.$$

If the system makes a transition from state 1 to state 2, then its conditional sojourn time in state 1 is $Y_{12} = L$, whereas in case of a transition to state 0, it stays a time $Y_{10} = Z$ in state 1. The unconditional sojourn time of the system in state 1 is

$$Y_1 = \min (L, Z).$$

Thus, by (9.23), the corresponding transition rates are

$$q_{12} = \lambda, \; q_{10} = \mu, \; \text{and} \; q_1 = \lambda + \mu.$$

When the system returns from state 1 to state 0, then it again spends time L in state 0, since the operating unit is 'as good as new' in view of the memoryless property of the exponential distribution.

a) *Survival probability* In this case, only the time to entering state 2 (system failure) is of interest. Hence, state 2 must be considered absorbing (Figure 9.2) so that

$$q_{20} = q_{21} = 0.$$

The survival probability of the system has the structure

$$\overline{F}_s(t) = P(L_s > t) = p_0(t) + p_1(t).$$

The corresponding system of differential equations (9.20) is

$$p_0'(t) = -\lambda p_0(t) + \mu p_1(t),$$
$$p_1'(t) = +\lambda p_0(t) - (\lambda + \mu) p_1(t), \qquad (9.24)$$
$$p_2'(t) = +\lambda p_1(t).$$

This system of differential equations will be solved on condition that both units are available at time $t = 0$. Combining the first two differential equations in (9.24) yields a homogeneous differential equation of the second order with constant coefficients for $p_0(t)$:

$$p_0''(t) + (2\lambda + \mu) p_0'(t) + \lambda^2 p_0(t) = 0.$$

The corresponding characteristic equation is

$$x^2 + (2\lambda + \mu) x + \lambda^2 = 0.$$

Its solutions are

$$x_{1,2} = -\left(\lambda + \frac{\mu}{2}\right) \pm \sqrt{\lambda\mu + \mu^2/4} \ .$$

Hence, since $p_0(0) = 1$, for $t \geq 0$,

$$p_0(t) = a \sinh \frac{c}{2} t \quad \text{with} \quad c = \sqrt{4\lambda\mu + \mu^2} \ .$$

Since $p_1(0) = 0$, the first differential equation in (9.24) yields $a = 2\lambda/c$ and

$$p_1(t) = e^{-\frac{2\lambda+\mu}{2}t} \left(\frac{\mu}{c} \sinh \frac{c}{2} t + \cosh \frac{c}{2} t\right), \quad t \geq 0 .$$

Thus, the survival probability of the system is

$$\bar{F}_s(t) = e^{-\frac{2\lambda+\mu}{2}} \left[\cosh \frac{c}{2} t + \frac{2\lambda+\mu}{c} \sinh \frac{c}{2} t\right], \quad t \geq 0 .$$

(For a definition of the hyperbolic functions sinh and cosh, see page 265). The mean value of the system lifetime L_s is most easily obtained from formula (2.52), page 64:

$$E(L_s) = \frac{2}{\lambda} + \frac{\mu}{\lambda^2} . \tag{9.25}$$

For the sake of comparison, in case of no replacement ($\mu = 0$), the system lifetime L_s has an Erlang distribution with parameters 2 and λ:

$$\bar{F}_s(t) = (1 + \lambda t) e^{-\lambda t}, \quad E(L_s) = 2/\lambda .$$

b) *Availability* If the replacement of failed units is continued after system failures, then the point availability

$$A(t) = p_0(t) + p_1(t)$$

of the system is of particular interest. In this case, the transition rate q_{21} from state 2 to state 1 is positive. However, q_{21} depends on the number $r = 1$ or $r = 2$ of mechanics which are in charge of the replacement of failed units. Assuming that a mechanic cannot replace two failed units at the same time, then (see Figure 9.3)

$$q_{21} = q_2 = r\mu.$$

For $r = 2$, the sojourn time of the system in state 2 is given by $Y_2 = \min(Z_1, Z_2)$, where Z_1 and Z_2 are independent and identically as Z distributed. Analogously, the sojourn time in state 1 is given by $Y_1 = \min(L, Z)$.

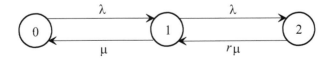

Figure 9.3 Transition graph for example 9.5 b)

Hence, the transition rates q_{10} and q_{12} have the same values as under a). The corresponding system of differential equations (9.20) becomes, when replacing the last differential equation by the normalizing condition (9.21),

$$p_0'(t) = -\lambda p_0(t) + \mu p_1(t),$$

$$p_1'(t) = +\lambda p_0(t) - (\lambda + \mu)p_1(t) + r\mu p_2(t),$$

$$1 = p_0(t) + p_1(t) + p_2(t).$$

The solution is left as an exercise to the reader.   □

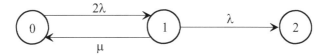

Figure 9.4 Transition graph for example 9.6 a)

Example 9.6 (*two-unit system, parallel redundancy*) Now assume that both units of the system operate at the same time when they are available. All other assumptions and the notation of the previous example are retained. In particular, the system is available if and only if at least one unit is available. In view of the initial condition $p_0(0) = 1$, the system spends

$$Y_0 = \min (L_1, L_2)$$

time units in state 0. Y_0 has an exponential distribution with parameter 2λ, and from state 0 only a transition to state 1 is possible. Therefore, $Y_0 = Y_{01}$ and

$$q_0 = q_{01} = 2\lambda.$$

When the system is in state 1, then it behaves as in example 9.5:

$$q_{10} = \mu, \quad q_{12} = \lambda, \quad q_1 = \lambda + \mu.$$

a) *Survival probability* As in the previous example, state 2 has to be thought of as absorbing: $q_{20} - q_{21} = 0$ (Figure 9.4). Hence, from (9.20) and (9.21),

$$p_0'(t) = -2\lambda p_0(t) + \mu p_1(t),$$

$$p_1'(t) = +2\lambda p_0(t) - (\lambda + \mu)p_1(t),$$

$$1 = p_0(t) + p_1(t) + p_2(t).$$

Combining the first two differential equations yields a homogeneous differential equation of the second order with constant coefficients for $p_0(t)$:

$$p_0''(t) + (3\lambda + \mu)p_0'(t) + 2\lambda^2 p_0(t) = 0.$$

The solution is

$$p_0(t) = e^{-\left(\frac{3\lambda+\mu}{2}\right)t}\left[\cosh\frac{c}{2}t + \frac{\mu-\lambda}{c}\sinh\frac{c}{2}t\right],$$

where

$$c = \sqrt{\lambda^2 + 6\lambda\mu + \mu^2}.$$

Furthermore,

$$p_1(t) = \frac{4\lambda}{c}e^{-\left(\frac{3\lambda+\mu}{2}\right)t}\sinh\frac{c}{2}t.$$

The survival probability of the system is

$$\overline{F}_s(t) = P(L_s > t) = p_0(t) + p_1(t).$$

Hence,

$$\overline{F}_s(t) = e^{-\left(\frac{3\lambda+\mu}{2}\right)t}\left[\cosh\frac{c}{2}t + \frac{3\lambda+\mu}{c}\sinh\frac{c}{2}t\right], \quad t \geq 0. \tag{9.26}$$

The mean system lifetime is

$$E(L_s) = \frac{3}{2\lambda} + \frac{\mu}{2\lambda^2}.$$

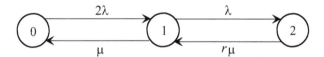

Figure 9.5 Transition graph for example 9.6 b)

For the sake of comparison, in the case without replacement ($\mu = 0$),

$$\overline{F}(t) = 2e^{-\lambda t} - e^{-2\lambda t}, \quad E(L_s) = \frac{3}{2\lambda}.$$

b) *Availability* If r ($r = 1$ or $r = 2$) mechanics replace failed units, then

$$q_2 = q_{21} = r\mu.$$

The other transition rates are the same as those under a) (Figure 9.5). The absolute state probabilities satisfy the system differential equations

$$p_0'(t) = -2\lambda p_0(t) + \mu p_1(t),$$

$$p_1'(t) = +2\lambda p_0(t) - (\lambda+\mu)p_1(t) + rp_2(t),$$

$$1 = p_0(t) + p_1(t) + p_2(t).$$

Solving this system of linear differential equations is left to the reader. ◻

9.3 STATIONARY STATE PROBABILITIES

If $\{\pi_j, j \in \mathbf{Z}\}$ is a stationary distribution of the Markov chain $\{X(t), t \geq 0\}$, then this special absolute distribution must satisfy Kolmogorov's equations (9.20). Since the π_j are constant, all the left-hand sides of these equations are equal to 0. Therefore, the system of linear differential equations (9.20) simplifies to a system of linear algebraic equations in the unknowns π_j:

$$0 = \sum_{k \in \mathbf{Z}, k \neq j} q_{kj} \pi_k - q_j \pi_j, \ j \in \mathbf{Z}. \tag{9,27}$$

This system of equations is frequently written in the form

$$q_j \pi_j = \sum_{k \in \mathbf{Z}, k \neq j} q_{kj} \pi_k, \quad j \in \mathbf{Z}. \tag{9.28}$$

This form clearly illustrates that the stationary state probabilities refer to an equilibrium state of the Markov chain:

> *The mean intensity per unit time of leaving state j, which is $q_j \pi_j$, is equal to the mean intensity per unit time of arriving at state j.*

According to assumption (9.21), only those solutions $\{\pi_j, j \in \mathbf{Z}\}$ of (9.27), which satisfy the normalizing condition, are of interest:

$$\sum_{j \in \mathbf{Z}} \pi_j = 1. \tag{9.29}$$

It is now assumed that the Markov chain is irreducible and positive recurrent. (Recall that an irreducible Markov chain with finite state space $\mathbf{Z}$ is always positive recurrent.) Then it can be shown that a unique stationary distribution $\{\pi_j, j \in \mathbf{Z}\}$ exists, which satisfies (9.27) and (9.29). Moreover, in this case the limits

$$p_j = \lim_{t \to \infty} p_{ij}(t)$$

exist and are independent of i. Hence, for any initial distribution, there exist the limits of the absolute state probabilities $\lim_{t \to \infty} p_j(t)$, and they are equal to p_j:

$$p_j = \lim_{t \to \infty} p_j(t), \ j \in \mathbf{Z}. \tag{9.30}$$

Furthermore, for all $j \in \mathbf{Z}$,

$$\lim_{t \to \infty} p_j'(t) = 0.$$

Otherwise, $p_j(t)$ would unboundedly increase as $t \to \infty$, contradictory to $p_j(t) \leq 1$. Hence, when passing to the limit as $t \to \infty$ in (9.20) and (9.21), the limits (9.30) are seen to satisfy the system of equations (9.27) and (9.29). Since this system has a unique solution, the limits p_j and the stationary probabilities π_j must coincide:

$$p_j = \pi_j, \ j \in \mathbf{Z}.$$

For a detailed discussion of the relationship between the solvability of (9.27) and the existence of a stationary distribution; see *Feller* (1968).

Continuation of Example 9.5 (*two-unit system, standby redundancy*) Since the sys- tem is available if at least one unit is available, its stationary availability is

$$A = \pi_0 + \pi_1.$$

When substituting the transition rates from Figure 9.3 into (9.27) and (9.29), the π_j are seen to satisfy the following system of algebraic equations

$$-\lambda \pi_0 + \quad \mu \pi_1 \quad = 0,$$
$$+\lambda \pi_0 - (\lambda + \mu) \pi_1 + r \pi_2 = 0,$$
$$\pi_0 + \quad \pi_1 + \pi_2 = 1.$$

Case $r = 1$

$$\pi_0 = \frac{\mu^2}{(\lambda + \mu)^2 - \lambda \mu}, \quad \pi_1 = \frac{\lambda \mu}{(\lambda + \mu)^2 - \lambda \mu}, \quad \pi_2 = \frac{\lambda^2}{(\lambda + \mu)^2 - \lambda \mu},$$

$$A = \pi_0 + \pi_1 = \frac{\mu^2 + \lambda \mu}{(\lambda + \mu)^2 - \lambda \mu}.$$

Case $r = 2$

$$\pi_0 = \frac{2\mu^2}{(\lambda + \mu)^2 + \mu^2}, \quad \pi_1 = \frac{2\lambda \mu}{(\lambda + \mu)^2 + \mu^2}, \quad \pi_2 = \frac{\lambda^2}{(\lambda + \mu)^2 + \mu^2},$$

$$A = \pi_0 + \pi_1 = \frac{2\mu^2 + 2\lambda \mu}{(\lambda + \mu)^2 + \mu^2}.$$

Continuation of Example 9.6 (*two-unit system, parallel redundancy*) Given the transition rates in Figure 9.5, the π_j are solutions of

$$-2\lambda \pi_0 + \quad \mu \pi_1 \quad = 0,$$
$$+2\lambda \pi_0 - (\lambda + \mu) \pi_1 + r \mu \pi_2 = 0,$$
$$\pi_0 + \quad \pi_1 + \pi_2 = 1.$$

Case $r = 1$

$$\pi_0 = \frac{\mu^2}{(\lambda + \mu)^2 + \lambda^2}, \quad \pi_1 = \frac{2\lambda \mu}{(\lambda + \mu)^2 + \lambda^2}, \quad \pi_2 = \frac{2\lambda^2}{(\lambda + \mu)^2 + \lambda^2},$$

$$A = \pi_0 + \pi_1 = \frac{\mu^2 + 2\lambda \mu}{(\lambda + \mu)^2 + \lambda^2}.$$

Case r = 2

$$\pi_0 = \frac{\mu^2}{(\lambda + \mu)^2}, \quad \pi_1 = \frac{2\lambda\mu}{(\lambda + \mu)^2}, \quad \pi_2 = \frac{\mu^2}{(\lambda + \mu)^2},$$

$$A = \pi_0 + \pi_1 = 1 - \left(\frac{\lambda}{\lambda + \mu}\right)^2.$$

Figure 9.6 shows a) the mean lifetimes and b) the stationary availabilities of the two-unit system for $r = 1$ as functions of $\rho = \lambda/\mu$. As anticipated, standby redundancy yields better results if switching a unit from a standby redundancy state to the operating state is absolutely reliable. With parallel redundancy, this switching problem does not exist, since an available spare unit is also operating.   □

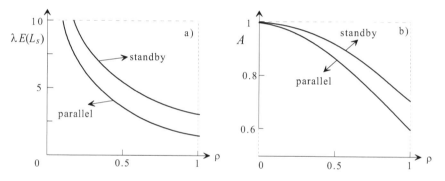

Figure 9.6 Mean lifetime a) and stationary availability b)

Example 9.7 A system has two different failure types: type 1 and type 2. After a type i-failure the system is said to be in failure state i; $i = 1, 2$. The time L_i to a type i-failure is assumed to have an exponential distribution with parameter λ_i, and the random variables L_1 and L_2 are assumed to be independent. Thus, if at time $t = 0$ a new system starts working, the time to its first failure is $Y_0 = \min(L_1, L_2)$. After a type 1-failure, the system is switched from failure state 1 into failure state 2. The time required for this is exponentially distributed with parameter ν. After entering failure state 2, the renewal of the system begins. A renewed system immediately starts working. The renewal time is exponentially distributed with parameter μ. This process continues to infinity.

All life- and renewal times as well as switching times are assumed to be independent. This model is, for example, of importance in traffic safety engineering: When the red signal in a traffic light fails (type 1-failure), then the whole traffic light is switched off (type 2-failure). That is, a *dangerous failure state* is removed by inducing a *blocking failure state*.

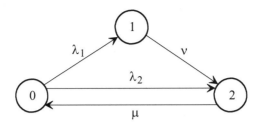

Figure 9.7 Transition graph for example 9.7

Consider the following system states

    0   system is operating

    1   type 1-failure state

    2   type 2-failure state

If $X(t)$ denotes the state of the system at time t, then $\{X(t),\ t \geq 0\}$ is a homogeneous Markov chain with state space $\mathbf{Z} = \{0, 1, 2\}$. Its transition rates are (Figure 9.7)

$$q_{01} = \lambda_1,\ \ q_{02} = \lambda_2,\ \ q_0 = \lambda_1 + \lambda_2,\ \ q_{12} = q_1 = v,\ \ q_{20} = q_2 = \mu.$$

Hence, the stationary state probabilities satisfy the system of algebraic equations

$$-(\lambda_1 + \lambda_2)\,\pi_0 + \mu\,\pi_2 = 0,$$

$$\lambda_1 \pi_0 - v\,\pi_1 = 0,$$

$$\pi_0 + \pi_1 + \pi_2 = 1.$$

The solution is

$$\pi_0 = \frac{\mu v}{(\lambda_1 + \lambda_2)v + (\lambda_1 + v)\mu},$$

$$\pi_1 = \frac{\lambda_1 \mu}{(\lambda_1 + \lambda_2)v + (\lambda_1 + v)\mu},$$

$$\pi_2 = \frac{(\lambda_1 + \lambda_2)v}{(\lambda_1 + \lambda_2)v + (\lambda_1 + v)\mu}. \qquad \square$$

9.4 SOJOURN TIMES IN PROCESS STATES

So far the fact has been used that independent, exponentially distributed times between changes of system states allow for modeling system behaviour by homogeneous Markov chains. Conversely, it can be shown that for any $i \in \mathbf{Z}$ the sojourn time of a homogeneous Markov chain $\{X(t),\ t \geq 0\}$ in state i also has an exponential distribution: By properties (9.8) and (9.15) of a homogeneous Markov chain,

$$P(Y_i > t | X(0) = i) = P(X(s) = i, \; 0 < s \le t | X(0) = i)$$

$$= \lim_{n \to \infty} P\left(X\left(\frac{k}{n}t\right) = i; \; k = 1, 2, ..., n \,\Big|\, X(0) = i\right)$$

$$= \lim_{n \to \infty} \left[p_{ii}\left(\frac{1}{n}t\right)\right]^n$$

$$= \lim_{n \to \infty} \left[1 - q_i \frac{t}{n} + o\left(\frac{1}{n}\right)\right]^n.$$

It follows that

$$P(Y_i > t | X(0) = i) = e^{-q_i t}, \quad t \ge 0, \tag{9.31}$$

since e can be represented by the limit

$$e = \lim_{x \to \infty} \left(1 + \frac{1}{x}\right)^x. \tag{9.32}$$

Thus, Y_i has an exponential distribution with parameter q_i.

Given $X(0) = i$, $X(Y_i) = X(Y_i + 0)$ is the state to which the Markov chain makes a transition on leaving state i. Let $m(nt)$ be the greatest integer m satisfying the in-equality $m/n \le t$ or, equivalently,

$$nt - 1 < m(nt) \le nt.$$

By making use of the geometric series, the joint probability distribution of the random vector $(Y_i, X(Y_i))$, $i \ne j$, can be obtained as follows:

$$P(X(Y_i) = j, \; Y_i > t | X(0) = i)$$

$$= P(X(Y_i) = j, \; X(s) = i \text{ for } 0 < s \le t | X(0) = i)$$

$$= \lim_{n \to \infty} \sum_{m=m(nt)}^{\infty} P\left(X\left(\frac{m+1}{n}\right) = j, \; Y_i \in \left[\frac{m}{n}, \frac{m+1}{n}\right) \Big| X(0) = i\right)$$

$$= \lim_{n \to \infty} \sum_{m=m(nt)}^{\infty} P\left(X\left(\frac{m+1}{n}\right) = j, \; X\left(\frac{k}{n}\right) = i \text{ for } 1 \le k \le m | X(0) = i\right)$$

$$= \lim_{n \to \infty} \sum_{m=m(nt)}^{\infty} \left[q_{ij}\frac{1}{n} + o\left(\frac{1}{n}\right)\right]\left[1 - q_i\frac{1}{n} + o\left(\frac{1}{n}\right)\right]^m$$

$$= \lim_{n \to \infty} \frac{\left[q_{ij}\frac{1}{n} + o\left(\frac{1}{n}\right)\right]}{q_i\frac{1}{n} + o\left(\frac{1}{n}\right)}\left[1 - q_i\frac{1}{n} + o\left(\frac{1}{n}\right)\right]^{m(nt)}.$$

Hence, by (9.32),

$$P(X(Y_i) = j, \; Y_i > t | X(0) = i) = \frac{q_{ij}}{q_i} e^{-q_i t}; \quad i \ne j; \; i, j \in \mathbf{Z}. \tag{9.33}$$

Passing to the marginal distribution of Y_i (i.e., summing the equations (9.33) with respect to $j \in \mathbf{Z}$) verifies (9.31).

Two other important conclusions are:

1) Letting $t = 0$ in (9.33) yields the one-step transition probability from state i into state j:

$$p_{ij} = P(X(Y_i + 0) = j | X(0) = i) = \frac{q_{ij}}{q_i}, \quad j \in \mathbf{Z}. \tag{9.34}$$

2) The state following state i is independent of Y_i (and, of course, independent of the history of the Markov chain before arriving at state i).

Knowledge of the transition probabilities p_{ij} suggests to observe a continuous-time Markov chain $\{X(t), t \geq 0\}$ only at those discrete time points at which state changes take place. Let X_n be the state of the Markov chain immediately after the nth change of state and $X_0 = X(0)$. Then $\{X_0, X_1, ...\}$ is a discrete-time homogeneous Markov chain with transition probabilities given by (9.34)

$$p_{ij} = P(X_n = j | X_{n-1} = i) = \frac{q_{ij}}{q_i}, \quad i, j \in \mathbf{Z}; \ n = 1, 2, ... \tag{9.35}$$

In this sense, the discrete-time Markov chain $\{X_0, X_1, ...\}$ is *embedded* in the continuous-time Markov chain $\{X(t), t \geq 0\}$. Embedded Markov chains can also be found in non-Markov processes. In these cases, they may facilitate the investigation of non-Markov processes. Actually, discrete-time Markov chains, which are embedded in arbitrary continuous-time stochastic processes, are frequently an efficient (if not the only) tool for analyzing these processes. Examples for the application of the *method of embedded Markov chains* to analyzing queueing systems are given in sections 9.7.3.2 and 9.7.3.3. Section 9.8 deals with semi-Markov chains, the framework of which is an embedded Markov chain.

9.5 CONSTRUCTION OF MARKOV SYSTEMS

In a *Markov system*, state changes are controlled by a Markov process. Markov systems, in which the underlying Markov process is a homogeneous, continuous-time Markov chain with state space $\mathbf{Z}$, are frequently special cases of the following basic model: The sojourn time of the system in state i is given by

$$Y_i = \min (Y_{i1}, Y_{i2}, ..., Y_{in_i}),$$

where the Y_{ij} are independent, exponential random variables with parameters

$$\lambda_{ij}; \ j = 1, 2, ..., n_i; \ i, j \in \mathbf{Z}.$$

A transition from state i to state j is made if and only if $Y_i = Y_{ij}$. If $X(t)$ as usual denotes the state of the system at time t, then, by the memoryless property of the exponential distribution, $\{X(t), t \geq 0\}$ is a homogeneous Markov chain with transition rates

$$q_{ij} = \lim_{h \to 0} \frac{p_{ij}(h)}{h} = \lambda_{ij}, \quad q_i = \sum_{j=1}^{n_i} \lambda_{ij}.$$

This representation of q_i results from (9.12) and (9.17). It reflects the fact that Y_i as the minimum of independent, exponentially distributed random variables Y_{ij} also has an exponential distribution, the parameter of which is obtained by summing the parameters of the Y_{ij}.

Example 9.8 (*repairman problem*) n machines with lifetimes $L_1, L_2, ..., L_n$ start operating at time $t = 0$. The L_i are assumed to be independent, exponential random variables with parameter λ. Failed machines are repaired. A repaired machine is 'as good as new'. There is one mechanic who can only handle one failed machine at a time. Thus, when there are $k > 1$ failed machines, $k - 1$ have to wait for repair. The repair times are assumed to be mutually independent and identically distributed as an exponential random variable Z with parameter μ. Life- and repair times are independent. Immediately after completion of its repair, a machine resumes its work.

Let $X(t)$ denote the number of machines which are in the failed state at time t. Then $\{X(t), t \geq 0\}$ is a Markov chain with state space $\mathbf{Z} = \{0, 1, ..., n\}$. The system stays in state 0 for a random time

$$Y_0 = \min (L_1, L_2, ..., L_n),$$

and then it makes a transition to state 1. The corresponding transition rate is

$$q_0 = q_{01} = \lambda n.$$

The system stays in state 1 for a random time

$$Y_1 = \min (L_1, L_2, ..., L_{n-1}, Z).$$

From state 1 it makes a transition to state 2 if $Y_1 = L_k$ for $k \in \{1, 2, ..., n-1\}$, and a transition to state 0 if $Y_1 = Z$. Hence,

$$q_{10} = \mu, \, q_{12} = (n-1)\lambda, \text{ and } q_1 = (n-1)\lambda + \mu.$$

In general (Figure 9.8),

$$q_{j-1,j} = (n-j+1)\lambda; \; j = 1, 2, ..., n,$$

$$q_{j+1,j} = \mu; \; j = 0, 1, ..., n-1,$$

$$q_{ij} = 0; \; |i-j| \geq 2,$$

$$q_j = (n-j)\lambda + \mu; \; j = 1, 2, ..., n,$$

$$q_0 = n\lambda.$$

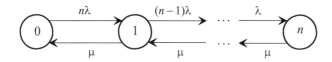

Figure 9.8 Transition graph for the repairman problem (example 9.8)

The corresponding system of equations (9.28) is

$$\mu\pi_1 = n\lambda\pi_0$$

$$(n-j+1)\lambda\pi_{j-1} + \mu\pi_{j+1} = ((n-j)\lambda + \mu)\pi_j \; ; \; j = 1, 2, ..., n-1$$

$$\mu\pi_n = \lambda\pi_{n-1}.$$

Beginning with the first equation, the stationary state probabilities are obtained by successively solving for the π_i:

$$\pi_j = \frac{n!}{(n-j)!} \rho^j \pi_0 \; ; \; j = 0, 1, ..., n,$$

where $\rho = \lambda/\mu$. From the normalizing condition (9.29),

$$\pi_0 = \left[\sum_{i=0}^{n} \frac{n!}{(n-i)!} \rho^i \right]^{-1}. \qquad \square$$

Erlang's Phase Method Systems with Erlang distributed sojourn times in their states can be transformed into Markov systems by introducing dummy states. This is due to the fact that a random variable, which is Erlang distributed with parameters n and μ, can be represented as a sum of n independent exponential random variables with parameter μ (formula (7.21), page 263). Hence, if the time interval, which the system stays in state i, is Erlang distributed with parameters n_i and μ_i, then this interval is partitioned into n_i disjoint subintervals (*phases*), the lengths of which are independent, identically distributed exponential random variables with parameter μ_i. By introducing the new states $j_1, j_2, ..., jn_i$ to label these phases, the original non-Markov system becomes a Markov system. In what follows, instead of presenting a general treatment of this approach, the application of *Erlang's phase* method is demonstrated by an example:

Example 9.9 (*two-unit system, parallel redundancy*) As in example 9.6, a two-unit system with parallel redundancy is considered. The lifetimes of the units are identically distributed as an exponential random variable L with parameter λ. The replacement times of the units are identically distributed as Z, where Z has an Erlang distribution with parameters $n = 2$ and μ. There is only one mechanic in charge of the replacement of failed units. All other assumptions and model specifications are as in example 9.6. The following system states are introduced:

0 both units are operating
1 one unit is operating, the replacement of the other one is in phase 1
2 one unit is operating, the replacement of the other one is in phase 2
3 no unit is operating, the replacement of the one being maintained is in phase 1
4 no unit is operating, the replacement of the one being maintained is in phase 2

The transition rates are (Figure 9.9):

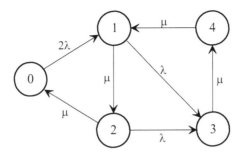

Figure 9.9 Transition graph for example 9.9

$$q_{01} = 2\lambda, \ q_0 = 2\lambda$$
$$q_{12} = \mu, \ q_{13} = \lambda, \ q_1 = \lambda + \mu$$
$$q_{20} = \mu, \ q_{23} = \lambda, \ q_2 = \lambda + \mu$$
$$q_{34} = \mu, \ q_3 = \mu$$
$$q_{41} = \mu, \ q_4 = \mu.$$

Hence the stationary state probabilities satisfy the following system of equations

$$\mu \pi_2 = 2\lambda \pi_0$$
$$2\lambda \pi_0 + \mu \pi_4 = (\lambda + \mu) \pi_1$$
$$\mu \pi_1 = (\lambda + \mu) \pi_2$$
$$\lambda \pi_1 + \lambda \pi_2 = \mu \pi_3$$
$$\mu \pi_3 = \mu \pi_4$$
$$1 = \pi_0 + \pi_1 + \pi_2 + \pi_3 + \pi_4.$$

The stationary probabilities $\pi_i^* = P('i$ units are failed') are of particular interest:

$$\pi_0^* = \pi_0, \ \pi_1^* = \pi_1 + \pi_2, \ \pi_2^* = \pi_3 + \pi_4.$$

With $\rho = E(Z)/E(L) = 2\lambda/\mu$ the π_i^* are

$$\pi_0^* = \left[1 + 2\rho + \tfrac{3}{2}\rho^2 + \tfrac{1}{4}\rho^3\right]^{-1},$$

$$\pi_1^* = \left[2\rho + \tfrac{1}{2}\rho^2\right]^{-1}\pi_0^*, \quad \pi_2^* = \left[\rho^2 + \tfrac{1}{4}\rho^3\right]^{-1}\pi_0^*.$$

The stationary system availability is given by $A = \pi_0^* + \pi_1^*$.   □

Unfortunately, applying Erlang's phase method to structurally complicated systems leads to rather complex Markov systems.

9.6 BIRTH AND DEATH PROCESSES

In this section, continuous-time Markov chains with property that only transitions to 'neighboring' states are possible, are discussed in more detail. These processes, called (continuous-time) birth and death processes, have proved to be an important tool for modeling queueing, reliability, and inventory systems. In the economical sciences, birth and death processes are among else used for describing the development of the number of enterprises in a particular area and of manpower fluctuations. In physics, flows of radioactive, cosmic, and other particles are modeled by birth and death processes. Their name, however, comes from applications in biology, where they have been used to stochastically model the development in time of the number of individuals in populations of organisms.

9.6.1 Birth Processes

A continuous-time Markov chain with state space $\mathbf{Z} = \{0, 1, ..., n\}$ is called a (pure) birth process if, for all $i = 0, 1, ..., n - 1$, only a transition from state i to $i + 1$ is possible. State n is absorbing if $n < \infty$.

Thus, the positive transition rates of a birth process are given by $q_{i,i+1}$. Henceforth they will be called birth rates and denoted as

$$\lambda_i = q_{i,i+1}, \quad i = 0, 1, ..., n - 1,$$

$$\lambda_n = 0 \text{ for } n < \infty.$$

The sample paths of birth processes are nondecreasing step functions with jump height 1. The homogeneous Poisson process with intensity λ is the simplest example of a birth process. In this case, $\lambda_i = \lambda$, $i = 0, 1, ...$. Given the initial distribution

$$p_m(0) = P(X(0) = m) = 1$$

(i.e., in the beginning the 'population' consists of m individuals), the absolute state probabilities $p_j(t)$ are equal to the transition probabilities $p_{mj}(t)$. In this case, the probabilities $p_j(t)$ are identically equal to 0 for $j < m$ and, according to (9.20), for $j \geq m$ they satisfy the system of linear differential equations

$$p'_m(t) = -\lambda_m p_m(t),$$

$$p'_j(t) = +\lambda_{j-1} p_{j-1}(t) - \lambda_j p_j(t); \quad j = m + 1, m + 2, ... \qquad (9.36)$$

$$p'_n(t) = +\lambda_{n-1} p_{n-1}(t), \quad n < \infty.$$

The solution of the first differential equation is

$$p_m(t) = e^{-\lambda_m t}, \quad t \geq 0. \qquad (9.37)$$

For $j = m + 1, m + 2, ...$, the differential equations (9.36) are equivalent to

$$e^{\lambda_j t}\left(p_j'(t) + \lambda_j p_j(t)\right) = \lambda_{j-1} e^{\lambda_j t} p_{j-1}(t)$$

or

$$\frac{d}{dt}\left(e^{\lambda_j t} p_j(t)\right) = \lambda_{j-1} e^{\lambda_j t} p_{j-1}(t).$$

By integration,

$$p_j(t) = \lambda_{j-1} e^{-\lambda_j t} \int_0^t e^{\lambda_j x} p_{j-1}(x)\, dx. \tag{9.38}$$

Formulas (9.37) and (9.38) allow the successive calculation of the probabilities $p_j(t)$ for $j = m+1, m+2, \dots$. For instance, on conditions $p_0(0) = 1$ and $\lambda_0 \neq \lambda_1$,

$$p_1(t) = \lambda_0\, e^{-\lambda_1 t} \int_0^t e^{\lambda_1 x}\, e^{-\lambda_0 x}\, dx$$

$$= \lambda_0\, e^{-\lambda_1 t} \int_0^t e^{-(\lambda_0 - \lambda_1)x}\, dx$$

$$= \frac{\lambda_0}{\lambda_0 - \lambda_1}\left(e^{-\lambda_1 t} - e^{-\lambda_0 t}\right), \quad t \geq 0.$$

If all the birth rates are different from each other, then this result and (9.38) yield by induction:

$$p_j(t) = \sum_{i=0}^{j} C_{ij}\, \lambda_i\, e^{-\lambda_i t}, \quad j = 0, 1, \dots,$$

$$C_{ij} = \frac{1}{\lambda_j} \prod_{k=0, k\neq i}^{j} \frac{\lambda_k}{\lambda_k - \lambda_i}, \quad 0 \leq i \leq j, \quad C_{00} = \frac{1}{\lambda_0}.$$

Linear Birth Process A birth process is called a *linear birth process* or a *Yule-Furry process* (see *Furry* (1937) and *Yule* (1924)) if its birth rates are given by

$$\lambda_i = i\lambda; \quad i = 0, 1, 2, \dots.$$

Since state 0 is absorbing, an initial distribution should not concentrate probability 1 on state 0. Linear birth processes occur, for instance, if in the interval $[t, t+h]$ each member of a population (bacterium, physical particle) independently of each other splits with probability $\lambda h + o(h)$ as $h \to 0$.

Assuming $p_1 = P(X(0) = 1) = 1$, the system of differential equations (9.36) becomes

$$p_j'(t) = -\lambda\left[j p_j(t) - (j-1) p_{j-1}(t)\right]; \quad j = 1, 2, \dots \tag{9.39}$$

with

$$p_1(0) = 1, \quad p_j(0) = 0; \quad j = 2, 3, \dots. \tag{9.40}$$

The solution of (9.39) given the initial distribution (9.40) is

$$p_i(t) = e^{-\lambda t}(1 - e^{-\lambda t})^{i-1}; \quad i = 1, 2, \dots.$$

Thus, $X(t)$ has a geometric distribution with parameter $p = e^{-\lambda t}$. Hence, the trend function of the linear birth process is

$$m(t) = e^{\lambda t}, \quad t \geq 0.$$

If $\mathbf{Z}$ is finite, then there always exists a solution of (9.36) which satisfies

$$\sum_{i \in \mathbf{Z}} p_i(t) = 1. \tag{9.41}$$

In case of an infinite state space $\mathbf{Z} = \{0, 1, ...\}$, the following theorem gives a necessary and sufficient condition for the existence of a solution of (9.36) with property (9.41). Without loss of generality, the theorem is proved on condition (9.40).

Theorem 9.2 (Feller-Lundberg) A solution $\{p_0(t), p_1(t), ... \}$ of the system of differential equations (9.36) satisfies condition (9.41) if and only if the series

$$\sum_{i=0}^{\infty} \frac{1}{\lambda_i} \tag{9.42}$$

diverges.

Proof Let

$$s_k(t) = p_0(t) + p_1(t) + \cdots + p_k(t).$$

Summing the middle equation of (9.36) from $j = 1$ to k yields

$$s_k'(t) = -\lambda_k p_k(t).$$

By integration, taking into account $s_k(0) = 1$,

$$1 - s_k(t) = \lambda_k \int_0^t p_k(x)\,dx. \tag{9.43}$$

Since $s_k(t)$ is monotonically increasing as $k \to \infty$, the following limit exists:

$$r(t) = \lim_{k \to \infty} (1 - s_k(t)).$$

From (9.43),

$$\lambda_k \int_0^t p_k(x)\,dx \geq r(t).$$

Dividing by λ_k and summing the arising inequalities from 0 to k gives

$$\int_0^t s_k(x)\,dx \geq r(t) \left(\frac{1}{\lambda_0} + \frac{1}{\lambda_1} + \cdots + \frac{1}{\lambda_k} \right).$$

Since $s_k(t) \leq 1$ for all $t \geq 0$,

$$t \geq r(t) \left(\frac{1}{\lambda_0} + \frac{1}{\lambda_1} + \cdots + \frac{1}{\lambda_k} \right).$$

If the series (9.42) diverges, then this inequality implies that $r(t) = 0$ for all $t > 0$. But this result is equivalent to (9.41).

Conversely, from (9.43),

$$\lambda_k \int_0^t p_k(x)\, dx \le 1$$

so that

$$\int_0^t s_k(x)\, dx \le \frac{1}{\lambda_0} + \frac{1}{\lambda_1} + \cdots + \frac{1}{\lambda_k}.$$

Passing to the limit as $k \to \infty$,

$$\int_0^t (1 - r(t))\, dt \le \sum_{i=0}^{\infty} \frac{1}{\lambda_i}.$$

If $r(t) \equiv 0$, the left-hand side of this inequality is equal to t. Since t can be arbitrarily large, the series (9.42) must diverge. This result completes the proof.   ∎

According to this theorem, it is theoretically possible that within a finite interval $[0, t]$ the population grows beyond all finite bounds. The probability of such an *explosive growth* is

$$1 - \Sigma_{i=0}^{\infty} p_i(t).$$

This probability is positive if the birth rates grow so fast that the series (9.42) converges. For example, an explosive growth would occur if

$$\lambda_i = i^2 \lambda; \quad i = 1, 2, \ldots$$

since

$$\sum_{i=1}^{\infty} \frac{1}{\lambda_i} = \frac{1}{\lambda} \sum_{i=1}^{\infty} \frac{1}{i^2} = \frac{\pi^2}{6\lambda} < \infty.$$

It is remarkable that an explosive growth occurs in an arbitrarily small time interval, since the convergence of the series (9.42) does not depend on t.

9.6.2 Death Processes

A continuous-time Markov chain with state space $\mathbf{Z} = \{0, 1, \ldots\}$ is called a (*pure*) *death process* if, for all $i = 1, 2, \ldots$ only transitions from state i to $i - 1$ are possible. State 0 is absorbing.

Thus, the positive transition rates of pure death processes are given by $q_{i,i-1}$, $i \ge 1$. In what follows, these transition rates will be called *death rates* and denoted as

$$\mu_0 = 0, \quad \mu_i = q_{i,i-1}; \quad i = 1, 2, \ldots.$$

The sample paths of such processes are non-increasing step functions. For pure death processes, on condition

$$p_n(0) = P(X(0) = n) = 1,$$

the system of differential equations (9.20) becomes

$$p'_n(t) = -\mu_n p_n(t)$$

$$p'_j(t) = -\mu_j p_j(t) + \mu_{j+1} p_{j+1}(t); \quad j = 0, 1, ..., n-1. \tag{9.44}$$

The solution of the first differential equation is

$$p_n(t) = e^{-\mu_n t}, \quad t \geq 0.$$

Integrating (9.44) yields

$$p_j(t) = \mu_{j+1} e^{-\mu_j t} \int_0^t e^{\mu_j x} p_{j+1}(x) \, dx; \quad j = n-1, ..., 1, 0. \tag{9.45}$$

Starting with $p_n(t)$, the probabilities

$$p_j(t), \quad j = n-1, n-2, ..., 0,$$

can be recursively determined from (9.45). For instance, assuming $\mu_n \neq \mu_{n-1}$,

$$p_{n-1}(t) = \mu_n e^{-\mu_{n-1} t} \int_0^t e^{-(\mu_n - \mu_{n-1})x} dx$$

$$= \frac{\mu_n}{\mu_n - \mu_{n-1}} (e^{-\mu_{n-1} t} - e^{-\mu_n t}).$$

More generally, if all the death rates are different from each other, then

$$p_j(t) = \sum_{i=j}^{n} D_{ij} \mu_i e^{-\mu_i t}, \quad 0 \leq j \leq n, \tag{9.46}$$

where

$$D_{ij} = \frac{1}{\mu_j} \prod_{\substack{k=j \\ k \neq i}}^{n} \frac{\mu_k}{\mu_k - \mu_i}, \quad j \leq i \leq n, \quad D_{nn} = \frac{1}{\mu_n}.$$

Linear Death Process A death process $\{X(t), t \geq 0\}$ is called a *linear death process* if for a positive parameter μ it has death rates

$$\mu_i = i\mu; \quad i = 0, 1, ...$$

Given the initial distribution

$$p_n(0) = P(X(0) = n) = 1,$$

the process stays in state n an exponentially with parameter $n\mu$ distributed time:

$$p_n(t) = e^{-n\mu t}, \quad t \geq 0.$$

Starting with $p_n(t)$, one obtains inductively from (9.45) or simply from (9.46):

$$p_i(t) = \binom{n}{i} e^{-i\mu t} (1 - e^{-\mu t})^{n-i}; \quad i = 0, 1, ..., n.$$

Hence, $X(t)$ has a binomial distribution with parameters n and $p = e^{-\mu t}$ so that the trend function of a linear death process is

$$m(t) = n e^{-\mu t}, \quad t \geq 0.$$

Example 9.10 A system consisting of n subsystems starts operating at time $t = 0$. The lifetimes of the subsystems are independent, exponentially with parameter λ distributed random variables. If $X(t)$ denotes the number of subsystems still working at time t, then $\{X(t), t \geq 0\}$ is a linear death process with death rates

$$\mu_i = i\lambda; \quad i = 0, 1, \dots.$$ ☐

9.6.3   Birth and Death Processes

9.6.3.1   Time-Dependent State Probabilities

A continuous-time Markov chain $\{X(t), t \geq 0\}$ with state space

$$\mathbf{Z} = \{0, 1, \dots, n\}, \ n \leq \infty,$$

is called a *birth and death process* if from any state i only a transition to state $i - 1$ or to state $i + 1$ is possible, provided that $i - 1 \in \mathbf{Z}$ and $i + 1 \in \mathbf{Z}$, respectively.

Therefore, the transition rates of a birth- and death process have property

$$q_{i,j} = 0 \quad \text{for } |i - j| > 1.$$

The transition rates $\lambda_i = q_{i,i+1}$ and $\mu_i = q_{i,i-1}$ are called *birth rates* and *death rates*, respectively. According to the restrictions given by the state space, $\lambda_n = 0$ for $n < \infty$ and $\mu_0 = 0$ (Figure 9.10). Hence, a birth process (death process) is a birth and death process, the death rates (birth rates) of which are equal to 0. If a birth and death process describes the number of individuals in a population of organisms, then, when arriving at state 0, the population is extinguished. Thus, without the possibility of immigration, state 0 is absorbing ($\lambda_0 = 0$).

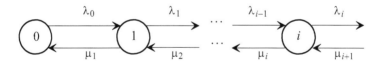

Figure 9.10 Transition graph of the birth- and death process

According to (9.20), the absolute state probabilities $p_j(t) = P(X(t) = j)$, $j \in \mathbf{Z}$, of a birth- and death process satisfy the system of linear differential equations

$$p_0'(t) = -\lambda_0 p_0(t) + \mu_1 p_1(t),$$

$$p_j'(t) = +\lambda_{j-1} p_{j-1}(t) - (\lambda_j + \mu_j) p_j(t) + \mu_{j+1} p_{j+1}(t), \quad j = 1, 2, \dots, \qquad (9.47)$$

$$p_n'(t) = +\lambda_{n-1} p_{n-1}(t) - \mu_n p_n(t), \quad n < \infty.$$

In the following two examples, the state probabilities $\{p_0(t), p_1(t), \dots\}$ of two important birth and death processes are determined via their respective z-transforms

$$M(t,z) = \sum_{i=0}^{\infty} p_i(t) z^i$$

under initial conditions of type

$$p_n(0) = P(X(0) = n) = 1.$$

In terms of the z-transform, this condition is equivalent to

$$M(0,z) \equiv z^n, \quad n = 0, 1, \dots.$$

Furthermore, partial derivatives of the z-transforms will be needed:

$$\frac{\partial M(t,z)}{\partial t} = \sum_{i=0}^{\infty} p_i'(t) z^i \quad \text{and} \quad \frac{\partial M(t,z)}{\partial z} = \sum_{i=1}^{\infty} i p_i(t) z^{i-1}. \tag{9.48}$$

Partial differential equations for $M(t,z)$ will be established and solved by applying the characteristic method.

Example 9.11 (linear birth and death process) $\{X(t), t \geq 0\}$ is called a *linear birth and death process* if it has transition rates

$$\lambda_i = i\lambda, \quad \mu_i = i\mu, \quad i = 0, 1, \dots$$

In what follows, this process is analyzed on condition that

$$p_1(0) = P(X(0) = 1) = 1.$$

Assuming $p_0(0) = 1$ would make no sense since state 0 is absorbing. The system of differential equations (9.20) becomes

$$p_0'(t) = \mu p_1(t),$$

$$p_j'(t) = (j-1)\lambda p_{j-1}(t) - j(\lambda+\mu)p_j(t) + (j+1)\mu p_{j+1}(t); \quad j = 1, 2, \dots. \tag{9.49}$$

Multiplying the j-th differential equation by z^j and summing from $j = 0$ to $j = \infty$, taking into account (9.48), yields the following linear, homogeneous partial differential equation for $M(t, z)$:

$$\frac{\partial M(t,z)}{\partial t} - (z-1)(\lambda z - \mu)\frac{\partial M(t,z)}{\partial z} = 0. \tag{9.50}$$

The corresponding (ordinary) *characteristic differential equation* is a *Riccati differential equation* with constant coefficients:

$$\frac{dz}{dt} = -(z-1)(\lambda z - \mu) = -\lambda z^2 + (\lambda+\mu)z - \mu. \tag{9.51}$$

a) $\lambda \neq \mu$ By separation of variables, (9.51) can be written in the form

$$\frac{dz}{(z-1)(\lambda z - \mu)} = -dt.$$

Integration on both sides of this equation yields

$$-\frac{1}{\lambda - \mu} \ln\left(\frac{\lambda z - \mu}{z - 1}\right) = -t + C.$$

The general solution $z = z(t)$ of the characteristic differential equation in implicit form is, therefore, given by

$$c = (\lambda - \mu)t - \ln\left(\frac{\lambda z - \mu}{z - 1}\right),$$

where c is an arbitrary constant. Thus, the general solution $M(t, z)$ of (9.50) has the structure

$$M(t, z) = f\left((\lambda - \mu)t - \ln\left(\frac{\lambda z - \mu}{z - 1}\right)\right),$$

where f can be any function with a continuous derivative. f can be determined by making use of the initial condition $p_1(0) = 1$ or, equivalently, $M(0, z) = z$. Since

$$M(0, z) = f\left(\ln\left(\frac{z - 1}{\lambda z - \mu}\right)\right) = z,$$

f must have structure

$$f(x) = \frac{\mu e^x - 1}{\lambda e^x - 1}.$$

Thus, $M(t, z)$ is

$$M(t, z) = \frac{\mu \exp\left\{(\lambda - \mu)t - \ln\left(\frac{\lambda z - \mu}{z - 1}\right)\right\} - 1}{\lambda \exp\left\{(\lambda - \mu)t - \ln\left(\frac{\lambda z - \mu}{z - 1}\right)\right\} - 1}.$$

After simplification, $M(t, z)$ becomes

$$M(t, z) = \frac{\mu\left[1 - e^{(\lambda - \mu)t}\right] - \left[\lambda - \mu e^{(\lambda - \mu)t}\right]z}{\left[\mu - \lambda e^{(\lambda - \mu)t}\right] - \lambda\left[1 - \mu e^{(\lambda - \mu)t}\right]z}.$$

This representation of $M(t, z)$ allows its expansion as a power series in z. The coefficient of z^j is the desired absolute state probability $p_j(t)$. Letting $\rho = \lambda/\mu$ yields

$$p_0(t) = \frac{1 - e^{(\lambda - \mu)t}}{1 - \rho e^{(\lambda - \mu)t}},$$

$$p_j(t) = (1 - \rho)\rho^{j-1} \frac{\left[1 - e^{(\lambda - \mu)t}\right]^{j-1}}{\left[1 - \rho e^{(\lambda - \mu)t}\right]^{j+1}} e^{(\lambda - \mu)t}, \quad j = 1, 2, \dots.$$

Since state 0 is absorbing, $p_0(t)$ is the probability that the population is extinguished at time t. Moreover,

$$\lim_{t\to\infty} p_0(t) = \begin{cases} 1 & \text{for } \lambda < \mu \\ \frac{\mu}{\lambda} & \text{for } \lambda > \mu \end{cases}.$$

Thus, for $\lambda > \mu$ the population will survive to infinity with positive probability μ/λ. If $\lambda < \mu$, the population will eventually disappear with probability 1. In the latter case, the distribution function of the lifetime L of the population is

$$P(L \le t) = p_0(t) = \frac{1 - e^{(\lambda-\mu)t}}{1 - \rho e^{(\lambda-\mu)t}}, \quad t \ge 0.$$

Hence, the population will survive the interval $[0, t]$ with probability

$$P(L > t) = 1 - p_0(t).$$

From this, applying formual (2.52), page 64,

$$E(L) = \frac{1}{\mu - \lambda} \ln\left(2 - \frac{\lambda}{\mu}\right).$$

The trend function $m(t) = E(X(t))$ is principally given by

$$m(t) = \sum_{j=0}^{\infty} j p_j(t).$$

By formulas (2.112), page 96, $m(t)$ can also be obtained from the z-transform:

$$m(t) = \left.\frac{\partial M(t,z)}{\partial z}\right|_{z=1}.$$

If only the trend function of the process is of interest, then here as in many other cases knowledge of the z-transform or the absolute state distribution is not necessary, since $m(t)$ can be determined from the respective system of differential equations (9.47). In this example, multiplying the j-th differential equation of (9.49) by j and summing from $j = 0$ to ∞ yields the following first-order differential equation:

$$m'(t) = (\lambda - \mu)m(t). \tag{9.52}$$

Taking into account the initial condition $p_1(0) = 1$, its solution is

$$m(t) = e^{(\lambda-\mu)t}.$$

By multiplying the j-th differential equation of (9.47) by j^2 and summing from $j = 0$ to ∞, a second order differential equation for $Var(X(t))$ is obtained. Its solution is

$$Var(X(t)) = \frac{\lambda + \mu}{\lambda - \mu}\left[1 - e^{-(\lambda-\mu)t}\right]e^{2(\lambda-\mu)t}.$$

Of course, since $M(t,z)$ is known, $Var(X(t))$ can be obtained from (2.112), too.

If the linear birth- and death process starts in states $s = 2, 3, ...$, no principal additional problems arise up to the determination of $M(t,z)$. But it will be more complicated to expand $M(t,z)$ as a power series in z.

The corresponding trend function, however, is easily obtained as solution of (9.52) with the initial condition $p_s(0) = P(X(0) = s) = 1$:

$$m(t) = s\,e^{(\lambda - \mu)t}, \quad t \ge 0.$$

b) $\lambda = \mu$ In this case, the characteristic differential equation (9.51) simplifies to

$$\frac{dz}{\lambda(z-1)^2} = -dt.$$

Integration yields

$$c = \lambda t - \frac{1}{z-1},$$

where c is an arbitrary constant. Therefore, $M(t,z)$ has structure

$$M(t,z) = f\left(\lambda t - \frac{1}{z-1}\right),$$

where f is a continuously differentiable function. Since $p_1(0) = 1$, f satisfies

$$f\left(-\frac{1}{z-1}\right) = z.$$

Hence, the desired function f is given by

$$f(x) = 1 - \frac{1}{x}, \quad x \ne 0.$$

The corresponding z-transform is

$$M(t,z) = \frac{\lambda t + (1 - \lambda t)z}{1 + \lambda t - \lambda t z}.$$

Expanding $M(t,z)$ as a power series in z yields the absolute state probabilities:

$$p_0(t) = \frac{\lambda t}{1 + \lambda t}, \quad p_j(t) = \frac{(\lambda t)^{j-1}}{(1 + \lambda t)^{j+1}}; \quad j = 1, 2, ..., \; t \ge 0.$$

An equivalent form of the absolute state probabilities is

$$p_0(t) = \frac{\lambda t}{1 + \lambda t}, \quad p_j(t) = [1 - p_0(t)]^2 [p_0(t)]^{j-1}; \quad j = 1, 2, ..., \; t \ge 0.$$

Mean value and variance of $X(t)$ are

$$E(X(t)) = 1, \quad Var(X(t)) = 2\lambda t.$$

This example shows that the analysis of apparently simple birth- and death processes requires some effort.   □

Example 9.12 Consider a birth- and death process with transition rates

$$\lambda_i = \lambda, \; \mu_i = i\,\mu; \quad i = 0, 1, ...$$

and initial distribution and $p_0(0) = P(X(0) = 0) = 1$.

The corresponding system of linear differential equations (9.47) is

$$p_0'(t) = \mu p_1(t) - \lambda p_0(t),$$

$$p_j'(t) = \lambda p_{j-1}(t) - (\lambda + \mu j) p_j(t) + (j+1)\mu p_{j+1}(t); \quad j = 1, 2, \dots . \tag{9.53}$$

Multiplying the j-th equation by z^j and summing from $j = 0$ to ∞ yields a homogeneous linear partial differential equation for the moment generating function:

$$\frac{\partial M(t,z)}{\partial t} + \mu(z-1)\frac{\partial M(t,z)}{\partial z} = \lambda(z-1) M(t,z). \tag{9.54}$$

The corresponding system of characteristic differential equations is

$$\frac{dz}{dt} = \mu(z-1), \quad \frac{dM(t,z)}{dt} = \lambda(z-1) M(t,z).$$

After separation of variables and subsequent integration, the first differential equation yields

$$c_1 = \ln(z-1) - \mu t$$

with an arbitrary constant c_1. By combining both differential equations and letting $\rho = \lambda/\mu$,

$$\frac{dM(t,z)}{M(t,z)} = \rho\, dz.$$

Integration yields

$$c_2 = \ln M(t,z) - \rho z,$$

where c_2 is an arbitrary constant. As a solution of (9.54), $M(t,z)$ satisfies

$$c_2 = f(c_1)$$

with an arbitrary continuous function f, i.e. $M(t,z)$ satisfies

$$\ln M(t,z) - \rho z = f(\ln(z-1) - \mu t).$$

Therefore,

$$M(t,z) = \exp\{f(\ln(z-1) - \mu t) + \rho z\}.$$

Since condition $p_0(0) = 1$ is equivalent to $M(0,z) \equiv 1$, f is implicitly given by

$$f(\ln(z-1)) = -\rho z.$$

Hence, the explicit representation of f is

$$f(x) = -\rho(e^x + 1).$$

Thus,

$$M(t,z) = \exp\left\{ -\rho\left(e^{\ln(z-1)-\mu t} + 1\right) + \rho z \right\}.$$

Equivalently,

$$M(t,z) = e^{-\rho(1-e^{-\mu t})} \cdot e^{+\rho(1-e^{-\mu t})z}.$$

Now it is easy to expand $M(t, z)$ as a power series in z. The coefficients of z^j are

$$p_j(t) = \frac{[\rho(1 - e^{-\mu t})]^j}{j!} e^{-\rho(1-e^{-\mu t})}; \quad j = 0, 1, \dots. \tag{9.55}$$

This is a Poisson distribution with intensity function $\rho(1 - e^{-\mu t})$. Therefore, this birth and death process has trend function

$$m(t) = \rho(1 - e^{-\mu t}).$$

For $t \to \infty$ the absolute state probabilities $p_j(t)$ converge to the stationary state probabilities:

$$\pi_j = \lim_{t \to \infty} p_j(t) = \frac{\rho^j}{j!} e^{-\rho}; \quad j = 0, 1, \dots.$$

If the process starts at a state $s > 0$, the absolute state probability distribution is not Poisson. In this case this distribution has a rather complicated structure, which will not be presented here. Instead, the system of linear differential equations (9.53) can be used to establish ordinary differential equations for the trend function $m(t)$ and the variance of $X(t)$. Given the initial distribution $p_s(0) = 1$, $s = 1, 2, \dots$, their respective solutions are

$$m(t) = \rho(1 - e^{-\mu t}) + s e^{-\mu t},$$

$$Var(X(t)) = (1 - e^{-\mu t})(\rho + s e^{-\mu t}).$$

The birth and death process considered in this example is of importance in queueing theory (section 9.7).   □

Example 9.13 (birth and death process with immigration) For positive parameters λ, μ, and ν, let transition rates be given by

$$\lambda_i = i\lambda + \nu, \quad \mu_i = i\mu; \quad i = 0, 1, \dots$$

If this model is used to describe the development in time of a population, then each individual will produce a new individual in $[t, t + \Delta t]$ with probability $\lambda \Delta t + o(\Delta t)$ or leave the population in this interval with probability $\mu \Delta t + o(\Delta t)$. Moreover, due to immigration from outside, the population will increase by one individual in $[t, t + \Delta t]$ with probability $\nu t + o(\Delta t)$. Thus, if $X(t) = i$, the probability that the population will increase or decrease by one individual in the interval $[t, t + \Delta t]$ is

$$(i\lambda + \nu) \Delta t + o(\Delta t) \text{ or } i\mu \Delta t + o(\Delta t),$$

respectively. These probabilities do not depend on t and refer to $\Delta t \to 0$. As in the previous example, state 0 is not absorbing.

The differential equations (9.47) become

$$p_0'(t) = \mu p_1(t) - \nu p_0(t),$$

$$p_j'(t) = (\lambda(j-1) + \nu)p_{j-1}(t) + \mu(j+1)p_{j+1}(t) - (\lambda j + \nu + \mu j)p_j(t).$$

Analogously to the previous examples, the z-transformation $M(t,z)$ of the probability distribution $\{p_0(t), p_1(t), ...\}$ is seen to satisfy the partial differential equation

$$\frac{\partial M(t,z)}{\partial t} = (\lambda z - \mu)(z-1)\frac{\partial M(t,z)}{\partial z} + \nu(z-1)M(t,z). \tag{9.56}$$

The system of the characteristic differential equations belonging to (9.56) is

$$\frac{dz}{dt} = -(\lambda z - \mu)(z-1),$$

$$\frac{dM(t,z)}{dt} = \nu(z-1)M(t,z).$$

From this, with the initial condition $p_0(0) = 1$ or, equivalently, $M(0,z) \equiv 1$, the solution is obtained analogously to the previous example

$$M(t,z) = \left\{ \frac{\lambda - \mu}{\lambda z + \lambda(1-z)e^{(\lambda-\mu)t} - \mu} \right\}^{\nu/\lambda} \quad \text{for} \quad \lambda \neq \mu,$$

$$M(t,z) = (1 + \lambda t)^{\nu/\lambda} \left\{ 1 - \frac{\lambda t z}{1 + \lambda t} \right\}^{-\nu/\lambda} \quad \text{for} \quad \lambda = \mu.$$

Generally it is not possible to expand $M(t,z)$ as a power series in z. But the absolute state probabilities $p_i(t)$ can be obtained by differentiation of $M(t,z)$:

$$p_i(t) = \left. \frac{\partial^i M(t,z)}{\partial z^i} \right|_{z=0} \quad \text{for} \quad i = 1, 2, ...$$

The trend function

$$m(t) = E(X(t)) = \left. \frac{\partial M(t,z)}{\partial z} \right|_{z=1}$$

of this birth and death process is

$$m(t) = \frac{\nu}{\lambda - \mu}\left[e^{(\lambda-\mu)t} - 1 \right] \quad \text{for} \quad \lambda \neq \mu, \tag{9.57}$$

$$m(t) = \nu t \quad \text{for} \quad \lambda = \mu.$$

If $\lambda < \mu$, the limit as $t \to \infty$ of the z-transform exists:

$$\lim_{t \to \infty} M(t,z) = \left(1 - \frac{\lambda}{\mu} \right)^{\nu/\lambda} \left(1 - \frac{\lambda}{\mu z} \right)^{-\nu/\lambda}.$$

For $\lambda < \mu$, the trend function (9.57) tends to a positive limit as $t \to \infty$:

$$\lim_{t \to \infty} m(t) = \frac{\nu}{\mu - \lambda} \quad \text{for} \quad \lambda < \mu. \qquad \square$$

9.6.3.2 Stationary State Probabilities

By (9.27), in case of their existence the stationary distribution $\{\pi_0, \pi_1, \dots\}$ of a birth and death process satisfies the following system of linear algebraic equations

$$\lambda_0 \pi_0 - \mu_1 \pi_1 = 0$$

$$\lambda_{j-1} \pi_{j-1} - (\lambda_j + \mu_j) \pi_j + \mu_{j+1} \pi_{j+1} = 0, \quad j = 1, 2, \dots \qquad (9.58)$$

$$\lambda_{n-1} \pi_{n-1} - \mu_n \pi_n = 0, \quad n < \infty.$$

This system is equivalent to the following one:

$$\mu_1 \pi_1 = \lambda_0 \pi_0$$

$$\mu_{j+1} \pi_{j+1} + \lambda_{j-1} \pi_{j-1} = (\lambda_j + \mu_j) \pi_j; \quad j = 1, 2, \dots \qquad (9.59)$$

$$\mu_n \pi_n = \lambda_{n-1} \pi_{n-1}, \quad n < \infty.$$

Provided its existence, it is possible to obtain the general solution of (9.58): Let

$$d_j = -\lambda_j \pi_j + \mu_{j+1} \pi_{j+1}; \quad j = 0, 1, \dots.$$

Then the system (9.58) simplifies to

$$d_0 = 0,$$

$$d_j - d_{j-1} = 0, \quad j = 1, 2, \dots$$

$$d_{n-1} = 0, \quad n < \infty.$$

Starting with $j = 0$, one successively obtains

$$\pi_j = \prod_{i=1}^{j} \frac{\lambda_{i-1}}{\mu_i} \pi_0; \quad j = 1, 2, \dots, n. \qquad (9.60)$$

1) If $n < \infty$, then the stationary state probabilities satisfy the normalizing condition

$$\sum_{i=0}^{n} \pi_i = 1.$$

Solving for π_0 yields

$$\pi_0 = \left[1 + \sum_{j=1}^{n} \prod_{i=1}^{j} \frac{\lambda_{i-1}}{\mu_i} \right]^{-1}. \qquad (9.61)$$

2) If $n = \infty$, then equation (9.61) shows that the convergence of the series

$$\sum_{j=1}^{\infty} \prod_{i=1}^{j} \frac{\lambda_{i-1}}{\mu_i} \qquad (9.62)$$

is necessary for the existence of a stationary distribution. A sufficient condition for the convergence of this series is the existence of a positive integer N such that

$$\frac{\lambda_{i-1}}{\mu_i} \le \alpha < 1 \quad \text{for all } i > N. \qquad (9.63)$$

Intuitively, this condition is not surprising: If the birth rates are greater than the corresponding death rates, the process will drift to infinity with probability 1. But this exludes the existence of a stationary distribution of the process. For a proof of the following theorem see *Karlin and Taylor* (1981).

Theorem 9.3 The convergence of the series (9.62) and the divergence of the series

$$\sum_{j=1}^{\infty} \prod_{i=1}^{j} \frac{\mu_i}{\lambda_i} \tag{9.64}$$

is sufficient for the existence of a stationary state distribution. The divergence of the series (9.64) is, moreover, sufficient for the existence of such a time-dependent solution $\{p_0(t), p_1(t), ...\}$ of (9.47) which satisfies the normalizing condition (9.21). ∎

Example 9.14 (*repairman problem*) The repairman problem introduced in example 9.8 is considered once more. However, it is now assumed that there are r mechanics for repairing failed machines, $1 \le r \le n$. A failed machine can be attended only by one mechanic. (For a modification of this assumption see example 9.15.) All other assumptions as well as the notation are as in example 9.8.

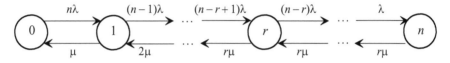

Figure 9.11 Transition graph of the general repairman problem

Let $X(t)$ denote the number of failed machines at time t. Then $\{X(t), t \ge 0\}$ is a birth and death process with state space $\mathbf{Z} = \{0, 1, ..., n\}$. Its transition rates are

$$\lambda_j = (n-j)\lambda, \quad 0 \le j \le n,$$

$$\mu_j = \begin{cases} j\mu, & 0 \le j \le r \\ r\mu, & r < j \le n \end{cases}$$

(Figure 9.11). Note that in this example the terminology 'birth and death rates' does not reflect the technological situation. If the *service rate* $\rho = \lambda/\mu$ is introduced, formulas (9.57) and (9.58) yield the stationary state probabilities

$$\pi_j = \begin{cases} \binom{n}{j} \rho^j \pi_0; & 1 \le j \le r \\ \dfrac{n!}{r^{j-r} r! (n-j)!} \rho^j \pi_0; & r \le j \le n \end{cases} \tag{9.65}$$

$$\pi_0 = \left[\sum_{j=0}^{r} \binom{n}{j} \rho^j + \sum_{j=r+1}^{n} \frac{n!}{r^{j-r} r! (n-j)!} \rho^j \right]^{-1}.$$

Policy 1:	$n=10,\ r=2$	Policy 2:	$n=5,\ r=1$
j	$\pi_{j,1}$	j	$\pi_{j,2}$
0	0.0341	0	0.1450
1	0.1022	1	0.2175
2	0.1379	2	0.2611
3	0.1655	3	0.2350
4	0.1737	4	0.1410
5	0.1564	5	0.0004
6	0.1173		
7	0.0704		
8	0.0316		
9	0.0095		
10	0.0014		

Table 9.1 Stationary state probabilities for example 9.14

A practical application of the stationary state probabilities (9.65) is illustrated by a numerical example: Let $n = 10$, $\rho = 0.3$, and $r = 2$. The efficiencies of the following two maintenance policies will be compared:

1) Both mechanics are in charge of the repair of any of the 10 machines.

2) The mechanics are assigned 5 machines each for the repair of which they alone are responsible.

Let $X_{n,r}$ be the random number of failed machines and $Z_{n,r}$ the random number of mechanics which are busy with repairing failed machines, dependent on the number n of machines and the number r of available mechanics. From Table 9.1, for policy 1,

$$E(X_{10,2}) = \Sigma_{j=1}^{10} j\,\pi_{j,1} = 3.902$$

$$E(Z_{10,2}) = 1 \cdot \pi_{1,1} + 2\,\Sigma_{j=2}^{10}\ \pi_{j,1} = 1.8296\,.$$

For policy 2,

$$E(X_{5,1}) = \Sigma_{j=1}^{5} j\,\pi_{j,2} = 2.011$$

$$E(Z_{5,1}) = 1 \cdot \pi_{1,2} + \Sigma_{j=2}^{5}\ \pi_{j,2} = 0.855\,.$$

Hence, when applying policy 2, the average number of failed machines out of 10 and the average number of busy mechanics out of 2 are

$$2\,E(X_{5,1}) = 4.022 \quad \text{and} \quad 2\,E(Z_{5,1}) = 1.710\,.$$

Thus, on the one hand, the mean number of failed machines under policy 1 is smaller than under policy 2, and, on the other hand, the mechanics are less busy under policy 2 than under policy 1. Hence, policy 1 should be preferred if no other relevant performance criteria have to be taken into account.   □

Example 9.15 The repairman problem of example 9.14 is modified in the following way: The available maintenance capacity of r units (which need not necessarily be human) is always fully used for repairing failed machines. Thus, if only one machine has failed, then all r units are busy with repairing this machine. If several machines are down, the full maintenance capacity of r units is uniformly distributed to the failed machines. This adaptation is repeated after each failure of a machine and after each completion of a repair. In this case, no machines have to wait for repair.

If j machines have failed, then the repair rate of each failed machine is

$$r\mu/j.$$

Therefore, the death rates of the corresponding birth and death process are constant, i.e., they do not depend on the system state:

$$\mu_j = j \cdot \frac{r\mu}{j} = r\mu; \quad j = 1, 2, \dots .$$

The birth rates are the same as in example 9.14:

$$\lambda_j = (n-j)\lambda; \quad j = 0, 1, \dots .$$

Thus, the stationary state probabilities are according to (9.60) and (9.61):

$$\pi_0 = \left[\sum_{j=1}^{n} \frac{n!}{(n-j)!} \left(\frac{\lambda}{r\mu} \right)^j \right]^{-1},$$

$$\pi_j = \frac{n!}{(n-j)!} \left(\frac{\lambda}{r\mu} \right)^j \pi_0; \quad j = 1, 2, \dots .$$

Comparing this result with the stationary state probabilities (9.65), it is apparent that in case $r = 1$ the uniform distribution of the repair capacity to the failed machines has no influence on the stationary state probabilities. This fact is not surprising, since in this case the available maintenance capacity of one unit (if required) is always fully used.   □

Many of the results presented so far in section 9.6 are due to *Kendall* (1948).

9.6.3.3 Nonhomogeneous Birth and Death Processes

Up till now, chapter 9 has been restricted to homogeneous Markov chains. They are characterized by transition rates which do not depend on time.

Nonhomogeneous Birth Processes 1) *Nonhomogeneous Poisson process* The most simple representative of a nonhomogeneous birth process is the nonhomogeneous Poisson process (page 274). Its birth rates are

$$\lambda_i(t) = \lambda(t); \quad i = 0, 1, \dots .$$

Thus, the process makes a transition from state i at time t to state $i+1$ in $[t, t+\Delta t]$ with probability $\lambda(t)\Delta t + o(\Delta t)$.

2) *Mixed Poisson process* If certain conditions are fulfilled, mixed Poisson processes (section 7.2.3) belong to the class of nonhomogeneous birth processes.

Lundberg (1964) proved that a birth process is a mixed Poisson process if and only if its birth rates $\lambda_i(t)$ have properties

$$\lambda_{i+1}(t) = \lambda_i(t) - \frac{d \ln \lambda_i(t)}{dt}; \quad i = 0, 1, \dots .$$

Equivalently, a pure birth process $\{X(t), t \geq 0\}$ with transition rates $\lambda_i(t)$ and with absolute state distribution

$$\{p_i(t) = P(X(t) = i); \quad i = 0, 1, \dots\}$$

is a mixed Poisson process if and only if

$$p_i(t) = \frac{t}{i} \lambda_{i-1}(t) p_{i-1}(t); \quad i = 1, 2, \dots ;$$

see also *Grandel* (1997).

Nonhomogeneous Linear Birth and Death Process In generalizing the birth and death process of example 9.11, now a birth and death process $\{X(t), t \geq 0\}$ is considered which has transition rates

$$\lambda_i(t) = \lambda(t) i, \quad \mu_i(t) = \mu(t) i; \quad i = 0, 1, \dots$$

and initial distribution

$$p_1(0) = P(X(0) = 1) = 1.$$

Thus, $\lambda(t)$ can be interpreted as the transition rate from state 1 into state 2 at time t, and $\mu(t)$ is the transition rate from state 1 into the absorbing state 0 at time t. According to (9.47), the absolute state probabilities $p_j(t)$ satisfy

$$p_0'(t) = \mu(t) p_1(t),$$

$$p_j'(t) = (j-1)\lambda(t) p_{j-1}(t) - j(\lambda(t) + \mu(t)) p_j(t) + (j+1)\mu(t) p_{j+1}(t); \quad j = 1, 2, \dots .$$

Hence, the corresponding z-transform $M(t, z)$ of

$$\{p_i(t) = P(X(t) = i); \quad i = 0, 1, \dots\}$$

is given by the partial differential equation (9.50) with time-dependent λ and μ:

$$\frac{\partial M(t, z)}{\partial t} - (z - 1)[\lambda(t) z - \mu(t)] \frac{\partial M(t, z)}{\partial z} = 0. \tag{9.66}$$

The corresponding characteristic differential equation is a differential equation of Riccati type with time-dependent coefficients (compare with (9.51)):

$$\frac{dz}{dt} = -\lambda(t) z^2 + [\lambda(t) + \mu(t)] z - \mu.$$

A property of this differential equation is that there exist functions $\varphi_i(x)$; $i = 1, 2, 3, 4,$

so that its general solution $z = z(t)$ can be implicitly written in the form

$$c = \frac{z\varphi_1(t) - \varphi_2(t)}{\varphi_3(t) - z\varphi_4(t)}.$$

Hence, for all differentiable functions $g(\cdot)$, the general solution of (9.66) has the form

$$M(t,z) = g\left(\frac{z\varphi_1(t) - \varphi_2(t)}{\varphi_3(t) - z\varphi_4(t)}\right).$$

From this and the initial condition $M(0,z) = z$ it follows that there exist two functions $a(t)$ and $b(t)$ so that

$$M(t,z) = \frac{a(t) + [1 - a(t) - b(t)]z}{1 - b(t)z}. \tag{9.67}$$

By expanding $M(t,z)$ as a power series in z,

$$p_0(t) = a(t),$$

$$p_i(t) = [1 - a(t)][1 - b(t)][b(t)]^{i-1}; \quad i = 1, 2, \dots . \tag{9.68}$$

Inserting (9.67) in (9.66) and comparing the coefficients of z yields a system of differential equations for $a(t)$ and $b(t)$:

$$(a'b - ab') + b' = \lambda(1 - a)(1 - b)$$

$$a' = \mu(1 - a)(1 - b).$$

The transformations $A = 1 - a$ and $B = 1 - b$ simplify this system to

$$B' = (\mu - \lambda)B - \mu B^2 \tag{9.69}$$

$$A' = -\mu A B. \tag{9.70}$$

The first differential equation is of Bernoulli type. Substituting in (9.69)

$$y(t) = 1/B(t)$$

gives a linear differential equation in y:

$$y' + (\mu - \lambda)y = \mu. \tag{9.71}$$

Since

$$a(0) = b(0) = 0,$$

y satisfies $y(0) = 1$. Hence the solution of (9.71) is

$$y(t) = e^{-\omega(t)}\left[\int_0^t e^{\omega(x)}\mu(x)\,dx + 1\right],$$

where

$$\omega(t) = \int_0^t [\mu(x) - \lambda(x)]\,dx.$$

From (9.70) and (9.71),

$$\frac{A'}{A} = -\mu B = -\frac{\mu}{y} = -\frac{y'}{y} - \omega'.$$

Therefore, the desired functions a and b are

$$a(t) = 1 - \frac{1}{y(t)} e^{-\omega(t)}$$

$$b(t) = 1 - \frac{1}{y(t)}, \quad t \geq 0.$$

With $a(t)$ and $b(t)$ known, the one-dimensional probability distribution (9.68) of the nonhomogeneous birth and death process $\{X(t), t \geq 0\}$ is completely characterized. In particular, the probability that the process is in the absorbing state 0 at time t is

$$p_0(t) = \frac{\int_0^t e^{\omega(x)} \mu(x) \, dx}{\int_0^t e^{\omega(x)} \mu(x) \, dx + 1}.$$

Hence, the process $\{X(t), t \geq 0\}$ will reach state 0 with probability 1 if the integral

$$\int_0^t e^{\omega(x)} \mu(x) \, dx \tag{9.72}$$

diverges as $t \to \infty$. A necessary condition for this is $\mu(x) \geq \lambda(x)$ for all $x \geq 0$.

Let L denote the first passage time of the process with regard to state 0, i.e.,

$$L = \inf_{t} \{t, \ X(t) = 0\}.$$

Since state 0 is absorbing, it is justified to call L the *lifetime* of the process. On condition that the integral (9.72) diverges as $t \to \infty$, L has distribution function

$$F_L(t) = P(L \leq t) = p_0(t), \quad t \geq 0.$$

Mean value and variance of $X(t)$ are

$$E(X(t)) = e^{-\omega(t)}, \tag{9.73}$$

$$Var(X(t)) = e^{-2\omega(t)} \int_0^t e^{\omega(x)} [\lambda(x) + \mu(x)] \, dx. \tag{9.74}$$

If the process $\{X(t), t \geq 0\}$ starts at $s = 2, 3, \ldots$, i.e., it has the initial distribution

$$p_s(0) = P(X(0) = s) = 1 \quad \text{for an } s = 2, 3, \ldots$$

then the corresponding z-transform is

$$M(t, z) = \left(\frac{a(t) + [1 - a(t) - b(t)]z}{1 - b(t)z} \right)^s.$$

In this case, mean value and variance of $X(t)$ are simply obtained by multiplying (9.73) and (9.74), respectively, by s.

9.7 APPLICATIONS TO QUEUEING SYSTEMS

9.7.1 Basic Concepts

One of the most important applications of continuous-time Markov chains is stochastic modeling of service facilities. The basic situation is the following: Customers arrive at a service system (queueing system) according to a random point process. If all servers are busy, an arriving customer either waits for service or leaves the system without having been served. Otherwise, an available server takes care of the customer. After random service times customers leave the system. The arriving customers constitute the *input* (*input flow, traffic, flow of demands*) and the leaving customers the *output* (*output flow*) of the queueing system. A queueing system is called a *loss system* if it has no waiting capacity for customers which do not find an available server on arriving at the system. These customers leave the system immediately after arrival and are said to be *lost*. A *waiting system* has unlimited waiting capacity for those customers who do not immediately find an available server and are willing to wait any length of time for service. A *waiting-loss system* has only limited waiting capacity for customers. An arriving customer is lost if it finds all servers busy and the waiting capacity fully occupied. A *single-server queueing system* has only one server, whereas a *multi-server queueing system* has at least two servers. 'Customers' or 'servers' need not be persons.

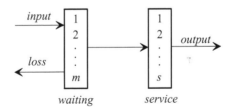

Figure 9.12 Scheme of a standard queueing system

Supermarkets are simple examples of queueing systems. Their customers are served at checkout counters. Filling stations also can be thought of as queueing systems with petrol pumps being the servers. Even a car park has the typical features of a waiting system. In this case, the parking lots are the 'servers' and the 'service times' are generated by the customers themselves. An anti-aircraft battery is a queueing system in the sense that it 'serves' the enemy aircraft. During recent years the stochastic modeling of communication systems, in particular computer networks, has stimulated the application of standard queueing models and the creation of new, more sophisticated ones. But the investigation of queueing systems goes back to the Danish engineer *A. K. Erlang* in the early 1900s, when he was in charge of designing telephone exchanges to meet criteria such as 'what is the mean waiting time of a customer before being connected' or 'how many lines (servers) are necessary to guarantee that with a given probability a customer can immediately be connected'?

The repairman problem considered in example 9.14 also fits into the framework of a queueing system. The successive failing of machines generates an input flow and the mechanics are the servers. This example is distinguished by a particular feature: each demand (customer) is produced by one of a finite number n of different sources 'inside the system', namely by one of the n machines. Classes of queueing systems having this particular feature are called *closed queueing systems*.

The global objective of queueing theory is to provide theoretical tools for the design and the quantitative analysis of service systems. Designing engineers of service systems need to make sure that the required service can be reliably delivered at minimal expense, since managers of service systems do not want to 'employ' more servers than necessary for meeting given performance criteria. Important criteria are:

1) The probability that an arriving customer finds an available server.

2) The mean waiting time of a customer for service.

3) The total sojourn time of a customer in the system.

It is common practice to characterize the structure of standard queueing systems by *Kendall's notation A/B/s/m*. In this code, A characterizes the input and B the service, s is the number of servers, and waiting capacity is available for m customers. Using this notation, standard classes of queueing systems are:

$A = M$ (*Markov*): Customers arrive in accordance with a homogeneous Poisson process (*Poisson input*).

$A = GI$ (*general independent*): Customers arrive in accordance with an ordinary renewal process (*recurrent input*).

$A = D$ (*deterministic*): The distances between the arrivals of neighbouring customers are constant (*deterministic input*).

$B = M$ (*Markov*) The service times are independent, identically distributed exponential random variables.

$B = G$ (*general*) The service times are independent, identically distributed random variables with arbitrary probability distribution.

For instance, $M/M/1/0$ is a loss system with Poisson input, one server, and exponential service times. $GI/M/3/\infty$ is a waiting system with recurrent input, exponential service times, and 3 servers. For queueing systems with an infinite number of servers, no waiting capacity is necessary. Hence their code is $A/B/\infty$.

In waiting systems and waiting-loss systems there are several ways of choosing waiting customers for service. These possibilities are called *service disciplines* (*queueing disciplines*). The most important ones are:

1) *FCFS* (*first come-first served*) Waiting customers are served in accordance with the order of their arrival. This discipline is also called *FIFO* (*first in-first out*), although 'first in' does not necessarily imply 'first out'.

2) LCFS (last come-first served) The customer, which arrived last, is served first. This discipline is also called *LIFO (last in-first out)*.

3) SIRO (service in random order) A server, when having finished with a customer, randomly picks one of the waiting customers for service.

There is a close relationship between service disciplines and priority (queueing) systems: In a *priority system* arriving customers have different *priorities* of being served. A customer with higher priority is served before a customer with lower priority, but no interruption of ongoing service takes place (*head of the line priority discipline*). When a customer with *absolute priority* arrives and finds all servers busy, then the service of a customer with lower priority has to be interrupted (*preemptive priority discipline*).

System Parameter and Assumptions In this chapter, if not stated otherwise, the interarrival times of customers are assumed to be independent and identically distributed as a random variable Y. The intensity of the input flow (mean number of arriving customers per unit time) is denoted as λ and referred to as *arrival rate* or *arrival intensity*. The service times of all servers, if not stated otherwise, are assumed to be independent and identically distributed as a random variable Z. The *service intensity* or *service rate* of the servers is denoted as μ, i.e. μ is the mean number of customers served per unit time by a server. Hence,

$$E(Y) = 1/\lambda \text{ and } E(Z) = 1/\mu.$$

The *traffic intensity* of a queueing system is defined as the ratio

$$\rho = \lambda/\mu.$$

Usually, the state, the system is in, is fully characterized by the number of customers $X(t)$, which are in the system at time t (waiting or being served). If the stochastic process $\{X(t), t \geq 0\}$ has eventually become stationary, then we say the queuing system is in the *steady state*. When the system is in the steady state, then the time dependence of its characteristic parameters, in particular of the state probabilities

$$P(X(t) = j); \ j = 0, 1, ... ,$$

has levelled out; they are constant. This will happen afer a sufficiently long operating time. In this case, the probability distribution of $X(t)$ does no longer depend on t so that $X(t)$ is simply written as X. In this case,

$$\{\pi_j = \lim_{t \to \infty} P(X(t) = j) = P(X = j); \ j = 0, 1, ..., s+m, \ s, m \leq \infty\}$$

is the stationary probability distribution of $\{X(t), t \geq 0\}$.

Let S denote the random number of busy servers in the steady state of the system. Then its *degree of server utilization* is

$$\eta = E(S)/s.$$

The coefficient η can be interpreted as the mean proportion of time a server is busy.

9.7.2   Loss Systems

9.7.2.1   *M/M/∞*-System

Strictly speaking, this system is neither a loss nor a waiting system. In this model, the stochastic process $\{X(t), t \geq 0\}$ is a homogeneous birth-and death process with state space $\mathbf{Z} = \{0, 1, ...\}$ and transition rates (see example 9.12)

$$\lambda_i = \lambda; \quad \mu_i = i\mu; \quad i = 0, 1, ...$$

The corresponding time-dependent state probabilities $p_j(t)$ of this queueing system are given by (9.55). The stationary state probabilities are obtained by passing to the limit as $t \to \infty$ in these $p_j(t)$ or by inserting the transition rates $\lambda_i = \lambda$ and $\mu_i = i\mu$ with $n = \infty$ into (9.60) and (9.61):

$$\pi_j = \frac{\rho^j}{j!} e^{-\rho}; \quad j = 0, 1, ... \tag{9.75}$$

This is a Poisson distribution with parameter $\rho = \lambda/\mu$. Hence, in the steady state the mean number of busy servers is equal to the traffic intensity of the system: $E(X) = \rho$.

5.7.2.2   *M/M/s/0*-System

In this case, $\{X(t), t \geq 0\}$ is a birth and death process with $\mathbf{Z} = \{0, 1, ..., s\}$ and

$$\lambda_i = \lambda; \quad i = 0, 1, ..., s - 1; \quad \lambda_i = 0 \text{ for } i \geq s,$$

$$\mu_i = i\mu; \quad i = 0, 1, ..., s.$$

Inserting these transition rates into the stationary state probabilities (9.60) and (9.61) with $n = s$ yields

$$\pi_0 = \left[\sum_{i=0}^{s} \frac{1}{i!} \rho^i \right]^{-1}; \quad \pi_j = \frac{1}{j!} \rho^j \pi_0; \quad j = 0, 1, ..., s. \tag{9.76}$$

The probability π_0 is called *vacant probability*. The *loss probability*, i.e., the probability that an arriving customer does not find an idle server, and, hence, leaves the system immediately, is

$$\pi_s = \frac{\frac{1}{s!} \rho^s}{\sum_{i=0}^{s} \frac{1}{i!} \rho^i}. \tag{9.77}$$

This is the famous *Erlang loss formula*. The following recursive formula for the loss probability as a function of s can easily be verified:

$$\pi_0 = 1 \text{ for } s = 0; \quad \frac{1}{\pi_s} = \frac{s}{\rho} \frac{1}{\pi_{s-1}} + 1; \quad s = 1, 2, ...$$

The mean number of busy servers is

$$E(X) = \sum_{i=1}^{s} i \pi_i = \sum_{i=1}^{s} i \frac{\rho^i}{i!} \pi_0 = \rho \sum_{i=0}^{s-1} \frac{\rho^i}{i!} \pi_0.$$

Combining this result with (9.76) and (9.77) yields

$$E(X) = \rho(1 - \pi_s).$$

Hence, the *degree of server utilization* is

$$\eta = \frac{\rho}{s}(1 - \pi_s).$$

Single-Server Loss System In case $s = 1$ vacant and loss probability are

$$\pi_0 = \frac{1}{1+\rho} \quad \text{and} \quad \pi_1 = \frac{\rho}{1+\rho}. \tag{9.78}$$

Since $\rho = E(Z)/E(Y)$,

$$\pi_0 = \frac{E(Y)}{E(Y) + E(Z)} \quad \text{and} \quad \pi_1 = \frac{E(Z)}{E(Y) + E(Z)}.$$

Hence, π_0 (π_1) is formally equal to the stationary availability (nonavailability) of a system with mean lifetime $E(Y)$ and mean renewal time $E(Z)$ the operation of which is governed by an alternating renewal process (formula (7.14), page 322).

Example 9.16 A 'classical application' (no longer of practical relevance) of loss models of type $M/M/s/0$ is a telephone exchange. Assume that the input (calls of subscribers wishing to be connected) has intensity $\lambda = 2\,[\text{min}^{-1}]$. Thus, the mean time between successive calls is $E(Y) = 1/\lambda = 0.5\,[\text{min}]$. On average, each subscriber occupies a line for $E(Z) = 1/\mu = 3\,[\text{min}]$.

1) What is the loss probability in case of $s = 7$ lines?

The corresponding traffic intensity is $\rho = \lambda/\mu = 6$. Thus, the loss probability equals

$$\pi_7 = \frac{\frac{1}{7!}6^7}{1 + 6 + \frac{6^2}{2!} + \frac{6^3}{3!} + \frac{6^4}{4!} + \frac{6^5}{5!} + \frac{6^6}{6!} + \frac{6^7}{7!}} = 0.185.$$

Hence, the mean number of occupied lines is

$$E(X) = \rho(1 - \pi_7) = 6(1 - 0.185) = 4.89,$$

and the degree of server (line) utilization is

$$\eta = \eta(7) = 4.89/7 = 0.698.$$

2) What is the minimal number of lines which have to be provided in order to make sure that at least 95% of the desired connections can be made?

The respective loss probabilities for $s = 9$ and $s = 10$ are

$$\pi_9 = 0.075 \quad \text{and} \quad \pi_{10} = 0.043.$$

Hence, the minimal number of lines required is $s = 10$. In this case, however, the degree of server utilization is smaller than with $s = 7$ lines:

$$\eta = \eta(10) = 0.574. \qquad \square$$

It is interesting and practically important that the stationary state probabilities of the queueing system $M/G/s/0$ also have the structure (9.76). That is, if the respective traffic intensities of the systems $M/M/s/0$ and $M/G/s/0$ are equal, then their stationary state probabilities coincide: for both systems they are given by (9.76). A corresponding result holds for the queueing systems $M/M/\infty$ and $M/G/\infty$. (Compare the stationary state probabilities (9.75) with the stationary state probabilities (7.37) (page 274) for the $M/G/\infty$-system.) Queueing systems having this property are said to be *insensitive* with respect to the probability distribution of the service. An analogous property can be defined with regard to the input. In view of (9.78), the $M/M/1/0$ -system is insensitive both with regard to arrival and service time distributions (*full insensitiveness*). A comprehensive investigation of the *insensitiveness* of queueing systems and other stochastic models is given in the handbook on queueing theory by *Gnedenko, König* (1983).

9.7.2.3 Engset's Loss System

Assume that n sources generate n independent Poisson inputs with common intensity λ, which are served by s servers, $s \le n$. The service times are independent, exponentially distributed random variables with parameter μ. As long as a customer from a particular source is being served, this source cannot produce another customer. (Compare to the repairman problem, example 9.14: during the repair of a machine, this machine cannot produce another demand for repair.) A customer which does not find an available server is lost. Let $X(t)$ denote the number of customers being served at time t. Then $\{X(t), t \ge 0\}$ is a birth- and death process with state space $\mathbf{Z} = \{0, 1, ..., s\}$. In case $X(t) = j$ only $n - j$ sources are active, that is they are able to generate customers. Therefore, the transition rates of this birth- and death process are

$$\lambda_j = (n-j)\lambda; \quad j = 0, 1, 2, ..., s-1,$$

$$\mu_j = j\mu; \quad j = 1, 2, ..., s.$$

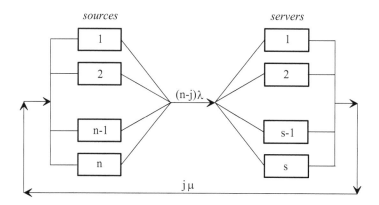

Figure 9.13 Engset's loss system in state $X(t) = j$

Inserting these transition rates into (9.60) and (9.61) with $n = s$ yields the stationary state distribution for Engset's loss system

$$\pi_j = \frac{\binom{n}{j}\rho^j}{\sum\limits_{i=0}^{s}\binom{n}{i}\rho^i}; \quad j = 0, 1, \ldots, s.$$

In particular, π_0 and the loss probability π_s are

$$\pi_0 = \frac{1}{\sum\limits_{i=0}^{s}\binom{n}{i}\rho^i}, \quad \pi_s = \frac{\binom{n}{s}\rho^s}{\sum\limits_{i=0}^{s}\binom{n}{i}\rho^i}; \quad j = 0, 1, \ldots, s.$$

Engset's loss system is just as the repairman problem considered in example 9.14, a closed queueing system.

9.7.3 Waiting Systems

9.7.3.1 *M/M/s/∞-System*

The Markov chain $\{X(t), t \geq 0\}$, which models this system, is defined as follows: If $X(t) = j$ with $0 \leq j \leq s$, then j servers are busy at time t. If $X(t) = j$ with $s > j$, then s servers are busy and $j - s$ customers are waiting for service. In either case, $X(t)$ is the total number of customers in the queueing system at time t. $\{X(t), t \geq 0\}$ is a birth and death process with state space $\mathbf{Z} = \{0, 1, \ldots\}$ and transition rates

$$\lambda_j = \lambda; \quad j = 0, 1, \ldots,$$

$$\mu_j = j\mu \text{ for } j = 0, 1, \ldots, s; \quad \mu_j = s\mu \text{ for } j > s. \tag{9.79}$$

In what follows it is assumed that

$$\rho = \lambda/\mu < s.$$

If $\rho > s$, then the arrival intensity λ of customers is greater than the maximum service rate μs of the system so that, at least in the longrun, the system cannot cope with the input, and the length of the waiting queue will tend to infinity as $t \to \infty$. Hence, no equilibrium (steady) state between arriving and leaving customers is possible. On the other hand, the condition is necessary and sufficient for the existence of a stationary state distribution, since in this case the corresponding series (9.62) converges and condition (9.63) is fulfilled.

Inserting the transition rates (9.79) into (9.60) yields

$$\pi_j = \frac{\rho^j}{j!}\pi_0 \text{ for } j = 0, 1, \ldots, s - 1,$$

$$\pi_j = \frac{\rho^j}{s! \, s^{j-s}}\pi_0 \text{ for } j \geq s. \tag{9.80}$$

The normalizing condition and the geometric series (formula (2.16), page 48) yields the vacant probability π_0:

$$\pi_0 = \left[\sum_{i=0}^{s-1} \frac{1}{i!} \rho^i + \frac{\rho^s}{(s-1)!(s-\rho)} \right]^{-1}.$$

The probability π_w that an arriving customer finds all servers busy is

$$\pi_w = \sum_{i=s}^{\infty} \pi_i.$$

π_w is called *waiting probability*, since it is the probability that an arriving customer must wait for service. Making again use of the geometric series yields a simple formula for π_w:

$$\pi_w = \frac{\pi_s}{1-\rho/s}. \tag{9.81}$$

In what follows, all derivations refer to the system in the steady state. If S denotes the random number of busy servers, then its mean value is

$$E(S) = \sum_{i=0}^{s-1} i\pi_i + s\pi_w. \tag{9.82}$$

From this,

$$E(S) = \rho. \tag{9.83}$$

(The details of the derivation of (9.83) are left as an exercise to the reader.) Also without proof: Formula (9.83) holds for any $GI/G/s/\infty$ -system. Hence the degree of server utilization in the $M/M/s/\infty$-system is $\eta = \rho/s$. By making use of (9.83), the mean value of the total number X of customers in the system is seen to be

$$E(X) = \sum_{i=1}^{\infty} i\pi_i = \rho \left[1 + \frac{s}{(s-\rho)^2} \pi_s \right]. \tag{9.84}$$

Let L denote the random number of customers waiting for service (queue length). Then the mean queue length is

$$E(L) = \sum_{i=s}^{\infty} (i-s)\pi_i = \sum_{i=s}^{\infty} i\pi_i - s\pi_w.$$

Combining this formula with (9.82)–(9.84) yields

$$E(L) = \frac{\rho s}{(s-\rho)^2} \pi_s. \tag{9.85}$$

Waiting Time Distribution Let W be the random time a customer has to wait for service if the service discipline *FCFS* is in effect. By the total probability rule

$$P(W > t) = \sum_{i=s}^{\infty} P(W > t | X = i)\pi_i. \tag{9.86}$$

If a customer enters the system when it is in state $X = i > s$, then all servers are busy so that the current output is a Poisson process with intensity $s\mu$. The random event '$W > t$' occurs if within t time units after the arrival of a customer the service of at most $i - s$ customers has been finished. Therefore, the probability that the service of precisely k customers, $0 \le k \le i - s$, will be finished in this interval of length t is

$$\frac{(s\,\mu t)^k}{k!}\,e^{-s\mu t}.$$

Hence,

$$P(W > t | X = i) = e^{-s\mu t} \sum_{k=0}^{i-s} \frac{(s\,\mu t)^k}{k!}$$

and, by (9.86)

$$P(W > t) = e^{-s\mu t} \sum_{i=s}^{\infty} \pi_i \sum_{k=0}^{i-s} \frac{(s\,\mu t)^k}{k!} = \pi_0 e^{-s\mu t} \sum_{i=s}^{\infty} \frac{\rho^i}{s!s^{i-s}} \sum_{k=0}^{i-s} \frac{(s\,\mu t)^k}{k!}.$$

By performing the index transformation $j = i - s$, changing the order of summation according to formula (2.115), page 99, and making use of both the exponential series and the geometric series (page 48) yield

$$P(W > t) = \pi_0 \frac{\rho^s}{s!} e^{-s\mu t} \sum_{j=0}^{\infty} \left(\frac{\rho}{s}\right)^j \sum_{k=0}^{j} \frac{(s\,\mu t)^k}{k!}$$

$$= \pi_s\, e^{-s\mu t} \sum_{k=0}^{\infty} \frac{(s\,\mu t)^k}{k!} \sum_{j=k}^{\infty} \left(\frac{\rho}{s}\right)^j$$

$$= \pi_s\, e^{-s\mu t} \sum_{k=0}^{\infty} \frac{(\lambda t)^k}{k!} \sum_{i=0}^{\infty} \left(\frac{\rho}{s}\right)^i = \pi_s\, e^{-s\mu t}\, e^{\lambda t}\, \frac{1}{1 - \rho/s}.$$

Hence, the distribution function of W is

$$F_W(t) = P(W \le t) = 1 - \frac{s}{s - \rho}\, \pi_s\, e^{-\mu(s-\rho)t}, \quad t \ge 0.$$

Note that $P(W > 0)$ is the waiting probability (9.81):

$$\pi_w = P(W > 0) = 1 - F_W(0) = \frac{s}{s - \rho}\, \pi_s.$$

The mean waiting time of a customer is

$$E(W) = \int_0^{\infty} P(W > t)\, dt = \frac{s}{\mu(s - \rho)^2}\, \pi_s. \tag{9.87}$$

A comparison of (9.85) and (9.87) yields *Little's formula* or *Little's law*:

$$E(L) = \lambda\, E(W). \tag{9.88}$$

Little's formula can be motivated as follows: The mean value of the sum of the waiting times arising in an interval of length τ is $\tau E(L)$. On the other hand, the same mean value is given by $\lambda \tau E(W)$, since the mean number of customers arriving in an interval of length τ is $\lambda \tau$. Hence,

$$\tau E(L) = \lambda \tau E(W),$$

which is Little's formula.

With $E(X)$ given by (9.84), an equivalent representation of Little's formula is

$$E(X) = \lambda\, E(T), \tag{9.89}$$

where T is the total sojourn time of a customer in the system, i.e., waiting plus service time $T = W + Z$. Hence, the mean value of T is

$$E(T) = E(W) + 1/\mu.$$

Little's formula holds for any $GI/G/s/\infty$–system. For a proof of this proposition and other 'Little type formulas' see *Franken et al.* (1981).

9.7.3.2 *M/G/1/∞*-System

In this single-server system, the service time Z is assumed to have an arbitrary probability density $g(t)$ and a finite mean $E(Z) = 1/\mu$. Hence, the corresponding stochastic process $\{X(t), t \geq 0\}$ describing the development in time of the number of customers in the system needs no longer be a homogeneous Markov chain as in the previous queuing models. However, there exists an embedded homogeneous discrete-time Markov chain, which can be used to analyze this system (see section 9.4).

The system starts operating at time $t = 0$. Customers arrive according to a homogeneous Poisson process with positive intensity λ. Let A be the random number of customers, which arrive whilst a customer is being served, and

$$\{a_i = P(A = i); \ i = 0, 1, ...\}$$

be its probability distribution. To determine the a_i, note that the conditional probability that during a service time of length $Z = t$ exactly i new customers arrive is

$$\frac{(\lambda t)^i}{i!}\, e^{-\lambda t}.$$

Hence,

$$a_i = \int_0^\infty \frac{(\lambda t)^i}{i!}\, e^{-\lambda t} g(t)\, dt, \ \ i = 0, 1, ...$$

This and the exponential series (page 48) yield the z-transform $M_A(z)$ of A:

$$M_A(z) = \sum_{i=0}^\infty a_i z^i = \int_0^\infty e^{-(\lambda - \lambda z)t} g(t)\, dt.$$

Consequently, if $\hat{g}(\cdot)$ denotes the Laplace transform of $g(t)$, then

$$M_A(z) = \hat{g}(\lambda - \lambda z). \tag{9.90}$$

By formula (2.112) (page 96), letting as usual $\rho = \lambda/\mu$, the mean value of A is

$$E(A) = \frac{dM_A(z)}{dz}\Big|_{z=1} = -\lambda\, \frac{d\hat{g}(r)}{dr}\Big|_{r=0} = \rho. \tag{9.91}$$

Embedded Markov Chain Let T_n be the random time point at which the nth customer leaves the system. If X_n denotes the number of customers in the system immediately after T_n, then $\{X_1, X_2, ...\}$ is a homogeneous, discrete-time Markov chain with state space $\mathbf{Z} = \{0, 1, ...\}$ and one-step transition probabilities

$$p_{ij} = P(X_{n+1} = j | X_n = i) = \begin{cases} a_j & \text{if } i = 0 \text{ and } j = 0, 1, 2, ... \\ a_{j-i+1} & \text{if } i-1 \leq j \text{ and } i = 1, 2, ... \\ 0 & \text{otherwise} \end{cases} \quad (9.92)$$

for all $n = 0, 1, ...$; $X_0 = 0$. This Markov chain is embedded in $\{X(t), t \geq 0\}$ since

$$X_n = X(T_n + 0); \quad n = 0, 1, ...$$

The discrete-time Markov chain $\{X_0, X_1, ...\}$ is irreducible and aperiodic. Hence, on condition $\rho = \lambda/\mu < 1$ it has a stationary state distribution $\{\pi_0, \pi_1, ...\}$, which can be obtained by solving the corresponding system of algebraic equations (8.9) (see page 342): Inserting the transition probabilities p_{ij} given by (9.92) into (8.9) gives

$$\pi_0 = a_0(\pi_0 + \pi_1),$$

$$\pi_j = \pi_0 a_j + \sum_{i=1}^{j+1} \pi_i a_{j-i+1}; \quad j = 1, 2, ... \quad (9.93)$$

Let $M_X(z)$ be the z-transform of the state X of the system in the steady state:

$$M_X(z) = \sum_{j=0}^{\infty} \pi_j z^j.$$

Then, multiplying (9.93) by z^j and summing up from $j = 0$ to ∞ yields

$$M_X(z) = \pi_0 \sum_{j=0}^{\infty} a_j z^j + \sum_{j=0}^{\infty} z^j \sum_{i=1}^{j+1} \pi_i a_{j-i+1}$$

$$= \pi_0 M_A(z) + M_A(z) \sum_{i=1}^{\infty} \pi_i z^{i-1} a_{j-i+1}$$

$$= \pi_0 M_A(z) + M_A(z) \frac{M_X(z) - \pi_0}{z}.$$

Solving this equation for $M_X(z)$ yields

$$M_X(z) = \pi_0 M_A(z) \frac{1-z}{M_A(z) - z}, \quad |z| < 1. \quad (9.94)$$

To determine π_0, note that

$$M_A(1) = M_X(1) = 1$$

and

$$\lim_{z \uparrow 1} \frac{M_A(z) - z}{1 - z} = \lim_{z \uparrow 1} \left(1 + \frac{M_A(z) - 1}{1 - z} \right) = 1 - \frac{dM_A(z)}{dz} \Big|_{z=1} \Big| = 1 - \rho.$$

Therefore, by letting $z \uparrow 1$ in (9.94),

$$\pi_0 = 1 - \rho. \tag{9.95}$$

Combining (9.90), (9.94), and (9.95) yields the *Formula of Pollaczek-Khinchin*:

$$M_X(z) = (1 - \rho) \frac{1 - z}{1 - \dfrac{z}{\hat{g}(\lambda - \lambda z))}}, \quad |z| < 1. \tag{9.96}$$

According to its derivation, this formula gives the z-transform of the stationary distribution of the random number X of customers in the system immediately after the completion of a customer's service. In view of the homogeneous Poisson input, it is even the stationary probability distribution of the 'original' Markov chain $\{X(t), t \geq 0\}$ itself. Thus, X is the random number of customers at the system in its steady state. Its probability distribution $\{\pi_0, \pi_1, ...\}$ exists and is solution of (9.93). Hence, numerical parameters as mean value and variance of the number of customers in the system in the steady state can be determined by (9.96) via formulas (2.112), page 96. For instance, the mean number of customers in the system is

$$E(X) = \frac{dM_X(z)}{dz}\Big|_{z=1}\Big| = \rho + \frac{\lambda^2[(E(Z))^2 + Var(Z)]}{2(1 - \rho)}. \tag{9.97}$$

Sojourn Time Let T be the time a customer spends in the system (sojourn time) if the *FCFS*-queueing discipline is in effect. Then T has structure

$$T = W + Z,$$

where W is the time a customer has to wait for service (waiting time). Let $F_T(t)$ and $F_W(t)$ be the respective distribution functions of T and W and $f_T(t)$ and $f_W(t)$ the corresponding densities with Laplace transforms $\hat{f}_T(r)$ and $\hat{f}_W(r)$. Since W and Z are independent,

$$\hat{f}_T(r) = \hat{f}_W(r)\hat{g}(r). \tag{9.98}$$

The number of customers in the system after the departure of a served one is equal to the number of customers which arrived during the sojourn time of this customer. Hence, analogously to the structure of the a_i, the probabilities π_i are given by

$$\pi_i = \int_0^\infty \frac{(\lambda t)^i}{i!} e^{-\lambda t} f_T(t)\, dt; \quad i = 0, 1, ...$$

The corresponding z-transform $M_X(z)$ of X or, equivalently, the z-transform of the stationary distribution $\{\pi_0, \pi_1, ...\}$ is (compare to the derivation of (9.90))

$$M_X(z) = \hat{f}_T(\lambda - \lambda z).$$

Thus, by (9.98),

$$M_X(z) = \hat{f}_W(\lambda - \lambda z)\hat{g}(\lambda - \lambda z).$$

This formula and (9.96) yields the Laplace transform of $f_W(r)$:

$$\hat{f}_W(r) = (1-\rho)\frac{r}{\lambda\hat{g}(r)+r-\lambda}.$$

By formulas (2.62) and (2.119), $E(W)$ and $Var(W)$ can be determined from $\hat{f}_W(r)$:

$$E(W) = \frac{\lambda[(E(Z))^2 + Var(Z)]}{2(1-\rho)}, \qquad (9.99)$$

$$Var(W) = \frac{\lambda^2[(E(Z))^2 + Var(Z)]^2}{4(1-\rho)^2} + \frac{\lambda E(Z^3)}{3(1-\rho)}.$$

The random number of busy servers S has the stationary distribution

$$P(S=0) = \pi_0 = 1-\rho, \quad P(S=1) = 1-\pi_0 = \rho$$

so that

$$E(S) = \rho.$$

The queue length is $L = X - S$. Hence, by (9.97),

$$E(L) = \frac{\lambda^2[(E(Z))^2 + Var(Z)]}{2(1-\rho)}. \qquad (9.100)$$

Comparing (9.99) and (9.100) verifies Little's formula (9.88):

$$E(L) = \lambda E(W).$$

Example 9.17 The use of the formula of Pollaczek-Khinchin is illustrated by assuming that Z has an exponential distribution:

$$g(t) = \mu e^{-\mu t}, \ t \geq 0.$$

By example 2.26 (page 101), the Laplace transform of $g(t)$ is

$$\hat{g}(r) = \frac{\mu}{r+\mu} \text{ so that } \hat{g}(\lambda - \lambda z) = \hat{g}(\lambda(1-z)) = \frac{\mu}{\lambda(1-z)+\mu}.$$

Inserting this in (9.96) gives

$$M_X(z) = (1-\rho)\mu\frac{1-z}{\mu - z[\lambda(1-z)+\mu]}$$

$$= (1-\rho)\mu\frac{1-z}{\mu(1-z)-z[\lambda(1-z)]} = (1-\rho)\mu\frac{1}{\mu-\lambda z}$$

so that by the exponential series (2.19) (page 48),

$$M_X(z) = (1-\rho)\frac{1}{1-\rho z} = \sum_{i=0}^{\infty}(1-\rho)\frac{\rho^i}{i!}z^i.$$

Hence, by the exponential series (2.19) (page 48),

$$p_i = (1-\rho)\frac{\rho^i}{i!}; \ i = 0, 1, \ldots.$$

This confirms the result (9.80) for the $M/M/s/\infty$-system with $s = 1$. □

9.7.3.3 *GI/M/1/∞*-System

In this single-server system, the interarrival times are given by an ordinary renewal process $\{Y_1, Y_2, ...\}$, where the Y_i are identically distributed as Y with probability density $f_Y(t)$ and finite mean value $E(Y) = 1/\lambda$. The service times are identically exponentially distributed with parameter μ. A customer leaves the system immediately after completion of its service. If an arriving customer finds the server busy, it joins the queue. The stochastic process $\{X(t), t \geq 0\}$, describing the development of the number of customers in the system in time, needs not be a homogeneous Markov chain. However, as in the previous section, an embedded homogeneous discrete-time Markov chain can be identified: The n th customer arrives at time

$$T_n = \Sigma_{i=1}^{n} Y_i; \quad n = 1, 2, ...$$

Let X_n denote the number of customers in the station immediately before arrival of the $(n + 1)$ th customer (being served or waiting). Then, $0 \leq X_n \leq n$, $n = 0, 1, ...$ The discrete-time stochastic process $\{X_0, X_1, ...\}$ is a Markov chain with parameter space $\mathbf{T} = \{0, 1, ...\}$ and state space $\mathbf{Z} = \{0, 1, ...\}$. Given that the system starts operating at time $t = 0$, the initial distribution of this discrete-time Markov chain is $P(X_0 = 0) = 1$.

For obtaining the transition probabilities of $\{X_0, X_1, ...\}$, let D_n be the number of customers leaving the station in the interval $[T_n, T_{n+1})$ of length Y_{n+1}. Then,

$$X_n = X_{n-1} - D_n + 1 \text{ with } 0 \leq D_n \leq X_n; \quad n = 1, 2, ...$$

By theorem 7.2, on condition $Y_{n+1} = t$, the random variable D_n has a Poisson distribution with parameter μt if the server is busy throughout the interval $[T_n, T_{n+1})$. Hence, for $i \geq 0$ and $1 \leq j \leq i + 1$,

$$P(X_n = j | X_{n-1} = i, Y_{n+1} = t) = \frac{(\mu t)^{i+1-j}}{(i+1-j)!} e^{-\mu t}; \quad n = 1, 2, ...$$

Consequently the one-step transition probabilities

$$p_{ij} = P(X_n = j | X_{n-1} = i); \quad i, j \in \mathbf{Z}; \quad n = 1, 2, ...$$

of the Markov chain $\{X_0, X_1, ...\}$ are

$$p_{ij} = \int_0^\infty \frac{(\mu t)^{i+1-j}}{(i+1-j)!} e^{-\mu t} f_Y(t) dt; \quad 1 \leq j \leq i + 1.$$

The normalizing condition yields

$$p_{i0} = 1 - \Sigma_{j=1}^{i+1} p_{ij}.$$

The transition probabilities p_{ij} do not depend on n so that $\{X_0, X_1, ...\}$ is a homogeneous Markov chain. It is *embedded* in the original state process $\{X(t), t \geq 0\}$ since

$$X_n = X(T_{n+1} - 0); \quad n = 0, 1, ...$$

Based on the embedded Markov chain $\{X_0, X_1, ...\}$, a detailed analysis of the queueing system $GI/M/1/\infty$ can be carried out analogously to the one of system $M/G/1/\infty$.

9.7.4 Waiting-Loss Systems

9.7.4.1 *M/M/s/m*-System

This system has s servers and waiting capacity for m customers, $m \geq 1$. A potential customer, which at arrival finds no idle server and the waiting capacity occupied, is lost, that is such a customer leaves the system immediately after arrival.

The number of customers $X(t)$ in the system at time t generates a birth- and death process $\{X(t), t \geq 0\}$ with state space $\mathbf{Z} = \{0, 1, ..., s + m\}$ and transition rates

$$\lambda_j = \lambda, \qquad 0 \leq j \leq s + m - 1,$$

$$\mu_j = \begin{cases} j\mu & \text{for} \quad 1 \leq j \leq s, \\ s\mu & \text{for} \quad s < j \leq s + m. \end{cases}$$

According to (9.60) and (9.61), the stationary state probabilities are

$$\pi_j = \begin{cases} \dfrac{1}{j!} \rho^j \pi_0 & \text{for} \quad 1 \leq j \leq s - 1, \\ \dfrac{1}{s! \, s^{j-s}} \rho^j \pi_0 & \text{for} \quad s \leq j \leq s + m. \end{cases}$$

$$\pi_0 = \left[\sum_{j=0}^{s-1} \frac{1}{j!} \rho^j + \sum_{j=s}^{s+m} \frac{1}{s! \, s^{j-s}} \rho^j \right]^{-1}.$$

The second series in π_0 can be summed up to obtain

$$\pi_0 = \begin{cases} \left[\displaystyle\sum_{j=0}^{s-1} \frac{1}{j!} \rho^j + \frac{1}{s!} \rho^s \frac{1-(\rho/s)^{m+1}}{1-\rho/s} \right]^{-1} & \text{for} \quad \rho \neq s, \\ \left[\displaystyle\sum_{j=0}^{s-1} \frac{1}{j!} \rho^j + (m+1) \frac{s^s}{s!} \right]^{-1} & \text{for} \quad \rho = s. \end{cases}$$

The *vacant probability* π_0 is the probability that there is no customer in the system and π_{s+m} is the *loss probability*, i.e., the probability that an arriving customer is lost (rejected). The respective probabilities π_f and π_w that an arriving customer finds a free (idle) server or waits for service are

$$\pi_f = \sum_{i=0}^{s-1} \pi_i, \quad \pi_w = \sum_{i=s}^{s+m-1} \pi_i.$$

Analogously to the loss system $M/M/s/0$, the mean number of busy servers is

$$E(S) = \rho (1 - \pi_{s+m}).$$

Thus, the degree of server utilization is

$$\eta = \rho (1 - \pi_{s+m})/s.$$

In the following example, the probabilities π_0 and π_{s+m}, which refer to a queueing system with s servers and waiting capacity for m customers, are denoted as

$$\pi_0(s,m) \text{ and } \pi_{s+m}(s,m),$$

respectively.

Example 9.18 A filling station has $s = 8$ petrol pumps and waiting capacity for $m = 6$ cars. On average, 1.2 cars arrive at the filling station per minute. The mean time a car occupies a petrol pump is 5 minutes. It is assumed that the filling station behaves like an $M/M/s/m$-queueing system. Since $\lambda = 1.2$ and $\mu = 0.2$, the traffic intensity is $\rho = 6$. The corresponding loss probability $\pi_{14} = \pi_{14}(8,6)$ is

$$\pi_{14}(8,6) = \frac{1}{8! \, 8^6} 6^{14} \pi_0(8,6) = 0.0167.$$

From the normalizing condition,

$$\pi_0(8,6) = \left[\sum_{j=0}^{7} \frac{1}{j!} 6^j + \frac{1}{8!} 6^8 \frac{1-(6/8)^7}{1-6/8} \right]^{-1}$$

$$= 0.00225.$$

Consequently, the average number of occupied petrol pumps is

$$E(S) = 6 \cdot (1 - 0.0167) = 5.9.$$

After having obtained these figures, the owner of the filling station considers 2 from the 8 petrol pumps superfluous and has them pulled down. It is assumed that this change does not influence the input flow so that cars continue to arrive with traffic intensity $\rho = 6$. The corresponding loss probability $\pi_{12} = \pi_{12}(6,6)$ becomes

$$\pi_{12}(6,6) = \frac{6^6}{6!} \pi_0(6,6) = 0.1023.$$

Thus, about 10% of all arriving cars leave the station without having filled up. To counter this drop, the owner provides waiting capacity for another 4 cars so that $m = 10$. The corresponding loss probability $\pi_{16} = \pi_{16}(6,10)$ is

$$\pi_{16}(6,10) = \frac{6^6}{6!} \pi_0(6,10) = 0.0726.$$

Formula

$$\pi_{6+m}(6,m) = \frac{6^6}{6!} \left[\sum_{j=0}^{5} \frac{1}{j!} 6^j + (m+1) \frac{6^6}{6!} \right]^{-1}$$

yields that additional waiting capacity for 51 cars has to be provided to equalize the loss caused by reducing the number of pumps from 8 to 6. So, the decision of the owner to pull down two of the pumps was surely not helpful. □

9.7.4.2 *M/M/s/∞*-System with Impatient Customers

Even if there is waiting capacity for arbitrarily many customers, some customers might leave the system without having been served. This happens when customers can only spend a finite time, their *patience time*, in the queue. If the service of a customer does not begin before its patience time expires, the customer leaves the system. For example, if somebody, whose long-distance train will depart in 10 minutes, has to wait 15 minutes to buy a ticket, then this person will leave the counter without a ticket. Real time monitoring and control systems have memories for data to be processed. But these data 'wait' only as long as they are up to date. Bounded waiting times are also typical for packed switching systems, for instance in computer-aided booking systems. Generally one expects that 'intelligent' customers adopt their behavior to the actual state of the queueing system. Of the many available models dealing with such situations, the following one is considered in some detail:

Customers arriving at an *M/M/s/∞*-system have independent, exponentially with parameter v distributed patience times. If $X(t)$ as usual denotes the number of customers in the system at time t, then $\{X(t), t \geq 0\}$ is a birth and death process with transition rates

$$\lambda_j = \lambda; \quad j = 0, 1, \ldots,$$

$$\mu_j = \begin{cases} j\mu & \text{for } j = 1, 2, \ldots, s, \\ s\mu + (j-s)v & \text{for } j = s, s+1, \ldots \end{cases}.$$

If $j \to \infty$, then $\mu_j \to \infty$, whereas the birth rate remains constant. Hence the sufficient condition for the existence of a stationary distribution stated in theorem 9.3 (page 419) is fulfilled. Once the queue length exceeds a certain level, the number of customers leaving the system is on average greater than the number of arriving customers per unit time. That is, the system is self-regulating, aiming at reaching the equilibrium state. Now formulas (9.60) and (9.61) yield the corresponding stationary state probabilities:

$$\pi_j = \begin{cases} \dfrac{1}{j!}\rho^j \pi_0 & \text{for } j = 1, 2, \ldots, s \\[2ex] \dfrac{\rho^s}{s!} \dfrac{\lambda^{j-s}}{\displaystyle\prod_{i=1}^{j-s}(s\mu+iv)} \pi_0 & \text{for } j = s+1, .s+2, \ldots \end{cases}$$

$$\pi_0 = \left[\sum_{j=0}^{s} \frac{1}{j!}\rho^j + \frac{\rho^s}{s!} \sum_{j=s+1}^{\infty} \frac{\lambda^{j-s}}{\displaystyle\prod_{i=1}^{j-s}(s\mu+iv)} \right]^{-1}.$$

Let L denote the random length of the queue in the steady state. Then,

$$E(L) = \sum_{j=s+1}^{\infty}(j-s)\pi_j.$$

Inserting the π_j yields after some algebra

$$E(L) = \pi_s \sum_{j=1}^{\infty} j\lambda^j \left[\prod_{i=1}^{j}(s\mu + i\nu) \right]^{-1}.$$

In this model, the *loss probability* π_ν is not strictly associated with the number of customers in the system. It is the probability that a customer leaves the system without having been served, because its patience time has expired. Therefore, $1 - \pi_\nu$ is the probability that a customer leaves the system after having been served. By applying the total probability rule with the exhaustive and mutually exclusive set of random events '$X = j$'; $j = s, s + 1, ...,$ one obtains

$$E(L) = \frac{\lambda}{\nu} \pi_\nu.$$

Thus, the mean queue length is directly proportional to the loss probability (compare to Little's formula (9.88)).

Variable Arrival Intensity Finite waiting capacities and patience times imply that in the end only a 'thinned flow' of potential customers will be served. Thus, it seems to be appropriate to investigate queueing systems, whose arrival (input) intensities depend on the state of the system. Those customers, however, which actually enter the system do not leave it without service. Since the tendency of customers to leave the system immediately after arrival increases with the number of customers in the system, the birth rates should decrease for $j \geq s$ as j tends to infinity. This property have, for example, for $\alpha \geq 0$ the birth rates

$$\lambda_j = \begin{cases} \lambda & \text{for } j = 0, 1, ..., s - 1, \\ \frac{s}{j+\alpha}\lambda & \text{for } j = s, s + 1, ... \end{cases}.$$

9.7.5   Special Single-Server Queueing Systems

9.7.5.1 System with Priorities

A single-server queueing system with waiting capacity for $m = 1$ customer is subject to two independent Poisson inputs 1 and 2 with respective intensities λ_1 and λ_2. The corresponding customers are called type 1- and type 2-customers. Type 1-customers have absolute (preemptive) priority, i.e. when a type 1- and a type 2-customer are in the system, the type 1-customer is being served. Thus, the service of a type 2-customer is interrupted as soon as a type 1-customer arrives. The displaced customer will occupy the waiting facility if it is empty. Otherwise it leaves the system. A waiting type 2-customer also has to leave the system when a type 1-customer arrives, since the newcomer will occupy the waiting facility. (Such a situation can only happen when a type 1-customer is being served.) An arriving type 1-customer is lost only then when both server and waiting facility are occupied by other type 1-customers.

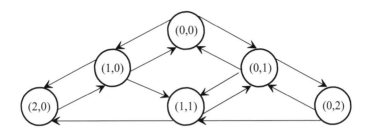

Figure 9.14 Transition graph for a single-server priority queueing system with $m = 1$

Thus, if only the number of type 1-customers in the system is of interest, then this priority queueing system becomes the waiting-loss system $M/M/s/1$ with $s = 1$, since type 2-customers have no impact on the service of type 1-customers at all. The service times of type 1- and type 2-customers are assumed to have exponential distributions with respective parameters μ_1 and μ_2. The state space of the system is represented in the form

$$\mathbf{Z} = \{(i,j);\ i,j = 0,1,2\},$$

where i denotes the number of type 1-customers and j the number of type 2-customers in the system. Note that if $X(t)$ denotes the system state at time t, the stochastic process $\{X(t), t \geq 0\}$ can be treated as a one-dimensional Markov chain, since scalars can be assigned to the six possible system states, which are given as two-component vectors. The Markov chain $\{X(t), t \geq 0\}$ is, however, not a birth- and death-process. Figure 9.14 shows its transition graph.

According to (9.28), the stationary state probabilities satisfy the system of equations

$$(\lambda_1 + \lambda_2) \pi_{(0,0)} = \mu_1 \pi_{(1,0)} + \mu_2 \pi_{(0,1)}$$

$$(\lambda_1 + \lambda_2 + \mu_1) \pi_{(1,0)} = \lambda_1 \pi_{(0,0)} + \mu_1 \pi_{(2,0)}$$

$$(\lambda_1 + \lambda_2 + \mu_2) \pi_{(0,1)} = \lambda_2 \pi_{(0,0)} + \mu_1 \pi_{(1,1)} + \mu_2 \pi_{(0,2)}$$

$$(\lambda_1 + \mu_1) \pi_{(1,1)} = \lambda_2 \pi_{(1,0)} + \lambda_1 \pi_{(0,1)} + \lambda_1 \pi_{(0,2)}$$

$$\mu_1 \pi_{(2,0)} = \lambda_1 \pi_{(1,0)} + \lambda_1 \pi_{(1,1)}$$

$$(\lambda_1 + \mu_2) \pi_{(0,2)} = \lambda_2 \pi_{(0,1)}$$

$$\pi_{(0,0)} + \pi_{(1,0)} + \pi_{(0,1)} + \pi_{(1,1)} + \pi_{(2,0)} + \pi_{(0,2)} = 1.$$

$m = 0$ Since there is no waiting capacity, each customer, notwithstanding its type, is lost if the server is busy with a type 1-customer. In addition, a type 2-customer is lost if, while being served, a type 1-customer arrives. The state space is

$$\mathbf{Z} = \{(0,0), (0,1), (1,0)\}.$$

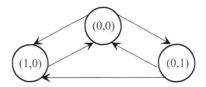

Figure 9.15 Transition graph for a 1-server priority loss system

Figure 9.15 shows the transition rates. The corresponding system (9.27) for the stationary state probabilities is

$$(\lambda_1 + \lambda_2)\pi_{(0,0)} = \mu_1\pi_{(1,0)} + \mu_2\pi_{(0,1)}$$

$$\mu_1\pi_{(1,0)} = \lambda_1\pi_{(0,0)} + \lambda_1\pi_{(0,1)}$$

$$1 = \pi_{(0,0)} + \pi_{(1,0)} + \pi_{(0,1)}.$$

The solution is

$$\pi_{(0,0)} = \frac{\mu_1(\lambda_1 + \mu_2)}{(\lambda_1 + \mu_1)(\lambda_1 + \lambda_2 + \mu_2)},$$

$$\pi_{(0,1)} = \frac{\lambda_2\mu_1}{(\lambda_1 + \mu_1)(\lambda_1 + \lambda_2 + \mu_2)}, \quad \pi_{(1,0)} = \frac{\lambda_1}{\lambda_1 + \mu_1}.$$

$\pi_{(1,0)}$ is the loss probability for type 1-customers. It is simply the probability that the service time of type 1-customers is greater than their interarrival time. On condition that at the arrival time of a type 2-customer the server is idle, this customer is lost if and only if during its service a type 1-customer arrives. The conditional probability of this event is

$$\int_0^\infty e^{-\mu_2 t}\lambda_1 e^{-\lambda_1 t}\,dt = \lambda_1\int_0^\infty e^{-(\lambda_1 + \mu_2)t}\,dt = \frac{\lambda_1}{\lambda_1 + \mu_2}.$$

Therefore, the (total) loss probability for type 2-customers is

$$\pi_l = \frac{\lambda_1}{\lambda_1 + \mu_2}\,\pi_{(0,0)} + \pi_{(0,1)} + \pi_{(1,0)}.$$

Example 9.19 Let $\lambda_1 = 0.1$, $\lambda_2 = 0.2$, and $\mu_1 = \mu_2 = 0.2$. Then the stationary state probabilities are

$$\pi_{(0,0)} = 0.2105, \quad \pi_{(0,1)} = 0.3073, \quad \pi_{(1,0)} = 0.0085,$$

$$\pi_{(1,1)} = 0.1765, \quad \pi_{(0,2)} = 0.2048, \quad \pi_{(2,0)} = 0.0924.$$

In case $m = 0$, with the same numerical values for the transition rates,

$$\pi_{(0,0)} = 0.4000, \quad \pi_{(1,0)} = 0.3333, \quad \pi_{(0,1)} = 0.2667.$$

The loss probability for type 2-customers is $\pi_l = 0.7333$.   □

9.7.5.2 *M/M/1/m*-System with Unreliable Server

If the implications of server failures on the system performance are not negligible, server failures have to be taken into account when building up a mathematical model. Henceforth, the principal approach is illustrated by a single-server queuing system with waiting capacity for m customers, Poisson input, and independent, identically distributed exponential service times with parameter μ. The lifetime of the server is assumed to have an exponential distribution with parameter α, both in its busy phase and in its idle phase, and the subsequent renewal time of the server is assumed to be exponentially distributed with parameter β. It is further assumed that the sequence of life- and renewal times of the server can be described by an alternating renewal process. When the server fails, all customers leave the system, i.e., the customer being served and the waiting customers if there are any are lost. Customers arriving during a renewal phase of the server are rejected, i.e., they are lost, too.

The stochastic process $\{X(t), t \geq 0\}$ describing the behaviour of the system is characterized as follows:

$$X(t) = \begin{cases} j & \text{if there are } j \text{ customers in the system at time } t;\ j = 0, 1, \ldots, m+1 \\ m+2 & \text{if the server is being renewed at time } t \end{cases}.$$

Its transition rates are (Figure 9.16):

$$q_{j,j+1} = \lambda;\quad j = 0, 1, \ldots, m$$
$$q_{j,j-1} = \mu;\quad j = 1, 2, \ldots, m+1 \tag{9.101}$$
$$q_{j,m+2} = \alpha;\quad j = 0, 1, \ldots m+1$$
$$q_{m+2,0} = \beta.$$

By (9.28), the stationary state probabilities satisfy the system of equations

$$(\alpha + \lambda)\pi_0 = \mu\pi_1 + \beta\pi_{m+2}$$
$$(\alpha + \lambda + \mu)\pi_j = \lambda\pi_{j-1} + \mu\pi_{j+1};\quad j = 1, 2, \ldots, m \tag{9.102}$$
$$(\alpha + \mu)\pi_{m+1} = \lambda\pi_m$$
$$\beta\pi_{m+2} = \alpha\pi_0 + \alpha\pi_1 + \cdots + \alpha\pi_{m+1}.$$

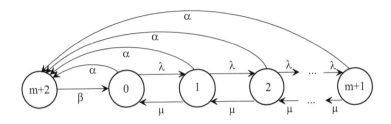

Figure 9.16 Transition graph of a queueing system with unreliable server

The last equation is equivalent to $\beta \pi_{m+2} = \alpha(1 - \pi_{m+2})$. Hence,

$$\pi_{m+2} = \frac{\alpha}{\alpha + \beta}.$$

Now, starting with the first equation in (9.102), the stationary state probabilities of the system $\pi_1, \pi_2, \dots, \pi_{m+1}$ can be successively determined. The probability π_0 is as usual obtained from the normalizing condition

$$\sum_{i=0}^{m+2} \pi_i = 1. \tag{9.103}$$

For the corresponding loss system ($m = 0$), the stationary state probabilities are

$$\pi_0 = \frac{\beta(\alpha + \mu)}{(\alpha + \beta)(\alpha + \lambda + \mu)}, \quad \pi_1 = \frac{\beta\lambda}{(\alpha + \beta)(\alpha + \lambda + \mu)}, \quad \pi_2 = \frac{\alpha}{\alpha + \beta}.$$

Modification of the Model It makes sense to assume that the server can only fail if it is busy. In this case,

$$q_{j,m+2} = \alpha \text{ for } j = 1, 2, \dots, m+1.$$

The other transition rates given by (9.101) remain valid. Thus, the corresponding transition graph is again given by Figure 9.16 with the arrow from node 0 to node $m + 2$ deleted. The stationary state probabilities satisfy the system of equations

$$\lambda \pi_0 = \mu \pi_1 + \beta \pi_{m+2}$$

$$(\alpha + \lambda + \mu) \pi_j = \lambda \pi_{j-1} + \mu \pi_{j+1} ; \ j = 1, 2, \dots, m \tag{9.104}$$

$$(\alpha + \mu) \pi_{m+1} = \lambda \pi_m$$

$$\beta \pi_{m+2} = \alpha \pi_1 + \alpha \pi_2 + \cdots + \alpha \pi_{m+1}.$$

The last equation is equivalent to $\beta \pi_{m+2} = \alpha(1 - \pi_0 - \pi_{m+2})$. It follows

$$\pi_{m+2} = \frac{\alpha}{\alpha + \beta}(1 - \pi_0).$$

Starting with the first equation in (9.104), the solution $\pi_0, \pi_1, \pi_2, \dots, \pi_{m+1}$ can be obtained as above. In case $m = 0$ the stationary state probabilities are

$$\pi_0 = \frac{\beta(\alpha + \mu)}{\beta(\alpha + \mu) + \lambda(\alpha + \beta)}, \quad \pi_1 = \frac{\lambda\beta}{\beta(\alpha + \mu) + \lambda(\alpha + \beta)}, \quad \pi_2 = \frac{\alpha\lambda}{\beta(\alpha + \mu) + \lambda(\alpha + \beta)}.$$

Comment It is interesting that this queueing system with unreliable server can be interpreted as a queueing system with priorities and absolutely reliable server. To see this, a failure of the server has to be declared as the arrival of a 'customer' with absolute priority. The service provided to this 'customer' consists in the renewal of the server. Such a 'customer' pushes away any other customer from the server, in this model even from the waiting facility. Hence it is not surprising that the theory of queueing systems with priorities also provides solutions for more complicated queuing systems with unreliable servers than the one considered in this section.

9.7.6 Networks of Queueing Systems

9.7.6.1 Introduction

Customers frequently need several kinds of service so that, after leaving one service station, they have to visit one or more other service stations in a fixed or random order. Each of these service stations is assumed to behave like the basic queueing system sketched in Figure 9.12. A set of queueing systems together with rules of their interactions is called a *network of queueing systems* or a *queueing network*. Typical examples are technological processes for manufacturing (semi-) finished products. In such a case the order of service by different queueing systems is usually fixed. Queuing systems are frequently subject to several inputs, i.e., customers with different service requirements have to be attended. In this case they may visit the service stations in different orders. Examples of such situations are computer and communication networks. Depending on whether and how data are to be provided, processed, or transmitted, the terminals (service stations) will be used in different orders. If technical systems have to be repaired, then, depending on the nature and the extent of the damage, service by different production departments within a workshop is needed. Transport and loading systems also fit into the scheme of queueing networks.

Using a concept from graph theory, the service stations of a queueing network are called *nodes*. In an *open queueing network* customers arrive from 'outside' at the system (external input). Each node may have its own external input. Once in the system, customers visit other nodes in a deterministic or random order before leaving the network. Thus, in an open network, each node may have to serve *external* and *internal customers*, where internal customers are the ones which arrive from other nodes. In *closed queueing networks* there are no external inputs into the nodes, and the total number of customers in the network is constant. Consequently, no customer departs from the network. Queueing networks can be represented by directed graphs. The directed edges between the nodes symbolize the possible transitions of customers from one node to another. The nodes in the network are denoted by $1, 2, ..., n$. Node i is assumed to have s_i servers; $1 \le s_i \le \infty$.

9.7.6.2 Open Queueing Networks

A mathematically exact analysis of queueing systems becomes extremely difficult or even impossible when dropping the assumptions of Poisson input and/or exponentially distributed service times. Hence, this section is restricted to a rather simple class of queueing networks, the *Jackson queueing networks*. They are characterized by four properties:

1) Each node has an unbounded waiting capacity.

2) The service times of all servers at node i are independent, identically distributed exponential random variables with parameter (intensity) μ_i. They are also independent of the service times at other nodes.

3) External customers arrive at node i in accordance with a homogeneous Poisson process with intensity λ_i. All external inputs are independent of each other and of all service times.

4) When the service of a customer at node i has been finished, the customer makes a transition to node j with probability p_{ij} or leaves the network with probability a_i. The *transition* or *routing matrix*

$$\mathbf{P} = ((p_{ij}))$$

is independent of the current state of the network and of its past.

Let $\mathbf{I}$ be the identity matrix. The matrix $\mathbf{I} - \mathbf{P}$ is assumed to be nonsingular so that the inverse matrix $(\mathbf{I} - \mathbf{P})^{-1}$ exists. According to the definition of the a_i and p_{ij},

$$a_i + \sum_{j=1}^n p_{ij} = 1 . \tag{9.105}$$

In a Jackson queueing network, each node is principally subjected to both external and internal input. Let α_j be the total input (arrival) intensity at node j. In the steady state, α_j must be equal to the total output intensity from node j. The portion of internal input intensity to node j, which is due to customers from node i, is $\alpha_i p_{ij}$. Thus,

$$\sum_{i=1}^n \alpha_i p_{ij}$$

is the total internal input intensity to node j. Consequently, in the steady state,

$$\alpha_j = \lambda_j + \sum_{i=1}^n \alpha_i p_{ij} ; \quad j = 1, 2, \dots, n . \tag{9.106}$$

By introducing vectors

$$\alpha = (\alpha_1, \alpha_2, \dots, \alpha_n) \text{ and } \lambda = (\lambda_1, \lambda_2, \dots, \lambda_n),$$

the relationship (9.106) can be written as

$$\alpha(\mathbf{I} - \mathbf{P}) = \lambda .$$

Since $\mathbf{I} - \mathbf{P}$ is assumed to be nonsingular, the vector of the total input intensities α is

$$\alpha = \lambda (\mathbf{I} - \mathbf{P})^{-1}. \tag{9.107}$$

Even under the assumptions stated, the total inputs at the nodes and the outputs from the nodes are generally nonhomogeneous Poisson processes.

Let $X_i(t)$ be the random number of customers at node i at time t. Its realizations are denoted as x_i; $x_i = 0, 1, \dots$. The random state of the network at time t is characterized by the vector $\mathbf{X}(t) = (X_1(t), X_2(t), \dots, X_n(t))$ with realizations $\mathbf{x} = (x_1, x_2, \dots, x_n)$. The set of all these vectors $\mathbf{x}$ forms the state space of the Markov chain $\{\mathbf{X}(t), t \geq 0\}$. Using set-theory notation, the state space is denoted as $\mathbf{Z} = \{0, 1, \dots\}^n$, i.e., $\mathbf{Z}$ is the set of all those n-dimensional vectors the components of which assume nonnegative integers. Since $\mathbf{Z}$ is countably infinite, this at first glance n-dimensional Markov chain becomes one-dimensional by arranging the states as a sequence.

To determine the transition rates of $\{\mathbf{X}(t), t \geq 0\}$, the n-dimensional vector $\mathbf{e}_i$ is introduced. Its ith component is a 1 and the other components are zeros:

$$\mathbf{e}_i = (0, 0, \dots, 0, 1, 0, \dots, 0). \tag{9.108}$$
$$\phantom{\mathbf{e}_i = (0,}1 \;\; 2 \;\; \cdots \;\;\; i \;\; \cdots \;\;\; n$$

Thus, $\mathbf{e}_i$ is the ith row of the identity matrix $\mathbf{I}$. Since the components of any state vector $\mathbf{x}$ are nonnegative integers, each $\mathbf{x}$ can be represented as a linear combination of all or some of the $\mathbf{e}_1, \mathbf{e}_2, \dots, \mathbf{e}_n$. In particular, $\mathbf{x} + \mathbf{e}_i$ $(\mathbf{x} - \mathbf{e}_i)$ is the vector which arises from $\mathbf{x}$ by increasing (decreasing) the ith component by 1. Starting from state $\mathbf{x}$, the Markov chain $\{\mathbf{X}(t), t \geq 0\}$ can make the following one-step transitions:

1) When a customer arrives at node i, the Markov chain makes a transition to state $\mathbf{x} + \mathbf{e}_i$.

2) When a service at node i is finished, $x_i > 0$, and the served customer leaves the network, the Markov chain makes a transition to state $\mathbf{x} - \mathbf{e}_i$.

3) When a service at node i with $x_i > 0$ is finished and the served customer leaves node i for node j, the Markov chain makes a transition to state $\mathbf{x} - \mathbf{e}_i + \mathbf{e}_j$.

Therefore, starting from state $\mathbf{x} = (x_1, x_2, \dots, x_n)$, the transition rates are

$$q_{\mathbf{x}, \mathbf{x} + \mathbf{e}_i} = \lambda_i$$

$$q_{\mathbf{x}, \mathbf{x} - \mathbf{e}_i} = \min(x_i, s_i) \mu_i a_i$$

$$q_{\mathbf{x}, \mathbf{x} - \mathbf{e}_i + \mathbf{e}_j} = \min(x_i, s_i) \mu_i p_{ij}, \;\; i \neq j.$$

In view of (9.105),

$$\sum_{j, j \neq i} p_{ij} = 1 - p_{ii} - a_i.$$

Hence, the rate of leaving state $\mathbf{x}$ is

$$q_{\mathbf{x}} = \sum_{i=1}^{n} \lambda_i + \sum_{i=1}^{n} \mu_i (1 - p_{ii}) \min(x_i, s_i).$$

According to (9.28), the stationary state probabilities

$$\pi_{\mathbf{x}} = \lim_{t \to \infty} P(\mathbf{X}(t) = \mathbf{x}), \;\; \mathbf{x} \in \mathbf{Z},$$

provided they exist, satisfy the system of equations

$$q_{\mathbf{x}} \pi_{\mathbf{x}} = \sum_{i=1}^{n} \lambda_i \pi_{\mathbf{x} - \mathbf{e}_i} + \sum_{i=1}^{n} a_i \mu_i \min(x_i + 1, s_i) \pi_{\mathbf{x} + \mathbf{e}_i}$$

$$+ \sum_{j=1}^{n} \sum_{\substack{i=1 \\ i \neq j}}^{n} a_i \mu_i \min(x_i + 1, s_i) p_{ij} \pi_{\mathbf{x} + \mathbf{e}_i - \mathbf{e}_j}. \tag{9.109}$$

In order to be able to present the solution of this system in a convenient form, recall that the stationary state probabilities of the waiting system $M/M/s_i/\infty$ with parameters α_i, μ_i, and $\rho_i = \alpha_i/\mu_i$ denoting in this order the intensity of the Poisson input, the service intensities of all servers, and the traffic intensity of the system are given by (see formula (9.80)),

$$
\varphi_i(j) = \begin{cases} \dfrac{1}{j!}\rho_i^j\,\varphi_i(0) & \text{for } j = 1,2,\ldots,s_i-1, \\[2mm] \dfrac{1}{s_i!\,s_i^{j-s_i}}\rho_i^j\,\varphi_i(0) & \text{for } j = s_i, s_i+1,\ldots, \end{cases} \qquad \rho_i < s_i,
$$

$$
\varphi_i(0) = \left[\sum_{j=0}^{s_i-1}\frac{1}{j!}\rho_i^j + \frac{\rho_i^{s_i}}{(s_i-1)!\,(s_i-\rho_i)}\right]^{-1}, \qquad \rho_i < s_i.
$$

(In the context of queueing networks, the notation $\varphi_i(\cdot)$ for the stationary state probabilities is common practice.) The stationary state probabilities of the queueing network are simply obtained by multiplying the corresponding state probabilities of the queuing systems $M/M/s_i/\infty$, $i = 1,2,\ldots n$:

> *If the vector of the total input intensities* $\boldsymbol{\alpha} = (\alpha_1,\alpha_2,\ldots,\alpha_n)$ *given by (9.106) satisfies the conditions* $\alpha_i < s_i\mu_i$, $i = 1,2,\ldots,n$, *then the stationary probability of state* $\mathbf{x} = (x_1,x_2,\ldots,x_n)$ *is*
>
> $$\pi_{\mathbf{x}} = \prod_{i=1}^{n}\varphi_i(x_i), \quad \mathbf{x} \in \mathbf{Z}. \tag{9.110}$$

Thus, the stationary state distribution of a Jackson queueing system is given in *product form*. This implies that each node of the network <u>behaves</u> like an $M/M/s_i/\infty$-system. However, the nodes need not be queueing systems of this type because the process $\{X_i(t), t \ge 0\}$ is usually not a birth and death process. In particular, the total input into a node need not be a homogeneous Poisson process. But the product form (9.110) of the stationary state probabilities proves that the queue lengths at the nodes in the steady state are independent random variables. There is a vast amount of literature dealing with assumptions under which the stationary distribution of a queueing network has the product form (see, for instance, *van Dijk* (1983)).

To verify that the stationary state distribution indeed has the product form (9.110), one has to substitute (9.110) into the system of equations (9.109). Using (9.105) and (9.106), one obtains an identity after some tedious algebra.

Queueing Networks with Feedback The simplest Jackson queueing network arises if $n = 1$. The only difference from the queueing system $M/M/s/\infty$ is that now a positive proportion of customers, who have departed from the network after having been served, will return and require further service. This leads to a queueing system with

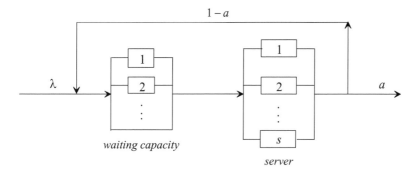

Figure 9.17 Queueing system with feedback

feedback (Figure 9.17). For instance, when servers have done a bad job, the affected customers will soon return to exercise possible guarantee claims. Formally, these customers remain in the network. Roughly speaking, a single-node Jackson queueing network is a mixture between an open and a closed waiting system. A customer leaves the system with probability a or reenters the system with probability $p_{11} = 1 - a$. If there is an idle server, then, clearly, the service of such a customer starts immediately. From (9.105) and (9.106), the total input rate α into the system satisfies

$$\alpha = \lambda + \alpha(1 - a).$$

(The index 1 is deleted from all system parameters.) Thus,

$$\alpha = \lambda/a.$$

Hence there exists a stationary distribution if

$$\lambda/a < s\mu \text{ or, equivalently, if } \rho = \lambda/\mu < a s.$$

In this case the stationary state probabilities are

$$\pi_j = \begin{cases} \dfrac{1}{j!}\left(\dfrac{\rho}{a}\right)^j \pi_0 & \text{for } j = 1, 2, \ldots, s - 1, \\[2ex] \dfrac{1}{s! s^{j-s}}\left(\dfrac{\rho}{a}\right)^j \pi_0 \text{ for} & j = s, s + 1, \ldots, \end{cases}$$

where

$$\pi_0 = \left[\sum_{j=1}^{s-1} \frac{1}{j!}\left(\frac{\rho}{a}\right)^j + \frac{\left(\frac{\rho}{a}\right)^s}{(s-1)!\left(s - \frac{\rho}{a}\right)} \right]^{-1}.$$

Interestingly, this is the stationary state distribution of the queueing system $M/M/s/\infty$ (without feedback), the input of which has intensity λ/a.

Sequential Queueing Networks In technological processes, the sequence of service is usually fixed. For example, a 'customer' may be a car being manufactured on an assembly line. Therefore, queueing systems switched in series, called *sequential queueing networks* or *tandem queueing networks*, are of considerable practical interest: External customers arrive only at node 1 (arrival intensity: λ_1). They subsequently visit in this order the nodes $1, 2, ..., n$ and then leave the network.

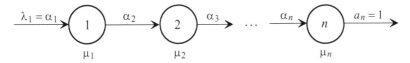

Figure 9.18 Sequential queueing network

The corresponding parameters are (Figure 9.18):

$$\lambda_i = 0; \quad i = 2, 3, ..., n$$

$$p_{i,i+1} = 1; \quad i = 1, 2, ..., n-1$$

$$a_1 = a_2 = \cdots = a_{n-1} = 0, \quad a_n = 1.$$

According to (9.106), the (total) input intensities of all nodes in the steady state must be the same:

$$\lambda_1 = \alpha_1 = \alpha_2 = \cdots = \alpha_n.$$

Hence, for single-server nodes ($s_i = 1$; $i = 1, 2, ..., n$), a stationary state distribution exists if

$$\rho_i = \lambda_1/\mu_i < 1; \quad i = 1, 2, ..., n,$$

or, equivalently, if

$$\lambda_1 < \min(\mu_1, \mu_2, ..., \mu_n).$$

Thus, the slowest server determines the efficiency of a sequential network. The stationary probability of state $\mathbf{x} = (x_1, x_2, ..., x_n)$ is

$$\pi_{\mathbf{x}} = \prod_{i=1}^{n} \rho_i^{x_i}(1 - \rho_i); \quad \mathbf{x} \in \mathbf{Z}.$$

The sequential network can be generalized by taking feedback into account. This is left as an exercise to the reader.   □

Example 9.20 Defective robots arrive at the admission's department of a maintenance workshop in accordance with a homogeneous Poisson process with intensity $\lambda = 0.2 [h^{-1}]$. In the admissions department (denoted as (1)) a first failure diagnosis is done. Depending on the result, the robots will have to visit other departments of the workshop. These are departments for checking and repairing the mechanics (2), electronics (3), and software (4) of the robots, respectively. The failure diagnosis in

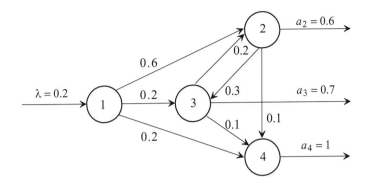

Figure 9.19 Maintenance workshop as a queueing network

the admissions department results in 60% of the arriving robots being sent to department (2) and 20% each to the departments (3) and (4). After having being maintained in department (2), 60% of the robots leave the workshop, 30% are sent to department (3), and 10% to department (4). After having being served by department (3), 70% of the robots leave the workshop, 20% are sent to department (2), and 10% are sent to department (4). After elimination of possible software failures all robots leave the workshop. A robot can be sent several times to one and the same department.

The following transition probabilities result from the transfer of robots between the departments:

$$p_{12} = 0.6, \; p_{13} = 0.2, \; p_{14} = 0.2,$$
$$p_{23} = 0.3, \; p_{24} = 0.1,$$
$$p_{32} = 0.2, \; p_{34} = 0.1.$$

The service intensities are assumed to be

$$\mu_1 = 1, \; \mu_2 = 0.45, \; \mu_3 = 0.4, \; \mu_4 = 0.1 \; [h^{-1}].$$

The graph plotted in Figure 9.19 illustrates the possible transitions between the departments. The edges of the graph are weighted by the corresponding transition probabilities. The system of equations (9.106) in the total input intensities is

$$\alpha_1 = 0.2$$
$$\alpha_2 = 0.6\,\alpha_1 \qquad\qquad + 0.2\,\alpha_3$$
$$\alpha_3 = 0.2\,\alpha_1 + 0.3\,\alpha_2$$
$$\alpha_4 = 0.2\,\alpha_1 + 0.1\,\alpha_2 + 0.1\,\alpha_3.$$

The solution is (after rounding)

$$\alpha_1 = 0.20, \quad \alpha_2 = 0.135, \quad \alpha_3 = 0.08, \quad \alpha_4 = 0.06.$$

The corresponding traffic intensities $\rho_i = \alpha_i / \mu_i$ are

$$\rho_1 = 0.2, \quad \rho_2 = 0.3, \quad \rho_3 = 0.2, \quad \rho_4 = 0.6.$$

From (9.110), the stationary probability of state $\mathbf{x} = (x_1, x_2, ..., x_n)$ for single-server nodes is

$$\pi_{\mathbf{x}} = \prod_{i=1}^{4} \rho^{x_i}(1 - \rho_i)$$

or $\qquad \pi_{\mathbf{x}} = 0.1792 \, (0.2)^{x_1} (0.3)^{x_2} (0.2)^{x_3} (0.6)^{x_4} ; \quad \mathbf{x} \in \mathbf{Z} = \{0, 1, ... \}^4.$

In particular, the stationary probability that there is no robot in the workshop is

$$\pi_{\mathbf{x}_0} = 0.1792,$$

where $\mathbf{x}_0 = (0, 0, 0, 0)$. Let X_i denote the random number of robots at node i in the steady state. Then the probability that, in the steady state, there is at least one robot in the admissions department is

$$P(X_1 > 0) = 0.8 \sum_{i=1}^{\infty} (0.2)^i = 0.2.$$

Analogously

$$P(X_2 > 0) = 0.3, \quad P(X_3 > 0) = 0.2, \quad \text{and} \quad P(X_4 > 0) = 0.6.$$

Thus, when there is a delay in servicing defective robots, the cause is most probably department (4) in view of the comparatively high amount of time necessary for finding and removing software failures. $\qquad\qquad\qquad\qquad\qquad\qquad\qquad\qquad\qquad\qquad$ □

9.7.6.3 Closed Queueing Networks

Analogously to the closed queueing system, customers cannot enter a *closed queueing network* 'from outside'. Customers which have been served at a node do not leave the network, but move to another node for further service. Hence, the number of customers in a closed queueing network is a constant N. Practical examples for closed queueing networks are multiprogrammed computer and communication systems.

When the service of a customer at node i is finished, then the customer moves with probability p_{ij} to node j for further service. Since the customers do not leave the network,

$$\sum_{j=1}^{n} p_{ij} = 1 ; \quad i = 1, 2, ..., n, \qquad\qquad (9.111)$$

where as usual n is the number of nodes. Provided the discrete Markov chain given by the transition matrix $\mathbf{P} = ((p_{ij}))$ and the state space $\mathbf{Z} = (1, 2, ..., n\}$ is irreducible, it has a stationary state distribution $\{\pi_1, \pi_2, ..., \pi_n\}$, which, according to (8.9), is the unique solution of the system of equations

$$\pi_j = \sum_{i=1}^{n} p_{ij} \pi_i ; \quad j = 1, 2, ..., n, \qquad\qquad (9.112)$$

$$1 = \sum_{i=1}^{n} \pi_i.$$

Let $X_i(t)$ be the random number of customers at node i at time t and

$$\mathbf{X}(t) = (X_1(t), X_2(t), \dots, X_n(t)).$$

The state space of the Markov chain $\{\mathbf{X}(t), t \geq 0\}$ is

$$\mathbf{Z} = \left\{ \mathbf{x} = (x_1, x_2, \dots, x_n) \text{ with } \sum_{i=1}^n x_i = N \text{ and } 0 \leq x_i \leq N \right\}, \qquad (9.113)$$

where the x_i are nonnegative integers. The number of elements (states) in $\mathbf{Z}$ is

$$\binom{n+N-1}{N}.$$

Let $\mu_i = \mu_i(x_i)$ be the service intensity of all servers at node i if there are x_i customers at this node, $\mu_i(0) = 0$. Then $\{\mathbf{X}(t), t \geq 0\}$ has the positive transition rates

$$q_{\mathbf{x}, \mathbf{x}-\mathbf{e}_i+\mathbf{e}_j} = \mu_i(x_i)p_{ij}; \quad x_i \geq 1, \ i \neq j,$$

$$q_{\mathbf{x}-\mathbf{e}_i+\mathbf{e}_j, \mathbf{x}} = \mu_j(x_j + 1)p_{ji}; \quad i \neq j, \ \mathbf{x}-\mathbf{e}_i+\mathbf{e}_j \in \mathbf{Z},$$

where the $\mathbf{e}_i$ are given by (9.108). From (9.111), the rate of leaving state $\mathbf{x}$ is

$$q_{\mathbf{x}} = \sum_{i=1}^n \mu_i(x_i)(1 - p_{ii}).$$

Hence, according to (9.28), the stationary distribution $\{\pi_{\mathbf{x}}, \mathbf{x} \in \mathbf{Z}\}$ of the Markov chain $\{\mathbf{X}(t), t \geq 0\}$ satisfies

$$\sum_{i=1}^n \mu_i(x_i)(1 - p_{ii})\pi_{\mathbf{x}} = \sum_{i,j=1, i \neq j}^n \mu_j(x_j + 1)p_{ji}\pi_{\mathbf{x}-\mathbf{e}_i+\mathbf{e}_j}, \qquad (9.114)$$

where $\mathbf{x} = (x_1, x_2, \dots, x_n) \in \mathbf{Z}$. In these equations, all $\pi_{\mathbf{x}-\mathbf{e}_i+\mathbf{e}_j}$ with $\mathbf{x}-\mathbf{e}_i+\mathbf{e}_j \notin \mathbf{Z}$ are equal to 0. Let $\varphi_i(0) = 1$ and

$$\varphi_i(j) = \prod_{k=1}^j \left(\frac{\pi_i}{\mu_i(k)} \right); \quad i = 1, 2, \dots, n; \ j = 1, 2, \dots, N.$$

Then the stationary probability of state $\mathbf{x} = (x_1, x_2, \dots, x_n) \in \mathbf{Z}$. is

$$\pi_{\mathbf{x}} = h \prod_{i=1}^n \varphi_i(x_i), \quad h = \left[\sum_{\mathbf{y} \in \mathbf{Z}} \prod_{i=1}^n \varphi_i(y_i) \right]^{-1} \qquad (9.115)$$

with $\mathbf{y} = (y_1, y_2, \dots, y_n)$. By substituting (9.115) into (9.114) one readily verifies that $\{\pi_{\mathbf{x}}, \mathbf{x} \in \mathbf{Z}\}$ is indeed a stationary distribution of the Markov chain $\{\mathbf{X}(t), t \geq 0\}$.

Example 9.21 Consider a closed sequential queueing network, which has a single server at each of its n nodes (Figure 9.20). There is only $N = 1$ customer in the system. When this customer is being served at a certain node, the other nodes are empty. Hence, with vectors $\mathbf{e}_i$ as defined by (9.108), the state space of the corresponding Markov chain $\{\mathbf{X}(t), t \geq 0\}$ is $\mathbf{Z} = \{\mathbf{e}_1, \mathbf{e}_2, \dots, \mathbf{e}_n\}$. The transition probabilities are

$$p_{i, i+1} = 1; \quad i = 1, 2, \dots, n-1; \ p_{n,1} = 1.$$

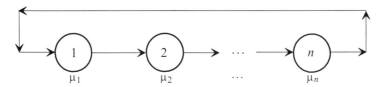

Figure 9.20 Closed sequential queueing network

The corresponding solution of (9.114) is a uniform distribution

$$\pi_1 = \pi_2 = \cdots = \pi_n = 1/n.$$

Let $\mu_i = \mu_i(1)$ be the service rate at node i. Then, for $i = 1, 2, ..., n,$

$$\varphi_i(0) = 1, \quad \varphi_i(1) = \frac{1}{n\mu_i}, \quad h = n\left[\sum_{i=1}^{n} \frac{1}{\mu_i}\right]^{-1}.$$

Hence, the stationary state probabilities (9.115) are

$$\pi_{\mathbf{e}_i} = \frac{1/\mu_i}{\sum\limits_{i=1}^{n} \dfrac{1}{\mu_i}}; \quad i = 1, 2, ..., n.$$

In particular, if $\mu_i = \mu$; $i = 1, 2, ..., n,$ then the states $\mathbf{e}_i$ have a uniform distribution:

$$\pi_{\mathbf{e}_i} = 1/n; \quad i = 1, 2, ..., n.$$

If there are $N \geq 1$ customers in the system and the μ_i do not depend on x_i, then the stationary state probabilities are

$$\pi_{\mathbf{x}} = \frac{(1/\mu_1)^{x_1} (1/\mu_2)^{x_2} \cdots (1/\mu_n)^{x_n}}{\sum\limits_{\mathbf{y} \in \mathbf{Z}} \prod\limits_{i=1}^{n} \left(\dfrac{1}{\mu_i}\right)^{y_i}},$$

where $\mathbf{x} = (x_1, x_2, ..., x_n) \in \mathbf{Z}$. Given $\mu_i = \mu$, $i = 1, 2, ..., n,$ the states have again a uniform distribution:

$$\pi_{\mathbf{x}} = \frac{1}{\binom{n+N-1}{N}}, \quad \mathbf{x} \in \mathbf{Z}. \qquad \square$$

Example 9.22 A computer system consists of two central processors 2 and 3, a disc drive 1, and a printer 4. A new program starts in the central processor 2. When this processor has finished its computing job, the computing phase continues in central processor 3 with probability α or the program goes to the disc drive with probability $1 - \alpha$. From the disc drive the program goes to central processor 3 with probability 1. From central processor 3 it goes to the central processor 2 with probability β or to the printer with probability $1 - \beta$. Here it terminates or goes back to central processor 2. When a program terminates, then another program (from outside) immediately joins the queue of central processor 2 so that there is always a fixed number of programs

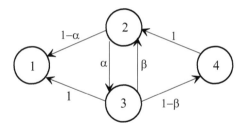

Figure 9.21 Computer system as a closed queueing network

in the system. Hence, a program formally goes from the printer to the central processor 2 with probability 1. If N denotes the constant number of programs in the system,
this situation represents a simple case of *multiprogramming* with N as the *level of
multiprogramming*. The state space **Z** of this system and the matrix **P** of the transition probabilities p_{ij} are

$$\mathbf{Z} = \{\mathbf{y} = (y_1, y_2, y_3, y_4); \; y_i = 0, 1, \dots, N; \; y_1 + y_2 + y_3 + y_4 = N\}$$

and

$$\mathbf{P} = \begin{pmatrix} 0 & 0 & 1 & 0 \\ 1-\alpha & 0 & \alpha & 0 \\ 0 & \beta & 0 & 1-\beta \\ 0 & 1 & 0 & 0 \end{pmatrix},$$

respectively (Figure 9.21). The corresponding solution of (9.114) is

$$\pi_1 = \frac{1-\alpha}{4-\alpha-\beta}, \quad \pi_2 = \pi_3 = \frac{1}{4-\alpha-\beta}, \quad \pi_4 = \frac{1-\beta}{4-\alpha-\beta}.$$

Let the service intensities of the nodes μ_1, μ_2, μ_3, and μ_4 be independent of the
number of programs at the nodes. Then,

$$\varphi_i(x_i) = \left(\frac{\pi_i}{\mu_i}\right)^{x_i}, \quad i = 1, 2, \dots, n.$$

Hence, the stationary probability of state

$$\mathbf{x} = (x_1, x_2, x_3, x_4) \text{ with } x_1 + x_2 + x_3 + x_4 = N$$

is

$$\pi_{\mathbf{x}} = \frac{h}{(4-\alpha-\beta)^N} \left(\frac{1-\alpha}{\mu_1}\right)^{x_1} \left(\frac{1}{\mu_2}\right)^{x_2} \left(\frac{1}{\mu_3}\right)^{x_3} \left(\frac{1-\beta}{\mu_4}\right)^{x_4}$$

with

$$h = \frac{(4-\alpha-\beta)^N}{\displaystyle\sum_{\mathbf{y} \in \mathbf{Z}} \left(\frac{1-\alpha}{\mu_1}\right)^{y_1} \left(\frac{1}{\mu_2}\right)^{y_2} \left(\frac{1}{\mu_3}\right)^{y_3} \left(\frac{1-\beta}{\mu_4}\right)^{y_4}}. \qquad \square$$

Application-oriented treatments of queueing networks are *Gelenbe, Pujolle* (1987),
Walrand (1988).

5.8 SEMI-MARKOV CHAINS

Transitions between the states of a continuous-time homogeneous Markov chain are controlled by its transition probabilities. According to section 9.4, the sojourn time in a state has an exponential distribution and depends on the current state, but not on the history of the process. Since in most applications the sojourn times in system states are non-exponential random variables, an obvious generalization is to allow arbitrarily distributed sojourn times whilst retaining the transition mechanism between the states. This approach leads to *semi-Markov chains*.

A semi-Markov chain with state space $\mathbf{Z} = \{0, 1, ...\}$ evolves in the following way: Transitions between the states are governed by a discrete-time homogeneous Markov chain $\{X_0, X_1, ...\}$ with state space $\mathbf{Z}$ and matrix of transition probabilities

$$\mathbf{P} = ((p_{ij})).$$

If the process starts at time $t = 0$ in state i_0, then the subsequent state i_1 is determined according to the transition matrix $\mathbf{P}$, while the process stays in state i_0 a random time $Y_{i_0 i_1}$. After that the state i_2, following state i_1, is determined. The process stays in state i_1 a random time $Y_{i_1 i_2}$ and so on. The random variables Y_{ij} are the *conditional sojourn times* of the process in state i given that the process makes a transition from i to j. They are assumed to be independent. Hence, immediately after entering a state at a time t, the further evolvement of a semi-Markov chain depends only on its state at this time point, but not on the evolvement of the process before t. The sample paths of a semi-Markov chain are piecewise constant functions which, by convention, are continuous on the right. In contrast to homogeneous continuous-time Markov chains, for predicting the development of a semi-Markov chain from a time point t, it is not only necessary to know its current state i, but also the 'age' of i at time t.

Let $T_0, T_1, ...$ denote the sequence of time points at which the semi-Markov chain makes a transition from one state to another (or to the same state). Then

$$X_n = X(T_n); \quad n = 0, 1, ... , \tag{9.116}$$

where $X_0 = X(0)$ is the initial state ($X_n = X(T_n + 0)$). Hence, the transition probabilities can be written in the following form

$$p_{ij} = P(X(T_{n+1}) = j | X(T_n) = i); \quad n = 0, 1, ...$$

In view of (9.116), the discrete-time stochastic process $\{X_0, X_1, ... \}$ is *embedded* in the (continuous-time) semi-Markov chain $\{X(t), t \geq 0\}$ (see page 401).

As already pointed out, the future development of a semi-Markov chain from a *jump point* T_n is independent of the entire history of the process before T_n. Let

$$F_{ij}(t) = P(Y_{ij} \leq t), \quad i, j \in \mathbf{Z},$$

denote the distribution function of the conditional sojourn time Y_{ij} of a semi-Markov chain in state i if the subsequent state is j. By the total probability rule, the *uncondi-tional sojourn time* Y_i of the chain in state i is

$$F_i(t) = P(Y_i \le t) = \sum_{j \in \mathbf{Z}} p_{ij} F_{ij}(t), \quad i \in \mathbf{Z}. \tag{9.117}$$

Special Cases 1) An alternating renewal process (page 319) is a semi-Markov chain with state space $\mathbf{Z} = \{0, 1\}$ and transition probabilities

$$p_{00} = p_{11} = 0 \text{ and } p_{01} = p_{10} = 1.$$

The states 0 and 1 indicate that the system is under renewal or operating, respectively. In this case, $F_{01}(\cdot)$ and $F_{10}(\cdot)$ are in this order the distribution functions of the re-newal time and the system lifetime.

2) A homogeneous Markov chain in continuous time with state space $\mathbf{Z} = \{0, 1, ...\}$ is a semi-Markov chain with the same state space and transition probabilities (9.34):

$$p_{ij} = \frac{q_{ij}}{q_i}, \quad i \ne j,$$

where q_{ij} (q_i) are the conditional (unconditional) transition rates of the Markov chain. By (9.31), the distribution function of the unconditional sojourn time in state i is

$$F_i(t) = 1 - e^{-q_i t}, \quad t \ge 0.$$

In what follows, semi-Markov processes are considered under the following three assumptions:

1) The embedded homogeneous Markov chain $\{X_0, X_1, ...\}$ has a unique stationary state distribution $\{\pi_0, \pi_1, ...\}$. By (8.9), this distribution is solution of

$$\pi_j = \sum_{i \in \mathbf{Z}} p_{ij} \pi_i, \quad \sum_{i \in \mathbf{Z}} \pi_i = 1. \tag{9.118}$$

As pointed out in section 8.3, a unique stationary state distribution exists if the Mar-kov chain is aperiodic, irreducible, and positive recurrent.

2) The distribution functions $F_i(t) = P(Y_i \le t)$ are nonarithmetic (see definition 5.3, page 216).

3) The mean sojourn times of the process in all states are finite:

$$\mu_i = E(Y_i) = \int_0^\infty [1 - F_i(t)] \, dt < \infty, \quad i \in \mathbf{Z}.$$

Note: In this section μ_i denotes no longer an intensity, but a mean sojourn time.

In what follows, a transition of the semi-Markov chain into state k is called a *k-tran-sition*. Let $N_k(t)$ be the random number of k-transitions occurring in $(0, t]$ and $H_k(t)$ its mean value: $H_k(t) = E(N_k(t))$. Then, for any $t > 0$,

$$\lim_{t \to \infty} [H_k(t + \tau) - H_k(t)] = \frac{\tau \pi_k}{\sum_{i \in \mathbf{Z}} \pi_i \mu_i}, \quad k \in \mathbf{Z}. \tag{9.119}$$

This relationship implies that after a sufficiently long time period the number of k-transitions in a given time interval does no longer depend on the position of this interval, but only on its length. Strictly speaking, the right-hand side of (9.119) gives the mean number of k-transitions in an interval of length τ once the process has reached its stationary regime, or, with other words, if it is in the steady state. The following formulas and the analysis of examples is based on (9.119), but the definition and properties of stationary semi-Markov chains will not be discussed in detail.

From (9.119), when the process is in the steady state, the mean number of k-transitions per unit time is

$$U_k = \frac{\pi_k}{\sum_{i \in \mathbf{Z}} \pi_i \mu_i}.$$

Hence the portion of time the chain is in state k is

$$A_k = \frac{\pi_k \mu_k}{\sum_{i \in \mathbf{Z}} \pi_i \mu_i}. \tag{9.120}$$

Consequently, in the longrun, the fraction of time the chain is in a set of states $\mathbf{Z}_0$, $\mathbf{Z}_0 \subseteq \mathbf{Z}$, is

$$A_{\mathbf{Z}_0} = \frac{\sum_{k \in \mathbf{Z}_0} \pi_k \mu_k}{\sum_{i \in \mathbf{Z}} \pi_i \mu_i}. \tag{9.121}$$

With other words, $A_{\mathbf{Z}_0}$ is the probability that a visitor, who arrives at a random time from 'outside', finds the semi-Markov chain in a state belonging to $\mathbf{Z}_0$.

Let c_k denote the cost, which is caused by a k-transition of the system. Then the mean total (transition) cost per unit time is

$$C = \frac{\sum_{k \in \mathbf{Z}} \pi_k c_k}{\sum_{i \in \mathbf{Z}} \pi_i \mu_i}. \tag{9.122}$$

Note that the formulas (9.119) to (9.122) depend only on the unconditional sojourn times of a semi-Markov chain in its states. This property facilitates their application.

Example 9.23 (*age renewal policy*) The system is renewed upon failure by an *emergency renewal* or at age τ by a *preventive renewal*, whichever occurs first.

To determine the stationary system availability, system states have to be introduced:
0   operating
1   emergency renewal
2   preventive renewal

Figure 9.22 Transition graph for example 9.23

Let L be the random system lifetime, $F(t) = P(L \leq t)$ its distribution function, and

$$\overline{F}(t) = 1 - F(t) = P(L > t)$$

its survival probability. Then the positive transition probabilities between the states are (Figure 9.22)

$$p_{01} = F(\tau), \quad p_{02} = \overline{F}(\tau), \quad p_{10} = p_{20} = 1.$$

Let Z_e and Z_p be the random times for emergency renewals and preventive renewals, respectively. Then the conditional sojourn times of the system in the states are

$$Y_{01} = L, \quad Y_{02} = \tau, \quad Y_{10} = Z_e, \quad Y_{20} = Z_p.$$

The unconditional sojourn times are

$$Y_0 = \min(L, \tau), \quad Y_1 = Z_e, \quad Y_2 = Z_p.$$

The system behaviour can be described by a semi-Markov chain $\{X(t), t \geq 0\}$ with state space $\mathbf{Z} = \{0, 1, 2\}$ and the transition probabilities and sojourn times given. The corresponding equations (9.118) in the stationary probabilities of the embedded Markov chain are

$$\pi_0 = \qquad \pi_1 + \pi_2$$

$$\pi_1 = F(\tau)\pi_0$$

$$1 = \qquad \pi_0 + \pi_1 + \pi_2.$$

The solution is

$$\pi_0 = 1/2, \quad \pi_1 = F(\tau)/2, \quad \pi_2 = \overline{F}(\tau)/2.$$

The mean sojourn times are

$$\mu_0 = \int_0^\tau \overline{F}(t)\, dt, \quad \mu_1 = d_e, \quad \mu_2 = d_p.$$

According to (9.120), the stationary availability $A_0 = A(\tau)$ of the system is

$$A(\tau) = \frac{\mu_0 \pi_0}{\mu_0 \pi_0 + \mu_1 \pi_1 + \mu_2 \pi_2}$$

or

$$A(\tau) = \frac{\int_0^\tau \overline{F}(t)\, dt}{\int_0^\tau \overline{F}(t)\, dt + d_e F(\tau) + d_p \overline{F}(\tau)}. \tag{9.123}$$

It is important that this result does not depend on the probability distributions of the renewal times Z_e and Z_p, but only on their mean values. An optimal renewal interval $\tau = \tau*$ satisfies the equation $dA(\tau)/d\tau = 0$ or

$$\lambda(\tau) \int_0^\tau \overline{F}(t)\, dt - F(\tau) = \frac{d}{1-d} \tag{9.124}$$

with $d = d_e/d_p$. A unique solution of this equation exists if $\lambda(t)$ is strictly increasing and $d_p < d_e$, i.e., $d < 1$. (Otherwise preventive renewals would not make sense.)

By coupling the equations (9.123) and (9.124) the corresponding maximal long-run availability $A(\tau^*)$ is seen to have structure

$$A(\tau^*) = \frac{1}{1 + (d_e - d_p)\lambda(\tau^*)}. \tag{9.125}$$

As a numerical special case, let L have a Rayleigh-distribution with parameter θ and renewal times $d_e = 10$ and $d_p = 2$. Then

$$F(t) = P(L \le t) = 1 - e^{-(t/\theta)^2}, \quad t \ge 0,$$

and, by formula (2.80), page 77, L has mean value

$$E(L) = \theta\sqrt{\pi/4}.$$

Since the corresponding failure rate is $\lambda(t) = 2t/\theta^2$, equation (9.123) becomes

$$2\frac{\tau}{\theta} \int_0^{\tau/\theta} e^{-x^2}\, dt + e^{-(\tau/\theta)^2} = 1.25.$$

The unique solution is $\tau^* = 0.5107 \cdot \theta$. This holds for any θ. (θ is a scale parameter.) By (9.125), the maximal stationary availability is

$$A(\tau^*) = \frac{\theta}{\theta + 8.1712},$$

whereas the stationary availability of the system without preventive renewals is smaller:

$$A = \frac{E(L)}{E(L) + d_e} = \frac{\theta}{\theta + 11.2838}.$$

If the renewal times are negligibly small, but the mean costs c_e and c_p for emergency and preventive renewals, respectively, are relevant, then, from (9.122), the mean renewal cost per unit time in the steady state are

$$K(\tau) = \frac{c_e\pi_1 + c_p\pi_2}{\mu_0\pi_0} = \frac{c_e F(\tau) + c_p \overline{F}(\tau)}{\int_0^\tau \overline{F}(t)\, dt}.$$

Analogously to the corresponding renewal times, c_e and c_p can be thought of mean values of arbitrarily distributed renewal costs. Since $K(\tau)$ has the same functional structure as $1/A(\tau) - 1$, maximizing $A(\tau)$ and minimizing $K(\tau)$ leads again to the same equation (9.124) if d is replaced with $c = c_p/c_e$.   □

Example 9.24 A series system consists of n subsystems $e_1, e_2, \ldots, e_n$. The lifetimes of the subsystems $L_1, L_2, \ldots, L_n$ are independent exponential random variables with parameters $\lambda_1, \lambda_2, \ldots, \lambda_n$. Let

$$G_i(t) = P(L_i \le t) = 1 - e^{-\lambda_i t}, \quad g_i(t) = \lambda_i e^{-\lambda_i t}, \quad t \ge 0; \; i = 1, 2, \ldots, n.$$

When a subsystem fails, the system interrupts its work. As soon as the renewal of the failed subsystem is finished, the system continues operating. Let μ_i be the average renewal time of subsystem e_i. As long as a subsystem is being renewed, the other subsystems cannot fail, i.e. during such a time period they are in the cold-standby mode. The following system states are introduced:

$X(t) = 0$ if the system is operating,

$X(t) = i$ if subsystem e_i is under renewal, $i = 1, 2, ..., n$.

Then $\{X(t), t \geq 0\}$ is a semi-Markov chain with state space $\mathbf{Z} = \{0, 1, ..., n\}$. The conditional sojourn times in state 0 of this semi-Markov chain are

$$Y_{0i} = L_i, \quad i = 1, 2, ..., n,$$

and its unconditional sojourn time in state 0 is

$$Y_0 = \min\{L_1, L_2, ..., L_n\}.$$

Thus, Y_0 has distribution function

$$F_0(t) = 1 - \bar{G}_1(t) \cdot \bar{G}_2(t) \cdots \bar{G}_n(t).$$

Letting $\lambda = \lambda_1 + \lambda_2 + \cdots + \lambda_n$ implies

$$F_0(t) = 1 - e^{-\lambda t}, \quad t \geq 0,$$

$$\mu_0 = E(Y_0) = 1/\lambda.$$

The system makes a transition from state 0 into state i with probability

$$p_{0i} = P(Y_0 = L_i)$$

$$= \int_0^\infty \bar{G}_1(x) \cdot \bar{G}_2(x) \cdots \bar{G}_{i-1}(x) \cdot \bar{G}_{i+1}(x) \cdots \bar{G}_n(x) g_i(x) \, dx$$

$$= \int_0^\infty e^{-(\lambda_1 + \lambda_2 + \cdots + \lambda_{i-1} + \lambda_{i+1} + \cdots + \lambda_n)x} \lambda_i e^{-\lambda_i x} \, dx = \int_0^\infty e^{-\lambda x} \lambda_i \, dx.$$

Hence,

$$p_{0i} = \frac{\lambda_i}{\lambda}, \quad p_{i0} = 1; \quad i = 1, 2, ..., n.$$

Thus, the system of equations (9.118) becomes

$$\pi_0 = \pi_1 + \pi_2 + \cdots + \pi_n,$$

$$\pi_i = \frac{\lambda_i}{\lambda} \pi_0; \quad i = 1, 2, ..., n.$$

In view of $\pi_1 + \pi_2 + \cdots + \pi_n = 1 - \pi_0$, the solution is

$$\pi_0 = \frac{1}{2}; \quad \pi_i = \frac{\lambda_i}{2\lambda}; \quad i = 1, 2, ..., n.$$

With these ingredients, formula (9.120) yields the stationary system availability

$$A_0 = \frac{1}{1 + \sum_{i=1}^n \lambda_i \mu_i}. \qquad \square$$

Example 9.25 Consider the loss system $M/G/1/0$ on condition that the server is subjected to failures: Customers arrive according to a homogeneous Poisson process with rate λ. Hence, their interarrival times are identically distributed as an exponential random variable Y with parameter λ. The server has random lifetime L_0 when being idle, and a random lifetime L_1 when being busy. L_0 is exponential with parameter λ_0, and L_1 is exponential with parameter λ_1. The service time Z has distribution function $B(t)$ with density $b(t)$. When at the time point of server failure a customer is being served, then this customer is lost, i.e., it has to leave the system. All occurring random variables are assumed to be independent. To describe the behavior of this system by a semi-Markov chain, three states are introduced:

State 0   The server is idle, but available.

State 1   The server is busy.

State 2   The server is under repair (not available)

To determine the steady state probabilities of the states 0, 1, and 2, the transition probabilities are needed:

$$p_{00} = p_{11} = p_{22} = p_{21} = 0, \quad p_{20} = 1$$

$$p_{01} = P(L_0 > Y) = \int_0^\infty e^{-\lambda_0 t} \lambda e^{-\lambda t} dt = \frac{\lambda}{\lambda + \lambda_0}$$

$$p_{02} = 1 - p_{01} = P(L_0 \le Y) = \frac{\lambda_0}{\lambda + \lambda_0}$$

$$p_{10} = P(L_1 > Z) = \int_0^\infty e^{-\lambda_1 t} b(t) dt$$

$$p_{12} = 1 - p_{10} = P(L_1 \le Z) = \int_0^\infty [1 - e^{-\lambda_1 t}] b(t) dt.$$

With these transition probabilities, the stationary state probabilities of the embedded Markov chain $\{X_0, X_1, ...\}$ can be obtained from (9.118):

$$\pi_0 = \frac{\lambda + \lambda_0}{2(\lambda + \lambda_0) + \lambda p_{12}}, \quad \pi_1 = \frac{\lambda}{2(\lambda + \lambda_0) + \lambda p_{12}}, \quad \pi_2 = \frac{\lambda_0 + \lambda p_{12}}{2(\lambda + \lambda_0) + \lambda p_{12}}.$$

The sojourn times in state 0, 1, and 2 are

$$Y_0 = \min(L_0, Y), \quad Y_1 = \min(L_1, Z), \quad Y_2 = Z.$$

Hence, the mean sojourn times are

$$\mu_0 = \frac{1}{\lambda + \lambda_0}, \quad \mu_1 = \int_0^\infty (1 - B(t)) e^{-\lambda_1 t} dt, \quad \mu_2 = E(Z).$$

With these parameters, the stationary state probabilities of the semi-Markov process are given by (9.120).   □

The time-dependent behaviour of semi-Markov chains is discussed, for instance, in *Kulkarni* (2010).

9.9 EXERCISES

9.1) Let $\mathbf{Z} = \{0, 1\}$ be the state space and

$$\mathbf{P}(t) = \begin{pmatrix} e^{-t} & 1 - e^{-t} \\ 1 - e^{-t} & e^{-t} \end{pmatrix}$$

the transition matrix of a continuous-time stochastic process $\{X(t), t \geq 0\}$. Check whether $\{X(t), t \geq 0\}$ is a homogeneous Markov chain.

9.2) A system fails after a random lifetime L. Then it waits a random time W for renewal. A renewal takes another random time Z. The random variables L, W, and Z have exponential distributions with parameters λ, ν, and μ, respectively. On completion of a renewal, the system immediately resumes its work. This process continues indefinitely. All life, waiting, and renewal times are assumed to be independent. Let the system be in states 0, 1, and 2 when it is operating, waiting, or being renewed. The transitions between the states are governed by a Markov chain $\{X(t), t \geq 0\}$.

(1) Draw the transition graph of $\{X(t), t \geq 0\}$ and set up a system of linear differential equations for the time-dependent state probabilities $p_i(t) = P(X(t) = i)$, $i = 0, 1, 2$.

(2) Use this system to derive an algebraic system of equations for the stationary state probabilities π_i of $\{X(t), t \geq 0\}$. Determine the stationary availability of the system.

9.3) Consider a 1-out-of-2 system, i.e., the system is operating when at least one of its two subsystems is operating. When a subsystem fails, the other one continues to work. On its failure, the joint renewal of both subsystems begins. On its completion, both subsystems resume their work at the same time. The lifetimes of the subsystems are identically exponential with parameter λ. The joint renewal time is exponential with parameter μ. All life- and renewal times are independent of each other. Let $X(t)$ be the number of subsystems operating at time t.

(1) Draw the transition graph of the Markov chain $\{X(t), t \geq 0\}$.

(2) Given the initial condition $P(X(0) = 2) = 1$, determine the time-dependent state probabilities $p_i(t) = P(X(t) = i)$, $i = 0, 1, 2$, and the stationary state distribution.

Hint Consider separately the cases $(\lambda + \mu + \nu)^2 (=)(<)(>) 4(\lambda\mu + \lambda\nu + \mu\nu)$.

9.4) A copy center has 10 copy machines of the same type which are in constant use. The times between two successive failures of a machine have an exponential distribution with mean value 100 *hours*. There are two mechanics who repair failed machines. A defective machine is repaired by only one mechanic. During this time, the second mechanic is busy repairing another failed machine, if there are any, or this mechanic is idle. All repair times have an exponential distribution with mean value 4 *hours*. All random variables involved are independent. Consider the steady state.

(1) What is the average percentage of operating machines?

(2) What is the average percentage of idle mechanics?

9.5) Consider the two-unit system with standby redundancy discussed in example 9.5 a) on condition that the lifetimes of the units are exponential with respective parameters λ_1 and λ_2. The other model assumptions listed in example 9.5 remain valid.

Model the system by a Markov chain and draw the transition graph.

9.6) Consider the two-unit system with parallel redundancy discussed in example 9.6 on condition that the lifetimes of the units are exponential with parameters λ_1 and λ_2, respectively. The other model assumptions listed in example 9.6 remain valid.

Model the behavior of the system by a Markov chain and draw the transition graph.

9.7) The system considered in example 9.7 is generalized as follows: If the system makes a direct transition from state 0 to the blocking state 2, then the subsequent renewal time is exponential with parameter μ_0. If the system makes a transition from state 1 to state 2, then the subsequent renewal time is exponential with parameter μ_1.

(1) Model the system by a Markov chain and draw the transition graph.

(2) What is the stationary probability that the system is blocked?

9.8) Consider a two-unit system with standby redundancy and one mechanic. All repair times of failed units have an Erlang distribution with parameters $n = 2$ and μ. Apart from this, the other model assumptions listed in example 9.5 remain valid.

(1) Model the system by a Markov chain and draw the transition graph.

(2) Determine the stationary state probabilities of the system.

(3) Sketch the stationary availability of the system as a function of $\rho = \lambda/\mu$.

9.9) Consider a two-unit parallel system (i.e., the system operates if at least one unit is operating). The lifetimes of the units have an exponential distributions with parameter λ. There is one repairman, who can only attend one failed unit at a time. Repairs times have an Erlang distribution with parameters $n = 2$ and $\lambda = 1/2$. The system arrives at the failed state as soon as a unit fails during the repair of the other one. All life and repair times are assumed to be independent.

(1) By using Erlang's phase method, determine the relevant state space of the system and draw the corresponding transition graph of the underlying Markov chain.

(2) Determine the stationary availability of the system.

9.10) When being in states 0, 1, and 2, a (pure) birth process $\{X(t), t \geq 0\}$ with state space $\mathbf{Z} = \{0, 1, 2, ...\}$ has the respective birth rates

$$\lambda_0 = 2, \ \lambda_1 = 3, \ \lambda_2 = 1.$$

Given $X(0) = 0$, determine the time-dependent state probabilities $p_i(t) = P(X(t) = i)$ for $i = 0, 1, 2$.

9.11) Consider a linear birth process with state space $\mathbf{Z} = \{0, 1, 2, ...\}$ and transition rates $\lambda_j = j\lambda$, $j = 0, 1, ...$

(1) Given $X(0) = 1$, determine the distribution function of the random time point T_3 at which the process enters state 3.

(2) Given $X(0) = 1$, determine the mean value of the random time point T_n at which the process enters state n, $n > 1$.

9.12) The number of physical particles of a particular type in a closed container evolves as follows: There is one particle at time $t = 0$. Its splits into two particles of the same type after an exponential random time Y with parameter λ (its lifetime). These two particles behave in the same way as the original one, i.e., after random times, which are identically distributed as Y, they split into 2 particles each, and so on. All lifetimes of the particles are assumed to be independent. Let $X(t)$ denote the number of particles in the container at time t.

Determine the absolute state probabilities $p_j(t) = P(X(t) = j)$; $j = 1, 2, ...$, of the stochastic process $\{X(t), t \geq 0\}$.

9.13) A death process with state space $\mathbf{Z} = \{0, 1, 2, ...\}$ has death rates

$$\mu_0 = 0, \ \mu_1 = 2, \text{ and } \mu_2 = \mu_3 = 1.$$

Given $X(0) = 3$, determine $p_j(t) = P(X(t) = j)$ for $j = 0, 1, 2, 3$.

9.14) A linear death process $\{X(t), t \geq 0\}$ has death rates $\mu_j = j\mu$; $j = 0, 1, ...$.

(1) Given $X(0) = 2$, determine the distribution function of the time to entering state 0 ('lifetime' of the process).

(2) Given $X(0) = n$, $n > 1$, determine the mean value of the time at which the process enters state 0.

9.15) At time $t = 0$ there are an infinite number of molecules of type a and $2n$ molecules of type b in a two-component gas mixture. After an exponential random time with parameter μ any molecule of type b combines, independently of the others, with a molecule of type a to form a molecule ab.

(1) What is the probability that at time t there are still j free molecules of type b in the container?

(2) What is the mean time till there are only n free molecules of type b left in the container?

9.16) At time $t = 0$ a cable consists of 5 identical, intact wires. The cable is subject to a constant load of $100kp$ such that in the beginning each wire bears a load of $20kp$. Given a load of $w\,kp$ per wire, the time to breakage of a wire (its lifetime) is exponential with mean value

$$\frac{1000}{w} \; [weeks].$$

When one or more wires are broken, the load of $100kp$ is uniformly distributed over the remaining intact ones. For any fixed number of wires, their lifetimes are assumed to be independent and identically distributed.

(1) What is the probability that all wires are broken at time $t = 50 \, [weeks]$?

(2) What is the mean time until the cable breaks completely?

9.17)* Let $\{X(t), t \geq 0\}$ be a death process with $X(0) = n$ and positive death rates $\mu_1, \mu_2, \ldots, \mu_n$.

Prove: If Y is an exponential random variable with parameter λ and independent of the death process, then

$$P(X(Y) = 0) = \prod_{i=1}^{n} \frac{\mu_i}{\mu_i + \lambda}.$$

9.18) A birth- and death process has state space $\mathbf{Z} = \{0, 1, \ldots, n\}$ and transition rates

$$\lambda_j = (n - j)\lambda \quad \text{and} \quad \mu_j = j\mu; \quad j = 0, 1, \ldots, n.$$

Determine its stationary state probabilities.

9.19) Check whether or under what restrictions a birth- and death process with transition rates

$$\lambda_j = \frac{j+1}{j+2}\lambda \quad \text{and} \quad \mu_j = \mu; \quad j = 0, 1, \ldots,$$

has a stationary state distribution.

9.20) A birth- and death process has transition rates

$$\lambda_j = (j+1)\lambda \quad \text{and} \quad \mu_j = j^2\mu; \; j = 0, 1, \ldots; \; 0 < \lambda < \mu.$$

Confirm that this process has a stationary state distribution and determine it.

9.21) Consider the following deterministic models for the mean (average) development of the size of populations:

(1) Let $m(t)$ be the mean number of individuals of a population at time t. It is reasonable to assume that a change of the population size, namely $dm(t)/dt$, is proportional to $m(t)$, $t \geq 0$, i.e., for a constant h the mean number $m(t)$ satisfies the differential equation

$$\frac{d m(t)}{d t} = h \, m(t).$$

a) Solve this differential equation assuming $m(0) = 1$.

b) Is there a birth and death process the trend function of which has the functional structure of $m(t)$?

(2) The mean population size $m(t)$ satisfies the differential equation

$$\frac{dm(t)}{dt} = \lambda - \mu\, m(t).$$

a) With a positive integer N, solve this equation under the initial condition

$$m(0) = N.$$

b) Is there a birth and death process the trend function of which has the functional structure of $m(t)$?

9.22) A computer is connected to three terminals (for example, measuring devices). It can simultaneously evaluate data records from only two terminals. When the computer is processing two data records and in the meantime another data record has been produced, then this new data record has to wait in a buffer, when the buffer is empty. Otherwise the new data record is lost. The buffer can store only one data record. The data records are processed according to the FCFS-queueing discipline. The terminals produce data records independently according to a homogeneous Poisson process with intensity λ. The processing times of data records from all terminals are independent, even if the computer is busy with two data records at the same time, and they have an exponential distribution with parameter μ. They are assumed to be independent of the input.

Let $X(t)$ be the number of data records in computer and buffer at time t.

(1) Verify that $\{X(t),\ t \ge 0\}$ is a birth and death process, determine its transition rates and draw the transition graph.

(2) Determine the stationary loss probability, i.e., the probability that in the steady state a data record is lost.

9.23) Under otherwise the same assumptions as in exercise 9.22, it is assumed that a data record, which has been waiting in the buffer a random *patience time*, will be deleted as being no longer up to date. The patience times of all data records are assumed to be independent, exponential random variables with parameter ν. They are also independent of all arrival and processing times of the data records.

(1) Draw the transition graph.

(2) Determine the stationary loss probability.

9.24) Under otherwise the same assumptions as in exercise 9.22, it is assumed that a data record will be deleted when its total sojourn time in the buffer and computer exceeds a random time Z, where Z has an exponential distribution with parameter α. Thus, the interruption of the current service of a data record is possible.

(1) Draw the corresponding transition graph.

(2) Determine the stationary loss probability.

9.25) A small filling station in a rural area provides diesel for agricultural machines. It has one diesel pump and waiting capacity for 5 machines. On average, 8 machines per hour arrive for diesel. An arriving machine immediately leaves the station without fuel if pump and all waiting places are occupied. The mean time a machine occupies the pump is 5 minutes. The station behaves like a $M/M/s/m$-queueing system.

(1) Determine the stationary loss probability.

(2) Determine the stationary probability that an arriving machine waits for diesel.

9.26) Consider a two-server loss system. Customers arrive according to a homogeneous Poisson process with intensity λ. A customer is always served by server 1 when this server is idle, i.e., an arriving customer goes only then to server 2, when server 1 is busy. The service times of both servers are iid exponential random variables with parameter μ. Let $X(t)$ be the number of customers in the system at time t.

Determine the stationary state probabilities of the stochastic process $\{X(t), t \geq 0\}$.

9.27) A two-server loss system is subject to a homogeneous Poisson input with intensity λ. The situation considered in exercise 9.26 is generalized as follows: If both servers are idle, a customer goes to server 1 with probability p and to server 2 with probability $1 - p$. Otherwise, a customer goes to the idle server (if there is any). The service times of the servers 1 and 2 are independent, exponential random variables with parameters μ_1 and μ_2, respectively. Arrival and service times are independent.

Describe the behaviour of the system by a suitable homogeneous Markov chain and draw the transition graph.

9.28) A single-server waiting system is subject to a homogeneous Poisson input with intensity $\lambda = 30 [h^{-1}]$. If there are not more than 3 customers in the system, the service times have an exponential distribution with mean $1/\mu = 2 [min]$. If there are more than 3 customers in the system, the service times are exponential with mean $1/\mu = 1$ [min]. All arrival and service times are independent.

(1) Show that there exists a stationary state distribution and determine it.

(2) Determine the mean length of the waiting queue in the steady state.

9.29) Taxis and customers arrive at a taxi rank in accordance with two independent homogeneous Poisson processes with intensities

$$\lambda_1 = 4 [h^{-1}] \text{ and } \lambda_2 = 3 [h^{-1}],$$

respectively. Potential customers, who find 2 waiting customers, do not wait for service, but leave the rank immediately. Groups of customers, who will use the same taxi, are considered to be one customer. On the other hand, arriving taxis, who find two taxis waiting, leave the rank as well.

What is the average number of customers waiting at the rank?

9.30) A transport company has 4 trucks of the same type. There are 2 maintenance teams for repairing the trucks after a failure. Each team can repair only one truck at a time and each failed truck is handled by only one team. The times between failures of a truck (lifetime) is exponential with parameter λ. The repair times are exponential with parameter μ. All life and repair times are assumed to be independent. Let $\rho = \lambda/\mu = 0.2$. What is the most efficient way of organizing the work:

(1) to make both maintenance teams responsible for the maintenance of all 4 trucks so that any team which is free can repair any failed truck, or

(2) to assign 2 definite trucks to each team?

9.31) Ferry boats and customers arrive at a ferry station in accordance with two independent homogeneous Poisson processes with intensities λ and μ, respectively. If there are k customers at the ferry station, when a boat arrives, then it departs with $min\,(k,n)$ passengers (n is the capacity of each boat). If $k > n$, then the remaining $k - n$ customers wait for the next boat. The sojourn times of the boats at the station are assumed to be negligibly small.

Model the situation by a suitable homogeneous Markov chain $\{X(t),\, t \geq 0\}$ and draw the transition graph.

9.32) The life cycle of an organism is controlled by shocks (e.g., accidents, virus attacks) in the following way: A healthy organism has an exponential lifetime L with parameter λ_h. If a shock occurs, the organism falls sick and, when being in this state, its (residual) lifetime S is exponential with parameter

$$\lambda_s, \ \lambda_s > \lambda_h.$$

However, a sick organism may recover and return to the healthy state. This occurs in an exponential time R with parameter μ. If during a period of sickness another shock occurs, the organism cannot recover and will die a random time D after the occurrence of the second shock. D is assumed to be exponential with parameter

$$\lambda_d, \ \lambda_d > \lambda_s.$$

The random variables L, S, R, and D are assumed to be independent.

(1) Describe the evolvement in time of the states the organism may be in by a Markov chain.

(2) Determine the mean lifetime of the organism.

9.33) Customers arrive at a waiting system of type $M/M/1/\infty$ with intensity λ. As long as there are less than n customers in the system, the server remains idle. As soon as the nth customer arrives, the server resumes its work and stops working only then, when all customers (including newcomers) have been served. After that the server again waits until the waiting queue has reached length n and so on. Let $1/\mu$ be the mean service time of a customer and $X(t)$ be the number of customers in the system at time t.

(1) Draw the transition graph of the Markov chain $\{X(t), t \geq 0\}$.

(2) Given that $n = 2$, compute the stationary state probabilities. Make sure they exist.

9.34) At time $t = 0$ a computer system consists of n operating computers. As soon as a computer fails, it is separated from the system by an automatic switching device with probability $1 - p$. If a failed computer is not separated from the system (this happens with probability p), then the entire system fails. The lifetimes of the computers are independent and have an exponential distribution with parameter λ. Thus, this distribution does not depend on the system state. Provided the switching device has operated properly when required, the system is available as long as there is at least one computer available. Let $X(t)$ be the number of computers which are available at time t. By convention, if, due to the switching device, the entire system has failed in $[0, t)$, then $X(t) = 0$.

(1) Draw the transition graph of the Markov chain $\{X(t), t \geq 0\}$.

(2) Given $n = 2$, determine the mean lifetime $E(X_s)$ of the system.

9.35) A waiting-loss system of type $M/M/1/2$ is subject to two independent Poisson inputs 1 and 2 with respective intensities λ_1 and λ_2, which are referred to as type 1- and type 2-customers. An arriving type 1-customer who finds the server busy and the waiting places occupied displaces a possible type 2-customer from its waiting place (such a type 2-customer is lost), but ongoing service of a type 2-customer is not interrupted. When a type 1-customer and a type 2-customer are waiting, then the type 1-customer will always be served first, regardless of the order of their arrivals. The service times of type 1- and type 2-customers are independent and have exponential distributions with respective parameters μ_1, and μ_2.

Describe the behavior of the system by a homogeneous Markov chain, determine the transition rates, and draw the transition graph.

9.36) A queueing network consists of two servers 1 and 2 in series. Server 1 is subject to a homogeneous Poisson input with intensity $\lambda = 5$ an hour. A customer is lost if server 1 is busy. From server 1 a customer goes to server 2 for further service. If server 2 is busy, the customer is lost. The service times of servers 1 and 2 are exponential with respective mean values

$$1/\mu_1 = 6 \, min \text{ and } 1/\mu_2 = 12 \, min.$$

All arrival and service times are independent.

What percentage of customers (with respect to the total input at server 1) is served by both servers?

9.37) A queueing network consists of three nodes (queueing systems) 1, 2, and 3, each of type $M/M/1/\infty$. The external inputs into the nodes have respective intensities

$$\lambda_1 = 4, \ \lambda_2 = 8, \text{ and } \lambda_3 = 12 \ [\text{customers per hour}].$$

The respective mean service times at the nodes are

$$4, 2, \text{ and } 1 \text{ [min]}.$$

After having been served by node 1, a customer goes to nodes 2 or 3 with equal probabilities 0.4 or leaves the system with probability 0.2. From node 2, a customer goes to node 3 with probability 0.9 or leaves the system with probability 0.1. From node 3, a customer goes to node 1 with probability 0.2 or leaves the system with probability 0.8. The external inputs and the service times are independent.

(1) Check whether this queueing network is a Jackson network.

(2) Determine the stationary state probabilities of the network.

9.38) A closed queueing network consists of 3 nodes. Each one has 2 servers. There are 2 customers in the network. After having been served at a node, a customer goes to one of the others with equal probability. All service times are independent random variables and have an exponential distribution with parameter μ.

What is the stationary probability to find both customers at the same node?

9.39) Depending on demand, a conveyor belt operates at 3 different speed levels 1, 2, and 3. A transition from level i to level j is made with probability p_{ij} with

$$p_{12} = 0.8, \ p_{13} = 0.2, \ p_{21} = p_{23} = 0.5, \ p_{31} = 0.4, \ p_{32} = 0.6.$$

The respective mean times the conveyor belt operates at levels 1, 2, or 3 between transitions are

$$\mu_1 = 45, \ \mu_2 = 30, \text{ and } \mu_3 = 12 \text{ [hours]}.$$

Determine the stationary percentages of time in which the conveyor belt operates at levels 1, 2, and 3 by modeling the situation as a semi-Markov chain.

9.40) The mean lifetime of a system is 620 hours. There are two failure types: Repairing the system after a type 1-failure requires 20 hours on average and after a type 2-failure 40 hours on average. 20% of all failures are type 2-failures. There is no dependence between the system lifetime and the subsequent failure type. Upon each repair the system is 'as good as new'. The repaired system immediately resumes its work. This process is continued indefinitely. Life- and repair times are independent.

(1) Describe the situation by a semi-Markov chain with 3 states and draw the transition graph of the underlying discrete-time Markov chain.

(2) Determine the stationary state probabilities of the system.

9.41)* Under otherwise the same model assumptions as in example 9.25, determine the stationary probabilities of the states 0, 1, and 2 introduced there on condition that the service time B is a constant μ; i.e., determine the stationary state probabilities of the loss system $M/D/1/0$ with unreliable server.

9.42) A system has two different failure types: type 1 and type 2. After a type i-fail-ure the system is said to be in failure state i; $i = 1, 2$. The time L_i to a type i-failure has an exponential distribution with parameter λ_i; $i = 1, 2$. Thus, if at time $t = 0$ a new system starts working, the time to its first failure is

$$Y_0 = \min (L_1, L_2).$$

The random variables L_1 and L_2 are assumed to be independent. After a type 1-fail-ure, the system is switched from failure state 1 into failure state 2. The respective mean sojourn times of the system in states 1 and 2 are μ_1 and μ_2. When in state 2, the system is being renewed. Thus, μ_1 is the mean switching time and μ_2 the mean renewal time. A renewed system immediately starts working, i.e., the system makes a transition from state 2 to state 0 with probability 1. This process continues to infinity. (For motivation, see example 9.7.)

(1) Describe the system behavior by a semi-Markov chain and draw the transition graph of the embedded discrete-time Markov chain.

(2) Determine the stationary availability of the system.

CHAPTER 10

Martingales

10.1 DISCRETE-TIME MARTINGALES

10.1.1 Definition and Examples

Martingales are important tools for solving prestigious problems in probability theory and its applications. Such problems occur in areas as random walks, point processes, mathematical statistics, actuarial risk analysis, and mathematics of finance. Heuristically, martingales are stochastic models for 'fair games' in a wider sense, i.e., games, in which each side has the same chance to win or to lose. In particular, *martingale* is the French word for that game, in which a gambler doubles her/his bet on every loss until he wins (Example 10.6). Martingales were introduced as a special class of stochastic processes by *J. Ville* und *P. Levy*. It was, however, *J. L. Doob*, who recognized their large theoretical and practical potential and began with their systematic investigation. Martingales as stochastic processes are defined for discrete and continuous parameter spaces **T**. Analogously to Markov processes, the terminology *discrete-time martingales* and *continuous-time martingales* is adopted. The definition of a martingale as given in this chapter heavily relies on the concept of the conditional mean value of a random variable given values of other random variables or, more generally, on the concept of the conditional mean value of a random variable given other random variables (see formulas $(3.61)-(3.64)$).

Definition 10.1 A stochastic process in discrete time $\{X_0, X_1, ...\}$ with state space **Z**, which satisfies

$$E(|X_n|) < \infty, \ n = 0, 1, 2, ...,$$

is called a (*discrete-time*) *martingale* if for all vectors $(x_0, x_1, ..., x_n)$ with $x_i \in \mathbf{Z}$ and $n = 0, 1, ...$

$$E(X_{n+1} | X_n = x_n, ..., X_1 = x_1, X_0 = x_0) = x_n. \tag{10.1}$$

Under the same assumptions, $\{X_0, X_1, ...\}$ is a (*discrete-time*) *supermartingale* if

$$E(X_{n+1} | X_n = x_n, ..., X_1 = x_1, X_0 = x_0) \le x_n, \tag{10.2}$$

and a (*discrete-time*) *submartingale* if

$$E(X_{n+1} | X_n = x_n, ..., X_1 = x_1, X_0 = x_0) \ge x_n. \tag{10.3}$$

●

If, for instance, the X_n are continuous random variables, then, in view of (3.54) (page 145), multiplying both sides of the (in-) equalities (10.1) to (10.3) by the joint density of the random vector $(X_0, X_1, ..., X_n)$ and integrating over its range yields

Martingale: $E(X_{n+1}) = E(X_n); \ n = 0, 1, ...,$

Supermartingale: $E(X_{n+1}) \leq E(X_n); \ n = 0, 1, ...,$

Submartingale: $E(X_{n+1}) \geq E(X_n); \ n = 0, 1, ...$

Thus, the trend function of a martingale is constant,

$$m = E(X_n) = E(X_0); \ n = 0, 1, ..., \tag{10.4}$$

whereas the trend functions of supermartingales (submartingales) are nonincreasing (increasing) in time. Despite its constant trend function, a martingale need not be a stationary process. Conditions (10.1) to (10.3) are obviously equivalent to

$$E(X_{n+1} - X_n | X_n = x_n, ..., X_1 = x_1, X_0 = x_0) = 0, \tag{10.5}$$

$$E(X_{n+1} - X_n | X_n = x_n, ..., X_1 = x_1, X_0 = x_0) \leq 0, \tag{10.6}$$

$$E(X_{n+1} - X_n | X_n = x_n, ..., X_1 = x_1, X_0 = x_0) \geq 0. \tag{10.7}$$

In particular, a stochastic process $\{X_0, X_1, ...\}$ with finite absolute first moments $E(|X_n|)$, $n = 0, 1, ...$ is a martingale if and only if it satisfies condition (10.5).

Since (10.1) is assumed to be true for all vectors $(x_0, x_1, ..., x_n)$ with $x_i \in \mathbf{Z}$, another, equivalent definition of a martingale is

$$E(X_{n+1} | X_n, ..., X_1, X_0) = X_n \ \text{ or } \ E(X_{n+1} - X_n | X_n, ..., X_1, X_0) = 0, \tag{10.8}$$

where the conditional (random) mean values are defined by formula (3.62) with $k = n$ and $Y = X_{n+1}$. The relations in (10.8) mean that they are true with probability 1. This definition applies analogously to super- and submartingales. From (10.8),

$$E(X_{n+2} | X_n, ..., X_1, X_0) = E[E(X_{n+2} | X_{n+1}, ..., X_1, X_0) | X_n, ..., X_1, X_0)]$$

$$= EX_{n+1} | X_n, ..., X_1, X_0) = X_n.$$

From this one gets by induction: $X_0, X_1, ..., X_n$ is a martingale if and only if for all positve integers m

$$E(X_{n+m} | X_n, ..., X_1, X_0) = X_n,$$

or, equivalently

$$E(X_{n+m} | X_n = x_n, ..., X_1 = x_1, X_0 = x_0) = x_n \ \text{ for all } \ (x_0, x_1, ..., x_n) \text{ with } x_i \in \mathbf{Z}.$$

If $\{X_0, X_1, ...\}$ is a martingale and X_n is interpreted as the random fortune of a gambler at time n, then, on condition $X_n = x_n$, the conditional mean fortune of the gambler at time $n + 1$ is also x_n, and this is independent on the development in time of the fortune of the gambler before n (*fair game* with regard both to the gambler and its opponent).

Example 10.1 (*sum martingale*) Let $\{Y_0, Y_1, ...\}$ be a sequence of independent random variables with $E(|Y_n|) < \infty$ for $n = 0, 1, 2, ...$ and $E(Y_i) = 0$ for $n = 1, 2, ...$ Then the sequence $\{X_0, X_1, ...\}$ defined by $X_n = Y_0 + Y_1 + \cdots + Y_n; \quad n = 0, 1, ...$ is a martingale. The proof is easily established:

$$E(X_{n+1}|X_n = x_n, ..., X_1 = x_1, X_0 = x_0)$$
$$= E(X_n + Y_{n+1}|X_n = x_n, ..., X_1 = x_1, X_0 = x_0)$$
$$= x_n + E(Y_{n+1}) = x_n.$$

The sum martingale $\{X_0, X_1, ...\}$ can be interpreted as a random walk on the real axis: X_n is the position of a particle after its nth jump or, in other words, its position at time n. The constant trend function of this martingale is

$$m = E(X_n) = E(Y_0); \quad n = 0, 1, ...$$ □

Example 10.2 (*product martingale*) Let $\{Y_0, Y_1, ...\}$ be a sequence of independent, positive random variables with $E(Y_0) < \infty$, $\mu = E(Y_i) < \infty$ for $i = 1, 2, ...,$ and

$$X_n = Y_0 Y_1 \cdots Y_n.$$

Then, for $n = 1, 2, ...$, since $X_{n+1} = X_n Y_{n+1}$,

$$E(X_{n+1}|X_n = x_n, ..., X_1 = x_1, X_0 = x_0)$$
$$= E(X_n Y_{n+1}|X_n = x_n, ..., X_1 = x_1, X_0 = x_0)$$
$$= x_n E(Y_{n+1}|X_n = x_n, ..., X_1 = x_1, X_0 = x_0)$$
$$= x_n E(Y_{n+1}) = x_n \mu.$$

Thus, $\{X_0, X_1, ...\}$ is a supermartingale for $\mu \le 1$ and a submartingale for $\mu \ge 1$. For $\mu = 1$, the random sequence $\{X_0, X_1, ...\}$ is a martingale with trend function

$$m = E(X_n) = E(Y_0), \quad n = 0, 1, ...$$

This martingale seems to be a realistic model for describing the development in time of share prices or other risky assets or derivates from these (for terminology see section 11.5.5.2) since, from historical experience, the share at a time point in the future is usually proportional to the current price. With this interpretation, $Y_n - 1$ is the relative change in the share price over the interval $[n, n + 1]$ with regard to X_n:

$$\frac{X_{n+1} - X_n}{X_n} = Y_n - 1; \quad n = 0, 1, ...$$ □

A further specification of the factors Y_i within the product martingale yields an exponential type martingale, which is considered in the following example.

Note For notational convenience, in this chapter (super-, sub-) martingales are sometimes denoted as $\{X_1, X_2, ...\}$ instead of $\{X_0, X_1, ...\}$.

Example 10.3 (*exponential martingale*) A special case of the product martingale is the *exponential martingale*. Let $\{Z_1, Z_2, ...\}$ be a sequence of independent, identically as Z distributed random variables, and θ be a real number with $w(\theta) = E(e^{\theta Z}) < \infty$.

A sequence of random variables $\{Y_1, Y_2, ...\}$ be defined as

$$Y_n = Z_1 + \cdots + Z_n; \quad n = 1, 2, ... \tag{10.9}$$

Then the sequence of random variables $\{X_1, X_2, ...\}$ with

$$X_n = \frac{e^{\theta Z_1}}{w(\theta)} \cdot \frac{e^{\theta Z_2}}{w(\theta)} \cdot \dots \cdot \frac{e^{\theta Z_n}}{w(\theta)} = \frac{e^{\theta Y_n}}{[w(\theta)]^n}; \quad n = 1, 2, ... \tag{10.10}$$

is a martingale. This follows immediately from example 10.2, since the factors $\frac{e^{\theta Z_i}}{w(\theta)}$ in (10.10) are independent and have mean value 1:

$$E\left(\frac{e^{\theta Z_i}}{w(\theta)}\right) = \frac{E(e^{\theta Z})}{w(\theta)} = \frac{w(\theta)}{w(\theta)} = 1.$$

In view of its structure, $\{X_1, X_2, ...\}$ is called an *exponential martingale*. If a parameter $\theta = \theta_0$ exists with $w(\theta_0) = 1$, then the exponential martingale simplifies to

$$\{X_1 = e^{\theta_0 Y_1}, X_2 = e^{\theta_0 Y_2}, ...\}. \qquad \square$$

Important special cases of the exponential martingale are:

1) Geometric Random Walk Let Z be a binary random variable with distribution

$$Z = \begin{cases} +1 & \text{with probability } p \\ -1 & \text{with probability } q \end{cases}, \quad q = 1 - p \neq 1/2,$$

then $\{Y_1, Y_2, ...\}$ given by (10.9) can be interpreted as a random walk, which starts at $Y_0 = 0$, and proceeds with steps of size 1 to the right or to the left, each with probabilities p and q, respectively, $0 < p < 1$. The sequence $\{e^{\theta Y_1}, e^{\theta Y_2}, ...\}$ is called a *geometric random walk*. In this case,

$$w(\theta) = E(e^{\theta Z}) = p e^{\theta} + q e^{-\theta}.$$

The geometric random walk is a martingale if $\theta = \ln [q/p]$ since then $w(\theta) = 1$, and the corresponding exponential martingale $\{X_1, X_2, ...\}$ has the structure $X_n = [q/p]^{Y_n}$ with trend function $m(n) = E(X_n) = 1$, $n = 1, 2, ...$

2) Discrete Black-Scholes Model A favorite model for describing the development of share prices, which are sampled at discrete time points 1, 2,..., is

$$X_n = S_1 \cdot S_2 \cdots S_n,$$

with $S_i = e^{Z_i}$ and independent, identically as $Z = N(\mu, \sigma^2)$ distributed Z_i, $i = 1, 2, ...$ S_i has a logarithmic normal distribution with parameters μ and σ^2 (page 84) and mean value $E(S_i) = e^{\mu + \sigma^2/2}$. Thus, $\{X_1, X_2, ...\}$ is a martingale iff $\mu = -\sigma^2/2$.

Example 10.4 (*branching process*) Consider the Galton-Watson branching process as introduced at page 370: Each member of the n th generation, $n = 0, 1, ...$, produces independently of each other a random number Y of offspring with mean value μ. Let X_{n+1} be the random number of offspring produced by the n th generation. Given $X_n = x_n$, the random variable X_{n+1} is independent of $X_0, X_1, ..., X_{n-1}$. Therefore,

$$E(X_{n+1} | X_n = x_n, ..., X_1 = x_1, X_0 = x_0) = \mu x_n. \qquad (10.12)$$

Hence, $\{X_0, X_1, ...\}$ is a martingale if $\mu = 1$, a supermartingale if $\mu \le 1$, and a submartingale if $\mu \ge 1$. Moreover, for any positive μ, the sequence $\{Z_0, Z_1, ...\}$ with $Z_n = X_n / \mu^n$ is a martingale. This can be verified as follows:

$$E(Z_{n+1} | Z_n = z_n, ..., Z_1 = z_1, Z_0 = z_0)$$

$$= E\left(\frac{X_{n+1}}{\mu^{n+1}} \,\middle|\, \frac{X_n}{\mu^n} = \frac{x_n}{\mu^n}, ..., \frac{X_1}{\mu^1} = \frac{x_1}{\mu^1}, \frac{X_0}{\mu^0} = \frac{x_0}{\mu^0}\right)$$

$$= \frac{1}{\mu^{n+1}} E(X_{n+1} | X_n = x_n, ..., X_1 = x_1, X_0 = x_0)$$

$$= \frac{1}{\mu^{n+1}} \mu x_n = \frac{x_n}{\mu^n} = z_n. \qquad \square$$

10.1.2 Doob-Type Martingales

In this section, the concept of a (super-, sub-) martingale $\{X_0, X_1, ...\}$ as introduced in definition 10.1 is generalized by conditioning with regard to another sequence of random variables $\{Y_0, Y_1, ...\}$. This, of course, only makes sense if $\{Y_0, Y_1, ...\}$ is somewhat related to $\{X_0, X_1, ...\}$. The following definition refers to the characterization of (super-, sub-) martingales by properties (10.5) to (10.7).

Definition 10.2 Let $\{X_0, X_1, ...\}$ and $\{Y_0, Y_1, ...\}$ be two discrete-time stochastic processes. If $E(|X_n|) < \infty$ for all $n = 0, 1, ...$, then the random sequence $\{X_0, X_1, ...\}$ is a *martingale with regard to* $\{Y_0, Y_1, ...\}$ or a *Doob-type martingale* if for all $(n + 1)$ -dimensional vectors $(y_0, y_1, ..., y_n)$ with y_i elements of the state space of $\{Y_0, Y_1, ...\}$ and for any $n = 0, 1, ...$,

$$E(X_{n+1} - X_n | Y_n = y_n, ..., Y_1 = y_1, Y_0 = y_0) = 0. \qquad (10.13)$$

Under otherwise the same assumptions, $\{X_0, X_1, ...\}$ is a *supermartingale with regard to* $\{Y_0, Y_1, ...\}$ if

$$E(X_{n+1} - X_n | Y_n = y_n, ..., Y_1 = y_1, Y_0 = y_0) \le 0,$$

and a *submartingale with regard to* $\{Y_0, Y_1, ...\}$ if

$$E(X_{n+1} - X_n | Y_n = y_n, ..., Y_1 = y_1, Y_0 = y_0) \ge 0. \qquad \bullet$$

Remark Most of the literature on martingales is measure-theoretically based. In this case, the definition of a martingale is usually done by means of the concept of a filtration. Loosely speaking, a *filtration* $\mathbf{F}_n$ contains all the information, which is available about the stochastic process $\{X_0, X_1, ...\}$ up to time point n. Generally, since with increasing time n the knowledge about the process increases, $\mathbf{F}_0 \subset \mathbf{F}_1 \subset \mathbf{F}_2 \subset \cdots$. Definition 10.1 uses the *natural filtration* $\mathbf{F}_n = \{X_0 = x_0, X_1 = x_1, ..., X_n = x_n\}$ for characterizing a martingale. Thus, the natural filtration is simply obtained by observing the process $\{X_0, X_1, ...\}$ up to time point n. Formally, $\mathbf{F}_n$ is the smallest σ-algebra generated by the events '$X_i = x_i$,' $i = 1, 2, ..., n$; see page 18. A filtration $\mathbf{F}_n$ may also contain other information than the natural filtration. In particular, in case of Doob-type martingales, our knowledge about the process $\{X_0, X_1, ...\}$ at time point n is given by the filtration $\mathbf{F}_n = \{Y_0 = y_0, Y_1 = y_1, ..., Y_n = y_n\}$. The value of X_n is fully determined by the filtration $\mathbf{F}_n$. In measure-theoretic terms, the random variable X_n is *measurable* with regard to $\mathbf{F}_n$. The random variable X_{n+1}, however, is not measurable with regard to $\mathbf{F}_n$. Thus, the martingale terminology can be unified by making use of the concept of a filtration:

A stochastic process $\{X_0, X_1, ...\}$ with $E(|X_n|) < \infty$ for all $n = 0, 1, ...$ is said to be a martingale with regard to the sequence of filtrations $\{\mathbf{F}_0, \mathbf{F}_1, ...\}$ if

$$E(X_{n+1}|\mathbf{F}_n) = x_n, \quad n = 0, 1, ...$$

Example 10.5 Let Y_i be the random price of a share at time i and S_i be the amount of share an investor holds in the interval $[i, i+1)$; $i = 0, 1, ..., S_i \geq 0$. Thus, at time $t = 0$ the total value of the investor's amount of shares is $X_0 = Y_0 S_0$ and in the interval $[i, i+1)$ the investor makes a 'profit' of $S_i(Y_{i+1} - Y_i)$. Hence, the investor's total profit up to time $t = n$ is

$$X_n = \sum_{i=0}^{n-1} S_i(Y_{i+1} - Y_i); \quad n = 1, 2, ... \tag{10.14}$$

It makes sense to assume that the investor's choice, what amount of share to hold in $[n, n+1)$, does not depend on the profit made in this and later intervals, but only on the profits made in the previous intervals. Hence, S_n is assumed to be fully determined by the $Y_0, Y_1, ..., Y_n$, i.e., the S_n are constant. Under this assumption, the sequence $\{X_1, X_2, ..., \}$ is a supermartingale with regard to $\{Y_0, Y_1, ...\}$ if $\{Y_0, Y_1, ...\}$ is a supermartingale. This is proved as follows:

$$E(X_{n+1} - X_n|Y_n = y_n, ..., Y_1 = y_1, Y_0 = y_0)$$

$$= E(S_n(Y_{n+1} - Y_n)|Y_n = y_n, ..., Y_1 = y_1, Y_0 = y_0)$$

$$= S_n E(Y_{n+1} - Y_n|Y_n = y_n, ..., Y_1 = y_1, Y_0 = y_0) \leq 0.$$

The last line makes use of the assumptions that given '$Y_n = y_n, ..., Y_1 = y_1, Y_0 = y_0$' the share amount S_n is a constant and that $\{Y_0, Y_1, ...\}$ is a supermartingale. Hence, no matter how well-considered the investor fixes the amount of share to be held in an interval, in the longrun she/he cannot expect to make positive profit if the share price develops unfavorably. (A supermartingale has a decreasing trend function.) □

Example 10.6 The structure of X_n given by (10.14) includes as a special case the winnings (losses) development when applying the *doubling strategy*: Jean bets €1 on the first game. If he wins, he gets 1. If he loses, his 'winning' are -1. Hoping to equalize the loss, Jean will bet €2 on the next game. If he wins, he will get €2 and, hence, will have made total winnings of €1. But if he loses he will have total 'winnings' of €-3, and will bet €4 on the next game and so on. After the first win Jean stops gambling. The following table shows the losses (winnings) development of Jean if he loses 5 times in a row and then wins:

game	1	2	3	4	5	6
result	loss	loss	loss	loss	loss	win
bet	1	2	4	8	16	32
'total winnings'	-1	-3	-7	-15	-31	+1

Generally, if Jean loses the first $n-1$ games and wins the nth game, then his bets are

$$S_i = 2^{i-1}, \quad i = 1, 2, ..., n,$$

and at this time point he quits the play with a win of €+1. Hence, at all future time points $n+1, n+2, ...$, Jean's total winnings stay constant at level €+1.

Let $Z_1, Z_2, ...$ be a sequence of independent random variables, identically distributed as Z, which indicate whether Jean has won the ith game or not:

$$Z_i = \begin{cases} +1 & \text{with probability } 1/2 \quad \text{(Jean wins)}, \\ -1 & \text{with probability } 1/2 \quad \text{(Jean loses)}. \end{cases} \tag{10.15}$$

In terms of the Z_i, the *stopping time* N of the play is defined as follows:

$$N = \min_{i=1,2,...} \{i, Z_i = 1\}.$$

Obviously N has the geometrical distribution (2.26) with $p = 1/2$:

$$p_k = P(N = k) = \left(\frac{1}{2}\right)^k, \quad k = 1, 2, ..., \quad \text{and } E(N) = 2.$$

Let X_n be the total winnings of Jean at time point n. To show that $\{X_1, X_2, ...\}$ is a martingale, equation (10.5) has to be verified:

$$E(X_n - X_{n-1} | X_{n-1} = x_{n-1}, ..., X_2 = x_2, X_1 = x_1) = 0. \tag{10.16}$$

Let $N = k$. Then the condition $'X_{n-1} = x_{n-1}, ..., X_2 = x_2, X_1 = x_1'$ in (10.16) can be deleted, since it is fully characterized by k and n. Three cases have to be considered:

1) $n < k$: $X_n = 2^0 Z_1 + 2^1 Z_2 + \cdots + 2^{n-1} Z_n = -1 - 2 - \cdots - 2^{n-1} = 1 - 2^n$.

   (in view of the geometric series (2.16), page 48)

2) $n = k$: $X_n = 2^0 Z_1 + 2^1 Z_2 + \cdots + 2^{n-1} Z_n + 2^n = -1 - 2 - \cdots - 2^{n-1} + 2^n = 1$.

3) $n > k$: $X_n = 1$ for all $n = k+1, k+2, ...$

Therefore,

$$E(X_n - X_{n-1}) = E(X_n - X_{n-1}|N > n)P(N > n) + E(X_n - X_{n-1}|N = n)P(N = n)$$

$$= -2^{n-1} \sum_{i=n+1}^{\infty} \left(\frac{1}{2}\right)^i + 2^{n-1}\left(\frac{1}{2}\right)^n$$

$$= -2^{n-1}\left(\frac{1}{2}\right)^{n+1} \sum_{i=0}^{\infty} \left(\frac{1}{2}\right)^i + \frac{1}{2}$$

$$= -2^{n-1}\left(\frac{1}{2}\right)^{n+1} 2 + \frac{1}{2} = 0,$$

which holds for all $n = 1, 2, \ldots$ (letting $X_0 = 0$). Thus, condition (10.16) is fulfilled so that $\{X_1, X_2, \ldots\}$ is a martingale. Hence, on average, Jean cannot make a profit when applying the doubling strategy. This theoretical result is not intuitive at all, since with increasing n the probability p_n that at least one of the Z_i in a series of n games assumes value 1 is $p_n = 1 - 2^{-n}$, and this probability tends to 1 very fast with increasing n. For being able to maintain the doubling strategy till a win, Jean must, however, have a sufficiently large (theoretically, an unlimited) amount of initial capital, since each bet size 2^i has a positive probability to occur (and the casino must allow arbitrarily large stakes). 'Large' is of course relative in this context, since if Jean starts gambling with an initial capital of €1 and his first bet size is one cent, then he can maximally maintain 6 bets so that his probability of winning one cent is $p_6 \approx 0.984$.

Now let us generalize the doubling strategy by assuming that the Z_i are given by

$$Z = \begin{cases} +1 & \text{with probability } p \\ -1 & \text{with probability } q \end{cases}, \quad q = 1 - p \neq 1/2. \tag{10.17}$$

Then, under otherwise the same assumptions, the mean value of $X_n - X_{n-1}$ becomes

$$E(X_n - X_{n-1}) = -2^{n-1} \sum_{i=n+1}^{\infty} pq^{i-1} + 2^{n-1}pq^{n-1}$$

$$= (2q)^{n-1}(p-q), \quad n = 1, 2, \ldots.$$

Thus, $\{X_1, X_2, \ldots\}$ is a supermartingale for $p \leq 1/2$ and a submartingale for $p \geq 1/2$. Even if $\{X_1, X_2, \ldots\}$ is a supermartingale, Jean can make money with the doubling strategy with any desired probability if his initial capital is large enough.

To establish the relationship of the doubling strategy to the previous example, let us introduce the notation $Y_i = Z_1 + Z_2 + \cdots + Z_i$. Then $Y_i - Y_{i-1} = Z_i$ so that

$$E(Y_i - Y_{i-1}) = p - q.$$

Thus, the sequence $(Y_1, Y_2, \ldots\}$ is a supermartingale if $p \leq 1/2$. (For extensions of this example see exercises 10.4 and 10.6.) $\square$

Example 10.7 At time $t = 0$ a population consists of 2 individuals, one of them is of type 1, the other one of type 2. An individual of type k splits into 2 individuals of type k, $k = 1, 2$. The splitting time is negligibly small. For all individuals, the time to splitting is a finite random variable. These times to splitting need not be identically distributed and/or independent. Let $t_1, t_2, \ldots$ be the sequence of time points, at which splittings occur. $\{t_1, t_2, \ldots\}$ is supposed to be a simple point process (page 255). $\{t_1, t_2, \ldots\}$ becomes a marked point process $\{(t_1, k_1), (t_2, k_2), \ldots\}$, where the marks $k_i = 1$ or $k_i = 2$ indicate whether an individual of type 1 or type 2 has split at time t_i. No deaths are assumed to occur so that we consider a special branching process.

After each splitting event the number of individuals in the population increases by 1. Hence, at time $t_n = t_n + 0$ (i.e., immediately after t_n) the population comprises a total number of $n + 2$ individuals, $n = 1, 2, \ldots$. It is assumed that at any time point each individual has the same probability to split. Let Y_n be the number of individuals of type 1 at time point t_n, $Y_0 = 1$. Then $\{Y_0, Y_1, \ldots\}$ is a nonhomogeneous Markov chain with state space $\{1, 2, \ldots\}$ and transition probabilities

$$p_{ii}(n) = P(Y_{n+1} = i | Y_n = i) = \frac{n + 2 - i}{n + 2},$$

$$p_{ii+1}(n) = P(Y_{n+1} = i | Y_n = i) = \frac{i}{n + 2}.$$

Note that the conditional mean value of Y_{n+1} on condition $Y_n = y_n$ is

$$E(Y_{n+1} | Y_n = y_n) = y_n + 0 \cdot p_{y_n y_n}(n) + 1 \cdot p_{y_n y_n+1}(n) \qquad (10.18)$$

$$= y_n + \frac{y_n}{n + 2}.$$

Now let X_n be the fraction of type 1-individuals in the population at time t_n:

$$X_n = \frac{Y_n}{n + 2}.$$

Then $\{X_0, X_1, \ldots\}$ is a martingale with respect to $\{Y_0, Y_1, \ldots\}$. To prove this, it is to show that

$$E(X_{n+1} | Y_n = y_n, \ldots, Y_1 = y_1, Y_0 = y_0) = x_n.$$

Since $\{Y_0, Y_1, \ldots\}$ is a Markov chain, the condition '$Y_n = y_n, \ldots, Y_1 = y_1, Y_0 = y_0$' can be replaced with '$Y_n = y_n$.' Hence, by (10.18),

$$E(X_{n+1} | Y_n = y_n) = E\left(\frac{Y_{n+1}}{n + 3} \middle| Y_n = y_n\right) = \frac{1}{n + 3} E(Y_{n+1} | Y_n = y_n)$$

$$= \frac{1}{n + 3}\left[y_n + \frac{y_n}{n + 2}\right] = \frac{y_n}{n + 2} = x_n.$$

In the literature, this population model is known as 'Pólya's urn scheme'. $\qquad \square$

Next, under rather strong additional conditions, a criterion is derived, which ensures that a Doob-type martingale is a martingale in the sense of definition 10.1. This derivation is facilitaed by the introduction of a new concept (*Kannan* (1979)).

Definition 10.3 Let $\{Y_0, Y_1, ...\}$ be a discrete-time Markov chain (not necessarily homogeneous) with state space $\mathbf{Z} = \{\cdots, -1, 0, +1, \cdots\}$ and transition probabilities

$$p_n(y, z) = P(Y_{n+1} = z | Y_n = y); \quad y, z \in \mathbf{Z}; \quad n = 0, 1, ...$$

A function $h(y, n); y \in \mathbf{Z}; n = 0, 1, ...$ is said to be *concordant* with $\{Y_0, Y_1, ...\}$ if it satisfies for all $y \in \mathbf{Z}$

$$h(y, n) = \sum_{z \in \mathbf{Z}} p_n(y, z) h(z, n + 1). \tag{10.19}$$

●

Theorem 10.1 Let $\{Y_0, Y_1, ...\}$ be a discrete-time Markov chain with state space

$$\mathbf{Z} = \{\cdots, -1, 0, +1, \cdots\}.$$

Then, for any function $h(y, n)$ which is concordant with $\{Y_0, Y_1, ...\}$,

a) the sequence of random variables $\{X_0, X_1, ...\}$ generated by

$$X_n = h(Y_n, n); \quad n = 0, 1, ...$$

is a martingale with regard to $\{Y_0, Y_1, ...\}$, and

b) the sequence $\{X_0, X_1, ...\}$ is a martingale.

Proof a) By the Markov property and the concordance of h with $\{Y_0, Y_1, ...\}$,

$$E(X_{n+1} - X_n | Y_n = y_n, ..., Y_1 = y_1, Y_0 = y_0)$$

$$= E(X_{n+1} | Y_n = y_n, ..., Y_1 = y_1, Y_0 = y_0) - E(X_n | Y_n = y_n, ..., Y_1 = y_1, Y_0 = y_0)$$

$$= E(h(Y_{n+1}, n+1) | Y_n = y_n) - E(h(Y_n, n) | Y_n = y_n)$$

$$= \sum_{z \in \mathbf{Z}} p_n(y_n, z) h(z, n+1) - h(y_n, n)$$

$$= h(y_n, n) - h(y_n, n) = 0.$$

This result shows that $\{X_0, X_1, ...\}$ is a martingale with regard to $\{Y_0, Y_1, ...\}$.

b) Let, for given $x_0, x_1, ..., x_n$, the random event A be defined as the 'martingale condition' $A = \{X_n = x_n, ..., X_1 = x_1, X_0 = x_0\}$. Since the X_n are fully determined by the $Y_0, Y_1, ..., Y_n$, there exists a set $\mathbf{Y}$ of vectors $\vec{y} = (y_n, ..., y_1, y_0)$ with property that the occurrence of any of the mutually disjoint random events

$$A_{\vec{y}} = \{Y_n = y_n, ..., Y_1 = y_1, Y_0 = y_0\}, \quad \vec{y} \in \mathbf{Y},$$

implies the occurrence of event A:

$$A = \bigcup_{\vec{y} \in \mathbf{Y}} A_{\vec{y}}.$$

Now the martingale property of $\{X_0, X_1, \ldots\}$ is easily established:

$$E(X_{n+1}|A) = \sum_{\vec{y} \in \mathbf{Y}} E\left(X_{n+1}\Big| A_{\vec{y}}\right) \frac{P(A_{\vec{y}})}{P(A)} = h(y_n, n) \sum_{\vec{y} \in \mathbf{Y}} \frac{P(A_{\vec{y}})}{P(A)}$$

$$= h(y_n, n) = x_n.$$

Hence, $\{X_0, X_1, \ldots\}$ is a martingale according to definition 10.1.   ∎

Example 10.8 (variance martingale) Let $\{Z_1, Z_2, \ldots\}$ be a sequence of independent, integer-valued random variables with probability distributions

$$q_i^{(n)} = P(Z_n = i), \quad i \in \mathbf{Z} = \{\cdots, -1, 0, +1, \cdots\},$$

and numerical parameters $E(Z_i) = 0$ and $E(Z_i^2) = \sigma_i^2$; $i = 1, 2, \ldots$.

With an integer-valued constant z_0, a discrete-time Markov chain $\{Y_0, Y_1, \ldots\}$ with state space $\mathbf{Z} = \{\cdots, -1, 0, +1, \cdots\}$ is introduced as $Y_n = z_0 + Z_1 + \cdots + Z_n$. Then,

$$E(Y_n) = z_0 \text{ for } n = 0, 1, \ldots \text{ and } Var(Y_n) = \sum_{i=1}^{n} \sigma_i^2 \text{ for } n = 1, 2, \ldots.$$

The function

$$h(y, n) = y^2 - \sum_{i=1}^{n} \sigma_i^2$$

is concordant with $\{Y_0, Y_1, \ldots\}$. To verify this, let $p_n(y, z)$ be the transition probabilities of $\{Y_0, Y_1, \ldots\}$ at time n. These transition probabilities are fully determined by the probability distribution of Z_{n+1} :

$$p_n(y, z) = P(Y_{n+1} = z | Y_n = y) = P(Z_{n+1} = z - y) = q_{z-y}^{(n+1)}; \quad y, z \in \mathbf{Z}.$$

Therefore,

$$\sum_{z \in \mathbf{Z}} p_n(y, z) h(z, n+1) = \sum_{z \in \mathbf{Z}} q_{z-y}^{(n+1)} h(z, n+1)$$

$$= \sum_{z \in \mathbf{Z}} q_{z-y}^{(n+1)} \left(z^2 - \sum_{i=1}^{n+1} \sigma_i^2\right) = \sum_{z \in \mathbf{Z}} q_{z-y}^{(n+1)} \left[(z - y + y)^2 - \sum_{i=1}^{n+1} \sigma_i^2\right]$$

$$= \sum_{z \in \mathbf{Z}} q_{z-y}^{(n+1)} (z-y)^2 + 2y \sum_{z \in \mathbf{Z}} q_{z-y}^{(n+1)} (z-y) + \sum_{z \in \mathbf{Z}} q_{z-y}^{(n+1)} y^2 - \sum_{i=1}^{n+1} \sigma_i^2$$

$$= \sigma_{n+1}^2 + 2y \cdot 0 + 1 \cdot y^2 - \sum_{i=1}^{n+1} \sigma_i^2 = y^2 - \sum_{i=1}^{n} \sigma_i^2 = h(y, n).$$

Hence, the function $h(y, n)$ is concordant with $\{Y_0, Y_1, \ldots\}$. Thus, by theorem 10.1, the random sequence $\{X_0, X_1, \ldots\}$ with X_n generated by

$$X_n = Y_n^2 - Var(Y_n) \tag{10.20}$$

is a martingale. It is called *variance martingale*.   □

10.1.3 Martingale Stopping Theorem and Applications

As pointed out in the beginning of this chapter, martingales are suitable stochastic models for fair games, i.e., the chances to win or to lose are equal. If one bets on a supermartingale, is it, nevertheless, possible to make money by finishing the game at the 'right time'? The decision, when to finish a game can, of course, only be made on the past development of the martingale (if no other information is available) and not on its future. Hence, a proper time for finishing a game seems to be a stopping time N for $\{X_0, X_1, ...\}$, where X_n is the gambler's net profit after the nth game. According to definition 4.2 (page 195), a stopping time for $\{X_0, X_1, ...\}$ is a positive, integer-valued random variable N with property that the occurrence of the event '$N = n$' is fully determined by the random variables $X_0, X_1, ..., X_n$ and, hence, does not depend on the $X_{n+1}, X_{n+2}, ...$ However, the *martingale stopping theorem* (also called *optional stopping theorem* or *optional sampling theorem*) excludes the possibility of winning in the longrun if finishing the game is controlled by a stopping time (see also examples 10.5 and 10.6).

Theorem 10.2 (*martingale stopping theorem*) Let N be a finite stopping time for the martingale $\{X_0, X_1, ...\}$, i.e. $P(N < \infty) = 1$. Then

$$E(X_N) = E(X_0) \tag{10.21}$$

if at least one of the following three conditions is fulfilled:

1) The stopping time N is bounded, i.e., there exists a finite constant C_1 so that, with probability 1, $N \leq C_1$. (Of course, in this case N is finite.)

2) There exists a finite constant C_2 with

$$\left| X_{\min(N,n)} \right| \leq C_2 \text{ for all } n = 0, 1, ...$$

3) $E(|X_N|) < \infty$ and $\lim_{n \to \infty} E(X_n | N > n) P(N > n) = 0.$ ∎

Remarks 1) When comparing formulas (10.4) and (10.21), note that in (10.21) N is a random variable.

2) Example 10.6 shows that (10.21) is not true for all martingales.

Example 10.9 (*Wald's identity*) Theorem 10.2 implies Wald's identity (4.74) on condition that N with $E(N) < \infty$ is a stopping time for a sequence of independent, identically as Y with $E(Y) < \infty$ distributed random variables $Y_1, Y_2, ... $ To see this, let

$$X_n = \sum_{i=1}^{n} (Y_i - E(Y)); \ n = 1, 2, ...$$

By example 10.1, the sequence $\{X_1, X_2, ...\}$ is a martingale. Therefore, theorem 10.2 is applicable with $E(X_1) = 0$:

$$E(X_N) = E\left(\sum_{i=1}^{N}(Y_i - E(Y))\right)$$

$$= E\left(\sum_{i=1}^{N} Y_i - N E(Y)\right) = E\left(\sum_{i=1}^{N} Y_i\right) - E(N)E(Y) = 0.$$

This proves Wald's identity:

$$E\left(\sum_{i=1}^{N} Y_i\right) = E(N)E(Y). \qquad (10.22)$$

□

Example 10.10 (*fair game*) Let $\{Z_1, Z_2, ...\}$ be a sequence of independent, identically as Z distributed random variables:

$$Z = \begin{cases} +1 & \text{with probability } 1/2 \\ -1 & \text{with probability } 1/2 \end{cases}.$$

Since $E(Z_i) = 0$, the sequence $\{Y_1, Y_2, ...\}$ defined by

$$Y_n = Z_1 + Z_2 + \cdots + Z_n; \ n = 1, 2, ...$$

is a martingale (example 10.1). Y_n is interpreted as the cumulative net profit (loss) of a gambler after the nth play if he bets one dollar on each play. The gambler finishes the game as soon he has won \$$a$ or lost \$$b$. Thus, the game will be finished at time

$$N = \min \{n; \ Y_n = a \text{ or } Y_n = -b\}. \qquad (10.23)$$

Obviously, N is a stopping time for the martingale $\{Y_1, Y_2, ...\}$. Note that this martingale is the symmetric random walk. Since $E(N)$ is finite, by equation (10.21),

$$0 = E(Y_1) = E(Y_N) = a P(Y_N = a) + (-b) P(Y_N = -b).$$

Combining this relationship with

$$P(Y_N = a) + P(Y_N = -b) = 1$$

yields the desired probabilities

$$P(Y_N = a) = \frac{b}{a+b}, \quad P(Y_N = -b) = \frac{a}{a+b}.$$

For determining $E(N)$, the variance martingale $\{X_1, X_2, ...\}$ with

$$X_n = Y_n^2 - Var(Y_n) = Y_n^2 - n$$

is used (formula (10.20)). By theorem 10.2,

$$E(X_1) = E(X_N) = E(Y_N^2) - E(N) = 0.$$

Therefore,

$$E(N) = E(Y_N^2) = a^2 P(Y_N = a) + b^2 P(Y_N = -b).$$

Thus, the mean duration of this fair game is

$$E(N) = a^2 \frac{b}{a+b} + b^2 \frac{a}{a+b} = a b. \qquad □$$

Example 10.11 (*unfair game*) Under otherwise the same assumptions as in the previous example, let

$$Z = \begin{cases} +1 & \text{with probability } p \\ -1 & \text{with probability } q \end{cases}, \quad q = 1 - p \neq 1/2. \tag{10.24}$$

Thus, the win and loss probabilities on a play are different. The mean value of Z_i is

$$E(Z_i) = p - q = 2p - 1.$$

The martingale $\{X_1, X_2, ...\}$ is defined as in the previous example:

$$X_n = \sum_{i=1}^{n} (Z_i - E(Z_i)); \quad n = 1, 2, ...$$

By introducing $Y_n = Z_1 + Z_2 + \cdots + Z_n$, the random variable X_n can be written as

$$X_n = Y_n - (p - q)n.$$

If this martingale is stopped at time N given by (10.23), equation (10.21) yields

$$E(X_N) = E(Y_N) - (p - q) E(N) = E(X_1) = 0, \tag{10.25}$$

or, equivalently,

$$a P(Y_N = a) + (-b) P(Y_N = -b) - (p - q) E(N) = 0.$$

For establishing another equation for the three unknowns

$$P(Y_N = a), \ P(Y_N = -b), \ \text{ and } \ E(N),$$

the exponential martingale (example 10.3) is used. If $\theta = \ln[q/p]$ with $q = 1 - p$, then, as pointed out in example 10.3, $E(e^{\theta Z_i}) = 1$ so that the geometric random walk $\{U_1, U_2, ...\}$ given by

$$U_n = \prod_{i=1}^{n} e^{\theta Z_i} = e^{\theta \sum_{i=1}^{n} Z_i} = e^{\theta Y_n}; \quad n = 1, 2, ...$$

is a martingale. Now, again by applying equation (10.21),

$$1 = E(U_1) = E(U_N) = e^{\theta a} P(Y_N = a) + e^{-\theta b} P(Y_N = -b). \tag{10.26}$$

Equations (10.25) and (10.26) together with $P(Y_N = a) + P(Y_N = -b) = 1$ yield the 'hitting probabilities'

$$P(Y_N = a) = \frac{1 - (p/q)^b}{(q/p)^a - (p/q)^b}, \quad P(Y_N = -b) = \frac{(q/p)^a - 1}{(q/p)^a - (p/q)^b},$$

and the mean duration of the game

$$E(N) = \frac{E(Y_N)}{p - q} = \frac{1}{p - q} \left(\frac{a[1 - (p/q)^b] - b[q/p^a - 1]}{(q/p)^a - (p/q)^b} \right).$$

By letting $n = b$ and $z = a + b$ one gets the result already obtained in example 8.3 (page 346, formula (8.20)) with elementary methods and without worrying whether the assumptions of theorem 10.2 are fulfilled. ◻

10.2 CONTINUOUS-TIME MARTINGALES

This section summarizes some results on continuous-time martingales. For simplicity and with regard to applications to Brownian motion processes in Chapter 11, their parameter space is restricted to $\mathbf{T} = [0, \infty)$, whereas the state space can be the whole real axis $\mathbf{Z} = (-\infty, +\infty)$ or a subset of it.

Definition 10.4 A stochastic process $\{X(t), t \geq 0\}$ with $E(|X(t)|) < \infty$ for all $t \geq 0$ is called a *martingale* if for all integers $n = 0, 1, ...,$ for every sequence $t_0, t_1, ..., t_n$ with $0 \leq t_0 < t_1 < \cdots < t_n$, for all vectors $(x_n, x_{n-1}, ..., x_0)$ with $x_i \in \mathbf{Z}$ and for any $t > t_n$,

$$E(X(t)|X(t_n) = x_n, ..., X(t_1) = x_1, X(t_0) = x_0) = x_n, \qquad (10.27)$$

●

Thus, for predicting the mean value of a martingale at a time t, only the last observation point before t is relevant. The development of the process before t_n contains no additional information with respect to its mean value at a time t, $t > t_n$. Hence, regardless how large the difference $t - t_n$ is, <u>on average</u> no increase/decrease of the process $\{X(t), t \geq 0\}$ can be predicted for the interval $[t_n, t]$.

Analogously to the definition of a discrete-time martingale via (10.8), a continuous-time martingale can be equivalently defined based on the formulas (3.61) and (3.62) at page 147: $\{X(t), t \geq 0\}$ is a continuos-time martingale if, with the notation and assumptions of theorem 10.2,

$$E(X(t)|X(t_n), \cdots, X(t_1), X(t_0)) = X(t_n). \qquad (10.28)$$

This property is frequently written in the more convenient forms

$$E(X(t)|X(y), y \leq s) = X(s), \ s < t, \qquad (10.29)$$

or

$$E(X(t) - X(s)|X(y), y \leq s) = 0, \ s < t. \qquad (10.30)$$

$\{X(t), t \geq 0\}$ is a *supermartingale* (*submartingale*) if in (10.27)–(10.30) the sign '=' is replaced with '≤' ('≥'). The trend function of a continuous-time martingale is constant:

$$m(t) = E(X(t)) \equiv m(0).$$

Example 10.12 Let $\{N(t), t \geq 0\}$ be a homogeneous Poisson process with intensity λ, $\lambda > 0$ (page 261). Then its trend function

$$m(t) = E(N(t)) = \lambda t$$

is increasing so that this process cannot be a martingale. The process $\{X(t), t \geq 0\}$, however, defined by

$$X(t) = N(t) - \lambda t$$

has trend function $m(t) \equiv 0$ and is indeed a martingale: For $s < t$,

$$E(X(t) - X(s)|X(y), y \leq s)$$

$$= E(N(t) - N(s) - \lambda(t-s)|N(y), y \leq s)$$

$$= E(N(t) - N(s)) - \lambda(t-s) = 0.$$

The condition '$N(y), y \leq s$' could be deleted, since the homogeneous Poisson process has independent increments. (Its development in $[0, s]$ has no influence on its development in $(s, t]$.) Of course, not every stochastic process $\{X(t), t \geq 0\}$ of structure $X(t) = Y(t) - E(Y(t))$ is a martingale.   □

Definition 10.5 (*stopping time*) A random variable L is a *stopping time* with regard to an (arbitrary) stochastic process $\{X(t), t \geq 0\}$ if for all $s > 0$ the occurrence of the random event '$L \leq s$' is fully determined by the evolvement of this process to time point s. Therefore, the occurrence of the random event '$L \leq s$' is independent of all $X(t)$ with $t > s$.   ●

Let '$I_{L>t}$' denote the indicator function for the occurrence of the event '$L > t$:'

$$I_{L>t} = \begin{cases} 1 & \text{if } L > t \text{ occurs}, \\ 0 & \text{otherwise} \end{cases}.$$

Theorem 10.3 (*martingale stopping theorem*) If $\{X(t), t \geq 0\}$ is a continuous-time martingale and L a finite stopping time for this martingale, then

$$E(X(L)) = E(X(0)) \tag{10.31}$$

if one of the following two conditions is fulfilled:

1) L is bounded,

2) $E(|X(L)|) < \infty$ and $\lim_{t \to \infty} E(|X(t)|I_{L>t}) = 0$.   ■

The interpretation of this theorem is the same as in case of the martingale stopping theorem for discrete-time martingales. For proofs of theorems 10.2 and 10.3 see, for instance, *Kannan* (1979), *Grimmett, Stirzaker* (2001), or *Rolski et al.* (1999).

Example 10.13 As an application of theorem 10.3, a proof of Lundberg's inequality (7.85) in actuarial risk analysis is given: Let $\{R(t), t \geq 0\}$ be the risk process under the assumptions made at page 294:

$$R(t) = x + \kappa t - C(t),$$

where x is the initial capital, κ the premium rate, and $\{C(t), t \geq 0\}$ the compound claim size process defined by

$$C(t) = \sum_{i=0}^{N(t)} M_i, \quad M_0 = 0,$$

where $\{N(t), t \geq 0\}$ is the homogeneous Poisson process with parameter $\lambda = 1/\mu$.

The claim sizes $M_1, M_2, ...$ are assumed to be independent and identically as M distributed random variables with finite mean $E(M)$ and distribution function and density

$$B(t) = P(M \le t), \quad b(t) = dB(t)/dt, \quad t \ge 0.$$

Let further

$$Y(t) = e^{-rR(t)} \text{ and } h(r) = E(e^{rM}) = \int_0^\infty e^{rt} b(t)dt$$

for any positive r with property $h(r) < \infty$. Then

$$E(Y(t)) = e^{-r(x+\kappa t)} E\left(e^{+rC(t)}\right)$$

$$= e^{-r(x+\kappa t)} \sum_{i=0}^\infty E(e^{+rC(t)}|N(t) = n) P(N(t) = n)$$

$$= e^{-r(x+\kappa t)} \sum_{i=0}^\infty [h(r)]^n \frac{(\lambda t)^n}{n!} e^{-\lambda t}$$

$$= e^{-r(x+\kappa t)} e^{\lambda t [h(r)-1]}.$$

Let

$$X(t) = \frac{Y(t)}{E(Y(t))} = e^{rC(t)-\lambda t [h(r)-1]}.$$

Since $\{C(t), t \ge 0\}$ has independent increments, the process $\{X(t), t \ge 0\}$ has independent increments as well. Hence, for $s < t$, since $E(X(t)) = 1$ for all $t \ge 0$,

$$E(X(t)|X(y), y \le s) = E(X(s) + X(t) - X(s)|X(y), y \le s)$$

$$= X(s) + E(X(t) - X(s)|X(y), y \le s)$$

$$= X(s) + E(X(t) - X(s)) = X(s) + 1 - 1 = X(s).$$

Thus, $\{X(t), t \ge 0\}$ is a martingale. Now, let

$$L = \inf_t \{t, R(t) < 0\}.$$

L is obviously a stopping time for the martingale $\{X(t), t \ge 0\}$. Therefore, for any finite $z > 0$, the truncated random variable $L \wedge z = \min(L, z)$ is a bounded stopping time for $\{X(t), t \ge 0\}$ (exercise 10.13). Hence, theorem 10.3 is applicable with the stopping time $L \wedge z$:

$$E(X(0)) = 1 = E(X(L \wedge z))$$

$$= E(X(L \wedge z|L < z) P(L < z) + E(X(L \wedge z|L \ge z)) P(L \ge z)$$

$$\ge E(X(L \wedge z|L < z) P(L < z)$$

$$= E(X(L |L < z) P(L < z)$$

$$= E(e^{rC(L)-\lambda L [h(r)-1]}| L < z) P(L < z).$$

The definitions of $R(t)$ and L imply $x + \kappa L < C(L)$. Thus, from the first and the last line of this derivation,

$$1 > E(e^{r(x+\kappa L)-\lambda L(h(r)-1)}|L < z)\, P(L < z),$$

or, equivalently,

$$1 > e^{rx} E(e^{[r\kappa - \lambda(h(r)-1)]L}|L < z)\, P(L < z). \tag{10.32}$$

If the parameter r is chosen in such a way that

$$r\kappa - \lambda[h(r) - 1] = 0, \tag{10.33}$$

then inequality (10.32) simplifies to

$$P(L < z) < e^{-rx}.$$

Since this inequality holds for all finite $z > 0$, it follows that

$$P(L < \infty) \le e^{-rx}. \tag{10.34}$$

The probability $P(L < \infty)$ is obviously nothing else but the ruin probability $p(x)$. On the other hand, in view of $\lambda = 1/\mu$, equation (10.33) is equivalent to equation (7.94), which defines the Lundberg coefficient r. When verifying this by partial integration of

$$E(e^{rM}) = \int_0^\infty e^{rx} b(t)dt,$$

note that the assumption $h(r) < \infty$ implies

$$\lim_{t\to\infty} e^{rt}\, \bar{B}(t) = 0.$$

Thus, (10.34) is indeed the Lundberg inequality (7.85) for the ruin probability. $\qquad\square$

10.3 EXERCISES

10.1) Let $Y_0, Y_1, \dots$ be a sequence of independent random variables, which are identically distributed as $N(0, 1)$. Are the stochastic sequences $\{X_0, X_1, \dots\}$ with

(1) $X_n = \sum_{i=0}^n Y_i^2$ (2) $X_n = \sum_{i=0}^n Y_i^3$ (3) $X_n = \sum_{i=0}^n |Y_i|$; $n = 0, 1, \dots$, martingales?

10.2) Let $Y_0, Y_1, \dots$ be a sequence of independent random variables with finite mean values. Show that the discrete-time stochastic process $\{X_0, X_1, \dots\}$ generated by

$$X_n = \sum_{i=0}^n (Y_i - E(Y_i))$$

is a martingale.

10.3) Let a discrete-time stochastic process $\{X_0, X_1, \dots\}$ be defined by

$$X_n = Y_0 \cdot Y_1 \cdot \dots \cdot Y_n,$$

where the random variables Y_i are independent and have a uniform distribution over the interval $[0, T]$. Under which conditions is $\{X_0, X_1, \dots\}$ (1) a martingale, (2) a submartingale, (3) a supermartingale?

10.4) Determine the mean value of the loss immediately before the win when applying the doubling strategy, i.e., determine $E(X_{N-1})$ (example 10.6).

10.5) Why is theorem 10.2 not applicable to the sequence of 'winnings' $\{X_1, X_2, ...\}$, which arises by applying the doubling strategy (example 10.6)?

10.6) Jean is not happy with the winnings he can make when applying the 'doubling strategy'. Hence, under otherwise the same assumptions and notations as in example 10.6, he triples his bet size after every lost game, starting again with €1.

(1) What is his winnings when he loses 5 games in a row and wins the 6th one?

(2) Is $\{X_1, X_2, ...\}$ a martingale?

10.7) Starting at value 0, the profit of an investor increases per week by $1 with probability p, $p > 1/2$, or decreases per week by one unit with probability $1-p$. The weekly increments of the investor's profit are assumed to be independent. Let N be the random number of weeks until the profit reaches for the first time a positive integer n. By means of Wald's equation, determine $E(N)$.

10.8) Starting at value 0, the fortune of an investor increases per week by $200 with probability 3/8, remains constant with probability 3/8, and decreases by $200 with probability 2/8. The weekly increments of the investor's fortune are assumed to be independent. The investor stops the 'game' as soon as he has made a total fortune of $2000 or a loss of $1000, whichever occurs first.

By using suitable martingales and applying the optional stopping theorem, determine

(1) the probability p_{2000} that the investor finishes the 'game' with a profit of $2000,

(2) the probability p_{-1000} that the investor finishes the 'game' with a loss of $1000,

(3) the mean duration $E(N)$ of the 'game.'

10.9) Let X_0 be uniformly distributed over $[0, T]$, X_1 be uniformly distributed over $[0, X_0]$, and, generally, X_{i+1} be uniformly distributed over $[0, X_i]$, $i = 0, 1, ...$.

Verify: The sequence $\{X_0, X_1, ...\}$ is a supermartingale with $E(X_k) = \frac{T}{2^{k+1}}$; $k = 0, 1, ...$.

10.10) Let $\{X_1, X_2, ...\}$ be a homogeneous discrete-time Markov chain with state space $\mathbf{Z} = \{0, 1, ..., n\}$ and transition probabilities

$$p_{ij} = P(X_{k+1} = j | X_k = i) = \binom{n}{j}\left(\frac{i}{n}\right)^j \left(\frac{n-i}{n}\right)^{n-j}; \quad i, j \in \mathbf{Z}.$$

Show that $\{X_1, X_2, ...\}$ is a martingale. (In Genetics, this martingale is known as the *Wright-Fisher model without mutation.*)

10.11) Show that if L is a stopping time for a stochastic process with discrete or continuous time and $0 < z < \infty,$ then

$$L \wedge z = \min(L, z)$$

is a stopping time for this process as well.

10.12) Let $\{N(t), t \geq 0\}$ be a nonhomogeneous Poisson process with intensity function $\lambda(t)$ and trend function

$$\Lambda(t) = \int_0^t \lambda(x)dx.$$

Check whether the stochastic process $\{X(t), t \geq 0\}$ with $X(t) = N(t) - \Lambda(t)$ is a martingale.

10.13) Show that every stochastic process $\{X(t), t \in \mathbf{T}\}$ satisfying

$$E(|X(t)|) < \infty, \ t \in \mathbf{T},$$

which has a constant trend function and independent increments, is a martingale.

10.14)* The ruin problem described in section 7.2.7 is modified in the following way: The risk reserve process $\{R(t), t \geq 0\}$ is only observed at the end of each year (or any other time unit). The total capital of the insurance company at the end of year n is

$$R(n) = x + \kappa n - \sum_{i=0}^{n} M_i; \ \ n = 1, 2, ...,$$

where x is the initial capital, κ is the constant premium income a year, and M_i is the total claim size the insurance company has to cover in year i, $M_0 = 0$. The random variables $M_1, M_2, ...$ are assumed to be independent and identically distributed as

$$M = N(\mu, \sigma^2) \text{ with } \kappa > \mu > 3\sigma.$$

Let $p(x)$ be the ruin probability of the company:

$$p(x) = P(\text{there is an } n = 1, 2, ... \text{ so that } R(n) < 0).$$

Show that

$$p(x) \leq e^{-2(\kappa - \mu)x/\sigma^2}, \ x \geq 0.$$

Hint Define $X_n = e^{-sR(n)}$, $n = 0, 1, ...,$ and select s such that $\{X_0, X_1, ...\}$ is a martingale. Apply theorem 10.2 with the stopping times $L = \min(n, R(n) < 0)$ and $L \wedge z, \ 0 < z < \infty.$

CHAPTER 11

Brownian Motion

11.1 INTRODUCTION

Tiny organic and inorganic particles when immersed in fluids move randomly along zigzag paths. In 1828, the English botanist *Robert Brown* published a paper in which he summarized his observations on this motion and tried to find its physical explanation. Originally, he was only interested in the behaviour of pollen in liquids in order to investigate the fructification process of phanerogams. However, at that time Brown could only speculate on the causes of this phenomenon and was at an early stage of his research even convinced that he had found an elementary form of life, which is common to all particles. Other early explanations refer to attraction and repulsion forces between particles, unstable conditions in the fluids in which they are suspended, capillary actions, and so on. Although the ceaseless, seemingly chaotic zigzag movement of microscopically small particles in fluids had already been observed before Brown, it is generally called *Brownian motion.*

The first approaches to mathematically modeling the Brownian motion were made by *L. Bachelier* (1900) and *A. Einstein* (1905). Both found the normal distribution to be an appropriate model for describing the Brownian motion and gave a physical explanation of the observed phenomenon: The chaotic movement of sufficiently small particles in fluids and in gases is due to the huge number of impacts with the surrounding molecules, even in small time intervals. (Assuming average physical conditions, there are about 10^{21} collisions per second between a particle and the surrounding molecules in a fluid.) More precisely, Einstein showed that water molecules could momentarily form a compact conglomerate which has sufficient energy to move a particle, when banging into it. (Note that the tiny particles are 'giants' compared with a molecule.) These bunches of molecules would hit the 'giant' particles from random directions at random times, causing its apparently irregular zigzag motion. Strangely, Einstein was obviously not aware of the considerable efforts, which had been made before him, to understand the phenomenon 'Brownian motion'. *N. Wiener* (1923), better known as the creator of the science of Cybernetics, was the first to present a general mathematical treatment of the Brownian motion. He defined and analyzed a stochastic process, which has served up till now as a stochastic model of Brownian motion. Henceforth, this process is called *Brownian motion process* or, if no misunderstandings are possible, simply *Brownian motion.* Frequently, mainly in the German literature, this process is also referred to as the *Wiener process.*

Nowadays the enormous importance of the Brownian motion is above all due to the fact that it is one of the most powerful tools in theory and applications of stochastic

modeling, whose role can be compared with that of the normal distribution in probability theory. The Brownian motion process is an essential ingredient into stochastic
calculus, plays a crucial role in mathematics of finance, is basic for defining one of
the most important classes of Markov processes, the *diffusion processes*, and for solving large sample estimation problems in mathematical statistics. Brownian motion
has fruitful applications in key disciplines as time series analysis, operations research,
communication theory (modeling signals and noise), and reliability theory (wear
modeling, accelerated life testing). This chapter only deals with the one-dimensional
Brownian motion.

Definition 11.1 (*Brownian motion*) A continuous-time stochastic process $\{B(t), t \geq 0\}$
with state space $\mathbf{Z} = (-\infty, +\infty)$ is called a (one-dimensional) *Brownian motion (process)* with parameter σ if it has the following properties:

1) $B(0) = 0$.

2) $\{B(t), t \geq 0\}$ has homogeneous and independent increments.

3) $B(t)$ has a normal distribution with

$$E(B(t)) = 0 \text{ and } Var(B(t)) = \sigma^2 t, \quad t > 0.$$ ●

Condition 1, namely $B(0) = 0$, is only a normalization and as an assumption not
really necessary. Actually, in what follows situations will arise in which a Brownian
motion is required to start at $B(0) = u \neq 0$. In such a case, the process retains property 2, but in property 3 assumption $E(B(t)) = 0$ has to be replaced with $E(B(t)) = u$.
The process $\{B_u(t), t \geq 0\}$ with $B_u(t) = u + B(t)$ is called a *shifted Brownian motion*.

In view of properties 2 and 3, the increment $B(t) - B(s)$ has a normal distribution with
mean value 0 and variance $\sigma^2 |t - s|$:

$$B(t) - B(s) = N(0, \sigma^2 |t - s|), \quad s, t \geq 0. \tag{11.1}$$

In applications of the Brownian motion to finance, the parameter σ is called *volatility*. σ^2 is also called *variance parameter* since

$$\sigma^2 = Var(B(1)). \tag{11.2}$$

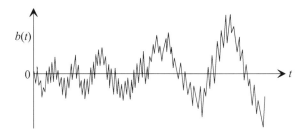

Figure 11.1 Sample path of the Brownian motion

Standard Brownian Motion If $\sigma = 1$, then $\{B(t), t \geq 0\}$ is called a *standard Brownian motion* and will be denoted as $\{S(t), t \geq 0\}$. For any Brownian motion with parameter σ,

$$B(t) = \sigma S(t).$$ (11.3)

Laplace Transform Since $B(t) = N(0, \sigma^2 t)$, because of formula (2.128), page 102, the Laplace transform of $B(t)$ is

$$E\left(e^{-\alpha B(t)}\right) = e^{+\frac{1}{2}\alpha^2 \sigma^2 t}.$$ (11.4)

11.2 PROPERTIES OF THE BROWNIAN MOTION

The first problem, which has to be addressed, is whether there exists a stochastic process having properties 1 to 3. An affirmative answer was already given by N. Wiener in 1923. In what follows, a constructive proof of the existence of the Brownian motion is given. This is done by showing that Brownian motion can be represented as the limit of a discrete-time random walk, where the size of the steps tends to 0 and the number of steps per unit time is speeded up.

Brownian Motion and Random Walk With respect to the physical interpretation of the Brownian motion, it is not surprising that there is a close relationship between Brownian motion and the random walk of a particle along the real axis. Modifying the random walk described in example 8.1, page 342, it is now assumed that after every Δt time units the particle jumps Δx length units to the right or to the left, each with probability 1/2. Thus, if $X(t)$ is the position of the particle at time t, then

$$X(t) = (X_1 + X_2 + \cdots + X_{[t/\Delta t]})\Delta x,$$ (11.5)

where $X(0) = X_0 = 0$ and

$$X_i = \begin{cases} +1 & \text{if the } i\text{th jump goes to the right,} \\ -1 & \text{if the } i\text{th jump goes to the left.} \end{cases}$$

As usual, $[t/\Delta t]$ denotes the greatest integer less than or equal to $t/\Delta t$. The random variables X_i are independent of each other and have probability distribution

$$P(X_i = 1) = P(X_i = -1) = 1/2 \quad \text{with} \quad E(X_i) = 0, \ Var(X_i) = 1.$$

Hence, formula (4.56) at page 187, applied to (11.5), yields

$$E(X(t)) = 0, \quad Var(X(t)) = (\Delta x)^2 [t/\Delta t].$$

With a positive constant σ, let $\Delta x = \sigma \sqrt{\Delta t}$. Then, taking the limit as $\Delta t \to 0$ in (11.5), a stochastic process in continuous time $\{X(t), t \geq 0\}$ arises which has trend and variance function

$$E(X(t)) = 0, \quad Var(X(t)) = \sigma^2 t.$$

Due to its construction, $\{X(t), t \geq 0\}$ has independent and homogeneous increments. Moreover, by the central limit theorem, $X(t)$ has a normal distribution for all $t > 0$. Therefore, the stochastic process of the 'infinitesimal random walk' $\{X(t), t \geq 0\}$ is a Brownian motion.

Even after Norbert Wiener, many amazing properties of the Brownian motion have been detected. Some of them will be considered in this chapter. The following theorem summarizes key properties of the Brownian motion.

Theorem 11.1 A Brownian motion $\{B(t), t \geq 0\}$ has the following properties:

a) $\{B(t), t \geq 0\}$ is mean-square continuous.

b) $\{B(t), t \geq 0\}$ is a martingale.

c) $\{B(t), t \geq 0\}$ is a Markov process.

d) $\{B(t), t \geq 0\}$ is a Gaussian process.

Proof a) From (11.1),

$$E((B(t) - B(s))^2) = Var(B(t) - B(s)) = \sigma^2 |t - s|. \tag{11.6}$$

Hence,

$$\lim_{h \to 0} E\left([B(t+h) - B(t)]^2\right) = \lim_{h \to 0} \sigma^2 |h| = 0.$$

Thus, the limit exists with regard to the convergence in mean-square (page 205).

b) Since a Brownian motion $\{B(t), t \geq 0\}$ has independent increments, for $s < t$,

$$E(B(t)|B(y), y \leq s)) = E(B(s) + B(t) - B(s)| B(y), y \leq s))$$

$$= B(s) + E(B(t) - B(s)|B(y), y \leq s))$$

$$= B(s) + E(B(t) - B(s)) = B(s) + 0 - 0 = B(s).$$

Therefore, $\{B(t), t \geq 0\}$ is a martingale.

c) Any stochastic process $\{X(t), t \geq 0\}$ with independent increments is a Markov process.

d) Let $t_1, t_2, ..., t_n$ be any sequence of real numbers with $0 < t_1 < t_2 < \cdots < t_n < \infty$. It has to be shown that for all $n = 1, 2, ...$ the random vector

$$(B(t_1), B(t_2), ..., B(t_n))$$

has an n-dimensional normal distribution. This is an immediate consequence of theorem 3.3 (page 149), since each $B(t_i)$ can be represented as a sum of independent, normally distributed random variables (increments) in the following way:

$$B(t_i) = B(t_1) + (B(t_2) - B(t_1)) + \cdots + (B(t_i) - B(t_{i-1})); \quad i = 2, 3, ..., n. \qquad \blacksquare$$

Theorem 11.2 Let $\{S(t), t \geq 0\}$ be the standardized Brownian motion. Then, for any constant $\alpha \neq 0$, the stochastic processes $\{Y(t), t \geq 0\}$ defined as follows are martingales:

a) $Y(t) = e^{\alpha S(t) - \alpha^2 t/2}$ (*exponential martingale*),

b) $Y(t) = S^2(t) - t$ (*variance martingale*).

Proof a) For $s < t$,

$$E(e^{\alpha S(t) - \alpha^2 t/2} | S(y), \, y \leq s) = E(e^{\alpha[S(s) + S(t) - S(s)] - \alpha^2 t/2} | S(y), \, y \leq s)$$

$$= e^{\alpha S(s) - \alpha^2 t/2} E(e^{\alpha[S(t) - S(s)]} | S(y), \, y \leq s)$$

$$= e^{\alpha S(s) - \alpha^2 t/2} E\left(e^{\alpha[S(t) - S(s)]}\right).$$

From (11.4) with $\sigma = 1$,

$$E\left(e^{\alpha[S(t) - S(s)]}\right) = e^{+\frac{1}{2}\alpha^2(t-s)}.$$

Hence,

$$E(e^{\alpha S(t) - \alpha^2 t/2} | S(y), \, y \leq s) = e^{\alpha S(s) - \alpha^2 s/2}. \tag{11.7}$$

b) For $s < t$, since $S(s)$ and $S(t) - S(s)$ are independent and $E(S(x)) = 0$ for all $x \geq 0$,

$$E(S^2(t) - t | S(y), \, y \leq s) = E([S(s) + S(t) - S(s)]^2 - t | S(y), \, y \leq s)$$

$$= S^2(s) + E\{2 S(s) [S(t) - S(s)] + [S(t) - S(s)]^2 - t | S(y), \, y \leq s\}$$

$$= S^2(s) + 0 + E\{[S(t) - S(s)]^2\} - t$$

$$= S^2(s) + (t - s) - t$$

$$= S^2(s) - s,$$

which proves the assertion. ∎

There is an obvious analogy between the exponential and the variance martingale defined in theorem 11.2 and corresponding discrete-time martingales considered in examples 10.3 and 10.8.

The relationship (11.7) can be used to generate further martingales: Differentiating (11.7) with regard to α once and twice, respectively, and letting $\alpha = 0$, 'proves' once more that $\{S(t), t \geq 0\}$ and $\{S^2(t) - t, t \geq 0\}$ are martingales. This algorithm produces more martingales by differentiating (11.7) $k = 3, 4, \ldots$ times. For instance, when differentiating (11.7) three and four times, respectively, the resulting martingales are

$$\{S^3(t) - 3 t S(t), t \geq 0\} \quad \text{and} \quad \{S^4(t) - 6 t S^2(t) + 3 t^2, t \geq 0\}.$$

Properties of the Sample Paths Since a Brownian motion is mean-square continuous, it is not surprising that its sample paths $b = b(t)$ are continuous functions in t. More exactly, the probability that a sample path of a Brownian motion is continuous is equal to 1. In view of this, it may come as a surprise that the sample paths of a Brownian motion are nowhere differentiable. This is here not proved either, but it is made plausible by using (11.6): For any sample path $b = b(t)$ and any sufficiently small, but positive Δt, the difference

$$\Delta b = b(t + \Delta t) - b(t)$$

is approximately equal to $\sigma \sqrt{\Delta t}$. Therefore,

$$\frac{\Delta b}{\Delta t} = \frac{b(t + \Delta t) - b(t)}{\Delta t} \approx \frac{\sigma \sqrt{\Delta t}}{\Delta t} = \frac{\sigma}{\sqrt{\Delta t}} .$$

Hence, for $\Delta t \to 0$, the difference quotient $\Delta b / \Delta t$ is likely to tend to infinity for any nonnegative t. Thus, it can be anticipated that the sample paths of a Brownian motion are nowhere differentiable; for proofs see, e.g., *Kannan* (1979). Another example for a continuous function, which is nowhere differentiable, is given in *Gelbaum and Olmstead* (1990).

The *variation* of a sample path (as well as of any real function) $b = b(t)$ in an interval $[0, \tau]$ with $\tau > 0$ is defined as the limit

$$\lim_{n \to \infty} \sum_{k=1}^{2^n} \left| b\left(\frac{k\tau}{2^n}\right) - b\left(\frac{(k-1)\tau}{2^n}\right) \right| . \tag{11.8}$$

A consequence of the non-differentiability of the sample paths is that this limit, no matter how small τ is, cannot be finite. Hence, any sample path of the Brownian motion is of *unbounded variation*. This property in its turn implies that the 'length' of a sample path over the finite interval $[0, \tau]$, and, hence, over any finite interval $[s, t]$ with $s < t$, is infinite. What geometric structure is such a sample path supposed to have? The most intuitive explanation is that the sample paths of any Brownian motion are strongly dentate (in the sense of the structure of leaves), but this structure must continue to the infinitesimal. This explanation corresponds to the physical interpretation of the Brownian motion: The numerous and rapid bombardments of particles in liquids or gases by the surrounding molecules cannot lead to a smooth sample path.

Unfortunately, the unbounded variation of the sample paths implies that particles move with an infinitely large velocity when dispersed in liquids or gases. Hence, the Brownian motion process cannot be a mathematically exact model for describing the movement of particles in these media. But it is definitely a good approximation. (For modeling the velocity of particles in liquids or gases the Ornstein-Uhlenbeck process has been developed; see page 511). However, as pointed out in the introduction, nowadays the enormous theoretical and practical importance of the Brownian motion within the theory of stochastic processes and their applications goes far beyond its being a mathematical model for describing the movement of microscopically small particles in liquids or gases.

11.3 MULTIDIMENSIONAL AND CONDITIONAL DISTRIBUTIONS

Let $\{B(t), t \geq 0\}$ be a Brownian motion and $f_t(x)$ the density of $B(t), t > 0$. From property 3 of definition 11.1,

$$f_t(x) = \frac{1}{\sqrt{2\pi t}\,\sigma}\, e^{-\frac{x^2}{2\sigma^2 t}}, \quad t > 0. \tag{11.9}$$

Since the Brownian motion is a Gaussian process, its multidimensional distributions are multidimensional normal distributions. To determine the parameters of this distribution, next the joint density $f_{s,t}(x_1, x_2)$ of $(B(s), B(t))$ will be derived.

Because of the independence of the increments of the Brownian motion and in view of $B(t) - B(s)$ having probability density $f_{t-s}(x)$, for small Δx_1 and Δx_2,

$$f_{s,t}(x_1, x_2)\Delta x_1 \Delta x_2 = P(x_1 \leq B(s) \leq x_1 + \Delta x_1, x_2 \leq B(t) \leq x_2 + \Delta x_2)$$

$$= P(x_1 \leq B(s) \leq x_1 + \Delta x_1, \ x_2 - x_1 \leq B(t) - B(s) \leq x_2 - x_1 + \Delta x_2 - \Delta x_1)$$

$$= f_s(x_1)\, f_{t-s}(x_2 - x_1)\Delta x_1 \Delta x_2.$$

Hence,

$$f_{s,t}(x_1, x_2) = f_s(x_1) f_{t-s}(x_2 - x_1). \tag{11.10}$$

(This derivation can easily be made rigorously.) Substituting (11.9) into (11.10) yields after some simple algebra

$$f_{s,t}(x_1, x_2) = \frac{1}{2\pi\sigma^2\sqrt{s(t-s)}}\, \exp\left\{-\frac{1}{2\sigma^2 s(t-s)}\left(tx_1^2 - 2s x_1 x_2 + s x_2^2\right)\right\}. \tag{11.11}$$

Comparing this density with the density of the bivariate normal distribution (3.24) at page 131 shows that the random vector $\{B(s), B(t)\}$ has a joint normal distribution with correlation coefficient

$$\rho = +\sqrt{s/t}, \quad 0 < s < t.$$

Therefore, the covariance function of the Brownian motion is

$$C(s, t) = Cov(B(s), B(t)) = \sigma^2 s, \ 0 < s < t. \tag{11.12}$$

In view of the independence of the increments of the Brownian motion, it is easier to directly determine the covariance function of $\{B(t), t \geq 0\}$: For $0 < s \leq t$,

$$C(s, t) = Cov(B(s), B(t)) = Cov(B(s), B(s) + B(t) - B(s))$$

$$= Cov(B(s), B(s)) + Cov(B(s), B(t) - B(s))$$

$$= Cov(B(s), B(s)) = \sigma^2 s.$$

Since the roles of s and t can be changed, for any s and t,

$$C(s, t) = \sigma^2 \min(s, t).$$

Let $0 < s < t$. By formula (3.19), page 128, the conditional density of $B(s)$ on condition $B(t) = b$ is

$$f_{B(s)}(x|B(t) = b) = \frac{f_{s,t}(x,b)}{f_t(b)}. \qquad (11.13)$$

Substituting (11.9) and (11.11) into (11.13) or by immediately making use of formula (3.25) at page 131,

$$f_{X(s)}(x|B(t) = b) = \frac{1}{\sqrt{2\pi \frac{s}{t}(t-s)}\,\sigma} \exp\left\{-\frac{1}{2\sigma^2 \frac{s}{t}(t-s)}\left(x - \frac{s}{t}b\right)^2\right\}. \qquad (11.14)$$

This is the density of a normally distributed random variable with parameters

$$E(B(s)|B(t) = b) = \frac{s}{t}b, \quad Var(B(s)|B(t) = b) = \sigma^2 \frac{s}{t}(t-s). \qquad (11.15)$$

For fixed t, one easily verifies that $Var(B(s)|B(t) = b)$ assumes its maximum at $s = t/2$.

Let $f_{t_1,t_2,\ldots,t_n}(x_1,x_2,\ldots,x_n)$ be the n-dimensional density of the random vector

$$(B(t_1), B(t_2), \ldots, B(t_n)) \quad \text{with} \quad 0 < t_1 < t_2 < \cdots < t_n < \infty.$$

From (11.10), by induction,

$$f_{t_1,t_2,\ldots,t_n}(x_1,x_2,\ldots,x_n) = f_{t_1}(x_1)f_{t_2-t_1}(x_2-x_1)\cdots f_{t_n-t_{n-1}}(x_n-x_{n-1}).$$

With $f_t(x)$ given by (11.9), the n-dimensional joint density becomes

$$f_{t_1,t_2,\ldots,t_n}(x_1,x_2,\ldots,x_n) \qquad (11.16)$$

$$= \frac{\exp\left\{-\frac{1}{2\sigma^2}\left[\frac{x_1^2}{t_1} + \frac{(x_2-x_1)^2}{t_2-t_1} + \cdots + \frac{(x_n-x_{n-1})^2}{t_n-t_{n-1}}\right]\right\}}{(2\pi)^{n/2}\,\sigma^n\,\sqrt{t_1(t_2-t_1)\cdots(t_n-t_{n-1})}}.$$

Transforming this density analogously to the two-dimensional case shows that (11.16) has the form (3.66), page 148. This proves once more that the Brownian motion is a Gaussian process.

The Brownian motion, as any Gaussian process, is completely determined by its trend and covariance function. Actually, since the trend function of a Brownian motion is identically zero, the Brownian motion is completely characterized by its covariance function. In other words, given σ^2, there is exactly one Brownian motion process with covariance function

$$C(s,t) = \sigma^2 \min(s,t).$$

Example 11.1 (*Brownian bridge*) The *Brownian bridge* $\{\bar{B}(t), t \in [0,1]\}$ is a stochastic process, which is given by the Brownian motion in the interval $[0,1]$ on condition that $B(1) = 0$:

$$\bar{B}(t) = B(t), \ 0 \le t \le 1; \ B(1) = 0.$$

Letting in (11.14) $b = 0$ and $t = 1$ yields the probability density of $\bar{B}(t)$:

$$f_{\bar{B}(t)}(x) = \frac{1}{\sqrt{2\pi t(1-t)}\,\sigma} \exp\left\{-\frac{x^2}{2\sigma^2 t(1-t)}\right\}, \quad 0 < t < 1.$$

Mean value and variance of $\bar{B}(t)$ are for $0 \le t \le 1$:

$$E(\bar{B}(t)) = 0,$$

$$Var(\bar{B}(t)) = \sigma^2 t(1-t).$$

The two-dimensional probability density of the random vector $(\bar{B}(s), \bar{B}(t))$ can be obtained from

$$f_{t_1,t_2}(x_1, x_2) = \frac{f_{t_1,t_2,t_3}(x_1, x_2, 0)}{f_{t_3}(0)}$$

with $t_1 = s$, $t_2 = t$, and $t_3 = 1$. Hence, for $0 < s < t < 1$, the joint density of the random vector $(\bar{B}(s), \bar{B}(t))$ is

$$f_{(\bar{B}(s),\bar{B}(t))}(x_1, x_2)$$

$$= \frac{\exp\left\{-\frac{1}{2\sigma^2}\left[\frac{t}{s(t-s)}x_1^2 - \frac{2}{t-s}x_1x_2 + \frac{1-s}{(t-s)(1-t)}x_2^2\right]\right\}}{2\pi\sigma^2\sqrt{s(t-s)(1-t)}}.$$

A comparison with (3.24), page 131, shows that the correlation and the covariance function of the Brownian bridge are

$$\rho(s,t) = \sqrt{\frac{s(1-t)}{t(1-s)}}, \quad 0 < s < t \le 1,$$

$$C(s,t) = \sigma^2 s(1-t), \quad 0 < s < t \le 1.$$

The Brownian bridge is a Gaussian process whose trend function is identically 0. Hence, it is uniquely determined by its covariance function.

The *geometric Brownian bridge* is defined as the stochastic process $\{Y(t), t \ge 0\}$ with

$$Y(t) = e^{\bar{B}(t)}, \ 0 \le t \le 1.$$

Both the Brownian bridge and the geometric Brownian bridge have some significance in modelling stochastically fluctuating parameters in mathematics of finance. □

11.4 FIRST PASSAGE TIMES

By definition, the Brownian motion $\{B(t),\ t \geq 0\}$ starts at $B(0) = 0$. The random time point $L(x)$, at which the process $\{B(t), t \geq 0\}$ reaches a given level x for the first time, is called the *first passage time* or the *first hitting time* of $\{B(t),\ t \geq 0\}$ with respect to level x. Since the sample paths of the Brownian motion are continuous functions, $L(x)$ is uniquely characterized by $B(L(x)) = x$ and can, therefore, be defined as

$$L(x) = \min_{t} \{t,\ B(t) = x\}, \quad x \in (-\infty, +\infty).$$

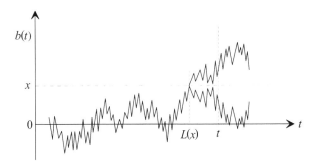

Figure 11.2 Illustration of the first passage time and the reflection principle

Next the probability distribution of $L(x)$ is derived on condition $x > 0$: Application of the total probability rule yields

$$P(B(t) \geq x) = P(B(t) \geq x | L(x) \leq t)\,P(L(x) \leq t) \tag{11.17}$$
$$+ P(B(t) \geq x | L(x) > t)\,P(L(x) > t).$$

The second term on the right-hand side of this formula vanishes, since, by definition of the first passage time,

$$P(B(t) \geq x | L(x) > t) = 0$$

for all $t > 0$. For symmetry reasons and in view of $B(L(x)) = x$,

$$P(B(t) \geq x | L(x) \leq t) = 1/2. \tag{11.18}$$

This situation is illustrated in Figure 11.2: Two sample paths of the Brownian motion, which coincide up to reaching level x and which after $L(x)$ are mirror symmetric with respect to the straight line $b(t) \equiv x$, have the same chance of occurring. (The probability of this event is, nonetheless, zero.) This heuristic argument is known as the *reflection principle*. Formulas (11.9), (11.17), and (11.18) yield

$$F_{L(x)}(t) = P(L(x) \leq t) = 2\,P(B(t) \geq x) = \frac{2}{\sqrt{2\pi t}\,\sigma} \int_{x}^{\infty} e^{-\frac{u^2}{2\sigma^2 t}}\,du. \tag{11.19}$$

For symmetry reasons, the probability distributions of the first passage times $L(x)$ and $L(-x)$ are identical for any x. Therefore, from (11.19),

$$F_{L(x)}(t) = \frac{2}{\sqrt{2\pi t}\,\sigma} \int_{|x|}^{\infty} e^{-\frac{u^2}{2\sigma^2 t}}\,du, \quad t > 0. \tag{11.20}$$

The relationship of the probability distribution of $L(x)$ to the normal distribution becomes visible when substituting $y^2 = u^2/(\sigma^2 t)$ in the integral of (11.20):

$$F_{L(x)}(t) = 2\left[1 - \Phi\left(\frac{|x|}{\sigma\sqrt{t}}\right)\right], \quad t > 0, \tag{11.21}$$

where as usual $\Phi(\cdot)$ is the distribution function of a standard normal random variable. Differentiation of (11.20) with respect to t yields the probability density of $L(x)$:

$$f_{L(x)}(t) = \frac{|x|}{\sqrt{2\pi}\,\sigma\,t^{3/2}} \exp\left\{-\frac{x^2}{2\sigma^2 t}\right\}, \quad t > 0. \tag{11.22}$$

Mean value $E(L(x))$ and variance $Var(L(x))$ do not exist, i.e., they are infinite.

The probability distribution determined by (11.21) or (11.22), respectively, is a special case of the *inverse Gaussian distribution* (page 513).

Maximum Let $M(t)$ be the maximal value the Brownian motion assumes in $[0, t]$:

$$M(t) = \max\{B(s),\ 0 \le s \le t\}. \tag{11.23}$$

The probability distribution of $M(t)$ is obtained as follows:

$$1 - F_{M(t)}(x) = P(M(t) \ge x) = P(L(x) \le t).$$

Hence, by (11.21), the distribution function of $M(t)$ is

$$F_{M(t)}(x) = 2\,\Phi\left(\frac{x}{\sigma\sqrt{t}}\right) - 1; \quad t > 0,\ x > 0, \tag{11.24}$$

The density of $M(t)$ one obtains by differentiation of (11.24) with regard to t:

$$f_{M(t)}(x) = \frac{2}{\sqrt{2\pi t}\,\sigma}\,e^{-x^2/(2\sigma^2 t)}; \quad t > 0,\ x > 0. \tag{11.25}$$

As a consequence from (11.24): For all finite x,

$$\lim_{t \to \infty} P(M(t) < x) = 0. \tag{11.26}$$

Therefore, with probability 1, $\lim_{t \to \infty} M(t) = \infty$. The unbounded growth of $M(t)$ is due to the linearly increasing variance $Var(B(t)) = \sigma^2 t$ of the Brownian motion as $t \to \infty$.

Contrary to the Brownian motion, its corresponding 'maximum process' $\{M(t),\ t \ge 0\}$ has nondecreasing sample paths. This process has applications among others in financial modeling and in reliability engineering (accelerated life testing, wear modeling).

Example 11.2 A sensor for measuring high temperatures gives an unbiased indication of the true temperature during its operating time. At the start, the measurement is absolutely correct. In the course of time, its accuracy deteriorates, but no systematic errors occur. Let $B(t)$ be the random deviation of the temperature indicated by the sensor at time t from the true temperature. Historical observations justify the assumption that $\{B(t), t \geq 0\}$ is a Brownian motion with parameter

$$\sigma = \sqrt{Var(B(1))} = 0.1 \ [^0C/day].$$

What is the probability that within a year (365 days) $B(t)$ exceeds the critical level $x = +5^0C$, i.e. the sensor reads at least once in a year 5^0C high? This probability is

$$F_{L(5)}(365) = P(L(5) < 365) = 2\left[1 - \Phi\left(\frac{5}{0.1\sqrt{365}}\right)\right]$$

$$= 2[1 - \Phi(2.617)] = 0.009.$$

If the accuracy of the sensor is allowed to exceed the critical value of $+5^0C$ with probability 0.05 during its operating time, then the sensor has to be exchanged by a new one after a time $t_{0.05}$ given by $P(L(-5) \leq t_{0.05}) = 0.05$. According to (11.21), $t_{0.05}$ satisfies equation

$$2\left[1 - \Phi\left(\frac{5}{0.1\sqrt{t_{0.05}}}\right)\right] = 0.05$$

or

$$\frac{5}{0.1\sqrt{t_{0.05}}} = \Phi^{-1}(0.975) = 1.96.$$

Thus, $t_{0.05} = 651$ [days]. ☐

The next example considers a first passage time problem with regard to the Brownian motion leaving an interval.

Example 11.3 Let $L(a, b)$ be the random time at which $\{B(t), t \geq 0\}$ for the first time hits either value a or value b:

$$L(a, b) = \min_{t} \{t, B(t) = a \text{ or } B(t) = b\}; \quad b < 0 < a.$$

Then the probability $p_{a,b}$ that $\{B(t), t \geq 0\}$ assumes value a before value b is

$$p_{a,b} = P(L(a) < L(b)) = P(L(a, b) = L(a))$$

(Figure 11.3) or, equivalently,

$$p_{a,b} = P(B(L(a, b)) = a).$$

To determine $p_{a,b}$, note that $L(a, b)$ is a stopping time for $\{B(t), t \geq 0\}$. In view of formula (11.24), $E(L(a, b))$ is finite. Hence, theorem 10.3 is applicable and yields

$$0 = E(B(L(a, b))) = a p_{a,b} + b(1 - p_{a,b}).$$

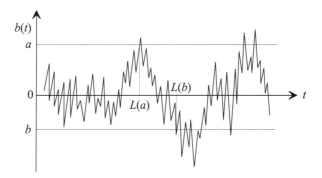

Figure 11.3 First passage times with regard to an interval

Therefore, the probability that the Brownian motion hits value a before value b is

$$p_{a,b} = \frac{|b|}{a+|b|}.$$ (11.27)

For determining the mean value of $L(a,b)$, the martingale $\{Y(t),\ t \geq 0\}$ with

$$Y(t) = \frac{1}{\sigma^2} B^2(t) - t$$

is used (theorem 11.2b) with $S(t) = B(t)/\sigma$. In this case, theorem 10.3 yields

$$0 = E\left(\frac{1}{\sigma^2} B^2(L(a,b))\right) - E(L(a,b)).$$

Hence,

$$E(L(a,b)) = E\left(\frac{1}{\sigma^2} B^2(L(a,b))\right) = \frac{1}{\sigma^2}\left[p_{a,b}\,a^2 + (1-p_{a,b})\,b^2\right]$$

so that, by (11.27),

$$E(L) = \frac{1}{\sigma^2}\,a\,|b|.$$ (11.28)

As an application of the situation considered in this example, assume that the total profit, which a speculator makes with a certain investment, develops according to a Brownian motion process $\{B(t),\ t \geq 0\}$, i.e., $B(t)$ is the cumulative 'profit', the speculator has achieved at time t (possibly negative). If the speculator stops investing after having achieved a profit of a or after having suffered a loss of b, then $p_{a,b}$ is the probability that he finishes with a profit of a.

With reference to example 11.2: The probability that the sensor reads $8^0 C$ high before it reads $4^0 C$ low is $4/(8+4) = 1/3$. Or, if in the same example the tolerance region for $B(t)$ is $[-5\,^0 C,\ 5\,C^0]$, then $B(t)$ on average leaves this region for the first time at time

$$E(L) = 25/0.01 = 2500 \ \text{[days]}.$$ □

11.5 TRANSFORMATIONS OF THE BROWNIAN MOTION

11.5.1 Identical Transformations

Transforming the Brownian motion leads to stochastic processes which are important in their own right, both from the theoretical and the practical point of view. Some transformations again lead to the Brownian motion. Theorem 11.3 compiles three transformations of this type (see *Lawler* (2006)).

Theorem 11.3 If $\{S(t),\ t \ge 0\}$ is the standard Brownian motion, then each of the following stochastic processes is also the standard Brownian motion:

(1) $\{X(t),\ t \ge 0\}$ with $X(t) = c\,S(t/c^2),\ c > 0,$

(2) $\{Y(t),\ t \ge 0\}$ with $Y(t) = S(t+h) - S(h),\ h > 0,$

(3) $\{Z(t),\ t \ge 0\}$ with $Z(t) = \begin{cases} t\,S(1/t) & \text{for } t > 0 \\ 0 & \text{for } t = 0 \end{cases}.$

Proof The theorem is proved by verifying properties 1) to 3) of definition 11.1. The processes (1) to (3) start at the origin: $X(0) = Y(0) = Z(0) = 0.$ Since the Brownian motion has independent, normally distributed increments, the processes (1) to (3) have the same property. Their trend functions are identically zero. Hence, it remains to show that the increments of the processes (1) to (3) are homogeneous. In view of (11.1), it suffices to prove that the variances of the increments of the processes (1) to (3) in any interval $[s, t]$ with $s < t$ are equal to $t - s.$ The following derivations make use of $E(S^2(t)) = t$ and formula (11.12).

(1)
$$Var(X(t) - X(s)) = E([X(t) - X(s)]^2)$$
$$= E(X^2(t)) - 2Cov(X(s), X(t)) + E(X^2(s))$$
$$= c^2\{E(S^2(t/c^2)) - 2Cov[S(s/c^2), S^2(t/c^2)] + E(S^2(s/c^2)\}$$
$$= c^2\left(\frac{t}{c^2} - 2\frac{s}{c^2} + \frac{s}{c^2}\right) = t - s.$$

(2)
$$Var(Y(t) - Y(s)) = E([S(t+h) - S(s+h)]^2)$$
$$= E\{S^2(t+h)) - 2\,Cov[S(s+h)\,S(t+h)] + E(S^2(s+h)\}$$
$$= (t+h) - 2(s+h) + (s+h) = t - s.$$

(3)
$$Var(Z(t) - Z(s)) = E\left([t\,S(1/t) - s\,S(1/s)]^2\right)$$
$$= t^2 E(S^2(1/t)) - 2\,s\,t\,Cov[S(1/s), S(1/t)] + s^2 E(S^2(1/s))$$
$$= t^2 \cdot \frac{1}{t} - 2\,s\,t \cdot \frac{1}{t} + s^2 \cdot \frac{1}{s} = t - s. \qquad \blacksquare$$

For any Brownian motion $\{B(t),\, t \geq 0\}$ (see, e.g., *Lawler* (2006)):

$$P\left(\lim_{t \to \infty} \frac{1}{t} B(t) = 0 \right) = 1. \tag{11.29}$$

If t is replaced with $1/t$, then taking the limit as $t \to \infty$ is equivalent to taking the limit as $t \to 0$. Hence,

$$P\left(\lim_{t \to 0} t B(1/t) = 0 \right) = 1. \tag{11.30}$$

A consequence of (11.29) is that any Brownian motion $\{B(t),\, t \geq 0\}$ crosses the t-axis with probability 1 at least once in the interval $[s, \infty)$, $s > 0$, and, hence, even countably infinite times. Since

$$\{t B(1/t),\, t \geq 0\}$$

is also a Brownian motion, it must have the same property. Therefore, for any $s > 0$, no matter how small s is, a Brownian motion $\{B(t), t \geq 0\}$ crosses the t-axis in $(0, s]$ countably infinite times with probability 1.

11.5.2 Reflected Brownian Motion

A stochastic process $\{X(t),\, t \geq 0\}$ defined by $X(t) = |B(t)|$ is called a *reflected Brownian motion* (reflected at the t-axis). Its trend and variance function are

$$m(t) = E(X(t)) = \frac{2}{\sqrt{2\pi t}\,\sigma} \int_0^\infty x e^{-\frac{x^2}{2\sigma^2 t}} dx = \sigma \sqrt{\frac{2t}{\pi}}, \quad t \geq 0,$$

$$Var(X(t)) = E(X^2(t)) - [E(X(t))]^2 = \sigma^2 t - \sigma^2 \frac{2t}{\pi} = (1 - 2/\pi)\sigma^2 t.$$

The reflected Brownian motion is a homogeneous Markov process with state space $\mathbf{Z} = [0, \infty)$. This can be seen as follows: For

$$0 \leq t_1 < t_2 < \cdots < t_n < \infty, \quad x_i \in \mathbf{Z},$$

taking into account the Markov property of the Brownian motion and its symmetric stochastic evolvement with regard to the t-axis,

$$P(X(t) \leq y | X(t_1) = x_1, X(t_2) = x_2, ..., X(t_n) = x_n)$$

$$= P(-y \leq B(t) \leq +y | B(t_1) = \pm x_1, B(t_2) = \pm x_2, ..., B(t_n) = \pm x_n)$$

$$= P(-y \leq B(t) \leq +y | B(t_n) = \pm x_n)$$

$$= P(-y \leq B(t) \leq +y | B(t_n) = x_n).$$

Hence, for $0 \leq s < t$, the transition probabilities

$$P(X(t) \leq y | X(s) = x)$$

of the reflected Brownian motion are determined by the increment of the Brownian motion in $[s, t]$ if it starts at time s at state x. Because this increment has an $N(x, \sigma^2\tau)$-distribution with $\tau = t - s$,

$$P(X(t) \le y | X(s) = x) = \frac{1}{\sqrt{2\pi\tau}\,\sigma} \int_{-y}^{y} e^{-\frac{(u-x)^2}{2\sigma^2\tau}} \, du,$$

or equivalently by

$$P(X(t) \le y | X(s) = x) = \Phi\left(\frac{y-x}{\sigma\sqrt{\tau}}\right) + \Phi\left(\frac{y+x}{\sigma\sqrt{\tau}}\right) - 1; \quad x, y \ge 0; \ \tau = t - s.$$

Since the transition probabilities depend on s and t only via $\tau = t - s$, the reflected Brownian motion is a homogeneous Markov process.

11.5.3 Geometric Brownian Motion

A stochastic process $\{X(t), t \ge 0\}$ with

$$X(t) = e^{B(t)} \tag{11.31}$$

is called *geometric Brownian motion*.

Unlike the Brownian motion, the sample paths of a geometric Brownian motion cannot become negative. Therefore and for analytical convenience, the geometric Brownian motion is a favourite tool in mathematics of finance for modeling share prices, interest rates, and so on.

According to (11.4), the Laplace transform of $B(t)$ is

$$\hat{B}(\alpha) = E(e^{-\alpha B(t)}) = e^{+\frac{1}{2}\alpha^2\sigma^2 t}. \tag{11.32}$$

Substituting in (11.32) the parameter α with a positive integer n yields the moments of $X(t)$:

$$E(X^n(t)) = e^{+\frac{1}{2}n^2\sigma^2 t}; \quad n = 1, 2, \dots. \tag{11.33}$$

In particular, mean value and second moment of $X(t)$ are

$$E(X(t)) = e^{+\frac{1}{2}\sigma^2 t}, \quad E(X^2(t)) = e^{+2\sigma^2 t}. \tag{11.34}$$

From (11.34) and (1.19):

$$Var(X(t)) = e^{t\sigma^2}(e^{t\sigma^2} - 1).$$

Although the trend function of the Brownian motion is constant, the trend function of the geometric Brownian motion is increasing:

$$m(t) = E(X(t)) = e^{\sigma^2 t/2}, \quad t \ge 0. \tag{11.35}$$

11.5.4 Ornstein-Uhlenbeck Process

As mentioned before, if the Brownian motion process would be the absolutely correct model for describing the movements of particles in liquids or gases, the particles had to move with an infinitely large velocity. To overcome this unrealistic assumption, *Uhlenbeck and Ornstein* (1930) developed a stochastic process for modeling the velocity of tiny particles in liquids and gases. Now this process is used as a stochastic model in other applications as well, e.g., in finance and population dynamics.

Definition 11.2 Let $\{B(t), t \geq 0\}$ be a Brownian motion with parameter σ. Then the stochastic process $\{U(t), t \in (-\infty, +\infty)\}$ defined by

$$U(t) = e^{-\alpha t} B(e^{2\alpha t}) \tag{11.36}$$

is said to be an *Ornstein-Uhlenbeck process* with parameters σ and α, $\alpha > 0$. ●

Thus, the stationary Ornstein-Uhlenbeck process arises from the nonstationary Brownian motion by time transformation and standardization.

The density of $U(t)$ is easily derived from (11.9):

$$f_{U(t)}(x) = \frac{1}{\sqrt{2\pi}\,\sigma} e^{-x^2/(2\sigma^2)}, \quad -\infty < x < \infty.$$

Thus, $U(t)$ has a normal distribution with parameters

$$E(U(t)) = 0, \quad Var(U(t)) = \sigma^2.$$

Hence, the trend function of the Ornstein-Uhlenbeck process is identically zero, and $U(t)$ is standard normal if $\{B(t), t \geq 0\}$ is the standard Brownian motion.

Since $\{B(t), t \geq 0\}$ is a Gaussian process, the Ornstein-Uhlenbeck process has the same property. (This is a corollary from theorem 3.3, page 149.) Hence, the multi-dimensional distributions of the Ornstein-Uhlenbeck process are multidimensional normal distributions. Moreover, there is a unique correspondence between the sample paths of the Brownian motion and the sample paths of the corresponding Ornstein-Uhlenbeck process. Thus, the Ornstein-Uhlenbeck process, as the Brownian motion, is a Markov process. The covariance function of the Ornstein-Uhlenbeck procss is

$$C(s, t) = \sigma^2 e^{-\alpha(t-s)}, \quad s \leq t. \tag{11.37}$$

This can be seen as follows: For $s \leq t$, in view of (11.12),

$$C(s, t) = Cov(U(s), U(t)) = E(U(s)U(t))$$

$$= e^{-\alpha(s+t)} E(B(e^{2\alpha s}) B(e^{2\alpha t}))$$

$$= e^{-\alpha(s+t)} Cov(B(e^{2\alpha s}), B(e^{2\alpha t})) = e^{-\alpha(s+t)} \sigma^2 e^{2\alpha s}$$

$$= \sigma^2 e^{-\alpha(t-s)}.$$

Corollary The Ornstein-Uhlenbeck process is weakly stationary. As a Gaussian process, it is also strongly stationary.

In contrast to the Brownian motion, the Ornstein-Uhlenbeck process has the following properties:

1) The increments of the Ornstein-Uhlenbeck process are not independent.

2) The Ornstein-Uhlenbeck process is mean-square differentiable.

11.5.5 Brownian Motion with Drift

11.5.5.1 Definitions and First Passage Times

Definition 11.3 A stochastic process $\{D(t), t \geq 0\}$ is called *Brownian motion with drift* if it has the following properties:

1) $D(0) = 0$,

2) $\{D(t), t \geq 0\}$ has homogeneous, independent increments,

3) Every increment $D(t) - D(s)$ has a normal distribution with mean value $\mu(t-s)$ and variance $\sigma^2|t-s|$. ●

An equivalent definition of the Brownian motion with drift is:

$\{D(t), t \geq 0\}$ is a Brownian motion with drift if and only if $D(t)$ has structure

$$D(t) = \mu t + B(t), \tag{11.38}$$

where $\{B(t), t \geq 0\}$ is the Brownian motion with variance parameter σ^2. The constant μ is called *drift parameter* or simply *drift*. Thus, a Brownian motion with drift arises by superimposing a Brownian motion on a deterministic function. This deterministic function is a straight line and coincides with the trend function of the Brownian motion with drift:

$$m(t) = E(D(t)) = \mu t.$$

If properties 2) and 3) are fulfilled, but the process starts at time $t = 0$ at level u, $u \neq 0$, then the resulting stochastic process $\{D_u(t), t \geq 0\}$ is called a *shifted Brownian motion with drift*. $D_u(t)$ has structure

$$D_u(t) = u + D(t).$$

The one-dimensional density functions of the Brownian motion with drift are

$$f_{D(t)}(x) = \frac{1}{\sqrt{2\pi t}\,\sigma} e^{-\frac{(x-\mu t)^2}{2\sigma^2 t}} \; ; \quad -\infty < x < \infty, \; t > 0. \tag{11.39}$$

Brownian motion processes with drift are, amongst other applications, used for modeling financial parameters, productivity criteria, cumulative maintenance costs, wear modeling as well as for modeling physical noise. Generally speaking, the Brownian motion with drift can successfully be applied to modeling situations in which causally

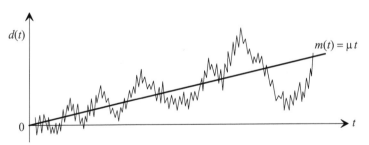

Figure 11.4 Sample path of a Brownian motion with positive drift

linear processes are permanently disturbed by random influences. In view of these applications it is not surprising that first passage times of Brownian motions with drift play an important role both with respect to theory and practice.

Let $L(x)$ be the first passage time of $\{D(t),\ t \ge 0\}$ with regard to level x. Then,

$$L(x) = \min_t\ \{t,\ D(t) = x\},\quad x \in (-\infty, +\infty).$$

Since every Brownian motion with drift has independent increments and is a Gaussian process, the following relationship between the probability densities of $D(t)$ and $L(x)$ holds (*Franz (1977)*):

$$f_{L(x)}(t) = \frac{x}{t} f_{D(t)}(x),\quad x > 0,\ \mu > 0.$$

Hence, the probability density of $L(x)$ is

$$f_{L(x)}(t) = \frac{x}{\sqrt{2\pi}\,\sigma\, t^{3/2}}\ \exp\left\{-\frac{(x-\mu t)^2}{2\sigma^2 t}\right\},\quad t > 0. \tag{11.40}$$

(See also *Scheike (1992)* for a direct proof of this result.) For symmetry reasons, the probability density of the first passage time $L(x)$ of a Brownian motion with drift starting at u can be obtained from (11.40) by replacing x there with $x - u$.

The probability distribution given by the density (11.40) is the *inverse Gaussian distribution* with parameters μ, σ^2, and x. (Replace in (2.89), page 85, the parameters α with x^2/σ^2 and β with $1/\mu$ to obtain (11.40)). Contrary to the first passage time of the Brownian motion, now mean value and variance of $L(x)$ exist:

$$E(L(x)) = \frac{x}{\mu},\quad Var(L(x)) = \frac{x\sigma^2}{\mu^3};\quad \mu > 0. \tag{11.41}$$

For $\mu = 0$, the density (11.40) simplifies to the first passage time density (11.20) of the Brownian motion. If $x < 0$ and $\mu < 0$, formula (11.40) yields the density of the corresponding first passage time $L(x)$ by substituting $|x|$ and $|\mu|$ for x and μ, respectively.

Let

$$F_{L(x)}(t) = P(L(x) \le t)\ \text{ and }\ \overline{F}_{L(x)}(t) = 1 - F_{L(x)}(t),\quad t \ge 0.$$

Assuming $x > 0$ and $\mu > 0$, integration of (11.40) yields

$$\bar{F}_{L(x)}(t) = \Phi\left(\frac{x - \mu t}{\sqrt{t}\,\sigma}\right) - e^{-2x\mu}\,\Phi\left(-\frac{x + \mu t}{\sqrt{t}\,\sigma}\right), \quad t > 0. \qquad (11.42)$$

If the second term on the right-hand side of (11.42) is sufficiently small, then one obtains an interesting result: The Birnbaum-Saunders distribution (7.159) at page 330 as a limit distribution of first passage times of compound renewal processes approximately coincides with the inverse Gaussian distribution.

After some tedious algebra, the Laplace transform of $f_{L(x)}(t)$ is seen to be

$$E\left(e^{-sL(x)}\right) = \int_0^\infty e^{-st} f_{L(x)}(t)\,dt = \exp\left\{-\frac{x}{\sigma^2}\left(\sqrt{2\sigma^2 s + \mu^2} - \mu\right)\right\}. \qquad (11.43)$$

Theorem 11.4 Let M be the absolute maximum of the Brownian motion with drift on the positive semiaxis $(0, \infty)$:

$$M = \max_{t \in (0,\infty)} D(t).$$

Then, for any positive x,

$$P(M > x) = \begin{cases} 1 & \text{for } \mu > 0, \\ e^{-2|\mu|x/\sigma^2} & \text{for } \mu < 0. \end{cases} \qquad (11.44)$$

Proof In view of (11.26), it is sufficient to prove (11.44) for $\mu < 0$. The exponential martingale $\{Y(t), t \geq 0\}$ with $Y(t) = e^{\alpha S(t) - \alpha^2 t/2}$ (theorem 11.2) is stopped at time $L(x)$. In view of

$$D(L(x)) = \mu L(x) + \sigma S(L(x)) = x,$$

the random variable $Y(L(x))$ can be represented as

$$Y(L(x)) = \exp\left\{\tfrac{\alpha}{\sigma}[x - \mu L(x)] - \alpha^2 L(x)/2\right\} = \exp\left\{\tfrac{\alpha}{\sigma} x - \left[\tfrac{\alpha\mu}{\sigma} + \alpha^2/2\right]L(x)\right\}.$$

Hence,

$$E(Y(L(x))) = e^{\frac{\alpha}{\sigma}x} E\left(\exp\left\{\tfrac{\alpha|\mu|}{\sigma} - \alpha^2/2\right\}L(x)\Big| L(x) < \infty\right) P(L(x) < \infty)$$

$$+ e^{\frac{\alpha}{\sigma}x} E\left(\exp\left\{\tfrac{\alpha|\mu|}{\sigma} - \alpha^2/2\right\}L(x)\Big| L(x) = \infty\right) P(L(x) = \infty).$$

Let $\alpha > 2|\mu|/\sigma$. Then the second term disappears and theorem 10.3 yields

$$1 = e^{\frac{\alpha}{\sigma}x} E\left(\exp\left\{\tfrac{\alpha|\mu|}{\sigma} - \alpha^2/2\right\}L(x)\,\Big|\, L(x) < \infty\right) P(L(x) < \infty).$$

Since $P(M > x) = P(L(x) < \infty)$, letting $\alpha \downarrow 2|\mu|/\sigma$ yields the desired result. ∎

Corollary The maximal value, which a Brownian motion with negative drift assumes in $(0, +\infty)$, has an exponential distribution with parameter

$$\lambda = \frac{2|\mu|}{\sigma^2}. \tag{11.45}$$

Example 11.4 (*Leaving an interval*) Analogously to example 11.3, let $L(a, b)$ be the first time point at which the Brownian motion with drift $\{D(t), t \geq 0\}$ hits either value a or value b, $b < 0 < a$.

Let $p_{a,b}$ be the probability that $\{D(t), t \geq 0\}$ hits level a before level b given $\mu \neq 0$:

$$p_{a,b} = P(L(a) < L(b)) = P(L(a, b) = a).$$

For establishing an equation in $p_{a,b}$, the exponential martingale in theorem 11.2 with

$$S(t) = (D(t) - \mu t)/\sigma$$

is stopped at time $L(a, b)$. From theorem 10.3,

$$1 = E\left(\exp\left\{ \frac{\alpha}{\sigma} (D(L(a,b)) - \mu\, L(a,b)) - \frac{\alpha^2 L(a,b)}{2} \right\} \right).$$

Equivalently,

$$1 = E\left(\exp\left\{ \frac{\alpha}{\sigma} (D(L(a,b))) - \left[\frac{\alpha\mu}{\sigma} + \frac{\alpha^2}{2} \right] L(a,b) \right\} \right).$$

Let $\alpha = -2\mu/\sigma$. Then,

$$1 = E\left(e^{\frac{\alpha}{\sigma}(D(L(a,b)))} \right) = p_{a,b} e^{-2\mu a/\sigma^2} + (1 - p_{a,b}) e^{-2\mu b/\sigma^2}.$$

Solving this equation for $p_{a,b}$ yields

$$p_{a,b} = \frac{1 - e^{-2\mu b/\sigma^2}}{e^{-2\mu a/\sigma^2} - e^{-2\mu b/\sigma^2}}. \tag{11.46}$$

If $\mu < 0$ and b tends to $-\infty$ in (11.46), then the probability $p_{a,b}$ becomes $P(L(a) < \infty)$, which proves once more formula (7.44) with $x = a$.

Generally, for a shifted Brownian motion with drift $\{D_u(t), t \geq 0\}$,

$$D_u(t) = u + D(t), \quad b < u < a, \quad \mu \neq 0,$$

formula (11.46) yields the corresponding probability $p_{a,b}(u)$ by replacing there a and b with $a - u$ and $b - u$, respectively (u can be negative):

$$p_{a,b}(u) = P(L(a) < L(b) | D_u(0)) = \frac{e^{-2\mu u/\sigma^2} - e^{-2\mu b/\sigma^2}}{e^{-2\mu a/\sigma^2} - e^{-2\mu b/\sigma^2}}. \qquad \square$$

Geometric Brownian Motion with Drift Let $\{D(t), t \geq 0\}$ be a Brownian motion with drift. Then the stochastic process $\{X(t), t \geq 0\}$ with

$$X(t) = e^{D(t)} \tag{11.47}$$

is called *geometric Brownian motion with drift.* If the drift μ is 0, then $\{X(t), t \geq 0\}$ is simply the *geometric Brownian motion* as defined by (11.31).

The Laplace transform of $D(t)$ is obtained by multiplying (11.4) by $e^{-t\mu\alpha}$:

$$E\left(e^{-\alpha D(t)}\right) = e^{-t\mu\alpha + \frac{1}{2}\sigma^2 t \alpha^2}. \tag{11.48}$$

Letting respective $\alpha = -1$ and $\alpha = -2$ yields the first and the second moment of $X(t)$:

$$E(X(t)) = e^{t(\mu+\sigma^2/2)}, \quad E(X^2(t)) = e^{2t\mu + 2\sigma^2 t}. \tag{11.49}$$

Therefore, by formula (2.62), page 67,

$$Var(X(t)) = e^{t(2\mu+\sigma^2)}(e^{t\sigma^2} - 1).$$

Since the inequalities $e^{D(t)} \geq x$ and $D(t) \geq \ln x$ are equivalent, the first passage time results obtained for the Brownian motion with drift can immediately be used for characterizing the first passage time behavior of the geometric Brownian motion with drift with level $\ln x$ instead of x, $x > 0$.

11.5.5.2 Application to Option Pricing

In financial modeling, Brownian motion and its transformations are used to describe the evolvement in time of prices of risky assets as shares, precious metals, crops, and combinations of them. *Derivatives* are financial products, which derive their values from one or more risky assets. Options belong to the most popular derivatives. An *option* is a contract, which entitles (but not obliges) its *holder* (*owner*) to either buy or sell a risky asset at a fixed, predetermined price, called *strike price* or *exercise price*. A *call* (*put*) *option* gives its holder the right to buy (to sell). An option has a finite or an infinite *expiration* or *maturity date*, which is determined by the contract. An *American option* can be exercised at any time point to its expiration; a *European option* can only be exercised at the time point of its expiration. So one can expect that an American option with finite expiration time τ is more expensive than a European option with the same expiration time if they are based on the same risky assets.

A basic problem in option trading is: What amount of money should a speculator pay to the *writer* (seller) of an option at the time of signing the contract to become holder of the option? Common sense tells that the writer will fix the option price at a level, which is somewhat higher than the mean payoff (profit) the speculator will achieve by acquiring this option. Hence, the following examples focus on calculating the mean (expected) payoff of a holder. For instance, if a European call option has the finite expiration date τ, strike price x_s, and the random price (value) of the underlying risky asset at time τ is $X(\tau)$, then the holder will achieve a positive random payoff of $X(\tau) - x_s$ if $X(\tau) > x_s$. If $X(\tau) \leq x_s$, then the owner will not exercise. This would make no financial sense, since in addition to the price the holder had to pay for accquiring the option, he/she would suffer a further loss of $x_s - X(\tau)$. In case of a European put option, the owner will exercise at time τ if $X(\tau) < x_s$ and make a random profit of $x_s - X(\tau)$. Thus, owners of European call or put options will achieve the random payoffs (Figure 11.5)

$$\max(X(\tau) - x_s, 0) \text{ and } \max(x_s - X(\tau), 0),$$

respectively. But, to emphasize it once more, 'payoff' in this context is not the net payoff of the holder, the holder's mean net profit is, if the model assumptions are correct, on average zero or even negative, since at the time of signing the contract he/she had to pay a price for becoming a holder.

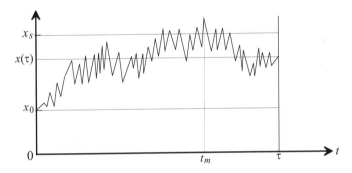

Figure 11.5 Payoff from a European option

Figure 11.5 illustrates the situation for a European option with expiration time τ. The underlying share price (risky asset) starts at the selling time of the option $t = 0$ at value x_0 per unit and ends at value $x(\tau)$. If a holder owns a European call option, he or she would not exercise, but for an owner of an American call option based on the same share there had been opportunities for making a profit (maximum payoff at time t_m). A holder of an European put option would have made a profit of $x_s - x(\tau)$.

Closely related to options is another kind of derivatives called *forward contracts*. A forward contract is an agreement between two parties, say Tom and Huckleberry. At time $t = 0$, Tom declares to buy a risky asset from Huckleberry at time τ for a certain price $Z(\tau)$, called *delivery price*. Huckleberry agrees both with the *maturity date* τ and the delivery price $Z(\tau)$, and they sign the contract. Different to options, Tom <u>must</u> buy at maturity date and Huckeleberry <u>must</u> sell at maturity, and no money changes hands between Tom and Huckleberry when signing the contract at time $t = 0$. If at the time of maturity the price $X(\tau)$ of the risky security exceeds the delivery price $Z(\tau)$, then Tom will win, otherwise Huckleberry. Determining the profit of Tom (Huckleberry) is quite analogous to determining the profit of the holder (price) of a European option. Related to forward contracts are *futures contracts*. They differ from each other mainly by administrative issues.

Another basic aspect in finance is *discounting*. Due to interest and inflation rates, the value, which a certain amount of money has today, will not be value which the same amount of money has tomorrow. In financial calculations, in particular in option pricing, this phenomenon is taken into account by a *discount factor*.

The following examples deal with option pricing under rather simplistic assumptions. For detailed and more general expositions, see, e.g., *Lin* (2006) and *Kijima* (2013).

Example 11.5 The price of a share at time t is given by a shifted Brownian motion $\{X(t) = D_{x_0}(t), t \geq 0\}$ with negative drift μ and volatility $\sigma = \sqrt{Var(B(1))}$:

$$X(t) = x_0 + D(t) = x_0 + \mu t + B(t). \tag{11.50}$$

Thus, x_0 is the initial price of the share: $x_0 = X(0)$. Based on this share, Huckleberry holds an American call option with strike price

$$x_s, \ x_s \geq x_0.$$

The option has no finite expiry date. Although the price of the share is on average decreasing, Huckleberry hopes to profit from random share price fluctuations. He makes up his mind to exercise the option at that time point, when the share price for the first time reaches value x with $x > x_s$. Therefore, if the holder exercises, his payoff will be $x - x_s$ (Figure 11.6). By following this policy, Huckleberry's mean payoff (gain) is

$$G(x) = (x - x_s)p(x) + 0 \cdot (1 - p(x)) = (x - x_s)p(x),$$

where $p(x)$ is the probability that the share price will ever reach level x. Equivalently, $p(x)$ is the probability that the Brownian motion with drift $\{D(t), t \geq 0\}$ will ever reach level $x - x_0$. Since the option has no finite expiration date, this probability is given by (11.44) if there x is replaced with $x - x_0$. Hence, Huckleberry's mean payoff is

$$G(x) = (x - x_s)e^{-\lambda(x - x_0)} \ \text{ with } \ \lambda = 2|\mu|/\sigma^2. \tag{11.51}$$

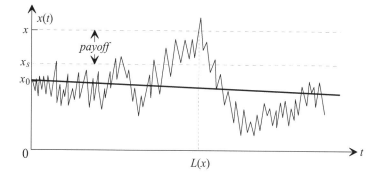

Figure 11.6 Payoff from random share price fluctuations

The condition $dG(x)/dx = 0$ yields the optimal value of x: Huckleberry will exercise as soon as the share price hits level

$$x^* = x_S + 1/\lambda.$$ (11.52)

The corresponding maximal mean payoff is

$$G(x^*) = \frac{1}{\lambda \, e^{\lambda (x_S - x_0) + 1}}.$$ (11.53)

Discounted Payoff Let the constant (risk free) discount rate α be positive. The discounted payoff from exercising the option at time t on condition that the share has at time t price x with $x > x_S$ is

$$e^{-\alpha t}(x - x_S).$$

Since under the policy considered, Huckleberry exercises the option at the random time $L_D(x - x_0)$, which is the first passage time of $\{D(t), t \geq 0\}$ with respect to level $x - x_0)$, his random discounted payoff is

$$e^{-\alpha L_D(x - x_0)}(x - x_S)$$

so that Huckleberry's mean discounted payoff is

$$G_\alpha(x) = (x - x_S)\int_0^\infty e^{-\alpha t} f_{L_D(x - x_0)}(t)\, dt,$$ (11.54)

where the density

$$f_{L_D(x - x_0)}(t)$$

is given by (11.40) with x replaced by $x - x_0$. The integral in (11.54) is equal to the Laplace transform of $f_{L_D(x - x_0)}(t)$ with parameter $s = \alpha$. Thus, from (11.43),

$$G_\alpha(x) = (x - x_S)\exp\left\{-\frac{x - x_0}{\sigma^2}\left(\sqrt{2\sigma^2\alpha + \mu^2} - \mu\right)\right\}.$$ (11.55)

The functional structures of the mean undiscounted payoff and the mean discounted payoff as given by 11.51) and (11.55), respectively, are identical. Hence the optimal parameters with respect to $G_\alpha(x)$ are again given by (11.52) and (11.53) with λ replaced by

$$\gamma = \frac{1}{\sigma^2}\left(\sqrt{2\sigma^2\alpha + \mu^2} - \mu\right).$$ (11.56)

Note that minimizing $G_\alpha(x)$ makes sense also for a positive drift parameter μ.   □

Example 11.6 Since for a negative drift parameter μ the sample paths of a stochastic process $\{X(t), t \geq 0\}$ of structure (11.50) eventually become negative with probability one, the share price model (11.50) has only limited application, in particular in cases of infinite expiration dates. Hence, in such a situation it seems to be more realistic to assume that the share price at time t is, apart from a constant factor, modeled by a geometric Brownian motion with drift:

$$X(t) = x_0 e^{D(t)}, \quad t \geq 0.$$

The other assumptions as well as the formulation of the problem and the notation introduced in the previous example remain valid. In particular, the price of the share at time $t = 0$ is again equal to x_0.

The random event $'X(t) \geq x'$ with $x > x_0$ is equivalent to

$$D(t) \geq \ln(x/x_0).$$

Therefore, by (11.44), the probability that the share price will ever reach level x is

$$p(x) = e^{-\lambda \ln(x/x_0)} = \left(\frac{x_0}{x}\right)^\lambda.$$

If the holder exercises the option as soon as the share price is x, his mean payoff is

$$G(x) = (x - x_s)\left(\frac{x_0}{x}\right)^\lambda. \tag{11.57}$$

The optimal level $x = x^*$ is

$$x^* = \frac{\lambda}{\lambda - 1} x_s. \tag{11.58}$$

To ensure that $x^* > x_s > 0$, an additional assumption has to be made:

$$\lambda = 2|\mu|/\sigma^2 > 1.$$

The corresponding maximal mean payoff is

$$G(x^*) = \left(\frac{\lambda - 1}{x_s}\right)^{\lambda - 1} \left(\frac{x_0}{\lambda}\right)^\lambda. \tag{11.59}$$

Discounted Payoff The undiscounted payoff $x - x_s$ is made when $\{D(t), t \geq 0\}$ hits level $\ln(x/x_0)$ for the first time, i.e., at time

$$L_D(\ln(x/x_0)).$$

Using this and processing as in the previous example, the mean discounted payoff is seen to be

$$G_\alpha(x) = (x - x_s)\left(\frac{x_0}{x}\right)^\gamma \tag{11.60}$$

with γ given by (11.56). The functional forms of the mean undiscounted payoff and (11.57) and (11.60) are identical. Hence, the corresponding optimal values x^* and $G_\alpha(x*)$ are given by (11.58) and (11.59) if in these formulas λ is replaced with γ. Note that condition $\gamma > 1$ is equivalent to

$$2(\alpha - \mu) > \sigma^2.$$

As in the previous example, a positive drift parameter μ need not be excluded.   □

Example 11.7 (*Formula of Black-Scholes-Merton*) A European call option is considered with strike price x_s and expiration date τ. The option is based on a risky asset the price of which, apart from a constant factor x_0, develops according to a geometric Brownian motion with drift $\{X(t),\, t \ge 0\}$:

$$X(t) = x_0\, e^{D(t)}, \quad t \ge 0.$$

The holder will buy if $X(\tau) > x_s$. Then, given a constant discount factor α, his random discounted payoff is

$$[e^{-\alpha\tau}(X(\tau) - x_s)]_+ = \max\,[e^{-\alpha\tau}(X(\tau) - x_s),\, 0].$$

Hence, the holder's mean discounted profit will be

$$G_\alpha(\tau, \mu, \sigma) = E([e^{-\alpha\tau}(X(\tau) - x_s)]_+).$$

In view of $D(\tau) = N(\mu\tau, \sigma^2\tau)$,

$$G_\alpha(\tau; \mu, \sigma) = e^{-\alpha\tau} \int_{\ln(x_s/x_0)}^{\infty} (x_0 e^y - x_s)\, \frac{1}{\sqrt{2\pi\sigma^2\tau}}\, \exp\left\{-\frac{1}{2\tau}\left(\frac{y - \mu\tau}{\sigma}\right)^2\right\} dy.$$

Substituting $u = \dfrac{y - \mu\tau}{\sigma\sqrt{\tau}}$ and letting

$$c = \frac{[\ln(x_s/x_0) - \mu\tau]}{\sigma\sqrt{\tau}}$$

yields

$$G_\alpha(\tau; \mu, \sigma) = x_0\, e^{(\mu-\alpha)\tau}\, \frac{1}{\sqrt{2\pi}} \int_c^{\infty} e^{u\sigma\sqrt{\tau}}\, e^{-u^2/2}\, du - x_s\, e^{-\alpha\tau}\, \frac{1}{\sqrt{2\pi}} \int_c^{\infty} e^{-u^2/2}\, du.$$

By substituting in the first integral $u = y + \sigma\sqrt{\tau}$ one obtains

$$\int_c^{\infty} e^{u\sigma\sqrt{\tau}}\, e^{-u^2/2}\, du = e^{\frac{1}{2}\sigma^2\tau} \int_{c-\sigma\sqrt{\tau}}^{\infty} e^{-y^2/2}\, dy.$$

Hence,

$$G_\alpha(\tau; \mu, \sigma)$$

$$= x_0\, e^{(\mu-\alpha+\sigma^2/2)\tau}\, \frac{1}{\sqrt{2\pi}} \int_{c-\sigma\sqrt{\tau}}^{\infty} e^{-y^2/2}\, dy - x_s\, e^{-\alpha\tau}\, \frac{1}{\sqrt{2\pi}} \int_c^{\infty} e^{-u^2/2}\, du$$

$$= x_0\, e^{(\mu-\alpha+\sigma^2/2)\tau}\, \Phi(\sigma\sqrt{\tau} - c) - x_s\, e^{-\alpha\tau}(\Phi(-c)).$$

At time t, the discounted price of the risky security is

$$X_\alpha(t) = e^{-\alpha t} X(t) = x_0\, e^{-(\alpha-\mu)t + \sigma S(t)},$$

where $\{S(t), t \geq 0\}$ is the standard Brownian motion. By theorem 11.2, the stochastic process $\{X_\alpha(t), t \geq 0\}$ is a martingale (exponential martingale) if $\alpha - \mu = \sigma^2/2$. On this condition, the mean discounted payoff of the holder is given by the *Formula of Black-Scholes-Merton*

$$\widetilde{G}_\alpha(\tau, \sigma) = x_0 \, \Phi(\sigma \sqrt{\tau} - c) - x_s \, e^{-\alpha \tau} \, \Phi(-c) \tag{11.61}$$

(*Black* and *Scholes* (1973), *Merton* (1973)). In this formula, the influence of the drift μ on the price development has been eliminated by the assumption that the discounted price of the risky asset develops according to a martingale. The formula of Black-Scholes-Merton gives the *fair price* of the option. This is partially motivated by the fact that a martingale has a constant trend function so that, on average, holder and writer of this option will neither lose nor win. $\square$

11.5.5.3 Application to Maintenance

In the following example, a functional of the Brownian motion will be used to model the random cumulative repair cost $X(t)$ a technical system causes over a time period $[0, t]$. The following basic situation is considered: A system starts working at time $t = 0$. As soon as $X(t)$ reaches level x, the system is replaced by an equivalent new one in negligibly small time. The cost of each replacement is c, and after a replacement a system is 'as good as new'. With regard to cost and length, all replacement cycles are independent of each other. Scheduling of replacements aims at minimizing the long-run total maintenance cost per unit time, in what follows referred to as *maintenance cost rate*.

Policy 1 The system is replaced by a new one as soon as the cumulative repair cost $X(t)$ reaches a given positive level x.

By the renewal reward theorem, i.e., by formula (7.148), page 325, the corresponding maintenance cost rate is

$$K_1(x) = \frac{x + c}{E(L_X(x))}. \tag{11.62}$$

Policy 1 basically needs the same input as the *economic lifetime policy*, which is introduced next for the sake of comparisons.

Policy 2 The system is replaced by a new one after reaching its *economic lifetime*, which is defined as that value $\tau = \tau^*$, which minimizes the average maintenance cost per unit time $K_2(\tau)$ if the system is always replaced by a new one after τ time units.

Again from the renewal reward theorem, $K_2(\tau)$ is given by

$$K_2(\tau) = \frac{E(X(\tau)) + c}{\tau}. \tag{11.63}$$

In this case a replacement cycle is has the constant length τ.

Example 11.8 The cost of a replacement is $\$10\,000$, and the cumulative repair cost $X(t)$ [in $\$$] has structure

$$X(t) = 0.1\,e^{D(t)}, \tag{11.64}$$

where $\{D(t), t \geq 0\}$ is the Brownian motion with drift parameter $\mu = 0.01\,[h^{-1}]$ and variance parameter $\sigma^2 = 0.0064$, i.e., in terms of the standard Brownian motion,

$$D(t) = 0.01t + 0.08\,S(t).$$

Policy 1 The stochastic repair cost process $\{X(t), t \geq 0\}$ reaches level x at that time point when the Brownian motion with drift $\{D(t), t \geq 0\}$ reaches level $\ln 10x$:

$$X(t) = x \iff D(t) = \ln 10x.$$

Hence, by formula (11.41), the mean value of the first passage time of the process $\{X(t), t \geq 0\}$ with regard to level x is

$$E(L_X(x)) = \frac{1}{0.01}\ln 10x = 100 \ln 10x.$$

The corresponding maintenance cost rate (11.62) is

$$K_1(x) = \frac{x + 10\,000}{100 \ln 10x}.$$

A limit $x = x^*$ minimizing $K_1(x)$ satisfies the necessary condition $dK_1(x)/dx = 0$ or

$$x \ln 10x - x = 10\,000.$$

The unique solution of this equation is (slightly rounded)

$$x^* = 1192.4\ [\$] \quad \text{so that}\quad K_1(x^*) = 11.92\,[\$/h]. \tag{11.65}$$

The mean length of an optimum replacement cycle is

$$E(L_X(x^*)) = 939\,[h].$$

Policy 2 Since by (11.49),

$$E(e^{D(t)}) = e^{(\mu + \sigma^2/2)t} = e^{0.0132\,t},$$

the corresponding maintenance cost rate (11.63) is

$$K_2(\tau) = \frac{10\,000 + 0.1\,e^{0.0132\,\tau}}{\tau}.$$

The optimal values are

$$\tau^* = 712\ [h] \quad \text{and}\quad K_2(\tau^*) = 15.74\,[\$/h]. \tag{11.66}$$

Thus, applying policy 1 instead of policy 2 reduces the maintenance cost rate by about 25%.

At first glance, a disadvantage of modelling repair cost processes by functionals of the Brownian motion is that these functionals generally are not monotone increasing. However, in this example $\{X(t), t \geq 0\}$ hits a level x for the first time at that time point when the Brownian motion with drift $\{D(t), t \geq 0\}$ reaches level $\ln 10x$. In other words, if a replacement cycle is given by the random interval $[0, L_D(y))$, then the processes $\{D(t), t \geq 0\}$ and $\{M(t), t \geq 0\}$ hit a positive level y for the first time at the same time point, namely $L_D(y)$. Hence, replacing $\{D(t), t \geq 0\}$ in the cumulative cost process $\{X(t), t \geq 0\}$ given by (11.64) with the 'maximum process' $\{M(t), t \geq 0\}$ defined by

$$M(t) = \max_{0 \leq y \leq t} D(y),$$

would, with regard to policy 1, yield the same the optimal values x^* and $K_1(x^*)$ as the ones given by formulas (11.65). The sample paths of $\{M(t), t \geq 0\}$ are nondecreasing and therefore, principally suitable for modelling the cumulative evolvement of repair costs. In the light of this it makes sense, and it is actually necessary to apply policy 2 to the cumulative repair cost process

$$X(t) = 0.1\, e^{M(t)}, \quad t \geq 0,$$

and to compare the results to (11.66). The probability distribution of $M(t)$ is given by the distribution of the first passage time $L(x) = L_D(x)$ since $P(L(x) \leq t) = P(M(t) > x))$. Hence, by (11.40)

$$P(M(t) > x) = \int_0^t \frac{x}{0.08\sqrt{2\pi}\, y^{1.5}} e^{-\frac{(x-0.01y)^2}{0.0128y}}\, dy.$$

Making use of formula (2.55), page 64, with $h(x) = e^x$ yields the corresponding maintenance cost rate in the form

$$K_2(\tau|M) = \frac{10\,000 + 0.1 \int_0^\infty x e^x \int_0^\tau \frac{1}{0.08\sqrt{2\pi}\, y^{1.5}} e^{-\frac{(x-0.01y)^2}{0.0128y}}\, dy dx}{\tau}.$$

The optimal values are

$$\tau^* = 696\,[h] \quad \text{and} \quad K_2(\tau^*|M) = 16.112.$$

They are quite close to the ones given by (11.66). As expected, $K_2(\tau^*|M) > K_2(\tau^*)$ with the respective τ^*-values. $\square$

11.5.6 Integrated Brownian Motion

If $\{B(t), t \geq 0\}$ is a Brownian motion, then its sample paths $b = b(t)$ are continuous. Hence, the integrals

$$b(t) = \int_0^t b(y)\, dy$$

exist for all sample paths. They are realizations of the *random integral*

$$U(t) = \int_0^t B(y)\, dy.$$

The stochastic process $\{U(t), t \ge 0\}$ is called *integrated Brownian motion*. This process can be a suitable model for situations, in which the observed sample paths seem to be 'smoother' than those of the Brownian motion. Analogously to the definition of the Riemann integral, for any n-dimensional vector $(t_1, t_2, ..., t_n)$ with

$$0 = t_0 < t_1 < \cdots < t_n = t \quad \text{and} \quad \Delta t_i = t_{i+1} - t_i; \quad i = 0, 1, 2, ..., n-1,$$

the random integral $U(t)$ is defined as the limit

$$U(t) = \lim_{\substack{n \to \infty \\ \Delta t_i \to 0}} \left\{ \sum_{i=0}^{n-1} [B(t_i + \Delta t_i) - B(t_i)] \Delta t_i \right\},$$

where passing to the limit refers to mean-square convergence. Therefore, the random variable $U(t)$, being the limit of a sum of normally distributed random variables, is itself normally distributed. More generally, by theorem 3.3, page 149, the integrated Brownian motion $\{U(t), t \ge 0\}$ is a Gaussian process. Hence, the integrated Brownian motion is uniquely characterized by its trend and covariance function. In view of

$$E\left(\int_0^t B(y) \, dy \right) = \int_0^t E(B(y)) \, dy = \int_0^t 0 \, dy \equiv 0,$$

the trend function of the integrated Brownian motion is identically equal to 0:

$$m(t) = E(U(t)) \equiv 0.$$

The covariance function

$$C(s, t) = Cov(U(s), U(t)) = E(U(s)U(t)), \quad s \le t,$$

of $\{U(t), t \ge 0\}$ is obtained as follows:

$$C(s, t) = E\left\{ \int_0^s B(y) \, dy \int_0^t B(z) \, dz \right\}$$

$$= E\left\{ \int_0^t \int_0^s B(y) B(z) \, dy \, dz \right\} = \int_0^t \int_0^s E(B(y) B(z)) \, dy \, dz.$$

Since $E(B(y), B(z)) = Cov(B(y)B(z)) = \sigma^2 \min(y, z)$, it follows that

$$C(s, t) = \sigma^2 \int_0^t \int_0^s \min(y, z) \, dy \, dz$$

$$= \sigma^2 \int_0^s \int_0^s \min(y, z) \, dy \, dz + \sigma^2 \int_s^t \int_0^s \min(y, z) \, dy \, dz$$

$$= \sigma^2 \int_0^s \left[\int_0^z y \, dy + \int_z^s z \, dy \right] dz + \sigma^2 \int_s^t \int_0^s y \, dy \, dz$$

$$= \sigma^2 \frac{s^3}{3} + \sigma^2 \frac{s^2}{2} (t - s).$$

Thus,

$$C(s, t) = \frac{\sigma^2}{6} (3t - s) s^2, \quad s \le t.$$

Letting $s = t$ yields the variance of $U(t)$:

$$Var(U(t)) = \frac{\sigma^2}{3} t^3.$$

11.6 EXERCISES

Note In all exercises, $\{B(t),\ t \geq 0\}$ is the Brownian motion with $Var(B(1)) = \sigma^2$.

11.1) Verify that the probability density $f_t(x)$ of $B(t)$,

$$f_t(x) = \frac{1}{\sqrt{2\pi t}\,\sigma}\, e^{-x^2/(2\sigma^2 t)}, \quad t > 0,$$

satisfies with a positive constant c the *thermal conduction equation*

$$\frac{\partial f_t(x)}{\partial t} = c\,\frac{\partial^2 f_t(x)}{\partial x^2}.$$

11.2) Determine the conditional probability density of $B(t)$ given $B(s) = y$, $0 \leq s < t$.

11.3)* Prove that the stochastic process $\{\bar{B}(t),\ 0 \leq t \leq 1\}$ given by $\bar{B}(t) = B(t) - t\,B(1)$ is the Brownian bridge.

11.4) Let $\{\bar{B}(t),\ 0 \leq t \leq 1\}$ be the Brownian bridge. Prove that the stochastic process

$$\{S(t), t \geq 0\} \text{ defined by } S(t) = (t+1)\bar{B}\!\left(\frac{t}{t+1}\right)$$

is the standard Brownian motion.

11.5) Determine the probability density of $B(s) + B(t)$, $0 \leq s < t$.

11.6) Let n be any positive integer. Determine mean value and variance of

$$X(n) = B(1) + B(2) + \cdots + B(n).$$

Hint Make use of formula (4.52), page 187.

11.7) Check whether for any positve τ the stochastic process $\{V(t),\ t \geq 0\}$ defined by

$$V(t) = B(t+\tau) - B(t)$$

is weakly stationary.

11.8) Let $X(t) = S^3(t) - 3t\,S(t)$. Prove that $\{X(t),\ t \geq 0\}$ is a continuous-time martingale, i.e., show that

$$E(X(t)|X(y),\ y \leq s) = X(s),\quad s < t.$$

11.9) Show by a counterexample that the Ornstein-Uhlenbeck process does not have independent increments.

11.10) (1) What is the mean value of the first passage time of the reflected Brownian motion $\{|B(t)|, t \geq 0\}$ with regard to a positive level x?
(2) Determine the distribution function of $|B(t)|$.

11.11)* Starting from $x = 0$, a particle makes independent jumps of length

$$\Delta x = \sigma \sqrt{\Delta t}$$

to the right or to the left every Δt time units. The respective probabilities of jumps to the right and to the left are

$$p = \frac{1}{2}\left(1 + \frac{\mu}{\sigma}\sqrt{\Delta t}\right) \text{ and } 1-p \text{ with } \sqrt{\Delta t} \leq \left|\frac{\sigma}{\mu}\right|, \quad \sigma > 0.$$

Show that as $\Delta t \to 0$ the position of the particle at time t is governed by a Brownian motion with drift with parameters μ and σ.

11.12) Let $\{D(t), t \geq 0\}$ be a Brownian motion with drift with paramters μ and σ. Determine $E\left(\int_0^t (D(s))^2\, ds\right)$.

11.13) Show that for $c > 0$ and $d > 0$

$$P(B(t) \leq ct + d \text{ for all } t \geq 0) = 1 - e^{-2cd/\sigma^2}.$$

Hint Make use of formula (11.29).

11.14) At time $t = 0$ a speculator acquires an American call option with infinite expiration time and strike price x_s. The price [in \$] of the underlying risky security at time t is given by $X(t) = x_0 e^{B(t)}$. The speculator makes up his mind to exercise this option at that time point, when the price of the risky security hits for the first time level x with $x > x_s \geq x_0$.
(1) What is the speculator's mean discounted payoff $G_\alpha(x)$ under a constant discount rate α?
(2) What is the speculator's payoff $G(x)$ without discounting?
In both cases, the cost of acquiring the option is not included in the speculator's payoff.

11.15) The price of a unit of a share at time point t is $X(t) = 10\, e^{D(t)}$, $t \geq 0$, where $\{D(t), t \geq 0\}$ is a Brownian motion process with drift parameter $\mu = -0.01$ and volatility $\sigma = 0.1$. At time $t = 0$ a speculator acquires an option, which gives him the right to buy a unit of the share at strike price $x_s = 10.5$ at any time point in the future, independently of the then current market value. It is assumed that this option has no expiry date. Although the drift parameter is negative, the investor hopes to profit from random fluctuations of the share price. He makes up his mind to exercise the option at that time point, when the expected difference between the actual share price x and the strike price x_s is maximal.
(1) What is the initial price of a unit of the share?
(2) Is the share price on average increasing or decreasing?
(3) Determine the corresponding share price which maximizes the expected profit of the speculator.

11.16) The value (in \$) of a share per unit develops, apart from the constant factor 10, according to a geometric Brownian motion $\{X(t),\, t \geq 0\}$ given by

$$X(t) = 10\, e^{B(t)}, \quad 0 \leq t \leq 120,$$

where $\{B(t),\, t \geq 0\}$ is the Brownian motion process with volatility $\sigma = 0.1$.

At time $t = 0$ a speculator pays \$17 for becoming owner of a unit of the share after 120 [*days*], irrespective of the then current market value of the share.

(1) What will be the mean undiscounted profit of the speculator at time point $t = 120$?

(2) What is the probability that the investor will lose some money when exercising at this time point?

In both cases, take into account the amount of \$17, which the speculator had to pay in advance.

11.17 The value of a share per unit develops according to a geometric Brownian motion with drift given by

$$X(t) = 10\, e^{0.2\,t + 0.1\, S(t)}, \quad t \geq 0,$$

where $\{S(t),\, t \geq 0\}$ is the standardized Brownian motion. An investor owns a European call option with running time $\tau = 1$ [year] and with strike price

$$x_S = \$12$$

on a unit of this share.

(1) Given a discount rate of $\alpha = 0.04$, determine the mean discounted profit of the holder of the option.

(2) For what value of the drift parameter μ do you get the fair price of the option?

(3) Determine this fair price.

11.18) The random price $X(t)$ of a risky security per unit at time t is

$$X(t) = 5\, e^{-0.01t + B(t) + 0.2|B(t)|},$$

where $\{B(t),\, t \geq 0\}$ is the Brownian motion with volatility

$$\sigma = 0.04.$$

At time $t = 0$ a speculator acquires the right to buy the share at price \$5.1 at any time point in the future, independently of the then current market value; i.e., the speculator owns an *American call option* with *strike price* $x_S = \$5.1$ on the share. The speculator makes up his mind to exercise the option at that time point, when the mean difference between the actual share price x and the strike price is maximal.

(1) Is the stochastic process $\{X(t),\, t \geq 0\}$ a geometric Brownian motion with drift?

(2) Is the share price on average increasing or decreasing?

(3) Determine the optimal actual share price $x = x^*$.

(4) What is the probability that the investor will exercise the option?

11.19) At time $t = 0$ a speculator acquires a European call option with strike price x_s and finite expiration time τ. Thus, the option can only be exercised at time τ at price x_s independently of its market value at time τ. The random price $X(t)$ of the underlying risky security develops according to

$$X(t) = x_0 + D(t),$$

where $\{D(t), t \geq 0\}$ is the Brownian motion with positive drift parameter μ and volatility σ. If $X(\tau) > x_s$, the speculator will exercise the option. Otherwise, the speculator will not exercise. Assume that

$$x_0 + \mu t > 3\sigma \sqrt{t}, \ 0 \leq t \leq \tau.$$

(1) What will be the mean undiscounted payoff of the speculator (cost of acquiring the option not included)?

(2) Under otherwise the same assumptions, what is the investor's mean undiscounted profit if

$$X(t) = x_0 + B(t) \text{ and } x_0 = x_s?$$

11.20) Show that

$$E(e^{\alpha U(t)}) = e^{\alpha^2 t^3/6}$$

for any constant α, where $U(t)$ is the integrated standard Brownian motion:

$$U(t) = \int_0^t S(x)\,dx, \ t \geq 0.$$

11.21)* For any fixed positive τ, let the stochastic process $\{V(t), t \geq 0\}$ be given by

$$V(t) = \int_t^{t+\tau} S(x)\,dx.$$

Is $\{V(t), t \geq 0\}$ weakly stationary?

11.22) Let $\{X(t), t \geq 0\}$ be the cumulative repair cost process of a system with

$$X(t) = 0.01 e^{D(t)},$$

where $\{D(t), t \geq 0\}$ is a Brownian motion with drift and parameters

$$\mu = 0.02 \text{ and } \sigma^2 = 0.04.$$

The cost of a system replacement by an equivalent new one is $c = 4000$.

(1) The system is replaced according to policy 1 (page 522). Determine the optimal repair cost limit x^* and the corresponding maintenance cost rate $K_1(x^*)$.

(2) The system is replaced according to policy 2 (page 522). Determine its economic lifetime τ^* based on the average repair cost development

$$E(X(t)) = 0.01 E(e^{D(t)})$$

and the corresponding maintenance cost rate $K_2(\tau^*)$.

(3) Analogously to example 11.8, apply replacement policy 2 to the cumulative repair cost process

$$X(t) = 0.01e^{M(t)}$$

with $M(t) = \max_{0 \leq y \leq t} D(y)$. Determine the corresponding economic lifetime of the system and the maintenance cost rate $K_2(\tau^*|M)$. Compare to the minimal maintenance cost rates determined under (1) and (2).

For part (3) of this exercise you need computer assistance.

CHAPTER 12

Spectral Analysis of Stationary Processes

12.1 FOUNDATIONS

Covariance functions of weakly stationary stochastic processes can be represented by their *spectral densities*. These *spectral representations* of covariance functions have proved a useful analytic tool in many technical and physical applications.

The mathematical treatment of spectral representations and the application of the results, particularly in electrotechnics and electronics, is facilitated by introducing the concept of a complex stochastic process: $\{X(t),\ t \in \mathbf{R}\}$ is a *complex stochastic process* if $X(t)$ has structure

$$X(t) = Y(t) + i\,Z(t), \quad \mathbf{R} = (-\infty, +\infty),$$

where $\{Y(t),\ t \in \mathbf{R}\}$ and $\{Z(t),\ t \in \mathbf{R}\}$ are two real-valued stochastic processes and $i = \sqrt{-1}$. Thus, the probability distribution of $X(t)$ is given by the joint probability distribution of the random vector $(Y(t), Z(t))$, $\mathbf{R} = (-\infty, +\infty)$. Trend and covariance function of $\{X(t),\ t \geq 0\}$ are defined by

$$m(t) = E(X(t)) = E(Y(t)) + i\,E(Z(t)), \tag{12.1}$$

$$C(s, t) = Cov\,(X(s), X(t)) = E\Big([X(s) - E(X(s))]\big[\,\overline{X(t) - E(X(t))}\,\big]\Big). \tag{12.2}$$

If $X(t)$ is real, then (12.1) and (12.2) coincide with (6.2) and (6.3), respectively.

Notation If $z = a + ib$ and $\bar{z} = a - ib$, then z and $\bar{z}$ are *conjugate complex numbers*. The *modulus* of z, denoted by $|z|$, is defined as $|z| = \sqrt{z\bar{z}} = \sqrt{a^2 + b^2}$.

A complex stochastic process $\{X(t),\ t \in \mathbf{R}\}$ is a *second-order process* if

$$E(|X(t)|^2) < \infty \text{ for all } t \in \mathbf{R}.$$

Analogously to the definition real-valued weakly stationary stochastic processes (see page 232), a second-order complex stochastic process $\{X(t),\ t \in \mathbf{R}\}$ is said to be *weakly stationary* if, with a complex constant m, it has the following properties:

1) $m(t) \equiv m$,

2) $C(s, t) = C(0, t - s)$.

In this case, $C(s, t)$ simplifies to a function of one variable:

$$C(s, t) = C(\tau),$$

where $\tau = t - s$.

Ergodicity If the complex stochastic process $\{X(t), t \in \mathbf{R}\}$ is weakly stationary, then one anticipates that, for any of its sample paths $x(t) = y(t) + iz(t)$, its constant trend function $m \equiv E(X(t))$ can be obtained by

$$m = \lim_{T \to \infty} \frac{1}{2T} \int_{-T}^{+T} x(t)\, dt. \qquad (12.3)$$

This representation of the trend function as an improper integral uses the full informa-tion, which is contained in one sample path of the process.

On the other hand, if n sample paths of the process

$$x_1(t), x_2(t), ..., x_n(t)$$

are each only scanned at one fixed time point t and if these values are obtained inde-pendently of each other, then m can be estimated by

$$m = \lim_{n \to \infty} \frac{1}{n} \sum_{k=1}^{n} x_k(t). \qquad (12.4)$$

The equivalence of formulas (12.3) and (12.4) allows a simple physical interpretation: the mean of a stationary stochastic process at a given time point is equal to its mean over the whole observation period ('time mean is equal to location mean'). With respect to their practical application, this is the most important property of *ergodic stochastic processes*. Besides the representation (12.2), for any sample path $x = x(t)$, the covariance function of an ergodic process can be obtained from

$$C(\tau) = \lim_{T \to \infty} \frac{1}{2T} \int_{-T}^{+T} [x(t) - m][\overline{x(t+\tau) - m}]\, dt. \qquad (12.5)$$

The exact definition of ergodic stochastic processes cannot be given here. In the engineering literature, the ergodicity of stationary processes is frequently simply defined by properties (12.3) and (12.5). The application of formula (12.5) is useful if the sample path of an ongoing stochastic process is being recorded continuously. The estimated value of $C(t)$ becomes the better the larger the observation period $[-T, +T]$.

Assumptions This chapter deals only with weakly stationary processes. Hence, the attribute 'weakly' is generally omitted. Moreover, without of loss of generality, the trend function of all processes considered is identically zero.

For this assumption, the representation (12.2) of the covariance function simplifies to

$$C(\tau) = C(t, t+\tau) = E(X(t)\overline{X(t+\tau)}). \qquad (12.6)$$

In what follows, *Euler's formula* is needed:

$$e^{\pm ix} = \cos x \pm \sin x. \qquad (12.7)$$

Solving for $\sin x$ and $\cos x$ yields

$$\sin x = \frac{1}{2i}\left(e^{ix} - e^{-ix}\right), \quad \cos x = \frac{1}{2}\left(e^{ix} + e^{-ix}\right). \qquad (12.8)$$

12.2 PROCESSES WITH DISCRETE SPECTRUM

In this section the general structure of stationary stochastic processes with discrete spectra is developed. Next the simple stochastic process $\{X(t), t \in \mathbf{R}\}$ with

$$X(t) = a(t) X \qquad (12.9)$$

is considered, where X is a complex random variable and $a(t)$ a complex function with $a(t) \neq$ constant. For $\{X(t), t \in \mathbf{R}\}$ to be stationary, the two conditions

$$E(X) = 0 \quad \text{and} \quad E(|X|^2) < \infty$$

are necessary. Moreover, because of (12.5) the function

$$E(X(t)\overline{X(t+\tau)}) = a(t)\,\overline{a(t+\tau)}\,E(|X|^2) \qquad (12.10)$$

is not allowed to depend on t. Letting $t = 0$, this implies

$$a(t)\,\overline{a(t)} = |a(t)|^2 = |a|^2 = \text{constant.}$$

Therefore, $a(t)$ has structure

$$a(t) = |a|\,e^{i\,\omega(t)}, \qquad (12.11)$$

where $\omega(t)$ is a <u>real</u> function. Substituting (12.11) into (12.10) shows that the difference $\omega(t+\tau) - \omega(t)$ does not depend on t. Thus, if $\omega(t)$ is assumed to be differentiable, then it satisfies the equation

$$d[\omega(t+\tau) - \omega(t)]/dt = 0,$$

or, equivalently,

$$\frac{d}{dt}\,\omega(t) = \text{constant.}$$

Hence, $\omega(t) = \omega t + \varphi$, where ω and φ are constants. (Note that for proving this result it is only necessary to assume the continuity of $\omega(t)$.) Thus,

$$a(t) = |a|\,e^{i(\omega t + \varphi)}.$$

If in (12.9) the random variable X is multiplied by $|a|e^{i\varphi}$ and $|a|e^{i\varphi}X$ is again denoted as X, then the desired result assumes the following form:

> A stochastic process $\{X(t), t \in \mathbf{R}\}$ defined by (12.9) is stationary if and only if
> $$X(t) = Xe^{i\omega t} \qquad (12.12)$$
> with $E(X) = 0$ and $E(|X|^2) < \infty$.

Letting $s = E(|X|^2)$, the corresponding covariance function is

$$C(\tau) = s\,e^{-i\omega\tau}.$$

Remark Apart from a constant factor, the parameter s is physically equal to the mean energy of the oscillation per unit time (mean power).

The real part $\{Y(t), t \in \mathbf{R}\}$ of the stochastic process $\{X(t), t \in \mathbf{R}\}$ given by (12.12) describes a cosine oscillation with random amplitude and phase. Its sample paths have, therefore, structure

$$y(t) = a \cos(\omega t + \varphi),$$

where a and φ are realizations of possibly dependent random variables A and Φ. The parameter ω is the *circular frequency* of the oscillation.

Generalizing the situation dealt with so far, a linear combination of two stationary processes of structure (12.12) is considered:

$$X(t) = X_1 e^{i\omega_1 t} + X_2 e^{i\omega_2 t}. \tag{12.13}$$

X_1 and X_2 are two complex random variables with mean values 0, whereas ω_1 and ω_2 are two constant real numbers with $\omega_1 \neq \omega_2$. The covariance function of the stochastic process $\{X(t), t \in \mathbf{R}\}$ defined by (12.13) is

$$C(t, t+\tau) = E(X(t)\overline{X(t+\tau)})$$

$$= E\left(\left[X_1 e^{i\omega_1 t} + X_2 e^{i\omega_2 t}\right]\left[\overline{X}_1 e^{-i\omega_1(t+\tau)} + \overline{X}_2 e^{-i\omega_2(t+\tau)}\right]\right)$$

$$= E\left(\left[X_1 \overline{X}_1 e^{-i\omega_1 \tau} + X_1 \overline{X}_2 e^{i(\omega_1 - \omega_2)t - i\omega_2 \tau}\right]\right)$$

$$+ E\left(\left[X_2 \overline{X}_2 e^{-i\omega_2 \tau} + X_2 \overline{X}_1 e^{i(\omega_2 - \omega_1)t - i\omega_1 \tau}\right]\right).$$

Thus, $\{X(t), t \in \mathbf{R}\}$ is stationary if and only if X_1 and X_2 are uncorrelated.

Note Two complex random variables X and Y with mean values 0 are said to be *uncorrelated* if they satisfy the condition $E(X\overline{Y}) = 0$ or, equivalently, $E(Y\overline{X}) = 0$, and *correlated* otherwise.

In this case, the covariance function of $\{X(t), t \in \mathbf{R}\}$ is given by

$$C(\tau) = s_1 e^{-i\omega_1 \tau} + s_2 e^{-i\omega_2 \tau}, \tag{12.14}$$

where

$$s_1 = E(|X_1|^2), \quad s_2 = E(|X_2|^2).$$

Generalizing equation (12.13) leads to

$$X(t) = \sum_{k=1}^{n} X_k e^{i\omega_k t} \tag{12.15}$$

with real numbers ω_k satisfying $\omega_j \neq \omega_k$ for $j \neq k$; $i, j = 1, 2, ..., n$. If the X_k are uncorrelated and have mean value 0, then it can be readily shown by induction that the process $\{X(t), t \in \mathbf{R}\}$ is stationary. Its covariance function is

$$C(\tau) = \sum_{k=1}^{n} s_k e^{-i\omega_k \tau}, \tag{12.16}$$

where

$$s_k = E(|X_k|^2); \quad k = 1, 2, ..., n.$$

In particular,

$$C(0) = E(|X(t)|^2) = \sum_{k=1}^{n} s_k. \tag{12.17}$$

The oscillation $X(t)$ given by (12.15) is an additive superposition of n harmonic oscillations. Its mean power is equal to the sum of the mean powers of these n harmonic oscillations.

Now let $X_1, X_2, \ldots$ be a countably infinite sequence of uncorrelated complex random variables with $E(X_k) = 0$; $k = 1, 2, \ldots$; and

$$\sum_{k=1}^{\infty} E\left(|X_k|^2\right) = \sum_{k=1}^{\infty} s_k < \infty. \tag{12.18}$$

Under these assumptions, the equation

$$X(t) = \sum_{k=1}^{\infty} X_k e^{i\omega_k t}, \quad \omega_j \neq \omega_k \quad \text{for } j \neq k, \tag{12.19}$$

defines a stationary process $\{X(t), t \in \mathbf{R}\}$ with covariance function

$$C(\tau) = \sum_{k=1}^{\infty} s_k e^{-i\omega_k \tau}. \tag{12.20}$$

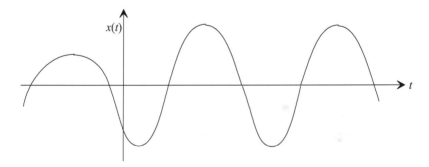

Figure 12.1 Sample path of a real narrow-band process

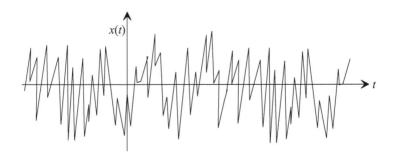

Figure 12.2 Sample path of a real wide-band process for large n

The sets $\{\omega_1, \omega_2, \ldots, \omega_n\}$ and $\{\omega_1, \omega_2, \ldots\}$ are said to be the *spectra* of the stochastic processes defined by (12.15) and (12.19), respectively. If all ω_k are sufficiently close to a single value ω, then $\{X(t), t \in \mathbf{R}\}$ is called a *narrow-band process* (Figure 12.1); otherwise it is called a *wide-band process* (Figure 12.2). Regarding convergence in mean-square, any stationary process $\{X(t), t \in \mathbf{R}\}$ can be sufficiently closely approximated to a stationary process of structure (12.15) in any finite interval $[-T \le t \le +T]$.

Later it will prove useful to represent the covariance function (12.20) in terms of the *delta function* $\delta(t)$. This function is defined as the limit

$$\delta(t) = \lim_{h \to 0} \begin{cases} 1/h & \text{for} \ -h/2 \le t \le +h/2 \\ 0 & \text{elsewhere} \end{cases}. \tag{12.21}$$

Symbolically,

$$\delta(t) = \begin{cases} \infty & \text{for} \ t = 0 \\ 0 & \text{elsewhere} \end{cases}.$$

The delta-function has a characteristic property, which is sometimes used as its definition: For any function $f(t)$,

$$\int_{-\infty}^{+\infty} f(t) \delta(t - t_0) \, dt = f(t_0). \tag{12.22}$$

The proof is easily established: If $F(t)$ is the antiderivative of $f(t)$, then

$$\int_{-\infty}^{+\infty} f(t) \delta(t - t_0) \, dt = \int_{-\infty}^{+\infty} f(t + t_0) \delta(t) \, dt$$

$$= \lim_{h \to 0} \left\{ \int_{-h/2}^{+h/2} f(t + t_0) \frac{1}{h} \, dt \right\}$$

$$= \frac{1}{2} \left\{ \lim_{h \to 0} \frac{F(t_0 + h/2) - F(t_0)}{h/2} + \lim_{h \to 0} \frac{F(t_0) - F(t_0 - h/2)}{h/2} \right\}$$

$$= \frac{1}{2} \{f(t_0) + f(t_0)\} = f(t_0).$$

Using property (12.22), the covariance function (12.20) can be written as

$$C(\tau) = \sum_{k=1}^{\infty} s_k \int_{-\infty}^{+\infty} e^{i \omega \tau} \delta(\omega - \omega_k) \, d\omega.$$

Symbolically,

$$C(\tau) = \int_{-\infty}^{+\infty} e^{i \omega \tau} s(\omega) \, d\omega, \tag{12.23}$$

where

$$s(\omega) = \sum_{k=1}^{\infty} s_k \, \delta(\omega - \omega_k). \tag{12.24}$$

The (generalized) function $s(\omega)$ is called the *spectral density* of the stationary process. Therefore, $C(\tau)$ is the *Fourier transform* of the spectral density of a stationary process with discrete spectrum.

Real Stationary Processes In contrast to a stochastic process of structure (12.12), a stationary process $\{X(t), t \in \mathbf{R}\}$ of structure (12.13), i.e.,

$$X(t) = X_1 e^{i\omega_1 t} + X_2 e^{i\omega_2 t},$$

can be real. To see this, let

$$X_1 = \frac{1}{2}(A + iB), \quad X_2 = \bar{X}_1 = \frac{1}{2}(A - iB), \quad \text{and } \omega_1 = -\omega_2 = \omega,$$

where A and B are two real random variables with mean values 0. Substituting these X_1 and X_2 into (12.13) yields (compare to Example 6.7, page 235)

$$X(t) = A \cos \omega t - B \sin \omega t.$$

If A and B are uncorrelated, then, letting $s = E(|X_1|^2) = E(|X_2|^2)$, the covariance function of $\{X(t), t \in \mathbf{R}\}$ is seen to be $C(\tau) = 2s \cos \omega\tau$. More generally, it can be shown that the process given by (12.15) with n terms defines a real stationary process if n is even and pairs of the X_k are complex conjugates.

12.3 PROCESSES WITH CONTINUOUS SPECTRUM

12.3.1 Spectral Representation of the Covariance Function

Let $\{X(t), t \in \mathbf{R}\}$ be a complex stationary process with covariance function $C(\tau)$. Then there exists a real, nondecreasing, and bounded function $S(\omega)$ so that

$$C(\tau) = \int_{-\infty}^{+\infty} e^{i\omega\tau} \, dS(\omega). \tag{12.25}$$

(This fundamental relationship is associated with the names of *Bochner, Khinchin* and *Wiener*; see, e.g., *Khinchin* (1934)). $S(\omega)$ is called the *spectral function* of the process. The definition of the covariance function implies that for all t

$$C(0) = S(\infty) - S(-\infty) = E(|X(t)|^2) < \infty.$$

Given $C(\tau)$, the spectral function is, apart from a constant c, uniquely determined. Usually c is selected in such a way that $S(-\infty) = 0$. If $s(\omega) = dS(\omega)/d\omega$ exists, then

$$C(\tau) = \int_{-\infty}^{+\infty} e^{i\omega\tau} s(\omega) \, d\omega. \tag{12.26}$$

The function $s(\omega)$ is called the *spectral density* of the process. Since $S(\omega)$ is nondecreasing and bounded, the spectral density has properties

$$s(\omega) \geq 0, \quad C(0) = \int_{-\infty}^{+\infty} s(\omega) \, d\omega < \infty. \tag{12.27}$$

Conversely, it can be shown that every function $s(\omega)$ with properties (12.27) is the spectral density of a stationary process.

Remark Frequently the function $f(\omega) = s(\omega)/2\pi$ is referred to as the spectral density. An advantage of this representation is that $\int_{-\infty}^{+\infty} f(\omega) \, d\omega$ is the mean power of the oscillation.

The set $\{\omega,\ s(\omega) > 0\}$ with its lower, upper marginal points

$$\inf_{\omega \in \mathbf{S}} \omega \quad \text{and} \quad \sup_{\omega \in \mathbf{S}} \omega$$

is said to be the (*continuous*) *spectrum* of the process. Its *bandwidth* w is defined as

$$w = \sup_{\omega \in \mathbf{S}} \omega - \inf_{\omega \in \mathbf{S}} \omega.$$

Note Here and in what follows mind the difference between w and ω.

Determining the covariance function is generally much simpler than determining the spectral density. Hence the inversion of the relationship (12.26) is of importance. It is known from the theory of the Fourier integral that this inversion is always possible if

$$\int_{-\infty}^{+\infty} |C(t)|\, dt < \infty. \tag{12.28}$$

In this case,

$$s(\omega) = \frac{1}{2\pi} \int_{-\infty}^{+\infty} e^{-i\omega t} C(t)\, dt. \tag{12.29}$$

The intuitive interpretation of assumption (12.28) is that $C(\tau)$ must sufficiently fast converge to 0 as $|\tau| \to \infty$. The stationary processes occurring in electroengineering and communication generally satisfy this condition. Integration of $s(\omega)$ over the interval $[\omega_1, \omega_2]$, $\omega_1 < \omega_2$, yields

$$S(\omega_2) - S(\omega_1) = \frac{i}{2\pi} \int_{-\infty}^{+\infty} \frac{e^{-i\omega_2 t} - e^{-i\omega_1 t}}{t} C(t)\, dt. \tag{12.30}$$

This formula is also valid if the spectral density does not exist. But in this case the additional assumption has to be made that at each point of discontinuity ω_0 of the spectral function the following value is assigned to $S(\omega)$:

$$S(\omega_0) = \frac{1}{2} [S(w_0 + 0) - S(\omega_0 - 0)].$$

Note that the delta function $\delta(t)$ satisfies condition (12.28). If $\delta(t)$ is substituted for $C(t)$ in (12.29), then formula (12.22) yields

$$s(\omega) = \frac{1}{2\pi} \int_{-\infty}^{+\infty} e^{-i\omega t} \delta(t)\, dt \equiv \frac{1}{2\pi}. \tag{12.31}$$

The formal inversion of this relationship according to (12.26) provides a complex representation of the delta function:

$$\delta(t) = \frac{1}{2\pi} \int_{-\infty}^{+\infty} e^{i\omega t}\, d\omega. \tag{12.32}$$

The time-discrete analogues to formulas (12.28) and (12.29) are

$$\sum_{t=-\infty}^{+\infty} |C(t)| < \infty, \quad s(\omega) = \frac{1}{2\pi} \sum_{t=-\infty}^{+\infty} e^{-it\omega} C(t). \tag{12.33}$$

Real Stationary Processes Since for any real stationary process $C(\tau) = C(\tau)$, the co-variance function can be written in the form

$$C(\tau) = [C(\tau) + C(-\tau)]/2.$$

Substituting (12.26) for $C(\tau)$ into this equation and using (12.8) yields

$$C(\tau) = \int_{-\infty}^{+\infty} \cos \omega\tau\, s(\omega)\, d\omega.$$

Because of $\cos \omega\tau = \cos(-\omega\tau)$, this formula can be written as

$$C(\tau) = 2\int_0^{+\infty} \cos \omega\tau\, s(\omega)\, d\omega. \tag{12.34}$$

Analogously, (12.29) yields the spectral density in the form

$$s(\omega) = \frac{1}{2\pi}\int_{-\infty}^{+\infty} \cos \omega t\, C(t)\, dt.$$

Since $s(\omega) = s(-\omega)$,

$$s(\omega) = \frac{1}{\pi}\int_{-\infty}^{+\infty} \cos \omega t\, C(t)\, dt. \tag{12.35}$$

Even in case of real processes it is, however, sometimes more convenient to use the formulas (12.26) and (12.29) instead of (12.34) and (12.35), respectively.

In many applications, the *correlation time* τ_0 is of interest. It is defined by

$$\tau_0 = \frac{1}{C(0)}\int_0^\infty C(t)\, dt. \tag{12.36}$$

If there is $|\tau| \le \tau_0$, then there is a significant correlation between $X(t)$ and $X(t+\tau)$. If $|\tau| > \tau_0$, then the correlation between $X(t)$ and $X(t+\tau)$ quickly decreases as $|\tau|$ tends to infinity.

Example 12.1 Let $\{..., X_{-1}, X_0, X_1, ...\}$ be the discrete white noise (purely random sequence) defined at page 246. Its covariance function is

$$C(\tau) = \begin{cases} \sigma^2 & \text{for} & \tau = 0 \\ 0 & \text{for} & \tau = \pm 1,\ \pm 2, ... \end{cases} \tag{12.37}$$

Hence, from (12.29),

$$s(\omega) = \sigma^2/2\pi.$$

Thus, the discrete white noise has a constant spectral density. This result is in accordance with (12.31), since the covariance function of the discrete white noise given by (12.37) is equivalent to $C(\tau) = \sigma^2\, \delta(\tau)$. $\qquad\square$

Example 12.2 The covariance function of the first-order autoregressive sequence has structure (page 249)

$$C(\tau) = c\, a^{|\tau|}; \quad \tau = 0, \pm 1, ...,$$

where a and c are real constants and $|a| < 1$. The corresponding spectral density is obtained from (12.33) as follows:

$$s(\omega) = \frac{1}{2\pi} \sum_{\tau=-\infty}^{\infty} C(\tau) e^{-i\tau\omega}$$

$$= \frac{c}{2\pi} \left[\sum_{\tau=-\infty}^{-1} a^{-\tau} e^{-i\tau\omega} + \sum_{\tau=0}^{\infty} a^{\tau} e^{-i\tau\omega} \right]$$

$$= \frac{c}{2\pi} \left[\sum_{\tau=1}^{\infty} a^{\tau} e^{i\tau\omega} + \sum_{\tau=0}^{\infty} a^{\tau} e^{-i\tau\omega} \right].$$

Hence,

$$s(\omega) = \frac{c}{2\pi} \left[\frac{a e^{i\omega}}{1 - a e^{i\omega}} + \frac{1}{1 - a e^{-i\omega}} \right]. \qquad \square$$

Example 12.3 The random telegraph signal considered in example 7.3 (page 265) has covariance function

$$C(\tau) = a e^{-b|\tau|}, \quad a > 0, \ b > 0. \tag{12.38}$$

Since condition (12.28) is satisfied, the corresponding spectral density $s(\omega)$ can be obtained from (12.29):

$$s(\omega) = \frac{1}{2\pi} \int_{-\infty}^{+\infty} e^{-i\omega t} a e^{-b|t|} \, dt$$

$$= \frac{a}{2\pi} \left\{ \int_{-\infty}^{0} e^{(b-i\omega)t} \, dt + \int_{0}^{\infty} e^{-(b+i\omega)t} \, dt \right\}$$

$$= \frac{a}{2\pi} \left\{ \frac{1}{b - i\omega} + \frac{1}{b + i\omega} \right\}$$

Hence,

$$s(\omega) = \frac{a b}{\pi (\omega^2 + b^2)}.$$

The corresponding correlation time is $\tau_0 = 1/b$.

This result is in line with Figure 12.3. Because of its simple structure, the covariance function (12.38) is sometimes even then applied if it only approximately coincides with the actual covariance function. $\qquad \square$

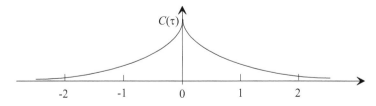

Figure 12.3 Covariance function for example 12.3

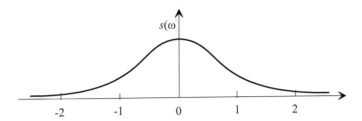

Figure 12.4 Spectral density for example 12.3

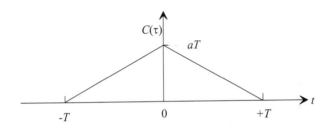

Figure 12.5 Covariance function for example 12.4

Example 12.4 Let

$$C(\tau) = \begin{cases} a(T-|\tau|) & \text{for } |\tau| \leq T \\ 0 & \text{for } |\tau| > T \end{cases}, \quad a > 0, \ T > 0. \qquad (12.39)$$

Figure 12.5 shows the graph of this covariance function. For example, the covariance function of the randomly delayed pulse code modulation considered in example 6.8 (page 236) has this structure (see Figures 6.4 and 6.5). The corresponding spectral density one gets by applying (12.29):

$$s(\omega) = \frac{a}{2\pi} \int_{-T}^{+T} e^{-i\omega t} (T-|t|)\, dt$$

$$= \frac{a}{2\pi} \left\{ T \int_{-T}^{+T} e^{-i\omega t}\, dt - \int_{0}^{+T} t\, e^{+i\omega t}\, dt - \int_{0}^{+T} t\, e^{-i\omega t}\, dt \right\}$$

$$= \frac{a}{2\pi} \left\{ \frac{2T}{\omega} \sin \omega T - 2 \int_{0}^{T} t \cos \omega t\, dt \right\}.$$

Hence,

$$s(\omega) = \frac{a}{\pi} \frac{1 - \cos \omega T}{\omega^2}.$$

Figure 12.6 shows the graph of $s(\omega)$.   □

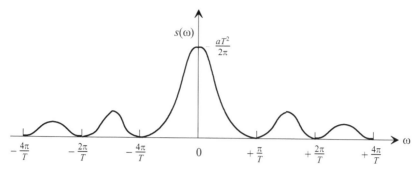

Figure 12.6 Spectral density for example 12.4

The previous examples should not give rise to the conjecture that for every function $f(\tau)$ which tends to zero as $|\tau| \to \infty$, a stationary stochastic process can be found with $f(\tau)$ being its covariance function. A slight modification of (12.39) yields a counterexample:

$$f(\tau) = \begin{cases} a\left(T - \tau^2\right) & \text{for } |\tau| \le T \\ 0 & \text{for } |\tau| > T \end{cases}, \quad a > 0, \ T > 0.$$

If this function is substituted for $C(\tau)$ in (12.29), then the resulting function $s(\omega)$ does not have properties (12.27). Therefore, $f(\tau)$ cannot be the covariance function of a stationary process.

Example 12.5 The stochastic processes considered in the examples 6.6 and 6.7 have covariance functions of the form

$$C(\tau) = a \cos \omega_0 \tau.$$

Using (12.8), the corresponding spectral density is obtained as follows:

$$s(\omega) = \frac{a}{2\pi} \int_{-\infty}^{+\infty} e^{-i\omega t} \cos \omega_0 t \, dt = \frac{a}{4\pi} \int_{-\infty}^{+\infty} e^{-i\omega t} \left(e^{i\omega_0 t} - e^{-i\omega_0 t} \right) dt$$

$$= \frac{a}{4\pi} \left\{ \int_{-\infty}^{+\infty} e^{i(\omega_0 - \omega)t} \, dt + \int_{-\infty}^{+\infty} e^{-i(\omega_0 + \omega)t} \, dt \right\}.$$

Applying (8.30) yields a symbolic representation of $s(\omega)$ (Figure 12.7):

$$s(\omega) = \frac{a}{2} \{ \delta(\omega_0 - \omega) + \delta(\omega_0 + \omega) \}. \tag{12.40}$$

Making use of (12.22), the corresponding spectral function is seen to be

$$S(\omega) = \begin{cases} 0 & \text{for } \omega \le -\omega_0, \\ a/2 & \text{for } -\omega_0 < \omega \le \omega_0, \\ a & \text{for } \omega > \omega_0. \end{cases}$$

Thus, the spectral function is piecewise constant (Figure 12.7).   □

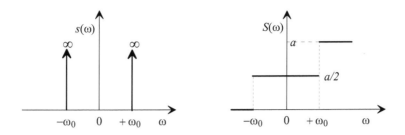

Figure 12.7 'Spectral density' and spectral function for example 12.5

Comment Since in example 12.5 the covariance function does not tend to zero as $|\tau| \to \infty$, the condition (12.28), which is necessary for applying (12.29), is not satisfied. This fact motivates the occurrence of the delta function in (12.40). Hence, (12.40) as well as (12.24) are symbolic representations of the spectral density. The usefulness of such symbolic representations based on the delta function will be illustrated later for a heuristic characterization of the white noise.

If $C_1(\tau)$ and $C_2(\tau)$ are the covariance functions of two stationary processes and

$$C(\tau) = C_1(\tau)\, C_2(\tau),$$

then it can be shown that there exists a stationary process with covariance function $C(\tau)$. The following example considers a stationary process, whose covariance function $C(\tau)$ is the product of the covariance functions of the stationary processes discussed in examples 12.3 and 12.5.

Example 12.6 Let $C(\tau)$ be given by the exponentially damped oscillation:

$$C(\tau) = a\, e^{-b|\tau|} \cos \omega_0 \tau, \qquad (12.41)$$

where $a > 0$, $b > 0$, and $\omega_0 > 0$. Thus, $C(\tau)$ satisfies condition (12.28) so that the corresponding spectral density can be obtained from (12.29):

$$s(\omega) = \frac{a}{\pi} \int_0^\infty e^{-bt} \cos(\omega t)\, \cos(\omega_0 t)\, dt$$

$$= \frac{a}{2\pi} \int_0^{+\infty} e^{-bt} [\cos(\omega - \omega_0)t + \cos(\omega + \omega_0)t]\, dt.$$

Therefore,

$$s(\omega) = \frac{a b}{2\pi} \left\{ \frac{1}{b^2 + (\omega - \omega_0)^2} + \frac{1}{b^2 + (\omega + \omega_0)^2} \right\}.$$

Functions of type (12.41) are frequently used to model covariance functions of stationary processes (possibly approximately), whose observed covariances periodically change their sign as τ increases. A practical example for such a stationary process is the fading of radio signals, which are recorded by radar. □

12.3.2 White Noise

In section 6.4.4 (page 246), the *discrete white noise* or a *purely random sequence* is defined as a sequence $\{X_1, X_2, ...\}$ of independent, identically distributed random variables X_i with parameters $E(X_i) = 0$ and $Var(X_i) = \sigma^2$. There is absolutely no problem with this definition.

Now let us assume that the indices i refer to time points $i\tau$, $i = 1, 2, ...$. What happens to the discrete white noise when τ tends to zero? Then, even for arbitrarily small τ, there will be no dependence between $X_{i\tau}$ and $X_{(i-1)\tau}$ as well as between $X_{i\tau}$ and $X_{(i+1)\tau}$. Hence, a continuous-time stochastic process $\{X(t), t \geq 0\}$, resulting from passing to the limit as $\tau \to 0$, must have the same covariance function as the discrete-time white noise (see formula (6.37), page 246):

$$C(\tau) = Cov(X(t), X(t+\tau)) = \begin{cases} \sigma^2 & \text{for } \tau = 0, \\ 0 & \text{for } \tau \neq 0, \end{cases} \qquad (12.42)$$

or, in terms of the delta-function, if the variance parameter σ^2 is written as $2\pi s_0$,

$$C(\tau) = 2\pi s_0 \delta(\tau). \qquad (12.43)$$

One cannot really think of a stochastic process in continuous time having this covariance function. Imagine $\{X(t), t \geq 0\}$ measures the temperature depending on time t at a location. Then the temperature at time point t would have no influence at the temperature one second later. Since there is no dependence between $X(t)$ and $X(t+\tau)$ for whatever small $|\tau|$, the continuous white noise is frequently said to be the 'most random process'.

By formulas (12.29) and (12.31), the spectral density belonging to (12.43) is

$$s(\omega) = \frac{1}{2\pi} \int_{-\infty}^{+\infty} e^{-i\omega t} 2\pi s_0 \delta(t) \, dt \equiv s_0$$

so that

$$\int_{-\infty}^{+\infty} s(\omega) \, d\omega = \infty.$$

Hence, a continuous-time white noise process cannot exist, since its spectral density only satisfies the first condition of (12.27). Nevertheless, the concept of white noise as an approximate statistical model is of great importance for various phenomena in electronics, electrical engineering, communication, time series, econometrics, and other disciplines. Its outstanding role in applications can be compared with the one of the point mass in mechanics, which also only exists in theory. (A mathematically exact definition of the white noise process is, however, possible on the fundament of stochastic calculus even if white noise does not exist in the real world.) Here, as a working basis, the following explanation of the continuous white noise is given:

The (continuous) white noise is a real, stationary, continuous-time stochastic process with constant spectral density.

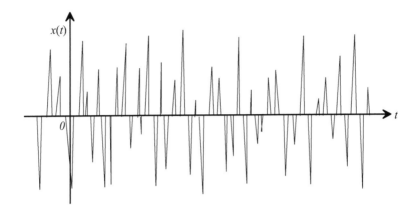

Figure 12.8 Illustration of a sample path of the white noise (time axis extremely stretched)

White noise can be thought of as a sequence of extremely sharp pulses, which occur after extremely short time intervals, and which have independent, identically distributed amplitudes with mean 0. The times in which the pulses rise and fall are so short that they cannot be registered by measuring instruments. Moreover, the response times of measurements are so large that during any response time a huge number of pulses occur, which cannot be registered (Figure 12.8).

Remark The term 'white noise' is due to a not fully justified comparison with the spectrum of the white light. This spectrum actually also has a wide-band structure, but its frequencies are not uniformly distributed over the entire bandwidth of the white light.

A stationary process$\{X(t), t \geq 0\}$ can be approximately considered a white noise process if the covariance between $X(t)$ and $X(t+\tau)$ tends to 0 extremely fast with increasing $|\tau|$. For example, if $X(t)$ denotes the the absolute value of the force which particles in a liquid are subjected to at time t (causing their Brownian motion), then this force arises from the about 10^{21} collisions per second between the particles and the surrounding molecules of the liquid (assuming average temperature, pressure and particle size). The process $\{X(t), t \geq 0\}$ is known to be weakly stationary with a covariance of type

$$C(\tau) = e^{-b|\tau|} \text{ with } b \geq 10^{19} \text{sec}^{-1}.$$

Hence, $X(t)$ and $X(t+\tau)$ are practically uncorrelated if

$$|\tau| \geq 10^{-18}.$$

A similar fast drop of the covariance function can be observed if $\{X(t), t \geq 0\}$ describes the fluctuations of the electromotive force in a conductor, which is caused by the thermal movement of electrons.

Example 12.7 Let $\{N(t), t \geq 0\}$ be a homogeneous Poisson process with intensity λ and $\{X(t), t \geq 0\}$ be a shot noise process (see example 6.5, page 229) defined by

$$X(t) = \sum_{i=1}^{N(t)} h(t - T_i),$$

where the function $h(t)$ quantifies the response of a system to the Poisson events arriving at time points T_i. In this example, the system is a vacuum tube, where a current impulse is initiated as soon as the cathode emits an electron. If e denotes the charge on an electron and if an emitted electron arrives at the anode after z time units, then the current impulse induced by an electron is known to be

$$h(t) = \begin{cases} \dfrac{\alpha e}{z^2} t & \text{for } 0 \leq t \leq z, \\ 0 & \text{elsewhere,} \end{cases}$$

where α is a tube-specific constant. $X(t)$ is, therefore, the total current flowing in the tube at time t. Now the covariance function of the process $\{X(t), t \geq 0\}$ can immediately be derived from the covariance function (7.32), page 272. The result is

$$C(s,t) = \begin{cases} \dfrac{\lambda (\alpha e)^2}{3z} \left[1 - \dfrac{3|t-s|}{2z} + \dfrac{|t-s|^3}{2z^3} \right] & \text{for } |t-s| \leq z, \\ 0 & \text{elsewhere.} \end{cases}$$

Since

$$\lim_{z \to 0} C(s,t) = \delta(s - t),$$

this shot noise process $\{X(t), t \geq 0\}$ behaves approximately as white noise if the transmission time z is sufficiently small. □

Band-Limited White Noise As already pointed out, a stationary process with constant spectral density over an unlimited bandwidth cannot exist. A stationary process, however, with spectral density

$$s(\omega) = \begin{cases} s_0 & \text{for } -w/2 \leq \omega \leq +w/2, \\ 0 & \text{otherwise,} \end{cases}$$

can (Figure 12.9 a). By making use of formulas (12.26) and (12.8), the corresponding covariance function is seen to be (Figure 12.9 b)

$$C(\tau) = \int_{-w/2}^{+w/2} e^{i\omega\tau} s_0 d\omega = 2s_0 \frac{\sin w\tau/2}{\tau}.$$

The mean power of such a process is proportional to $C(0) = s_0 w$, since

$$\lim_{x \to 0} \frac{\sin x}{x} = 1.$$

The parameter w is the bandwidth of its spectrum. With increasing w the band-limited white noise process behaves increasingly like a white noise. □

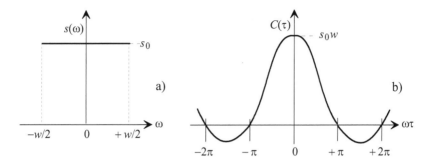

Figure 12.9 Spectral density and covariance function of the band-limited white noise

12.4 EXERCISES

12.1) Define the stochastic process $\{X(t), t \in \mathbf{R}\}$ by

$$X(t) = A \cos(\omega t + \Phi),$$

where A and Φ are independent random variables with $E(A) = 0$ and Φ is uniformly distributed over the interval $[0, 2\pi]$.

Check whether the covariance function of the weakly stationary process $\{X(t), t \in \mathbf{R}\}$ can be obtained from the limit relation (12.5).

The covariance function of a slightly more general process has been determined in example 6.6 at page 235.

12.2) A weakly stationary, continuous-time process has covariance function

$$C(\tau) = \sigma^2 e^{-\alpha|\tau|}\left(\cos \beta\tau - \frac{\alpha}{\beta} \sin \beta|\tau|\right).$$

Prove that its spectral density is given by

$$s(\omega) = \frac{2\sigma^2 \alpha \omega^2}{\pi(\omega^2 + \alpha^2 + \beta^2 - 4\beta^2\omega^2)}.$$

12.3) A weakly stationary continuous-time process has covariance function

$$C(\tau) = \sigma^2 e^{-\alpha|\tau|}\left(\cos \beta\tau + \frac{\alpha}{\beta} \sin \beta|\tau|\right).$$

Prove that its spectral density is given by

$$s(\omega) = \frac{2\sigma^2 \alpha (\alpha^2 + \beta^2)}{\pi(\omega^2 + \alpha^2 - \beta^2 + 4\alpha^2\beta^2)}.$$

12.4) A weakly stationary continuous-time process has covariance function

$$C(\tau) = a^{-b\tau^2} \text{ for } a > 0, \ b > 0.$$

Prove that its spectral density is given by

$$s(\omega) = \frac{a}{2\sqrt{\pi b}} \, e^{-\frac{\omega^2}{4b}}.$$

12.5) Define a weakly stationary stochastic process $\{V(t), \ t \geq 0\}$ by

$$V(t) = S(t+1) - S(t),$$

where $\{S(t), \ t \geq 0\}$ is the standard Brownian motion process.

Prove that its spectral density is proportional to

$$\frac{1 - \cos\omega}{\omega^2}.$$

12.6) A weakly stationary, continuous-time stochastic process has spectral density

$$s(\omega) = \sum_{k=1}^{n} \frac{\alpha_k}{\omega^2 + \beta_k^2}, \quad \alpha_k > 0.$$

Prove that its covariance function is given by

$$C(\tau) = \pi \sum_{k=1}^{n} \frac{\alpha_k}{\beta_k} e^{-\beta_k |\tau|}, \quad \alpha_k > 0.$$

12.7) A weakly stationary, continuous-time stochastic process has spectral density

$$s(\omega) = \begin{cases} 0 & \text{for } |\omega| < \omega_0 \text{ or for } |\omega| > 2\omega_0, \\ a^2 & \text{for } \quad \omega_0 \leq |\omega| \leq 2\omega_0, \end{cases} \quad \omega_0 > 0.$$

Prove that its covariance function is given by

$$C(\tau) = 2a^2 \sin(\omega_0\tau)\left(\frac{2\cos\omega_0\tau - 1}{\tau}\right).$$

REFERENCES

Allen, L.J.S. (2011). *An Introduction to Stochastic Processes with Applications to Biology.* Chapman & Hall/CRC, Boca Raton, London, New York. 2nd ed.

Andél, J. (1984). *Statistische Analyse von Zeitreihen.* Akademie-Verlag. Berlin.

Asmussen, S. (2000). *Ruin Probabilities.* World Scientific. Singapore, London.

Bachelier, L. (1900). Théorie de la spéculation. *Ann. Scient. de l' cole Normale Supér.* 3. 21-86.

Beichelt, F. (1997). *Stochastische Prozesse für Ingenieure.* Teubner Verlag. Stuttgart. English translation (2002) with P. Fatti: *Stochastic Processes and their Applications.* Taylor and Francis. London, New York.

Beichelt, F. (2006). *Stochastic Processes in Science, Engineering, and Finance.* Chapman & Hall/CRC. Boca Raton, London, New York.

Beichelt, F. and P. Tittmann (2012). *Reliability and Maintenance-Networks and Systems.* Chapman & Hall/CRC. Boca Raton, London, New York.

Bernoulli, J. (1713). *Ars Conjectandi.* Thurnisorium. Basel. (Ed. by N. Bernoulli).

Brandt, A., Franken, P., and B. Lisek (1990). *Stationary Stochastic Models.* Wiley. New York.

Brown, R. (1828). A brief account of microscopial observations made in the months of June, July, and August, 1827, on particles contained in the pollen of plants, and on the general existence of active molecules in organic and inorganic bodies. *Phil. Mag. Series* 2. 161.

Brzeniak, Z. and T. Zastawniak (1999). *Basic Stochastic Processes.* Springer. New York.

Chatfield, C. (2012). *The Analysis of Time Series.* Chapman & Hall/CRC. Boca Raton, New York, London. 6th ed.

Chung, K. L. (1960). *Markov Chains with Stationary Transition Probabilities.* Springer-Verlag. Heidelberg.

Cramér, H. and M.R. Leadbetter (1967). *Stationary and Related Stochastic Processes.* John Wiley & Sons, New York.

Dubourdieu, J. (1938). Remarques relatives a la théorie mathmatique del' assurance accidents. *Bull. Trim. de l'Inst. des Actuaires Français* 49. 76.

Durrett, R. (2012). *Essentials of Stochastic Processes.* Springer. New York. 2nd ed.

Durrett, R. (2015). *Branching Process Models of Cancer.* Springer. New York.

Einstein, A. (1905). Über die von der molekularkinetischen Theorie der Wärme geforderte Bewegung von in ruhenden Flüssigkeiten suspendierten Teilchen. *Annalen der Physik* 17. 549.

Feller, W. (1968). *An Introduction to Probability Theory and its Applications*, Vol. I. John Wiley & Sons, New York. 3rd ed.

Feller, W. (1971). *An Introduction to Probability Theory and its Applications*. Vol. II. John Wiley & Sons. New York.

Franken, P. et. al. (1981). (Authors are Franken, König, Arndt, and Schmidt). *Queues and Point Processes*. Akademie-Verlag. Berlin.

Franz, J. (1977). Niveaudurchgangszeiten zur Charakterisierung sequentieller Schätzverfahren. *Mathem. Operationsforschung und Statisik* 8. 499-508.

Furry, W. (1937). On fluctuation phenomena in the passage of high energy electrons through lead. *Phys. Rev.* 52. 569.

Galton, F. and H.W. Watson (1875). On the probability of the extinction of families. *Journ. Anthropol. Soc. London* (*Royal Anthropol. Inst. G. B. Ireland*) 4. 138-144.

Gardner, W. A. (1989). *Introduction to Random Processes with Applications to Signals and Systems*. McGraw-Hill Publishing Company. New York.

Gelbaum, B. and B. Olmstead (1990). *Theorems and Counterexamples in Mathematics*. Springer. New York.

Gelenbe, E. and G. Pujolle (1987). *Introduction to Queueing Networks*. Wiley & Sons. New York.

Gnedenko, B.W. and D. König (1983). *Handbuch der Bedienungstheorie* I and II. Akademie-Verlag. Berlin.

Grandell, J. (1991). *Aspects of Risk Theory*. Springer-Verlag. New York. Berlin.

Grandell, J. (1997). *Mixed Poisson Processes*. Chapman & Hall. London.

Grimmett, G.R. and D.R. Stirzaker (2001). *Probability and Random Processes*. Oxford University Press, Oxford. 3rd ed.

Gut, A. (1990). Cumulative shock models. *Advances Appl. Probability* 22. 504.

Haccou, P., Jagers, P. and V.A. Vatutin (2011). *Branching Processes*. Cambridge University Press. Cambridge.

Hardy, G. H. (1908). Mendelian proportions in a mixed population. Science, 49-50.

Harris, T.E. (1963). *The Theory of Branching Processes*. Springer. Berlin.

Helstrom, C. W. (1984). *Probability and Stochastic Processes for Engineers*. Macmillan Publishing Company. New York, London.

Heyde, C.C. and E. Seneta (1972). The simple branching process, a turning point test and a fundamental identity: a historical note on I.J. Bienaym. *Biometrika* 59, 680.

Jones, P.W. and P. Smith (2010). *Stochastic Processes-An Introduction*. Chapman & Hall/CRC. Boca Raton, London, New York. 2nd ed.

Kaas, R. et al. (2004). (Authors are Kaas, Goovaerts, Dhaene, and Denuit). *Modern Actuarial Risk Theory*. Springer. New York.

Kannan, D. (1979). *An Introduction to Stochastic Processes.* North Holland. New York, London.

Karlin, S. (1966). *A First Course in Stochastic Processes.* Academic Press. New York, London.

Karlin, S. and H.M. Taylor (1981). *A Second Course in Stochastic Processes.* Academic Press, New York, London.

Karlin, S. and H.M. Taylor (1994). *An Introduction to Stochastic Modeling.* Academic Press, New York, London.

Khinchin, A. Ya. (1934). Korrelationstheorie der stationären stochastischen Prozesse. *Mathematische Annalen* 109. 415-458.

Kijima, M. (2013). *Stochastic Processes with Applications to Finance.* Chapman & Hall/CRC. Boca Raton, London, New York. 2nd ed.

Kimmel, M. and D. Axelrod (2015). *Branching Processes in Biology.* Springer. New York.

Kirkwood, J.R. (2015). *Markov Processes.* Chapman & Hall/CRC. Boca Raton.

Kolmogorov, A. (1933). *Grundbegriffe der Wahrscheinlichkeitsrechnung.* Springer. Berlin. English edition (1950): *Foundations of the Theory of Probability.* Chelsea. New York.

Kulkarni, V.G. (2010). *Modeling and Analysis of Stochastic Systems.* Chapman & Hall/CRC. Boca Raton, New York, London. 2nd ed.

Lai, C.-D. and M. Xie. (2006). Stochastic Aging and Dependence for Reliability. Springer. New-York.

Lawler, G. (2006). *Introduction to Stochastic Processes.* Chapman & Hall/CRC. Boca Raton, New York, London. 2nd ed.

Liao, M. (2014). *Applied Stochastic Processes.* Chapman & Hall/CRC, Boca Raton, New York, London.

Lin, X.S. (2006). *Introductory Stochastic Analysis for Finance and Insurance.* Wiley. Hoboken, New Jersey.

Lomax, K.S. (1954). Business failures: another example of the analysis of failure data. *Journ. Amer. Stat. Assoc.*, 49, 847-852.

Lotka, A.J. (1931). The extinction of families. *Journ. Wash. Acad. Sci.* 21, 377-453.

Lundberg, O. (1964). *On Random Processes and their Application to Sickness and Accident Statistics.* Almqvist och Wiksell. Uppsala.

Madsen, H. (2008). *Time Series Analysis.* Chapman & Hall/CRC. Boca Raton.

Matthes, K., Kerstan, J., and J. Mecke (1974). *Unbegrenzt teilbare Punktprozesse.* Akademie-Verlag, Berlin. (*Infinitely Divisible Point Processes.* Wiley. 1978.)

Moser, L.F. (1839). *Die Gesetze der Lebensdauer.* Veit-Verlag. Berlin.

Nisbet, R.M. and W.S.C. Gurney (1982). *Modelling Fluctuating Populations.* Wiley. Chichester, New York.

Prado, R. and M. West (2010). *Time Series-Modeling, Computation, and Inference.* Chapman & Hall/CRC. Boca Raton, New York, London.

Rolski, T. et al. (1999). (Authors are Rolski, Schmidli, Schmidt, Teugels). *Stochastic Processes for Insurance and Finance.* John Wiley & Sons. New York.

Rosin, E. and E. Rammler (1933). The laws governing the fineness of powdered coal. *J. Inst. Fuel* 7, 29.

Ross, S. M. (2010). *Introduction to Probability Models.* Academic Press. San Diego. 10th ed.

Scheike, T. H. (1992). A boundary-crossing result for Brownian motion. *Journ. Appl. Prob.* 29, 448.

Schottky, W. (1918). Über spontane Stromschwankungen in verschiedenen Elektrizitätsleitern. *Annalen der Physik* 57. 541-567.

Seshradi, V. (1999). *The Inverse Gaussian Distribution.* Springer. New York.

Sevastyanov, B.A. (1971). *Branching processes* (in Russian). Mir. Moscow.

Smoluchowski, M.V. (1916). Drei Vorträge über Diffusion, Brownsche Bewegung und Koagulation von Kolloidteilchen. *Physikalische Zeitschrift,* 17, 557-585.

Snyder, D.L. (1975). *Random Point Processes.* John Wiley. New York, London.

Steffensen, J.F. (1930). Om Sandsyndligheden for at Afkommet uddor. *Matematisk Tidsskrift.,* B, 19-23.

Stigman, K. (1995). *Stationary Marked Point Processes.* Chapman & Hall. New York.

Taylor, H.M. and S. Karlin (1998). *An Introduction to Stochastic Modeling.* Academic Press. New York

Tijms, H.C. (1994). *Stochastic Models-An Algorithmic Approach.* Wiley. New York.

Uhlenbeck, G.E. and L.S. Ornstein (1930). On the Theory of Brownian motion. *Phys. Rev.* 36. 823-841.

van Dijk, N. (1993). *Queueing Networks and Product Forms.* Wiley, New York.

von Collani ed. (2003). *Defining the Science of Stochastics.* Volume 1 in the *Sigma-Series of Stochastics* (eds.: Beichelt, F. and J. Sheil). Heldermann. Lemgo.

Walrand, J. (1988). *An Introduction to Queueing Networks.* Prentice Hall. Englewood Cliffs.

Wiener, N. (1923). Differential space. *J. Math. Phys. Mass. Inst. Techn.* 2. 131.

Willmot, G.E. and X. S. Lin (2001). *Lundberg Approximations for Compound Distributions with Insurance Applications.* Springer, New York.

Yule, G.U. (1924). A mathematical theory of evolution, based on the conclusions of Dr. J.C. Willis, F.R.S. *Phil. Trans. Royal Soc. London,* B, 213, 21-87.

INDEX

A

absorbing state 344, 350
absorption probabilities 345
 -mean time to 345, 366
aging 87, 88
alternating renewal process 320
a posteriori probabilities 26
a priori probabilities 26
arithmetic distribution 216
arithmetic random variable 216
arrival intensity (rate) 427
autocorrelation function 226
autocovariance function 226
autoregressive mean average models 251
autoregressive sequence 249, 537
availability 321, 174
 -interval 321
 -long-run 322
 -point 321
 -stationary 322

B

bandwith 240, 241
Bayes' theorem 26
Bernoulli distribution 49
beta distribution 76
beta function 76
binary random variable 49
binomial coefficient 13
binomial distribution 51, 98
 -negative 53
Birnbaum-Saunders distribution 330
birth and death processes 364, 405, 410
 -continuous time 405
 -linear 411
 -with immigration 416
 -discrete-time 364

-(pure) birth process 405
-(pure) death process 408
-nonhomogeneous 421
bivariate exponential distribution 132
bivariate normal distribution 131, 184
Black-Scholes-Merton, formula of 522
Blackwell's renewal theorem 311
branching process 479
Brownian bridge 503
Brownian motion 496
-and random walk 497
-first passage time 504
-geometric 510
-integrated 524
-reflected 509
-shifted 496
-standard Brownian motion 497
-with drift 512
Buffon's needle problem 16

C

Campbell's theorem 272
Cauchy distribution 74
central limit theorem 208
central tendency 63
Chapman-Kolmogorov equations 341, 384
characteristic function 102
Chevalier de Méré, problem of 4, 30
circular frequency 532
claim arrival process 292
claim size process 292
classes of states of a Markov chain
-absorbing 386
-closed 350, 386
-equivalence 351

-essential   352
-minimal   350
-null   355
-recurrent   355
 -positive   355
-transient   355
coefficient of variation   47, 67
concordant function   484
conditional density   127, 128
conditional distribution   127
conditional mean value   128
conditional probability   24
congruence method   167
contingency table   105
convergence criteria   204, 205
convolution   8, 100, 179, 300
correlated   137, 532
 -positively, negatively   137
correlation coefficient   137
correlation function   226
correlation time   537
counting process   257, 300
 -simple   258
covariance   135
covariance function   226
 -of a complex stochastic process   529
 -spectral representation   532, 535
Cramér-Lundberg approximation   297
cumulative repair cost   522

D
de Morgan, rules of   10
density   59, 121
 -conditional   128
 -joint   121, 145
derivatives   516
DFR   89
DFRA   90
Dirichlet's formula   99, 101

discounting   518
discrete Black-Scholes model   478
discrete-time Markov chain   339
 -homogeneous   340
 -irreducible   350
 -reducible   350
distribution function   42, 59
Doob-type martingales   479
doubling strategy   481

E
Ehrenfest's diffusion model   343
embedded Markov chain   401, 435, 458
Engset's loss system   430
equilibrium state   342
ergodicity (of a complex process)   530
ergodic state (of a Markov chain)   355
Erlang distribution   75, 190, 301, 307
Erlang's loss formula   428
Erlang's phase method   403
events
 -certain   1
 -deterministic   1, 10
 -elementary   8
 -impossible   1, 10
 -simple   8
events, random   10
 -disjoint   10
 -exclusive   10, 26
 -exhaustive   26
 -independent   28, 29
exaustive set of random events   26
exercise price   516
expected value   46, 63, 64
expiration date   516
explosive growth   384
exponential distribution   60, 75
 -bivariate   132
exponential martingale   478, 499

extinction, probability of   373
  -mean time to   369
  -probability of   373

F
faculty   13
failure probability   86
failure rate   88
  -integrated   88
fair division   5
fair game   487
fair price   522
favorable cases   12
feedback   451
filtration   239, 480
first passage time   329, 355, 504, 513
flow (of demands)   425
force of mortality   88
formula of Bayes   26
formula of Pollaczek-Khinchin   436
formula of total probability   26
forward contracts   517
frequency distribution   44
fundamental renewal theorem   312

G
gambler's ruin problem   346
gamma distribution   75
gamma function   75
Gaussian distribution   78
Gaussian error integral   80
Gaussian process   233, 498
generating function   95
  -moment   95, 96, 100, 103
  -probability   95, 96
geometric Brownian motion   510
  -with drift   374
geometric distribution   50, 97
  -modified   375

  -truncated   375
Gumbel distribution   133, 138, 149

H
half-life period   2
Hardy-Weinberg law   348
hazard function   88
hitting time   359
homogeneous Poisson process   261
hypergeometric distribution   53

I
IFR   89
IFRA   90
image   101
impatient customer   441
inclusion-exclusion formula   19
increment (of a stochastic process)   232
  -homogeneous increments   232
  -independent increments   233
independence   118, 122
  -of random events   28, 29
  -of random variables   118, 122
inequality (-ties)   199
  -for moments   202
  -for probabilities   199
initial distribution   341
  -stationary   342, 385
initial reserve (capital)   293
input (flow) (queueing system)   425
insensitivity   430
integral equations of renewal type   305
intensity   259, 261, 396
intensity function   274
interarrival times   263
interval probability   59, 60
interval reliability   317, 318
inverse Gaussian distribution 85, 505, 513
irreducible Markov chain   350

J+K

Jackson queueing network	447
joint probability density	121, 145
joint probability distribution	181, 120
	145
Kendall's notation	282
Kolmogorov's differential equations	388
-backward	388
-forward	388
Kronecker symbol	340

L

Landau order symbol	89, 261
Laplace distribution	66
Laplace random experiment	12
Laplace transform	99, 497, 514
Laplace transformation	99
laws of large numbers	206
-weak	206, 207
-strong	207, 208
limit theorems	204, 208
-limit theorem of Moivre-Laplace	210
-local limit theorems	214
linear birth- and death process	411
linear birth process	406
linear death process	409
linear filter	239
linear regression	133
Little's formula (law)	433
logarithmic normal distribution	83
logistic distribution	84
Lomax distribution	93
loss probability	428, 431, 439, 442
Lundberg coefficient	297
Lundberg inequality	298, 490

M

maintenance cost rate	290, 522
marginal distribution	118, 121, 145

marked point process	256
-random	260, 285, 287, 324
Markov(ian) property	233, 339
Markov chain	339, 383
-continuous-time	383
-discrete-time	339
-distribution of a	385
-absolute	342, 385
-finite (multi)-dimensional	341
-one-dimensional	342, 385
-embedded	401, 435
-irreducible	350, 360
-semi-Markov	458
-sojourn times of a	399
-stationary	342, 385
Markov process	233, 498
Markov system	401
Marshall-Olkin distribution	132, 138
-n-dimensional	149
martingale	475, 498
-continuous-time	475, 489
-discrete-time	475
-Doob-type	479
-exponential	478, 499
-product	477
-sub	475, 476, 489
-sum-	477
-super	475, 476, 489
-variance	485, 499
martingale stopping theorem	486, 490
maturity date	516
maximum of independent random variables	175
maximum of the Brownian motion	505
-of the Brownian motion with drift	514
mean-square continuous	234
mean value	46, 63, 162, 170, 187
measurable space	18
median	65

memoryless property 87
mid-square method 168
minimal repair 289
minimum
 -of independent random variables 177
mixed Poisson process 278
mixture of distributions 92
mixture of random variables 92
mode 66
moment 46, 47, 65, 96, 100
moving averages 240, 246, 248, 349
multidimensional random variable 117
multimodal 66
multinomial criterion 280

N

narrow-band process 534
negative binomial distribution 53, 95, 282
network of queueing systems 447
noise (random) 225, 230, 246, 512, 542
normal distribution 78, 191, 301, 307
 -bivariate 132, 184
 -n-dimensional 148
normalization 72

O

option, call-, put-, 516
 -American 516
 -European 516
option pricing 516
Ornstein-Uhlenbeck process 511
output (of a queueing system) 425

P

paradox of the renewal theory 318
parallel system 176
pareto distribution 74, 93
patience time 441
percentile 65

period (of a state) 353
point availability 321
Poisson distribution 56, 97, 98
Poisson events 263
Poisson process(es) 261
 -compound 287, 293
 -doubly stochastic 283
 -homogeneous 261, 385
 -mixed 278
 -nonhomogeneous 274
 -superposition of 284
 -thinning of 285
Pollaczek-Khinchin formula 436
Pólya process 281
Pólya's urn scheme 483
power distribution 73
preimage 101
preemptive priority 427
premium income 293
premium rate 294
priority queueing systems 442
probability 12
 -axiomatic definition 17
 -classical definition 12
 -conditional 24
 -geometric definition 15
probability density 59
probability distribution
 -conditional 119, 127
 -of a random variable 41, 59
 -of a random vector 120
 -of a stochastic process 225
 -of a sum 179, 189
 -standardized 72
 -truncated 71
probability histogram 44
probability mass function 43, 59
probability space 18

product martingale 477
pulse code modulation 228
  -randomly delayed 236
purely random process (sequence) 246

Q

quantile 65
queueing network 447
  -closed 454
  -Jackson 447
  -open 447
  -sequential 452
  -with feedback 450
queueing system 425
  -closed 426, 419, 431
  -insensitive 430
  -loss 428
  -multi-server 425
  -priority 442
  -single-server 425, 429
  -waiting 431
  -waiting-loss 439
  -with unreliable server 445

R

random experiment 7
random function 223
random integral 524
randomly delayed pulse code modulation 236, 539
random numbers 163
random number generators 167
  -congruence method 167
  -mid-square method 168
random point process 255
  -marked 256
  -simple 258, 261
  -stationary 258
random sample 188

random telegraph signal 265, 538
random variable(s) 40, 41
  -binary 49
  -bivariate 117, 120
  -compound 260, 293
  -continuous 42, 59
  -discrete 59
  -functions of 155
  -independent 118, 122, 145
  -maximum of 175
  -minimum of 177
  -mixture of 92
  -nonnegative 86
  -product of 172
  -range of 40
  -ratio of 173
  -standardized 72
  -two-dimensional 117
random vector 117, 120
  -n-dimensional 144
random walk
  -symmetric 342, 497
  -unbounded 357
  -with absorbing barrier(s) 344, 365, 368
  -with reflecting barrier(s) 343
Rayleigh distribution 77
realization 40, 224
  -of a random variable 40
  -of a stochastic process 224
recurrence time 256, 315
  -backward 315
  -forward 315
recurrence time (Markov chain) 355
recurrent Markov chain 357
reflected Brownian motion 509
reflection principle 504
relative frequency 3, 20, 45
renewal 299
  -cycle 325

-density 303
-equations 304
-function 302
-bounds on the 308
-point (of a Markov chain) 355
renewal counting process 300
renewal process 299, 300
-alternating 320
-compound 324
-first passage time 329
-delayed 306
-ordinary 299
-stationary 318
renewal theorem(s) 325
repairman problem 402, 419
replacement cycle 290
residual lifetime 86
retransformation 101
risk analysis 292, 313, 327
risk reserve 293
risky asset 516
Rosin-Rammler distribution 78, 84
routing matrix 448
ruin probability 293

S

safety loading 295
sample path 224, 500
sample space 8, 39
Schottky effect 230
second-order stochastic process 231
semi-Markov chain 458
series system 178
service disciplines 426
service intensity (rate) 427
set 9
-countably infinite 9
-discrete 9
-nondenumerable 9

shifted Brownian motion 496
-with drift 512
shot noise 223, 269
sigma algebra 18
skewness 72
smoothing techniques 239
-Epanechnikov kernel 241
-exponential 241
-geometrical 241
sojourn time 399, 436, 458
spectral density 535
spectral function 535
spectrum 531
-continuous 535, 541
-discrete 531
standard deviation 47, 67
standardization 72
standard normal distribution 80
state (of a discrete-time Markov chain)
-absorbing 350
-accessible 351
-aperiodic 353
-ergodic 355
-essential 352
-inessential 352
-null, positive recurrent 355
-recurrent 355
-transient 355
stationarity of a stochastic process 230
-of a point process 258
-strongly 230, 232
-weakly 232
stationary state probabilities 385, 396
steady state 427
stochastic process 224
-complex 529
-Gaussian 233
-Markov 233
-mean-square continuous 234

-second-order   231, 529
-strongly stationary   230
-weakly stationary   232, 529
-with homogeneous increments   232
 -with independent increments   233
stochastic regularities   2, 7
stochastics   5
stopping time   195, 490
strike price   516
strongly stationary   230
structure distribution   93
submartingale   476, 489
sum martingale   477
sum of random variables   179, 181, 186
                                            194
supermartingale   476, 489
superposition of Poisson processes   285
survival probability   86, 293
symmetric function (density)   66

T
thinning of Poisson processes   285
time series   237
total probability rule   124
traffic (flow)   425
traffic intensity   427
trajectory   224
transition graph   350, 390
transition matrix   340, 383, 384, 448
transition probabilities   340, 383, 387
 -m-step   340
transition rates   387, 389
 -conditional   388
 -unconditional   388
transition times   389

trend estimation (in time series)   243
trend function   226, 238
 -of a complex stochastic process   529
truncation   71

U+V
uncorrelated   137, 532
unfair game   488
uniform distribution   44, 49, 73, 267
 -two-dimensional   126
unimodal   66
universe   8
unreliable server   455, 427
vacant probability   428, 439
variability   47, 67
variable input intensity   442
variance   47, 67
 -of a sum   184, 187
variance martingale   485, 499
variation   500
volatility   496

W+Y+Z
waiting probability   432, 439
waiting time   432, 433
Wald's identity   194, 486
Weibull distribution   77, 292
white noise   542
 -band-limited   544
 -continuous   542
 -discrete   246
wide-band process   534
Wiener process (Brownian motion)   495
Yule-Furry process   406
z-transformation   96, 179

Printed and bound by CPI Group (UK) Ltd, Croydon, CR0 4YY

24/10/2024

01778283-0017